Inpatient Neuropsychology

Erin K. Bailey
Joel E. Kamper • Tracy S. Kretzmer
Editors

Inpatient Neuropsychology

Springer

Editors
Erin K. Bailey
James A. Haley Veterans' Hospital, Mental Health
and Behavioral Science Service (MS/BSS)
University of South Florida, Morsani College of
Medicine, Department of Psychiatry and
Behavioral Sciences
Tampa, FL, USA

Joel E. Kamper
James A. Haley Veterans' Hospital, Mental Health
and Behavioral Science Service (MS/BSS)
Tampa, FL, USA

Tracy S. Kretzmer
James A. Haley Veterans' Hospital, Mental Health
and Behavioral Science Service (MS/BSS)
Tampa, FL, USA

ISBN 978-3-032-05879-9 ISBN 978-3-032-05880-5 (eBook)
https://doi.org/10.1007/978-3-032-05880-5

This Springer imprint is published by the registered company Springer Nature Switzerland AG
The registered company address is: Gewerbestrasse 11, 6330 Cham, Switzerland

Dedication

This book is dedicated to our patients and their families, who we have had the privilege of serving.

Thank you for allowing us to be a part of your journey.

PREFACE

The demand for integrated healthcare services has increased over the past several decades, though in no other environment than the inpatient setting is collaboration and communication among healthcare providers most essential. As experts in brain-behavior relationships, the neuropsychological skill set adds distinct value to integrated systems of care, as evidenced by the growth of employment opportunities in hospital-based settings. However, this expansion of practice compels a transformative shift in how we, as a profession, approach our role as a neuropsychologist. Whether working in acute rehabilitation or functioning as a consultation & liaison (C&L) provider in the hospital setting, inpatient practice demands application of our unique expertise in novel ways.

This text was largely conceived from necessity. Apart from literature in acquired brain injury, there is limited aggregated guidance on how neuropsychologists can truly integrate alongside medical providers in environments where extensive, psychometrically driven evaluation is often impractical. Similarly, while postdoctoral training in neuropsychology requires interdisciplinary consultation as a core competency, it is unclear how many inpatient neuropsychology training opportunities are available that allow for nuanced development of this skill. Among those that are afforded postdoctoral experience in inpatient settings, this is usually confined to hospital systems that have historically respected and utilized neuropsychological services, such as academic medical centers and Veterans Affairs hospitals. Consequently, early career practitioners may be ill-equipped to advocate for and create an effective practice in hospital settings or with medical teams unfamiliar with our inimitable clinical acumen. Equally important is the ability of the neuropsychologist to fluidly navigate a fast-paced environment where having proficiency of medical knowledge and disease is paramount.

Across years of consulting listservs, discussions with mentors, receiving anecdotal feedback from medical colleagues, reviewing available literature, and continually adjusting our practice through trial-and-error, we retained a goal to develop a book focused exclusively on inpatient practice for neuropsychologists. In the following chapters, we aim to provide a framework for how neuropsychologists can create, or expand, inpatient services. Further, to address the lack of comprehensive resources for those working in C&L settings, we endeavor to contribute practice considerations for how neuropsychologists can effectively function as inpatient consultants. While not exhaustive, we hope that this guidance will increase comfort and dexterity enacting the full spectrum of our high-quality skill set.

To achieve our goal, we enlisted authorship from experts in our field and divided content into three focus areas that sequentially build proficiency of inpatient practice. Our first section, Foundational Competencies, discusses topics essential for any neuropsychologist working in an inpatient setting. This includes professional advocacy, navigation of administration and billing considerations, and foundational medical knowledge, as well as best practices for conducting bedside examinations and individualized care for a wide spectrum of patients. The next section, Clinical Syndromes and Applied Practice, reviews pathology more often encountered in inpatient settings, including delirium and hydrocephalus, among other medical conditions. Here, we hope the reader finds beneficial guidance for elucidating differential diagnoses in cases of increased medical complexity. Our final section, Advanced Practice Considerations, covers spheres of practice that may be less familiar to those who have primarily functioned as outpatient diagnosticians, but are often required of neuropsychologists in inpatient settings. Topics include behavioral management of neurologically disordered populations and assessment of decisional capacities, as well as evaluation and treatment of psychosocial and psychiatric comorbidities that can impact treatment and prognosis.

In our personal experiences, we have directly witnessed the benefit to patients, families, and hospital systems afforded by providing inpatient neuropsychology services. Moreover, we have firsthand knowledge of how appreciated and respected these services are by our colleagues and administrators. Though many of the concepts and information are applicable to a wide spectrum of providers and specialties, we hope that this text motivates and inspires our fellow neuropsychologists to pursue inpatient practice. Ultimately, it is only through our own efforts and advocacy that we can showcase the incremental value to patient care imparted by the inclusion of neuropsychologists in integrated care settings.

Tampa, FL, USA

Erin K. Bailey
Joel E. Kamper
Tracy S. Kretzmer

Acknowledgments

To my family and friends, whose love and support encourages me every day.
To my patients, who teach me the true meaning of resilience and hope.
And to our military veterans, whose sacrifices remind us of the power of service.
Thank you.
–EKB

To my wife, family, and friends for their support and love.
To my mentors for their dedication and inspiration.
And to my patients for giving me the opportunity to help them.
–JEK

To my family and friends for supporting me always.
To my mentors for believing in me.
And to my patients and their families for teaching and inspiring me
–TSK

The editorial team would also like to express our sincere appreciation to Elizabeth Kamper for contribution of her exquisite art.

CONTENTS

About the Editors

Erin K. Bailey, Ph.D., ABPP-CN is a clinical neuropsychologist at the James A. Haley Veterans' Hospital in Tampa, FL, where she serves as the Assistant Training Director for the Psychology Predoctoral Internship Program. She is also an Assistant Professor in the Department of Psychiatry and Behavioral Neurosciences at the University of South Florida, Morsani College of Medicine and Vice Chair of the USF Social-Behavioral Institutional Review Board. She is board certified by the American Board of Professional Psychology in Clinical Neuropsychology and is a diplomate of the American Academy of Clinical Neuropsychology. Dr. Bailey serves as an ethics consultant for the JAHVH Health Care Ethics Consultation Service and as neuropsychology representative to the JAHVH Dementia Committee. Dr. Bailey provides training and supervision for psychology interns, neuropsychology residents, and psychiatry fellows in the outpatient neuropsychology and memory disorders clinics, as well as the inpatient C&L neuropsychology service—where she serves as co-director. Her professional interests include dementia, decisional capacity, and healthcare ethics.

Joel E. Kamper, Ph.D., ABPP-CN is a clinical neuropsychologist at the James A. Haley Veterans' Hospital (JAHVH) in Tampa, FL, where he serves as the Assistant Training Director for the Neuropsychology Postdoctoral Residency Program. A diplomate of the American Academy of Clinical Neuropsychology, he has provided C&L neuropsychology services at JAHVH since starting the program in 2015 and currently serves as its co-director. He also maintains a robust outpatient practice and is involved in training and supervision for predoctoral interns, postdoctoral residents, and psychiatry fellows. Dr. Kamper has a passion for professional education and sits on the boards of multiple national neuropsychology practice and training organizations. He is additionally involved in regulation and licensure as a volunteer member of several committees tasked with EPPP development through the Association of State and Provincial Psychology Boards (ASPPB). His professional interests include differential diagnosis, rare and complex diseases, and delirium.

Tracy S. Kretzmer, Ph.D., ABPP-CN is a clinical neuropsychologist at the James A. Haley Veterans' Hospital in Tampa, FL. She completed her doctorate in Medical Psychology at the University of Alabama, Birmingham, and is a diplomate of the American Academy of Clinical Neuropsychology. She has worked primarily in Physical Medicine and Rehabilitation inpatient settings both as a clinician and a program director. For the past 20 years she has also provided training and supervision for predoctoral interns and postdoctoral residents. Dr. Kretzmer participates in several professional organizations and brain injury special interest groups. Her research and clinical interests have focused on traumatic brain injury, mental health interventions, access to evidence-based care for individuals with cognitive disabilities, behavioral management, and TBI comorbidities and outcomes.

Contributors

Martina Azar Brown University Health, Providence, RI, USA

Michelle Boston-Babicz Madonna Ptak Memory Disorders Clinic, BayCare Medical Group, Clearwater, FL, USA

Erin K. Bailey James A. Haley Veterans' Hospital, Mental Health and Behavioral Science Service (MS/BSS), University of South Florida, Morsani College of Medicine, Department of Psychiatry and Behavioral Sciences, Tampa, FL, USA

Sheryl Berardinelli Jefferson Moss-Magee Rehabilitation, Elkins Park, PA, USA

Cady Block Department of Psychiatry and Psychology, Mayo Clinic Florida, Jacksonville, FL, USA

Joseph J. Boscarino Mental Health and Behavioral Science Service (MH/BSS), James A. Haley Veterans' Hospital, Tampa, FL, USA

Emanuel M. Boutzoukas Mental Health and Behavioral Science Service (MH/BSS), James A. Haley Veterans' Hospital, Tampa, FL, USA

Sarah W. Clark Mental Health Service, Richmond VA Medical Center, Richmond, VA, USA

Erica Cotton (née Sieg) Department of Psychiatry and Behavioral Sciences, Department of Neurology, Northwestern University Feinberg School of Medicine, Chicago, IL, USA

Erica L. Dawson Department of Neurology, The Ohio State University Wexner Medical Center, Columbus, OH, USA

Michelle Fox Mental Health Service, Lt. Col. Luke Weathers, Jr. VA Medical Center, Memphis, TN, USA

Adam Fusick Mental Health and Behavioral Science Service (MH/BSS), James A. Haley Veterans' Hospital, Tampa, FL, USA
Department of Psychiatry and Behavioral Sciences, Morsani College of Medicine, University of South Florida, Tampa, FL, USA

Douglas Johnson-Greene Department of Physical Medicine and Rehabilitation, University of Miami Health System, Miami, FL, USA

Emily Hallowell Mental Health & Behavioral Sciences Service, VA Providence Healthcare System, Providence, RI, USA
Department of Psychiatry and Human Behavior, Warren Alpert Medical School of Brown University, Providence, RI, USA

Robin Hanks Department of Rehabilitation Psychology and Neuropsychology, Detroit Medical Center, Detroit, MI, USA
Department of Physical Medicine and Rehabilitation, Wayne State University, Detroit, MI, USA

Kyle J. Jennette Department of Psychiatry, University of Illinois at Chicago – College of Medicine, Chicago, IL, USA
Department of Neurology and Rehabilitation, University of Illinois at Chicago – College of Medicine, Chicago, IL, USA

Joel E. Kamper James A. Haley Veterans' Hospital, Mental Health and Behavioral Science Service (MS/BSS), Tampa, FL, USA

Robert Kanser Department of Physical Medicine and Rehabilitation, University of North Carolina School of Medicine, Chapel Hill, NC, USA

Rachel Keelan Department of Rehabilitation Psychology and Neuropsychology, Detroit Medical Center, Detroit, MI, USA
Department of Physical Medicine and Rehabilitation, Wayne State University, Detroit, MI, USA

Tracy S. Kretzmer James A. Haley Veterans' Hospital, Mental Health and Behavioral Science Service (MS/BSS), Tampa, FL, USA

Andrea Kurasz Mental Health and Behavioral Science Service (MH/BSS), James A. Haley Veterans' Hospital, Tampa, FL, USA

Daniel W. Klyce Mental Health Service, Richmond VA Medical Center, Richmond, VA, USA
Department of Physical Medicine and Rehabilitation, Virginia Commonwealth University, Richmond, VA, USA
Psychology Service, Sheltering Arms Institute, Richmond, VA, USA

Brittany Lang Department of Neurology, University of Utah School of Medicine, Salt Lake City, UT, USA

Karina Loyo Department of Rehabilitation Psychology/Neuropsychology, TIRR Memorial Hermann Health System, Houston, TX, USA

Bani Malhotra Department of Physical Medicine and Rehabilitation, Virginia Commonwealth University, Richmond, VA, USA

Claire McGrath Bancroft, Mt. Laurel, NJ, USA

Thomas Oswald James A. Haley Veterans' Hospital, Mental Health and Behavioral Science Service (MS/BSS), Tampa, FL, USA

Jennifer Pender Mental Health and Polytrauma Services, John D. Dingell Department of Veteran Affairs Medical Center, Detroit, MI, USA

Matthew S. Phillips Department of Psychiatry, University of Illinois at Chicago – College of Medicine, Chicago, IL, USA

Summar Reslan Mental Health and Polytrauma Services, John D. Dingell Department of Veteran Affairs Medical Center, Detroit, MI, USA

David J. Rothman Department of Physical Medicine and Rehabilitation, Virginia Commonwealth University, Richmond, VA, USA

Jason R. Soble Department of Psychiatry, University of Illinois at Chicago – College of Medicine, Chicago, IL, USA
Department of Neurology and Rehabilitation, University of Illinois at Chicago – College of Medicine, Chicago, IL, USA

Megan Stockamp Department of Psychiatry and Behavioral Sciences, Morsani College of Medicine, University of South Florida, Tampa, FL, USA

Greg A. Sullivan Mental Health and Behavioral Science Service (MH/BSS), James A. Haley Veterans' Hospital, Tampa, FL, USA
Department of Psychiatry and Behavioral Neurosciences, Morsani College of Medicine, University of South Florida, Tampa, FL, USA

Devin M. Ulrich Division of Movement Disorders, Department of Neurology, Rush University Medical Center, Chicago, IL, USA

Ryan Van Patten Center for Neurorestoration & Neurotechnology, VA Providence Healthcare System, Providence, RI, USA
Mental Health & Behavioral Sciences Service, VA Providence Healthcare System, Providence, RI, USA
Department of Psychiatry and Human Behavior, Warren Alpert Medical School of Brown University, Providence, RI, USA

Emily N. Vanderbleek Mental Health and Behavioral Science Service (MH/BSS), James A. Haley Veterans' Hospital, Tampa, FL, USA

Daniel S. Weitzner Madonna Ptak Memory Disorders Clinic, BayCare Medical Group, Clearwater, FL, USA

Amanda Zwilling Department of Rehabilitation Psychology and Neuropsychology, Detroit Medical Center, Detroit, MI, USA
Department of Physical Medicine and Rehabilitation, Wayne State University, Detroit, MI, USA

Part I

Foundational Competencies

The Practice of Inpatient Neuropsychology

1

Tracy S. Kretzmer, Erica Cotton (née Sieg), and Thomas Oswald

ACRONYMS AND ABBREVIATIONS

APA	American Psychological Association
C&L	Consultation and Liaison
CARF	Commission on Accreditation of Rehabilitation Facilities
DoD	Department of Defense
ED	Emergency Department
FNQ-R	The Family Needs Questionnaire-Revised
IPC	Infection Prevention and Control
IRB	Institutional Review Board
LST	Life-Sustaining Treatment
NIH	National Institute of Health
PI	Program Improvement
PPACA/ACA	Patient Protection and Affordable Care Act
QI	Quality Improvement
TBI	Traumatic Brain Injury
TJC	The Joint Commission
VA	Veterans Affairs (*the Department of*)
WHO	World Health Organization

INTRODUCTION

Over the past several decades, the practice of neuropsychology has expanded across various clinical settings, particularly within hospital, rehabilitation, and other medical settings. This is not unexpected given the changing landscape of healthcare following the passing of the Patient Protection and Affordable Care Act (PPACA/ACA, Public Law No: 111-148, Mar 23 2013), which emphasizes an integrated healthcare approach (Kubu et al. 2016). Today, many neuropsychologists work in hospitals and medical centers, where opportunities to demonstrate the benefits of neuropsychological services beyond traditional outpatient assessment are vast (Sweet et al. 2021). While recent surveys examining the practices of neuropsychologists working in integrated healthcare systems have provided important insights, findings specific to inpatient settings are not

T. S. Kretzmer (✉) · T. Oswald
Mental Health and Behavioral Science Service (MH/BSS), James A. Haley Veterans' Hospital, Tampa, FL, USA
e-mail: Tracy.Kretzmer@va.gov; Thomas.Oswald@va.gov

E. Cotton (née Sieg)
Department of Psychiatry and Neurology, Northwestern University & Northwestern Memorial Hospital, Chicago, IL, USA
e-mail: Esieg@nm.org

E. K. Bailey et al. (eds.), *Inpatient Neuropsychology*, https://doi.org/10.1007/978-3-032-05880-5_1

routinely examined (Kubu et al. 2016; Sweet et al. 2021). As a result, the number of neuropsychologists currently employed in inpatient settings, the range of patient populations served, types of practice approaches utilized, and the various roles and responsibilities performed are not fully understood.

Further, many neuropsychologists complete their training primarily, or almost exclusively, in outpatient clinics. Even in programs offering training experiences with inpatient populations, few programs truly provide competency in inpatient practice. For those who have never worked in inpatient settings, it is not unusual to feel unsure about how best to effectively and efficiently address the needs of patients, families, and treatment teams. In fact, many inpatient neuropsychologists report a steep learning curve when first initiating inpatient care, often having to employ an array of assessment and intervention skills, including aspects of health, rehabilitation, and medical psychology. While several helpful texts exist that discuss inpatient assessment approaches (Kessler 2006; Pavol 2019), resources for managing other common roles and responsibilities are limited (Johnson-Greene 2018; Blair and Gorman 2003). In fact, a recent informal survey revealed 80% of respondents (N = 69) were interested in developing a professional support group for discussing inpatient clinical practice, with roughly half also seeking routine discussions on the business of inpatient services (American Psychological Association [APA], Division 22 Listserv 2011). Although knowledge-based and applied competencies in neuropsychology are well established, there is no current organizational guidance specific to the practice of inpatient neuropsychology and few empirical or practical resources to reference.

Of the resources that exist for inpatient practice, most are focused on rehabilitation. This is not unexpected given that many inpatient neuropsychologists practice in acute care and rehabilitation units. These programs often include neuropsychologists as integrated team members due, at least in part, to accreditation standards (i.e., The Joint Commission [TJC], Commission on Accreditation of Rehabilitation Facilities [CARF]) that historically required neuropsychologists to address neurocognitive, behavioral, and emotional sequelae of acute injury. Although, due to cost, neuropsychologists are no longer exclusive in meeting this requirement (e.g., now including occupational therapy, speech therapy), the initial enaction of these standards provided an extended opportunity for neuropsychologists to gain proficiency working in the inpatient setting. As a result, both formative and practical guidance have been published describing team integration, essential roles, and assessment and treatment approaches in rehabilitation settings (Johnson-Greene 2018; Butt and Caplan 2010; Blair and Gorman 2003). However, there is paucity of literature offering guidance in non-rehabilitation, inpatient settings, such as general medical, oncology, acute psychiatric, and long-term/residential programs. To address this gap, this chapter will outline common roles and responsibilities relevant to inpatient neuropsychological practice, including consultation, integrated team membership, training, research, and organizational stewardship. We will also discuss ways to promote the inclusion of neuropsychological practice in inpatient settings and reinforce the unique value added by these services.

ESTABLISHING VALUE

Since the publication of Prigatano and Pliskin's hallmark book (2003) advocating the need to empirically examine the efficacy and utility of neuropsychological services, numerous studies have demonstrated the beneficial impact of outpatient neuropsychological evaluations. For example, consumer satisfaction of outpatient neuropsychological evaluations is consistently rated highly among physicians (Tremont et al. 2002; Temple et al. 2006), as well as patient and family stakeholders (Westervelt et al. 2007; Watt and Crowe 2017). Neuropsychological feedback in outpatient settings has demonstrated efficacy in improving patients' understanding of their diagnosis, their ability to cope, and their overall quality of life (Rosado et al. 2017). Donders' (2020) extensive clinical review further delineated the benefits of outpatient neuropsychological evaluations in the prediction and management of patient outcomes across a diverse range of patient populations.

Neuropsychological services also provide economic value to larger healthcare systems (Donders 2020; Glen et al. 2020). Examples of "incremental value" or "return on investment" may include improvements in healthcare utilization rates, costs, or lengths of stay. This is particularly important given the continued changes in healthcare economics and reimbursement rates, as well as the propensity of neuropsychologists employed in large healthcare systems (Sweet et al. 2021). Several studies investigating the economic value of outpatient services have found lower rates of healthcare utilization following completion of an outpatient neuropsychological evaluation, including reduced frequency in emergency department (ED) visits, fewer hospital admissions, and lower healthcare costs (VanKirk et al. 2013; Janecek et al. 2018).

Competencies fundamental to neuropsychological practice are not environmentally dependent; therefore, the value associated with outpatient services can be assumed to similarly apply in inpatient settings. However, few studies have investigated the clinical benefits or economic impact of inpatient neuropsychological services, where differences in practice warrant empirical examination to better elucidate the unique value added. These differences include testing environments, the number of measures administered, ability to engage, modification of standardized measurements, and—depending upon whether the inpatient provider serves as a consultant or integrated team member—the number of sessions and types of services provided. Patient-centered outcomes may also differ due to the provision of neuropsychological services earlier in the disease course, including cognitive assessments, psychoeducation, recommended strategies that promote treatment engagement, and triaging to outpatient neuropsychological services for follow-up care (Colvin et al. 2022).

Empirical Study

To date, only two studies have investigated the impact of inpatient neuropsychological services. This is concerning, as not being able to fully convey the specific utility and incremental value of inpatient services to healthcare system, consumer and/or payor stakeholder(s) may serve as barriers to the continued growth of the field. Indeed, recent surveys of inpatient providers show a desire for robust empirical study of this practice (Petranovich et al. 2023; Kullgren et al. 2015). In an early study by Bishop et al. (2003), the clinical utility of inpatient neuropsychological evaluations in guiding patient care was assessed in relation to whether discharge recommendations influenced eventual discharge disposition. Of the 100 inpatient charts reviewed, 78% of hospital discharge summaries included information about the neuropsychological evaluation, 68% included specific recommendations from the neuropsychological report, and 48% referred to specific neurocognitive results from the evaluation. In addition, when neuropsychological reports included specific discharge placement recommendations, actual discharge placement was consistent with neuropsychological recommendation 80% of the time.

A more recent study by Sieg et al. (2019) examined the impact of inpatient neuropsychological assessment on healthcare costs and utilization. Findings indicated that patients deemed to have high levels of post-discharge needs—as based on the results of their inpatient neuropsychological evaluation—accumulated higher inpatient costs over the following year compared to patients with low or moderate post-discharged needs. Further, patients who refused to undergo an inpatient neuropsychological evaluation had greater healthcare utilization, including ED visits and readmissions at 1 year. The authors hypothesize several reasons for patient refusal including disengagement in patients' own healthcare or being representative of the acuity of their illness. Ultimately, this research supports the value of inpatient neuropsychology services at a systems-level by helping identify populations at risk for mounting costs and healthcare utilization, as well as informing avenues for early intervention on a patient-level.

Quality and Program Improvement

While generalizable research efforts are needed, it is also important to identify the value of inpatient neuropsychological services specific to individual facilities, as this allows inpatient neuropsychologists to demonstrate their value to local leadership. Tracking both discipline-specific data (e.g., RVUs, number of consults, referral sources, etc.) and patient-centered outcomes through quality improvement (QI) or performance improvement (PI) activities can be just as important to hospital leadership. QI and PI activities do not require the same investment of time and effort as clinical research (i.e., grant funding, institutional review, personnel management; see Fig. 1.1), though are beneficial in guiding program development, negotiating local workload expectations, improving reimbursement rates, reducing length of stay, and/or obtaining appropriate levels of staff support.

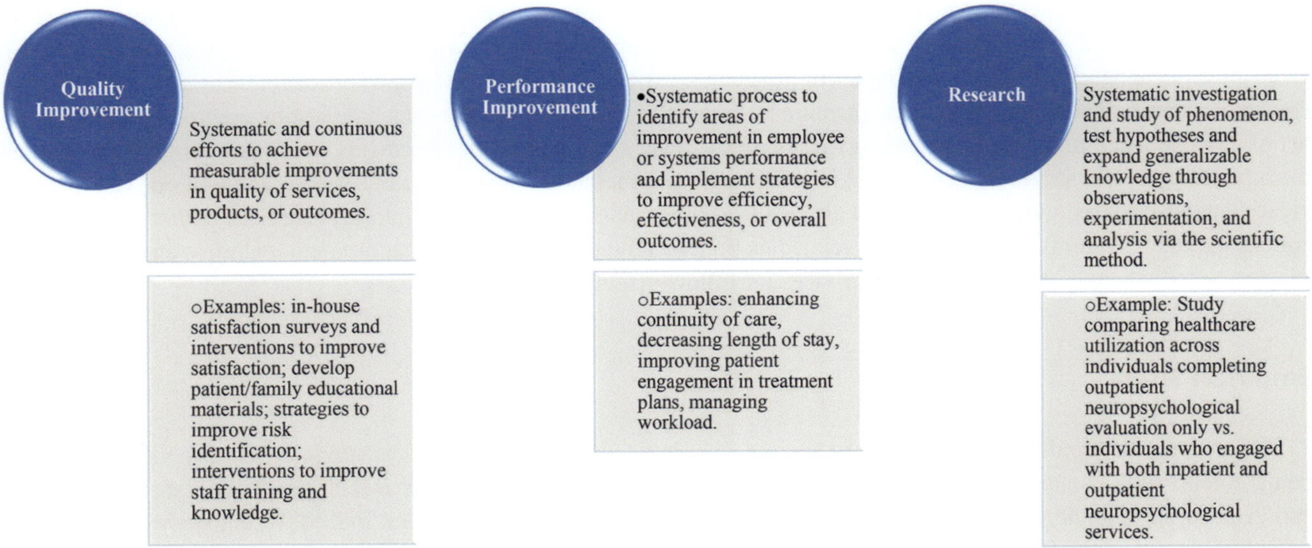

Fig. 1.1 Quality improvement, performance improvement, and research

ROLES OF THE INPATIENT NEUROPSYCHOLOGIST

Neuropsychologists who provide care to hospitalized patients may witness notable decline or amazing recovery. Compared to outpatient settings, this rewarding work is fast paced and highly dynamic. Patients often suffer from acute medical illness or injury, necessitating a flexible approach that tends to lack the predictability of scheduled outpatient appointments or the structure of a traditional cognitive evaluation. To be successful in these environments, inpatient neuropsychologists must learn to adapt assessment and treatment protocols, communicate and collaborate with diverse disciplines, and rapidly complete work products due to urgent consult needs. For most, care delivery extends beyond the bounds of discrete episodes of assessment and feedback, allowing for a deeper understanding of the experiences of patients and their families.

Within inpatient settings, neuropsychologists often provide services as either an integrated team member, working side-by-side with a range of healthcare disciplines, or as an independent consultant referred to evaluate in specific cases. Each role is distinct (though not necessarily mutually exclusive) and dictates subsequent decisions related to workflow, time allocation, productivity standards, documentation expectations, collaboration boundaries, office space, and billing.

The Inpatient Consultant

Consultation is a benchmark training competency in the field of psychology (APA 2011) and a core skill identified by the Houston Conference Guidelines (Hannay 1998) for neuropsychology specialty practice. However, the practice of consultation in the inpatient setting differs significantly from outpatient care. The medical and/or psychiatric acuity of hospitalized patients not only demands advanced clinical expertise, but also strong interpersonal communication skills and timely assessment. An effective inpatient neuropsychologist must integrate a vast knowledge base spanning medicine, neuropsychology, and psychology, while also collaborating with (and managing expectations of) diverse referral sources.

Managing Referrals

In a consultation and liaison (C&L) role, inpatient neuropsychologists often provide coverage across large portions of a medical facility. As such, referral sources are diverse, spanning multiple medical teams and various levels of patient complexity, each with specific needs and unique demands for the purpose and rapidity of services. Referrals consequently demand timely responses to prevent delays in care prior to discharge, as most inpatient admissions are brief (i.e., days). This equates to a considerable pace for consulting neuropsychologists, the navigation of which requires self-confidence, assertive communication skills, and excellent time management. Inpatient consultants also need to be adept at establishing referral parameters to maintain efficiency of workflow while simultaneously maintaining professional relationships. Referral sources should be aware of what services a consultant can reliably provide versus services that are more appropriate for another clinician (e.g., psychiatrist) and/or an outpatient referral following discharge. This can be accomplished through initial education by the consultant to referring providers.

Communicating Findings

Referrals are made when providers need assistance establishing and confirming a diagnosis, understanding the severity of cognitive impairment, or clarifying how to proceed with treatment (Pavol 2019; Karol and Sturm 2017; Tremont et al. 2002). An effective consultant delivers on these referrals by clarifying the referral question, generating a targeted clinical approach, and detailing useful recommendations. This is often done in direct communication with the referring provider, as opposed to only being conveyed through the final report.

Recommendations made in these settings frequently address decision-making abilities, safety considerations, and anticipated level of supervision upon discharge (Johnson-Greene 2018; Pavol 2019). Results of the assessment can also yield insight into the potential barriers for patients' return to driving and/or work (Coleman et al. 2002; Sherer et al. 2002). Indecisive impressions, overly comprehensive evaluations, and templated recommendations are usually not helpful in this setting. If results are inconclusive, at least some conceptualization and conclusions can still be provided—even if based on qualitative observations and clinical judgment—and if further medical work-up should be recommended to the treatment team.

Patient feedback in the inpatient setting allows for a unique opportunity to educate on the complex interplay of medical, psychological, and cognitive symptoms. In this role, the neuropsychology consultant can aid in conveying a holistic picture of a patient's circumstance, in a manner that is easily understood. While those new to inpatient work may have an initial preference to defer feedback to the medical team, research shows our feedback is highly valued by referring providers and that satisfaction with our service—including making recommendations directly to the patient and/or team—prompts re-consultation (Temple et al. 2006).

Evaluative Approaches

Inpatient consultants must maintain a breadth of applicable knowledge to provide competent services throughout a medical facility, including varying medical, psychiatric, and rehabilitation units (Karol and Sturm 2017). This requires fluency with a wide range of medical and psychiatric conditions and treatments, many of which may not be included in formal neuropsychology training or have direct impact on cognitive functioning. Similarly, proper assessment in the inpatient environment necessitates a broad, flexible skillset. Evaluations are time-limited and rarely exhaustive given illness acuity and anticipated change in medical and cognitive status over time. Evaluations may include assessment of capacity or delirium, cognitive screening, or brief neuropsychological evaluations (1–2 hours). Services are typically provided at bedside; therefore, testing measures should be selected accordingly. Recommended measures include those that can be easily transported, scored, and adapted to accommodate for confounding variables, such as testing interruptions, physical injury, sensory loss, or reduced cognitive endurance. Inpatient consultants also need to be skilled at quickly establishing rapport which can aid in facilitating engagement and motivation.

The Integrated Team Member

In contrast to independent consultants, who provide services across patient populations and medical teams, inpatient neuropsychologists may also function as integrated members within a specific unit or program. Teams are typically defined based on specific patient populations and, for this reason, close collaboration with other disciplines, contributing to shared goals, participating in team meetings, and maintaining continuity of patient care are required of this role. To be successful, integrated team members must align with the policies, culture, and long-term objectives of the unit or program while also maintaining specialized expertise. This balance applies to patient care as well as program development or quality assurance activities in service of the individual team and applicable accrediting body (e.g., CARF, TJC).

Clinical work unique to this role includes activities that are both non-traditional and, frequently, non-billable. For example, integrated team members may spend extended time on the unit, observing patients in informal and other treatment settings to better understand how cognitive impairments might impede functioning. Providing or participating in group sessions, formal education, and "hallway" consultations help foster team cohesion and align shared goals. In accordance with this ethos, the integrated team role for a neuropsychologist is as much about team functioning as it is clinical work. Strategies for how best to integrate into inpatient teams will be explored in the following section.

Table 1.1 Types of integrated healthcare teams

Team Model	Description	Challenges
Multidisciplinary	Separate roles Distinct treatment goals	Fragmented care Communication Conflicting Priorities
Interdisciplinary	Partially shared roles Integrated treatment goals	Collaboration Lack of Common Language
Transdisciplinary	Fully overlapping roles Unified treatment plan	Reimbursement Communication Conflicts Ethics Competency

Integrated Care Team Models

There are several team approaches utilized within inpatient settings. These include multidisciplinary, interdisciplinary, and transdisciplinary models (Table 1.1), the differences between which are helpful to understand when attempting to get a better sense of team culture. While some teams neatly embody one of these models, others may reflect a combination of multiple dynamics depending on program needs, resources, stage of development, and/or program leadership. For further readings on team approaches, we recommend reviewing *Practical Psychology in Medical Rehabilitation* and *Handbook of Rehabilitation Psychology, 2nd edition* (Dunn 2017; Butt and Caplan 2010).

Multidisciplinary teams include a group of individuals with diverse expertise who work collaboratively to provide comprehensive patient care. This approach ensures patients receive all services necessary to address multiple concerns in a singular setting. Although the team works toward the common goal of improving patient outcomes, each provider practices independently to develop discipline-specific treatment plans. There is no overlap of duties or treatment, although team members share their findings during rounds or team staff meetings. Despite the unifying goal, the inherent challenge of this model is fragmented care. Without open and routine communication, collaborative care can deteriorate and lead to conflicting expectations (Butt and Caplan 2010).

Interdisciplinary teams operate under the premise of shared responsibilities and a degree of role overlap. This overlap of roles often requires staff to develop skillsets outside of their own discipline. For example, a neuropsychologist may need to understand more about a medical procedure or a treatment task used by nursing to better support patient care. Interdisciplinary team staff meetings often include discussion of assessment findings, identification of treatment priorities, and determination of task responsibilities. The inherent challenge with this model is collaboration. Team members must effectively and openly communicate to negotiate priorities, reconcile potential role conflicts, and resolve differences about treatment decisions. In environments with high workload demands, sufficient time for true interdisciplinary collaboration may be limited. Settings that prioritize interdisciplinary care often require support from program and hospital administration to ensure providers are allotted protected time to engage in such activities.

Transdisciplinary teams posit that each provider assumes roles and responsibilities that significantly overlap with those of other disciplines. The elimination of disciplinary lines is intended to promote a unified treatment plan implemented by all, regardless of training or discipline. This model is rare, given significant, inherent challenges. There are financial reimbursement concerns, high communication demands, risk of personality conflicts, mandates to share training and skills, competency and ethical concerns, and the uncomfortable requirement to divest from discipline-specific training.

Regardless of the specific model used, teams optimally include consistent staff for each program or unit (e.g., oncology team) and a plan to ensure transmission of clinical information across all shifts. There is a collective commitment to a mission as well as a set of values, beliefs, and norms. Discipline-specific care is rendered in service of promoting team cohesion and solidarity.

Team Dynamics

Historically, inpatient treatment teams, within medical settings, have had a vertical structure of leadership, with the physician leading the team (medical model). However, the optimal team structure enables fluid leadership that encourages collaboration and flexibly accommodates prioritization of certain disciplines as dictated by patient need. For example, a physician may defer team leadership to the neuropsychologist when, after thorough medical work-up, a functional neurologic symptom disorder is suspected. Similarly, ideal team leadership considers (and if needed, addresses) any circumstances that pose a risk for staff conflict ("splitting") or team fragmentation. It is not uncommon for patients or families to align with specific providers/disciplines whose treatment or recommendations are optimistic for improvement, while simultaneously rejecting others that address unpleasant aspects of their condition. Interventions in such circumstances can include intentionally presenting recommendations and treatment plans as a unified team as well as patient and family education. Teams with strong dynamics make decisions that are collaborative, collegial, and mutual. Inclusive decision-making ultimately benefits patients and families, but also serves to reduce staff burnout by ensuring all team members have a purpose and voice.

Team Integration

Integration into an interdisciplinary treatment team happens gradually as the result of ongoing, deliberate intent. Understanding team dynamics, such as structure, processes, leadership, and communication style can only be obtained over time. Akin to meeting a new patient, establishment of rapport with fellow staff can help build credibility and ease any initial apprehension on the part of the provider and the team. Figure 1.2 offers tips that promote successful integration when first joining a team. Of particular importance is embodying the values of a care team, such as being helpful and supportive of colleagues. To this end, the inpatient neuropsychologist may elect to accept, or say "yes", to initial referrals that may otherwise be reduplicative, benign, or misinformed, and defer gentle boundary-setting or education until after establishing a foundation of trust and reliability. Similarly, delaying advocation for clinic change and instead creating a working list of ideas that can be eventually developed and implemented may help avoid harmful initial misattributions of forcefulness and arrogance.

In addition, Blair and Gorman (2003) offer several "survival tips" for neuropsychologists working in rehabilitation settings, though these remain applicable for all types of inpatient practice (Table 1.2).

Once a sound, working relationship has been established, the neuropsychologist can take a more active role in team decision-making and providing professional feedback regarding the optimal use of neuropsychological services within the specific setting. This is vital as neuropsychologists, with expertise in both interpersonal and brain-behavior relationships, are often asked to occupy numerous roles beyond that of a diagnostician (Fig. 1.3).

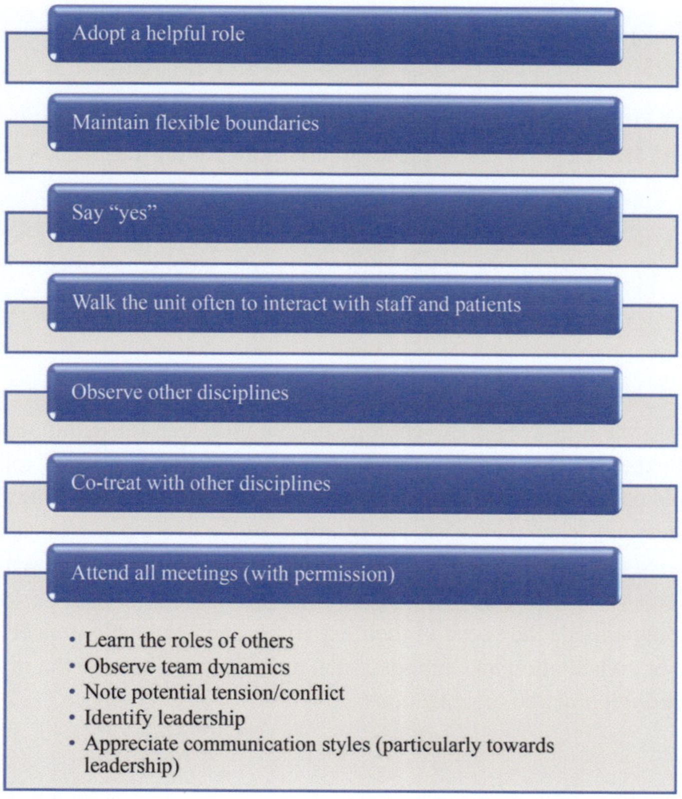

Fig. 1.2 Integrating into the treatment team

Table 1.2 Blair and Gorman's (2003) strategies for successful team integration

Strategy	Examples
Join the System First	Respect your colleagues by learning the culture/milieu before attempting to make changes.
Learn the Language of the Unit	Each treatment team will utilize different medical and treatment terminology based on the patient population (i.e. rehabilitation vs. oncology).
Know Where You Fit	Understand the hierarchy and where you fit it. Most inpatient providers have not worked with neuropsychologists.
Become the Glue – Translate, Integrate, Communicate	Teams may have diverse opinions about treatment approaches. Helping team members communicate more effectively with each other, and patients/family is a foundational skill of neuropsychologists.
Learn and Demonstrate the Art of Professional Deferral	By deferring to others on "peripheral issues" team members are more likely to accept your ideas about "central issues." Balancing the need to emphasis the correctness of a clinical "fact" vs. reinforcing a relationship with other team members.
Remain Alert to Ethical Challenges	Managing patient confidentiality within a treatment team setting.
Do Not Over Test	Due to acuity, limited length of stay, patient fatigue, reimbursement rates, other providers evaluating/treating cognition, etc. call for the need to limit testing batteries.
Deal Productively with Administration	Constructively respond to admin requests, demonstrate your value, and that you are an asset to your patients and team.
Understand Basic Coding and Reimbursement Issues	Do not assume that administration/billing department understands reimbursement issues for neuropsychology.
Do Not Forget You are a Neuro-*psychologist*	Do not only emphasize the "neuro" part. A good clinical neuropsychologist requires you to be both a good clinical psychologist and neuropsychologist. You need both skills sets equally when integrated into teams.

Adapted from Blair & Gorman, 2003

Ensuring team members understand how neuropsychology can be most helpful within that specific setting will serve to delineate clear boundaries and expectations, thereby decreasing the probability of interdisciplinary conflict. However, some degree of conflict is an inevitable part of working within a team (Fig. 1.4). The neuropsychologist should be adept at identifying and managing these disagreements when they arise. For example, as the overlap of clinical services with other disciplines is common, preemptive discussions of scope of practice and promoting shared knowledge will help ensure that each field of expertise is utilized in the patients' best interests. These discussions are important because feelings of territoriality, or concerns that a colleague is practicing outside the scope of their expertise, can negatively impact team dynamics. In such cases, it is helpful to have resources for consultation and support, such as colleagues outside the immediate team or other professional mentorship developed through practice organizations.

Clinical Services

Within an integrated team, neuropsychologists are typically consulted for every new admission to the medical or rehabilitation unit. Moreso than outpatient practice, working with inpatients may include interventions, such as provision of brief, supportive therapy and psychoeducation addressing the impact of medical and psychiatric conditions on brain health

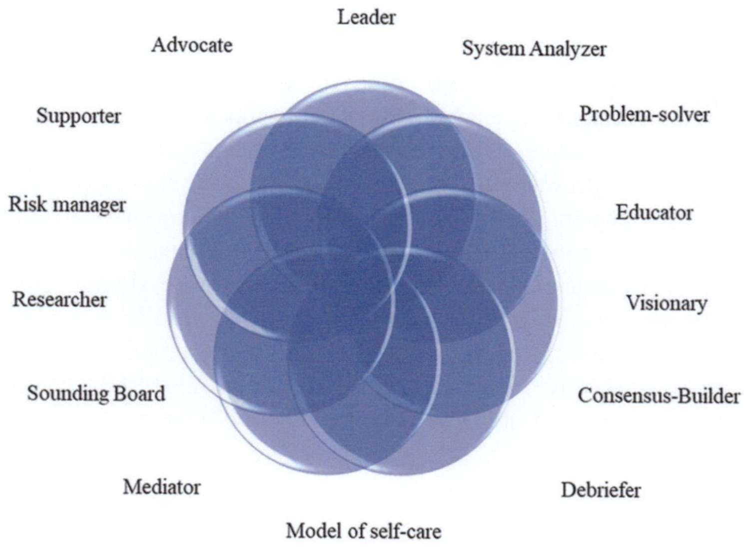

Adapted from Brown & Folen, 2005

Fig. 1.3 Additional roles of the inpatient neuropsychologist

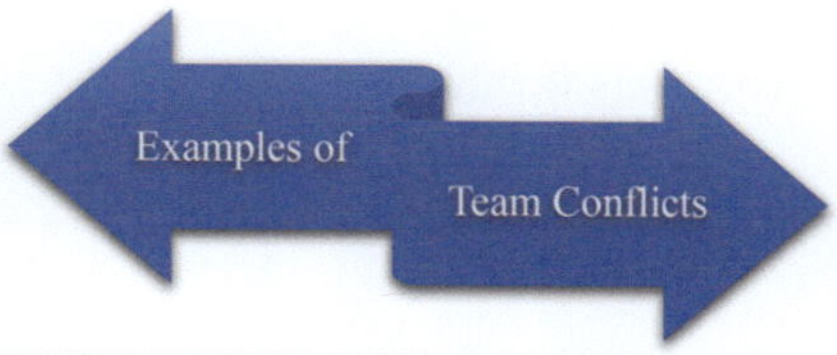

Decisions regarding patient advocacy versus staff education.
Addressing poor team functioning contributing to maladaptive patient behaviors.
Ethical concerns and capacity issues.
Intersectionality clashes between patients/caregivers and staff.
Patient confidentiality versus team communications.
Scheduling conflicts and care interruptions.
Maligned expectations regarding roles and responsibilities.

Fig. 1.4 Examples of team conflicts in the inpatient setting

(e.g., sleep, pain, nutrition, substance misuse, grief/adjustment, anxiety, depression). Fortunately, the nature of inpatient admission allows for numerous opportunities throughout the day for flexible delivery of services compared to outpatient settings. This flexibility is essential to accommodate and prioritize changing clinical needs. For example, patients may experience emotional lability or behavioral outbursts that require immediate intervention.

Regardless of whether the referral is for cognitive evaluation, serial assessment, or intervention, most services will start by collecting relevant patient history focusing on recent medical or psychiatric changes contributing to a patient's mental status. Premorbid functioning and modifiable factors potentially contributing to cognitive efficiency should also be explored. Mood and safety should be assessed as part of this process, and should a patient be determined to be high risk for self- or other-directed harm, team members should be notified directly so immediate steps can be taken to protect the well-being of the patient. Depending upon the severity of injury, collaterals may be the initial source of background information. Preliminary cognitive and psychological screeners are typically administered during the intake visit or shortly thereafter. This information is used to establish a neuropsychological plan of care for the patient's admission that also includes recommendations for treatment providers on how best to manage and approach current cognitive, behavioral, and/or psychological sequelae. One particular benefit of delaying a fuller evaluation until after an initial intake is that any necessary accommodations for illness acuity, environmental restrictions, and patient limitations due to physical, sensory, and/or cognitive impairments can be identified prior to formal cognitive evaluation.

Unlike neuropsychologists who function as consultants, those embedded within treatment teams need not fully address the referral question in this initial visit. Instead, patients can be followed throughout their admission which allows for information and clinical data to be collected longitudinally, updating a patient's diagnosis, prognosis, and treatment plan as indicated. At times, the ability to provide neuropsychological services will be limited by acuity and/or severity factors. In these situations, maintaining communication and rapport with the patient (or family) and monitoring progress during routine medical rounds remains essential. Continued engagement also enables the neuropsychologist to monitor for any acute cognitive or emotional changes. Once a patient is deemed appropriate for more comprehensive cognitive examination, these evaluations still tend to be more targeted and shorter compared to other settings (Sweet et al. 2021). Additional opportunities to collect data include co-treatment with other disciplines or observation of the patient across settings and providers. This information can enhance conceptualization and assist with treatment and discharge recommendations. Interventions focused on cognitive rehabilitation or behavioral syndromes (e.g., insight awareness development, anger management) may also be implemented at this stage of recovery.

Hybrid Roles

An inpatient neuropsychologist may decide to formulate a hybrid position that includes independent consultation to other areas of the inpatient facility, while remaining integrated with a specific team or unit. These role divisions should be intentional and advertised accordingly. Time and work parameters of each role should be negotiated in advance with the clinical supervisor or hospital leadership so that expansions or reductions of duties can be titrated as needed. We recommend consideration of the prevalence of neurological impairments within each medical unit to assist in defining these roles.

WORKING WITH FAMILIES

One of the most important duties of the inpatient neuropsychologist is working with families and care-partners, as they play a crucial role in the inpatient process. Their ability to cope with a loved one's illness or injury and engage in the treatment process is a strong predictor of patient outcomes (Rashid et al. 2023; Norton et al. 2013). Given the interconnectedness, interdependence, shared emotional connections, and agreed-upon roles and rules family exist within, inclusion of family members in the treatment process, in addition to addressing their own reactions and needs, is vital. Inpatient neuropsychologists need to establish a working alliance with family members.

Depending upon the patient population and setting, inpatient neuropsychologists may work as frequently (or more) with family members as with the patient. Family services can include both patient- and caregiver-centric recommendations, including education about diagnosis and prognosis, neuropsychological feedback, supportive therapy, as well as recommended strategies to manage behavioral, cognitive, and emotional changes. These services are useful in reinforcing the essential role of family as care partners, encouraging family-team collaboration, and alleviating potential anxiety (Gontkovsky and Golden 2010). To illustrate this point, McLaughlin and Carey (1993) found that family members endorsing higher rates of distress upon admission experienced more conflict with the treatment team, poorer adjustment to the patient's disability, and longer lengths of stay. Supportive interventions aimed at the caregiver experience can also help build the resilience necessary for self-care. This may include reminders to attend to their own critical needs or providing community resources in preparation for discharge or transition to the next level of care.

Just as patients' reactions and adjustments evolve across the course of their hospitalization, family members are similarly navigating their own reactions and adjustment, while simultaneously providing support and assistance to their loved one. It is important to meet families where they are along the adjustment process (Lezak 1986). Cavallo and Kay (2005) describe 3 main stages of family adjustment following brain injury, that are nevertheless applicable across patient populations. These stages are:

- *The Acute Phase:* Family focus remains on issues of survival while long-term impact of physical and neurological damage is minimized. Normal routines are disregarded so all resources can be directed to their loved one.
- *The Rehabilitation Phase:* Once medically stable and intensive rehabilitation is initiated, families hold high expectations for recovery. Continued treatment engagement keeps the possibility of unlimited recovery open. Families begin to adjust in preparation for receiving their loved one back into the family system.
- *The Integration Phase:* Lengthiest and most difficult phase, with high variability across families. As rehabilitation care ends, the injured person is gradually reintegrating into the community and the family system. Expectations for complete

recovery begin to recede as the reality of permanent physical, cognitive, and/or behavioral change becomes apparent. Family systems attempt to find/develop a new balance, so that all family members can proceed with their own lives.

Understanding what adjustment phase families are in will help determine treatment approaches. For example, during acute hospitalizations, emotional distress is often high as families attempt to navigate the uncertainty of their loved one's survival or recovery. During this period, it is not unusual for family members to experience shock, grief, trauma, and overwhelm. For patients who stabilize and recover, family stress levels often decrease as uncertainty and understanding of the condition improves. Distress and anxiety may increase again, as families prepare to discharge home due to worry about their ability to care for their loved one sufficiently. Helping families manage their distress is an important role for inpatient neuropsychologists as it has both short- and long-term impacts. Without treatment, chronic caregiver distress may progress to long-term symptoms of depression, fatigue, posttraumatic stress disorder (PTSD), and anxiety (Kreutzer et al. 1994; Shirasaki et al. 2024). This is associated with a broad range of poor patient outcomes, including higher rates of healthcare utilization and disease progression (Rashid et al. 2023; Norton et al. 2013).

Family needs also evolve along the disease continuum, with health information tending to be desired during the early or acute stages of hospitalization (Kreutzer et al. 2018; Sinnakaruppan and Williams 2001). Neuropsychologists excel in how best to deliver complex information in these situations. For example, information should be repeated and delivered in a way that limits esoteric terminology and medical jargon. In addition to tailoring verbal education and feedback, materials in the form of handouts or videos should be provided to help family members retain new information; while also ensuring they have resources to review as needed later. Recommendations for delivering feedback, such as providing information and resources, can be found in Fig. 1.5.

While family needs evolve throughout the course of admission, most treatment teams emphasize family education and skills development, while other crucial needs may go unrecognized or be minimally addressed. Kreutzer et al. (2018) examined both met and unmet needs of families during inpatient brain injury rehabilitation. Findings demonstrated the most common unmet needs fell within the domains of emotional and instrumental support. Examples of unmet emotional needs included wanting to discuss feelings about their loved one with someone who has gone through the same experience, helping

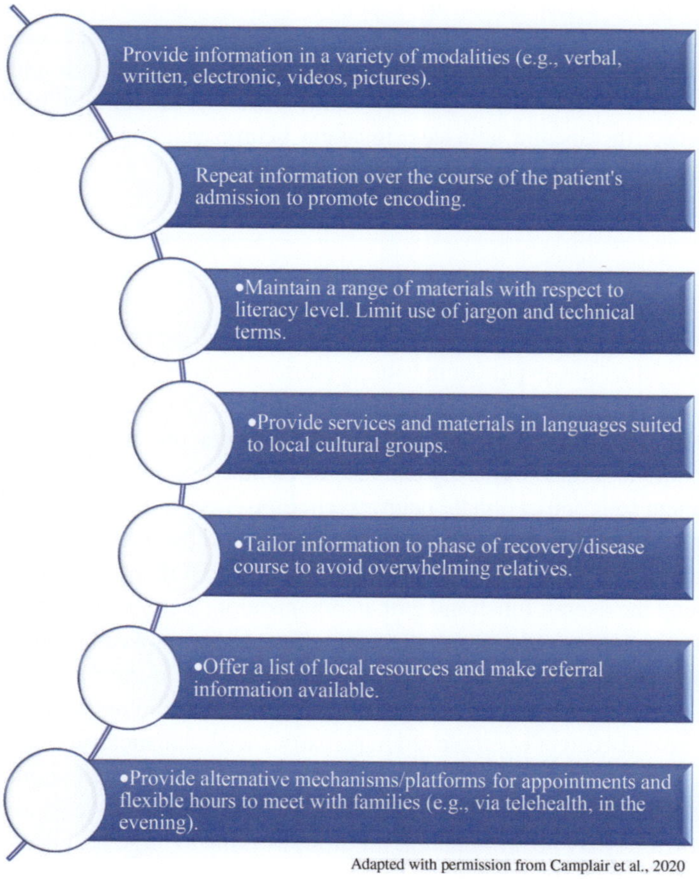

Fig. 1.5 Strategies to optimize feedback

prepare for the worst, helping with negative feelings (such as resentment) toward their loved one, and addressing doubts and fears about the future. Unmet instrumental needs included needing assistance with household activities, time with friends, not having a break from problems, and paying attention to their own needs while their loved one was hospitalized.

To this end, self-report measures can be useful for identifying the full spectrum of caregiver needs. One such example is The Family Needs Questionnaire-Revised (FNQ-R; Kreutzer 1988). Given the impact of family engagement on patient outcomes, identifying perceived needs prior to discharge is essential. As family needs will vary by setting and disease type, it is important for the inpatient neuropsychologist to have an extensive understanding of the literature for each patient population, as specific setting needs, standardized measurement tools, common unmet needs, and identifiable future needs will help inform an effective plan of care.

Special Considerations

Conveying Prognosis

Depending upon the patient population you are working with, the role of hope for recovery may vary. As such, discussing prognosis so families can set realistic expectations for the future and develop plans for next steps is critical. This will naturally vary depending on the characteristics, severity, and mechanism of the disease or injury. For those with poor medical prognosis, such as terminal cancer or neurodegenerative disease, future planning may include liaising with palliative care or hospice services, completing Advanced Directives, and addressing anticipatory grief. For patients who are in the early stages of recovery from an acute neurological illness or injury, discussion of prognosis may include adjustment to disability, predicted future care needs, and preemptive problem-solving of likely impact to functional status. Being able to discuss realistic prognosis while managing hope can be challenging, particularly as you first learn to develop this skill.

Beliefs and Culture

Provision of neuropsychological services should include consideration of different value and belief systems, cultural backgrounds, level of acculturation, experiences and comfort with healthcare systems, and available financial and social resources. While all families experience grief and distress when a loved one suffers a severe illness, how emotional distress is managed and expressed may differ across cultures (Simpson et al. 2000). Understanding and respecting each family's culture and unique needs is crucial to helping them manage expectations, take part in training and care, and facilitate the transition home or to the next level of care (Camplair et al. 2020; Watson et al. 2022). Finally, by identifying cultural and psychological factors that impact therapeutic alliances, inpatient neuropsychologists can provide guidance to treatment teams on how best to facilitate positive partnerships with families.

MANAGING BURNOUT

The various roles that an inpatient neuropsychologist can pursue are exciting and rewarding. That said, the patient population is inherently vulnerable. Most are experiencing the acute to subacute effects of serious illnesses or injuries and adjusting to an acquired disability. Similarly, loved ones often shoulder the burden of their family members' illness, while also navigating their own emotional reactions. In many cases, we are interacting with individuals at one of the worst times of their lives. Prolonged exposure to others' suffering can lead to compassion fatigue and burnout. As such, inpatient care can be a challenging environment for compassionate neuropsychologists as emotional reactions are normal.

Burnout is typically rooted in chronic workplace stress and characterized by emotional exhaustion, cynicism, and a sense of inefficacy (Maslach and Leiter 2017; The World Health Organization [WHO] 2019). It negatively impacts work effectiveness and productivity, decreasing the overall functioning of an organization. Burnout also contributes to poor quality of care, poor patient satisfaction, worse patient outcomes, job dissatisfaction, and employee turnover (Maslach 2017). There is an increased risk of burnout-related syndromes among healthcare professionals if normal emotional reactions are not balanced with self-care and organizational support (Kinman and Teoh 2017). Neuropsychologists are equally at risk, with high rates of burnout among those who are early career or employed in pediatric, academic medical centers, and Veterans' Affairs (VA) settings (Shapiro et al. 2025). These same groups also demonstrate lower levels of professional fulfillment than their neuropsychology counterparts. Identified contributions to this difference include perceived level of autonomy, job flexibility, and workplace culture—factors of which may be more salient for those working in large healthcare systems compared to private practice.

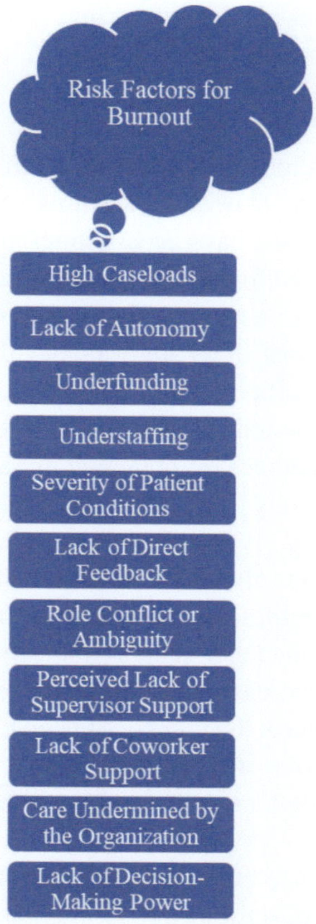

Fig. 1.6 Risk factors for burnout

Given the negative impact on the provider, their patients, and the healthcare system at large, it is important for inpatient neuropsychologists to monitor themselves and their colleagues for signs and symptoms of burnout and advocate to address any modifiable factors that may increase risk (Fig. 1.6). Unrecognized or unaddressed burnout can erode team dynamics, communication, and work culture. Appreciating these risks and intentionally implementing effective coping strategies are paramount to maintaining professional productivity and patient safety.

Strategies for preventing burnout include anticipating the psychological impact of providing care, understanding your own stress response, and developing a personalized self-care plan that fits into your lifestyle, personality, resources, and interests. At least some interventions should be tailored for regular use throughout the workday in short intervals, such as brief breaks away from one's desk, taking a walk, or having conversations with colleagues during meals. Equally important are strategies that promote a healthy lifestyle outside of work, such as obtaining adequate sleep/diet/exercise, human connection, pursuit of interests, and use of positive thinking (Saban et al. 2013). However, ultimate effective management of burnout requires both individual and systems-level intervention (Feigon et al. 2018; Shapiro et al. 2025).

NON-CLINICAL ROLES

Research Activities

It is unclear how many inpatient providers also engage in funded research. As clinical workload and reimbursement demands rise, neuropsychologists are increasingly becoming exclusive in their role as clinician or researcher. An exception is academic medical centers (particularly those with an established history of grant funding and research activity), where neuropsychologists are still routinely employed with clear expectations for productivity across both domains. This inevitably presents challenges managing both research/grant deadlines while also prioritizing necessary clinical care, especially if supportive resources such as psychometricians and/or administrative staff are not available. Despite these obstacles, engaging in both activities can be quite rewarding, as research provides a greater range of activities, allows for contribution to patient care across multiple tiers of influence, and often leads to career advancement.

Institutional Review Board

Inpatient neuropsychologists hired with expectations to conduct research are often required to obtain grant funding to offset salary disbursement from fewer billable hours. However, obtaining grant funding for inpatient research has unique challenges. At a minimum, the neuropsychologist must have access to an institutional review board (IRB) and, ideally, a grants office or administrative grants support staff. The former ensures the research is completed ethically, and the latter two ensure the neuropsychologist (and intended staff) have a way to get the funds they are given for their research. If an employing institution does not have a dedicated IRB, external IRBs are available, depending on region and purpose. Also, while most major academic institutions have large grants offices, this may not be true for smaller or private institutions. Rather, these organizations may have well-established administrative pathways to receive various types of external funding (governmental, private, charitable, etc.), which are good resources for obtaining initial support.

Grant Funding

Institutional and Private Grants: Larger institutions often have various internal mechanisms for small-scale grant funding (i.e., generally between $1000 and $50,000). The purpose of these grants is to collect pilot data or pay for a portion of an otherwise completed project (e.g., higher-level statistician support for existing data, data-pull from electronic medical records). These funding mechanisms generally have easy application processes and a brief duration cycle (1–2 years). However, given the scale of these funding opportunities, the number of people able to be supported is generally very limited. Most awards of this type only cover a small portion of a primary investigator's salary (5–20%, in general), or some combination of small salary coverage for the primary investigator (~10%) and a greater allotment for a research assistant (~20%), as they are generally less expensive to employ.

Grant funding can also be obtained through private organizations that range in purpose and mission, such targeted for specific diseases (professional or advocate-oriented disease societies), charitable organizations, geographic location, populations, or aims (e.g., prevention, intervention, outcome measurements). These outside funding sources can offer a range of funding opportunities from small-scale projects like those offered by institutions, to very large projects akin to a government-funded project. Consistent with this, application length and complexity typically mirror the size of the funding award. These awards can be excellent alternatives to governmental funding; however, the award cycles or opportunities may be more sporadic or limited.

Federal Grants: Governmental funding is one of the most competitive application processes. The National Institute of Health (NIH) offers a range of application opportunities from training grants to multi-center and center-development grants. Recipients of NIH grants are generally those on academic tracks such as a full or nearly full-time research focus. Recipients typically have robust catalogues of publications and histories of receiving similar awards prior to a large federal grant. Federal funding is generally intended to cover multiple grant-related personnel including the primary investigator, research assistants, mentors, co-investigators, and sub-investigators. Larger projects may even include multiple sites or centers, with budgets upwards of $1000,000 to $5,000,000. Consistent with the significantly higher funding award, the application process is usually rigorous, complex, and lengthy. As such, most large academic institutions have dedicated grant staff who assist the primary investigator in grant planning, form completion, checklists, and submission of the prepared documents. Final grant application packets often exceed 100 pages and are commonly closer to 300 pages.

Government agencies, such as the Department of Defense (DoD) or the VA, also have grant funding opportunities available to those practicing within these settings for the purpose of research that supports active-duty military and veteran populations. Like other institutional and governmental grants, there is often a range of intended funds, project sizes, and purpose. These grants have a corresponding range of length and complexity of applications matching award size, with larger award sizes having similar expectations for demonstrations of prior academic success. There are also career development awards within these practice settings to encourage relatively newer investigators to pursue research in areas prioritized by the given institution.

Training

Training experiences in inpatient neuropsychology are not readily available at all institutions. In those that do offer inpatient training, the degree of exposure and quality of experience received by trainees is not well described. Limited data that

Knowledge of:

□Roles, responsibilities, skills and values/attitudes of different disciplines in a range of settings (e.g. healthcare, education, forensic, policy, community).
□Different systems relevant to neuropsychological practice and strategies to negotiate these systems effectively.
□Interprofessional collaboration and its role in contributing to optimal care.
□General and setting-specific contributions and limitations of neuropsychological expertise.
□Functional impact of neuropsychological disorders on patient abilities to negotiate systems (e.g. healthcare, employment, education) and relevant management strategies.

Demonstrate the ability to:

□Collaborate effectively with individuals and systems across disciplines, adapt communications to context appropriately, address questions about neurobehavioral functioning, and contribute to integrated understanding.
□Work cooperatively across systems, validating shared values and demonstrating mutual respect.
□Work collaboratively, providing neuropsychological expertise in clinical and research teams.

Adapted from Roper, et al. (2018) & Rey-Casserly, et al. (2012)

Fig. 1.7 Foundational competencies relevant to integrated care

are available suggest most inpatient training occurs at the postdoctoral level (Baum et al. 2022). However, these data are specific to pediatric rehabilitation settings and the generalizability of these findings is unknown. At present, there is no current consensus on specific training requirements for inpatient neuropsychology. While the training competencies of consultation and interdisciplinary systems address many of the skills needed for inpatient settings, the extent and efficacy of these competencies for inpatient practice remain unclear (Smith and CNS. 2019). Roper et al. (2018) outlined competencies important for working in integrated healthcare environments across graduate, intern, post-doctoral, and post-licensure stages (Fig. 1.7; Roper et al. 2018; Rey-Casserly et al. 2012).

Ultimately, neuropsychologists working in integrated settings, whether inpatient or outpatient, will require skill in building interprofessional relationships, effective communication, and collaborative decision-making. To this end, Roper et al. (2018) offer three broad targets for competency-based training:

- Foundational competencies related to interdisciplinary systems and integrated care.
- Functional competencies related to general neuropsychological practice.
- Knowledge and skills specific to the team's treatment setting.

As the practice of inpatient neuropsychology expands, so too will training in these settings. In creating specific training requirements for neuropsychological trainees, drawing on competencies from other professions and subspecialties can enhance the breadth of knowledge necessary to effectively function in a specialized setting.

ORGANIZATIONAL STEWARDSHIP

Hospital Education

In many medical settings, mental health conditions and the subsequent impact on recovery, treatment adherence, and engagement are not fully appreciated. Inpatient neuropsychologists have a unique opportunity to provide education about a broad range of mental health and cognitive sequelae, including how symptoms may impact care delivery, compliance, or treatment engagement. Provision of education can help advertise and describe the services offered by neuropsychology. Educational opportunities can range from multi-hour workshops, to seminars, to brief, 10-min in-services. In addition, Griffiths et al. (2014) found that staff express preference for on-the-job training alongside experienced colleagues or specialists. As experts in cognition and behavior, neuropsychologists are uniquely qualified for providing education and skill development in the management of confusion/delirium, cognitive impairment, and mental health problems. Education that includes skill development can improve staff confidence and self-efficacy, which is associated with lower staff burnout and turnover.

Hospital Committees and Response Teams

Due to the broad range of skills and competencies, neuropsychologists are often invited to serve on hospital committees and/or specialized response teams. These may include Hospital Ethics, Quality Improvement, Disaster Planning, IRB, and Life-Sustaining Treatment (LST) committees. Hospitals may also develop workgroups or multidisciplinary response teams to address additional needs specific to that facility. Often, these teams require additional training or skills specific to the needed intervention (e.g., managing disruptive behaviors, family-staff ruptures). Negotiating with supervisors or hospital administration to allow for protected time to engage in these hospital activities is always recommended.

Community Outreach

It is not uncommon for clinical providers to become active in the stakeholder communities analogous to the clinical populations served (e.g., epilepsy, stroke, cancer). Participating with local and national organizations is a great way to garner additional support for patients and foster awareness within the community. Inpatient neuropsychologists working in accredited programs (e.g., CARF, TJC) are often asked to participate in community outreach activities. These include education, such as community talks that raise awareness about a specific disease process, sharing prevention strategies, and providing resources to help the public gain access to evidence-based care. Community outreach can also include specialty input to help guide service development, such as offering general training or material to staff, advocates, or patient and family support groups run by local organizations.

SUMMARY

The role of an inpatient neuropsychologist is both intellectually stimulating and deeply rewarding, offering unique opportunities to make meaningful contributions to care during some of the most critical and vulnerable moments of patients' lives. By providing clarity during a time of medical or psychiatric uncertainty, advocating for cognitive and psychological well-being, and collaborating closely with multidisciplinary providers, inpatient neuropsychologists play a pivotal role in enhancing diagnostic accuracy and treatment planning. Compared with traditional outpatient settings, inpatient practice offers a dynamic environment that is fast paced, less predictable, and often includes a broader range of professional and clinical activities. Whether working as a consultant or as an integrated team member, collaboration, effective communication, and efficiency are key to success. Unfortunately, there are limited resources available that provide guidance on the day-to-day practices of working in inpatient settings. Towards this aim, professional surveys specific to inpatient neuropsychology practices and available training opportunities are needed to better prepare a future workforce. In addition, continued empirical examination to further quantify the unique value added by inpatient neuropsychological services on patient outcomes and healthcare costs will be essential for continued growth in this practice sector.

REFERENCES

American Psychological Association. Revised competency benchmarks for professional psychology; 2011, June.

Baum KT, Smith-Paine J, Tlustos SJ, Johnson A, Petranovich C. Clinical and training practices: a survey of pediatric neuropsychologists serving inpatient rehabilitation. Child Neuropsychol. 2022;28(4):510–34. https://doi.org/10.1080/09297049.2021.1993809.

Bishop CL, Temple RO, Tremont G, Westervelt HJ, Stern RA. Utility of the neuropsychological evaluation in an acute medical hospital. Clin Neuropsychol. 2003;17(4):468–73. https://doi.org/10.1076/clin.17.4.468.27944.

Blair KL, Gorman PW. Survival tips for the neuropsychologist in an inpatient rehabilitation setting. Rehabil Psychol. 2003;48(4):310–3. https://doi.org/10.1037/0090-5550.48.4.310.

Brown KS, Folen RA. Psychologists as leaders of multidisciplinary chronic pain management teams: a model for health care delivery. Prof Psychol Res Pract. 2005;36(6):587.

Butt L, Caplan B. The rehabilitation team. In: Frank RG, Rosenthal M, Caplan B, editors. Handbook of rehabilitation psychology. 2nd ed. American Psychological Association; 2010. p. 451–7. https://doi.org/10.1037/15972-032.

Camplair PS, Butler RW, Lezak MD. Providing psychological services to families of brain-injured adults and children in the present health-care environment. In: Clinical neuropsychology and cost outcome research. Psychology Press; 2020. p. 83–107.

Cavallo MM, Kay T. The family system. In: Silver JM, McAllister TW, Yudofsky SC, editors. Textbook of traumatic brain injury. American Psychiatric Publishing, Inc; 2005. p. 533–58.

Coleman RD, Rapport LJ, Ergh TC, Hanks RA, Ricker JH, Millis SR. Predictors of driving outcome after traumatic brain injury. Arch Phys Med Rehabil. 2002;83(10):1415–22. https://doi.org/10.1053/apmr.2002.35111.

Colvin MK, Roebuck-Spencer TM, Sperling SA, Acheson S, Bailie JM, Espe-Pfeifer P, Glen T, Bragg T, Bott N, Hilsabeck RC. Linking patient-centered outcomes to neuropsychological practice: a conceptual framework and opportunities for research. Arch Clin Neuropsychol. 2022;37(6):1091–102. https://doi.org/10.1093/arclin/acac015.

De Hert S. Burnout in healthcare workers: prevalence, impact and preventative strategies. Local Regional Anesthes. 2020;Volume 13:171–83.

Donders J. The incremental value of neuropsychological assessment: a critical review. Clin Neuropsychol. 2020;34(1):56–87. https://doi.org/10.1080/13854046.2019.1575471.

Dunn M. Interdisciplinary teams. In: Budd MA, Hough S, Wegener ST, Stiers W, editors. Practical psychology in medical rehabilitation. Springer International Publishing/Springer Nature; 2017. p. 483–7. https://doi.org/10.1007/978-3-319-34034-0_53.

Feigon M, Block C, Guidotti Breting L, Boxley L, Dawson E, Cobia D. Work–life integration in neuropsychology: A review of the existing literature and preliminary recommendations. The Clinical Neuropsychologist, 2018;32(2):300–317.

Glen T, Hostetter G, Roebuck-Spencer TM, Garmoe WS, Scott JG, Hilsabeck RC, Arnett P, Espe-Pfeifer P. Return on investment and value research in neuropsychology: a call to arms. Arch Clin Neuropsychol. 2020;35(5):459–68. https://doi.org/10.1093/arclin/acaa010.

Gontkovsky ST, Golden CJ. Neuropsychology within the inpatient rehabilitation environments; 2010.

Griffiths A, Knight A, Harwood R, Gladman JR. Preparation to care for confused older patients in general hospitals: a study of UK health professionals. Age Ageing. 2014;43(4):521–7.

Hannay HJ. Proceedings of the Houston conference on specialty education and training in clinical neuropsychology, September 3–7, 1997, University of Houston Hilton and conference center. Arch Clin Neuropsychol. 1998;13(2):157–8.

Janecek JK, Dorociak KE, Piper LE, Kelleher T, Pliskin NH, Gowhari M, Molokie RE. Integration of neuropsychology services in a sickle cell clinic and subsequent healthcare use for pain crises. Clin Neuropsychol. 2018;33(7):1195–211. https://doi.org/10.1080/13854046.2018.1535664.

Johnson-Greene D. Clinical neuropsychology in integrated rehabilitation care teams. Arch Clin Neuropsychol. 2018;33(3):310–8.

Karol RL, Sturm L. Models of consultation. In: Budd M, Hough S, Wegener S, Stiers W, editors. Practical psychology in medical rehabilitation. Cham: Springer; 2017.

Kessler HR. The bedside neuropsychological examination. 2nd ed. American Psychological Association; 2006.

Kinman G, Teoh K. They took off their uniform when they got home but couldn't remove the Armour. Psychologist. 2017;30:58–61.

Kreutzer JS. Family needs questionnaire. Richmond: Rehabilitation Research and Training Center on Severe Traumatic Brain Injury, Medical College of Virginia; 1988.

Kreutzer JS, Gervasio AH, Camplair PS. Patient correlates of caregivers' distress and family functioning after traumatic brain injury. Brain Inj. 1994;8(3):211–30. https://doi.org/10.3109/02699059409150974.

Kreutzer JS, Marwitz JH, Klyce DW, Schaaf KPW, Sima AP, Welch AM, Niemeier JP. Family needs on an inpatient brain injury rehabilitation unit: a quantitative assessment. J Head Trauma Rehabil. 2018;33(4):228–36.

Kubu CS, Ready RE, Festa JR, Roper BL, Pliskin NH. The times they are a changin': neuropsychology and integrated care teams. Clin Neuropsychol. 2016;30(1):51–65.

Kullgren KA, Tsang KK, Ernst MM, Carter BD, Scott EL, Sullivan SK. Inpatient pediatric psychology consultation-liaison practice survey: corrected version. Clin Practice Pediatr Psychol. 2015;3(4):340–51.

Lezak MD. Psychological implications of traumatic brain damage for the patient's family. Rehabil Psychol. 1986;31(4):241.

Maslach CS. Job burnout. Annu Rev Psychol. 2001;52(1):397–422.

Maslach C. Finding solutions to the problem of burnout. Consult Psychol J Practice Res. 2017;69(2):143.

Maslach C, Leiter MP. New insights into burnout and health care: strategies for improving civility and alleviating burnout. Med Teach. 2017;39(2):160–3.

McLaughlin A, Carey J. The adversarial alliance: developing therapeutic relationships between families and the team in brain injury rehabilitation. Brain Inj. 1993;7(1):45–51.

Norton MC, Clark C, Fauth EB, Piercy KW, Pfister R, Green RC, et al. Caregiver personality predicts rate of cognitive decline in a community sample of persons with Alzheimer's disease. The Cache County dementia progression study. Int Psychogeriatr. 2013;25(10):1629–37.

Patient Protection and Affordable Care Act (PPACA/ACA), Public Law No: 111–148, Mar 23, 2010, *amended by* Health Care and Education Reconciliation Act, Public Law No: 111–152, 124 Stat. 1029 (Mar 30, 2010) (codified as amended in sections of 26, 42 U.S.C. (2013)).

Pavol MA. Inpatient neuropsychological assessment in older adults. In: Ravdin LD, Katzen HL, editors. Handbook on the neuropsychology of aging and dementia. Cham: Springer; 2019.

Petranovich C, Johnson A, Smith-Paine J, Tlustos SJ, Baum KT. A survey of pediatric neuropsychologists serving inpatient rehabilitation, part II: billing, time allocation and tracking, and professional identity and perceptions. Child Neuropsychol. 2023;29(3):445–56. https://doi.org/10.1080/09297049.2022.2097652.

Rashid R, Aamodt WW, Horn S, Dahodwala N. Association of Caregiver Depression Risk with patient outcomes in Parkinson disease. JAMA Netw Open. 2023;6(8):e2327485. https://doi.org/10.1001/jamanetworkopen.2023.27485.

Rey-Casserly C, Roper BL, Bauer RM. Application of a competency model to clinical neuropsychology. Prof Psychol Res Pract. 2012;43(5):422.

Roper BL, Block CK, Osborn K, Ready RE. Education and training for clinical neuropsychologists in integrated care settings. Arch Clin Neuropsychol. 2018;33(3):263–8. https://doi.org/10.1093/arclin/acx141.

Rosado DL, Buehler S, Botbol-Berman E, Feigon M, León A, Luu H, Pliskin NH. Neuropsychological feedback services improve quality of life and social adjustment. The Clinical Neuropsychologist, 2017;32(3):422–435. https://doi.org/10.1080/13854046.2017.1400105.

Saban KL, Hogan TP, DeFrino D, Evans CT, Bauer ED, Louise-Bender Pape T, Steiner M, Proescher EJ, Vlasses FR, Smith B. Burnout and coping strategies of polytrauma team members caring for veterans with traumatic brain injury. Brain Inj. 2013;27(3):301–9.

Shapiro DN, Anderson C, Berrios-Siervo G, MacDonald B, Quinton TL, Wilson M. Well-being, burnout, and professional fulfillment among neuropsychologists and trainees in neuropsychology: factors that contribute and recommendations for improvement. Clin Neuropsychol. 2025:1–16.

Sherer M, Novack TA, Sander AM, Struchen MA, Alderson A, Thompson RN. Neuropsychological assessment and employment outcome after traumatic brain injury: a review. Clin Neuropsychol. 2002;16(2):157–78.

Shirasaki K, Hifumi T, Nakanishi N, Nosaka N, Miyamoto K, Komachi MH, Otani N. Postintensive care syndrome family: a comprehensive review. Acute medicine & surgery, 2024;11(1):e939.

Sieg E, Mai Q, Mosti C, Brook M. The utility of neuropsychological consultation in identifying medical inpatients with suspected cognitive impairment at risk for greater hospital utilization. Clin Neuropsychol. 2019;33(1):75–89. https://doi.org/10.1080/13854046.2018.1465124.

Simpson G, Mohr R, Redman A. Cultural variations in the understanding of traumatic brain injury and brain injury rehabilitation. Brain Inj. 2000;14(2):125–40.

Sinnakaruppan I, Williams DM. Family carers and the adult head-injured: a critical review of carers' needs. Brain Inj. 2001;15(8):653–72. https://doi.org/10.1080/02699050010025759.

Smith G, CNS. Education and training in clinical neuropsychology: recent developments and documents from the clinical neuropsychology synarchy. Arch Clin Neuropsychol. 2019;34(3):418–31.

Sweet JJ, Klipfel KM, Nelson NW, Moberg PJ. Professional practices, beliefs, and incomes of US neuropsychologists: the AACN, NAN, SCN 2020 practice and "salary survey". Clin Neuropsychol. 2021;35(1):7–80.

Temple RO, Carvalho J, Tremont G. A national survey of physicians' use of and satisfaction with neuropsychological services. Arch Clin Neuropsychol. 2006;21(5):371–82. https://doi.org/10.1016/j.acn.2006.05.002.

Tremont G, Westervelt HJ, Javorsky DJ, Podolanczuk A, Stern RA. Referring physicians' perceptions of the neuropsychological evaluation: how are we doing? Clin Neuropsychol. 2002;16(4):551–4. https://doi.org/10.1076/clin.16.4.551.13902.

VanKirk KK, Horner MD, Turner TH, Dismuke CE, Muzzy W. Hospital service utilization is reduced following neuropsychological evaluation in a sample of U.S. veterans. Clin Neuropsychol. 2013;27(5):750–61. https://doi.org/10.1080/13854046.2013.783122.

Watson WD, Lahey S, Baum KT, Hamner T, Koterba CH, Alvarez G, Slomine BS. The role of the Neuropsychologist across the stages of recovery from acquired brain injury: a summary from the pediatric rehabilitation Neuropsychology collaborative. Child Neuropsychology, 2022;29(2):299–320. https://doi.org/10.1080/09297049.2022.2086691.

Watt S, Crowe SF. Examining the beneficial effect of neuropsychological assessment on adult patient outcomes: a systematic review. Clin Neuropsychol. 2017;32(3):368–90. https://doi.org/10.1080/13854046.2017.1414885.

Westervelt HJ, Brown LB, Tremont G, Javorsky DJ, Stern RA. Patient and family perceptions of the neuropsychological evaluation: how are we doing? Clin Neuropsychol. 2007;21:263–73.

World Health Organization [WHO]. Burn-Out an "Occupational Phenomenon": International Classification of Diseases; 2019, May.

Clinical Practice, Billing, and Marketing of Neuropsychology Services in Inpatient Settings

2

Erica L. Dawson, Douglas Johnson-Greene, and Cady Block

ABBREVIATIONS

AACN	American Academy of Clinical Neuropsychology
AAMC	Association of American Medical Colleges
AHRQ	Agency for Healthcare Research and Quality
AMA	American Medical Association
AMGA	American Medical Group Association
APA	American Psychological Association
CMS	Centers for Medicare and Medicaid Services
COVID-19	Coronavirus Disease 2019
CPT	Current Procedural Terminology
EEG	Electroencephalogram
fMRI	Functional Magnetic Resonance Imaging
HR	House of Representatives (*bill originating in*)
HRB	Harvard Business Review
ICD	International Classification of Diseases
MIPS	Merit-Based Incentive Payment System
MOU	Memorandum of Understanding
MRI	Magnetic Resonance Imaging
NAN	National Academy of Neuropsychology
PM&R	Physical Medicine and Rehabilitation
SCN	Society for Clinical Neuropsychology
US	United States
VA	Veterans Affairs (*the Department of*)
WHO	World Health Organization
wRVU	Work Relative Value Unit

E. L. Dawson (✉)
Department of Neurology, The Ohio State University Wexner Medical Center, Columbus, OH, USA
e-mail: Erica.Dawson@osumc.edu

D. Johnson-Greene
Department of Physical Medicine and Rehabilitation, University of Miami Health System, Miami, FL, USA
e-mail: Djgreene@med.miami.edu

C. Block
Department of Psychiatry and Psychology, Mayo Clinic Florida, Jacksonville, FL, USA
e-mail: Block.Cady@mayo.edu

E. K. Bailey et al. (eds.), *Inpatient Neuropsychology*, https://doi.org/10.1007/978-3-032-05880-5_2

INTRODUCTION

Neuropsychologists play an important role in inpatient settings, providing essential and complementary assessment and treatment services to persons with acute illness and rehabilitation needs, many of whom experience cognitive and behavioral challenges (Novack et al. 2010). We are uniquely positioned to integrate objective cognitive data with emotional and behavioral symptoms, as well as other contextual information, to inform diagnostic considerations and patient care. Inpatient neuropsychological evaluations are often used for treatment and discharge planning, as well as anticipating safety issues and impairment of essential functional skills upon discharge, such as the ability to make informed medical decisions.

There are unique practice considerations in inpatient settings compared to outpatient settings. For example, inpatient evaluations tend to be much shorter, in part due to the evolving nature of many illnesses, which translates into time-limited applicability (Johnson-Greene 2018; Sweet et al. 2021). These targeted assessments are also typically intended to answer specific questions, such as the capacity to consent to medical treatment or anticipated need for supervision following discharge. Another important goal of inpatient evaluations is to produce cognitive baseline data for future evaluations when there are questions regarding disease progression, improvement of function, or prediction of treatment responsiveness (Bercaw et al. 2011). Lastly, inpatient settings necessitate the provision of peripheral services beyond the evaluation, including intervention and education to patients, family members, and other care providers. Despite these important roles and the significant number of neuropsychologists who report practicing in inpatient settings, little has been written about the business and billing aspects of neuropsychological practice in this context.

Some business questions that immediately become apparent are the correspondence between the professional standards of care, the capability of providing the desired services, and being reimbursed for one's time. In this chapter, we will explore the business of neuropsychology in inpatient settings, with an emphasis on different models of service delivery, billing practices, barriers to reimbursement, and marketing considerations. It is important to note that citations were referenced where available at the time of this publication. The remainder of this chapter's content was derived from experiences of the authors, all of whom operate within academic medical and rehabilitation settings in the United States. Some information in this chapter may be less applicable to institutions that are governmental, resourced differently, or operate within the context of a socialized or other type of healthcare system. Additional information on working within a larger hospital system, with third-party or community stakeholders, and pertaining to issues of equity/diversity/inclusion in the inpatient setting, is outside the scope of this chapter but can be found elsewhere.

INPATIENT NEUROPSYCHOLOGY PRACTICE

Practice Settings

Inpatient neuropsychological evaluations occur in many settings, two of which are acute inpatient and subacute rehabilitation hospital units. Approximately 10% of neuropsychologists are housed directly within a Physical Medicine and Rehabilitation (PM&R) department, where inpatient services are commonly expected (Sweet et al. 2021). A much higher percentage are employed within Psychiatry or Neurology departments, in which outpatient services are the predominant expectation (Sweet et al. 2021). Outside of academic or Veterans Affairs (VA) medical centers, some neuropsychologists in private practice (or with affiliated appointments in non-traditional healthcare systems) forge relationships with local hospitals, neurorehabilitation centers, or nursing facilities to diversify their practices. Regardless of the setting, the relationship the provider has with the institution can range from an intermittent consultant to a permanent service line team member. Considerations regarding business models associated with these various practice settings and roles are discussed below.

Assessment Models

The inpatient setting, by definition, necessitates a tightly focused evaluation given the medical acuity of patients, time constraints, increased likelihood of changes in clinical presentation over time, and elevated risk for interruptions. Inpatient neuropsychological reports tend to be brief and typically must be completed rapidly (e.g., 24 h or less) to be accessible in a fast-paced environment. Test batteries are often truncated to cognitive and mood screening plus an abbreviated clinical interview. In some circumstances, brief bedside cognitive screening measures can be paired with select standardized measures to generate a quick snapshot of functioning, such as in the case of assessing for delirium or medical decision-making capacity. In other cases, such as when establishing a cognitive baseline for future comparison, test batteries that enable assessment of a range of cognitive abilities may be most useful.

Billing Practices and Productivity

Essential Terminology

Current Procedural Terminology (CPT®) codes refer to specific procedures or services conducted by healthcare professionals, such as diagnostic interviewing and testing conducted by a technician versus a professional (American Medical Association [AMA] 2024). CPT codes are paired with diagnostic International Classification of Diseases (ICD) codes (World Health Organization [WHO] 1993) to indicate the type and purpose of each service (Borman 2017). While various diagnostic code systems are in use, ICD-10 is most prevalent in hospitals within the United States as of this writing. The Centers for Medicare and Medicaid Services (CMS) is a federal health care agency that is often referenced by other insurance agencies when considering which combinations of CPT codes and ICD codes should be reimbursed, as CMS typically reimburses only those services that appear "reasonable, necessary, and/or appropriate" based on current standards of care (Barisa 2017).

Productivity Measurement

Billing practices for inpatient evaluations vary depending on the type of services provided. The primary goal is to use CPT billing codes that accurately reflect all activities necessary to generate the final product while also maximizing total productivity (See Tables 2.1, 2.2, and 2.3 for commonly used CPT codes). The American Psychological Association (APA 2019a, b, c) offers several resources and webinars pertaining to billing for neuropsychological services, which are referenced at the end of this chapter.

Standards for determining clinician productivity vary across institutions, with some measuring hours billed or net revenue generated, and others adhering to published *work* Relative Value Unit (wRVU) billable benchmarks (Sweet et al. 2021). The standard selected is site-specific, based on our experience, and may not be translatable across academic medical centers, non-profit hospitals, VA medical centers, and for-profit institutions. Hours billed easily equates with various types of clinical activities in terms of time spent, whereas the wRVU system is more complicated. wRVUs for each CPT code are established and updated periodically by CMS and are purported to capture the level of technical skill and perceived effort associated with specific services (Childers and Maggard-Gibbons 2020). However, among psychologists, a primary limitation of using a single wRVU target is the puzzling, yet marked, difference in attainable hourly wRVU associated with intervention versus assessment services, particularly for evaluations completed without assistance of a psychometrist (which is often the case in inpatient settings). For example, if you calculate the average hourly wRVU associated with neuropsychological evaluation CPT codes (e.g., 96,116, 96,132/96133), you will find that neuropsychological evaluation is "worth" considerably less wRVU per hour than an hour of outpatient psychotherapy (90837). Therefore, if neuropsychologists and psychologists have the same wRVU target, that discrepancy results in many more hours of billable work necessary for neuropsychologists to "catch up" with their colleagues who perform mostly intervention services. (See Dawson and Speelman 2023; Steel et al. 2023 for more information on this subject.)

To our knowledge, the only national agency that offers published, neuropsychologist-specific benchmark data is Sullivan Cotter, a healthcare analytic and advisory firm that generates proprietary benchmarks based on its own survey data. Other organizations, such as the American Medical Group Association (AMGM), the Medical Group Management Association (AGMA), Vizient Inc., and the Association of American Medical Colleges (AAMC), do not offer neuropsychology-specific data. Furthermore, some benchmarks are notoriously generic and pertain to "allied health professional." While closer in parity with psychologists who conduct mostly intervention services, Sullivan Cotter neuropsychology-specific benchmarks were still not equivalent hour-for-hour at the time of this publication. Neuropsychologists can also reference the joint American Academy of Clinical Neuropsychology (AACN)/National Academy of Neuropsychology (NAN)/Society of Clinical Neuropsychology (SCN) 2020 Practice and "Salary Survey," which is published approximately every five years and details professional practice survey data for neuropsychologists across different settings, career stages, and geographic regions (Sweet et al. 2021). The salary survey offers parameters of productivity targets and salary which may be useful for negotiation purposes. However, despite being supported by three national professional neuropsychology associations, these data may not always be recognized by hospital administrations due to rigid adherence to proprietary benchmark data.

Table 2.1 Assessment codes commonly used by neuropsychologists in inpatient services

Definition	Activities	Code	Time Required	wRVU Value
Psychological diagnostic evaluation	Biopsychosocial assessment with patient history, mental status, and recommendations	90791	Not a time-based code	3.84
Neurobehavioral status exam	Clinical interview, test administration, scoring/interpretation, and report writing	96116	Initial 60 minutes	1.86
		96121	Each additional 60 minutes	1.71
Psychological testing by professional	Includes psychological testing, *by the professional*, scoring/interpretation, report writing, clinical decision making, treatment planning, and feedback	96130	Initial 60 minutes	2.56
		96131	Each additional 60 minutes	1.96
Neuropsychological evaluation by professional	Includes neuropsychological evaluation, *by the professional* (i.e., test interpretation, report writing, clinical decision making, treatment planning, and feedback)	96132	Initial 60 minutes	2.56
		96133	Each additional 60 minutes	1.96
Test administration and scoring by professional	Includes psychological or neuropsychological testing, *by the professional*, for two or more tests (any method)	96136	Initial 30 minutes	0.55
		96137	Each additional 30 minutes	0.46
Test administration and scoring by technician	Includes psychological or neuropsychological testing, *by the technician,* for two or more tests (any method)	96138	Initial 30 minutes	0.00
		96139	Each additional 30 minutes	0.00

(continued)

Table 2.1 (continued)

Automated testing	Includes a single automated test (for a given condition, monitoring progression of disease or condition) administered via electronic platform that produces an automated report only	96146	Not a time-based code	0.00
Health and Behavior assessment	Includes health behavior assessment or reassessment including a health-focused clinical interview, behavioral observations, clinical decision making, test administration, scoring/interpretation, and report	96156	Not a time-based code	2.20

Note: wRVU totals reflect 2024 values.

CPT® Codes for Inpatient Evaluation

When selecting which code(s) to bill for an evaluation in any setting, providers should be mindful to use CPT code(s) that most accurately reflect the amount of time and effort spent on *all* tasks necessary to complete the assessment (e.g., interview, record review, interview of staff or caregivers, testing, documentation, consultation with the patient's treatment providers, etc.). For specific assessment CPT codes and associated definitions, time requirements, and wRVU values, please refer to Table 2.1.

CPT code 90791 (*Psychiatric Diagnostic Evaluation*) is not time-based but is associated with higher overall wRVU compared to other CPT codes typically billed by neuropsychologists. In inpatient settings, 90791 may be appropriate for a clinical interview coupled with brief/screening assessments (e.g., a cognitive screener and/or standardized questionnaires); such an evaluation may take somewhere around an hour to complete (but can take longer). The Health and Behavior assessment code adopted in 2020 (i.e., 96156) is another option, although it is associated with lower wRVU values compared to 90791 and may not be considered a reimbursable code by some health insurance companies. For evaluations that include measures which are more time-consuming to administer and score, the use of time-based codes is recommended, as this better captures the entirety of a neuropsychological evaluation. 90791 in combination with 96136/96137 (*Psychological or Neuropsychological Testing by Provider*) may be permissible by some insurance carriers and institutional compliance offices, but resistance is not uncommon in the authors' experiences. This is likely (at least, in part) due to information shared by the APA (2019a, c), when the neuropsychology CPT codes changed, which typically pairs 96116/96121 (*Neurobehavioral Status Examination*) with neuropsychological assessment codes. There can also be insurance issues when there is a mismatch between CPT codes and diagnostic codes, such as if a medical diagnosis is required but 90791 is used for the interview. Notably, neuropsychological interviews involve elements of both neurobehavioral status examination and psychiatric diagnostic evaluation, and psychological disorders represent medical phenomena, so distinguishing these types of interviews can feel arbitrary. See Table 2.2 for a list of psychology intervention codes and their associated wRVUs.

For longer inpatient evaluations, codes traditionally used for outpatient evaluations (e.g., 96116/96121, 96132/96133, and 96136/96137 or 96138/96139) may be most appropriate because they better capture the time needed to synthesize multiple sources of information and formulate informed recommendations. That said, the use of these codes is often limited to once-per-calendar year, per-patient unless there has been a significant change in the patient's status that would warrant re-evaluation (e.g., return to work following a medical event). Repeat evaluation might be feasible if the first evaluation was completed in an inpatient setting and subsequent evaluations will be conducted in an outpatient setting, especially if the inpatient charges sometimes fall under per diem reimbursement (see below). However, repeat testing within the same calendar year must be employed judiciously to reduce the risk of a health insurance review or audit for justification of medical necessity or complete denial or coverage.

Table 2.2 Intervention codes used by neuropsychologists in inpatient services

Definition	Activities	Code	Time Required	wRVU Value
Psychotherapy intervention	Individual intervention	90832	30 minutes	1.78
		90834	45 minutes	2.35
		90837	60 minutes	3.47
	Crisis intervention, includes urgent assessment and history of a life-threatening, complex, or distressing crisis, mental status exam, and disposition	90839	60 minutes	3.28
		90840	Each additional 30 minutes	1.57
	Family intervention, without patient present	90846	50 minutes	2.51
	Family intervention, with patient present	90847	50 minutes	2.62
	Multi-family intervention	90849	45-60 minutes	0.62
	Group intervention (other than multi-family)	90853	45-60 minutes	0.62
Health and behavior intervention	Individual intervention	96158	Initial 30 minutes	1.52
		96159	Each additional 15 minutes	0.52
	Group intervention	96164	Initial 30 minutes	0.22
		96165	Each additional 15 minutes	0.10
	Family intervention, with patient present	96167	Initial 30 minutes	1.62
		96168	Each additional 15 minutes	0.58

(continued)

Table 2.2 (continued)

	Family intervention, without patient present	96170	Initial 30 minutes	0.00
		96171	Each additional 15 minutes	0.00
Caregiver behavior management training	Includes contact with caregivers in a multifamily setting, without patient present. Written note must document specific behavioral management/modification procedures, strategies, and interventions to improve the patient's functioning	96202	Initial 60 minutes	0.43
		96203	Each additional 15 minutes	0.12

Note: wRVU totals reflect 2024 values. Intervention services without the patient present (90846, 96171) are typically not covered by Medicare. If use of therapy techniques involved, consider 90847 for family psychotherapy (with patient present). If used for physical health diagnosis and education, consider health & behavior codes 96167-96168 for family intervention service. Can add-on interactive complexity code 90785 (with 90791, 90832, 90834, 90837, or 90853) if one of four criteria are met:1) need to manage maladaptive communication among participants, such as high anxiety or reactivity, repeated questions, or disagreements; 2) caregiver emotions/behaviors that interfere with their understanding and ability to assist in treatment plan implementation; 3) evidence or disclosure of an event leading to mandated report such as abuse or neglect – as long as discussion of the event and report were initiated with patient and other visit participants; and/or 4) use of play equipment or other physical devices to communicate with the patient in the event of undeveloped or lost language communication skills.

Special Considerations

Use of Technology: In the context of the COVID-19 pandemic, many neuropsychologists started conducting assessments using a range of telehealth practices (Hewitt et al. 2022). Within inpatient settings, some hospitals began incorporating electronic tablets into routine patient care to reduce contact and minimize transmission of infection. Tablets remain useful in the evaluation and treatment of patients under specific isolation precautions, and some providers have started using tablets for outpatient evaluations as well (Germine et al. 2019). Services incorporating technology are typically billable without modifiers and using the same codes as with any other evaluation as long as both the patient and provider are co-located (APA 2019b).

Provision of Feedback: Same-day feedback is typically billed as part of the original evaluation using time-based neuropsychological evaluation codes. In contrast, feedback occurring on a separate date, presumably preceding discharge, may be billed separately using an interventional code such as 90837 (*Psychotherapy, 60 Min*), particularly when such feedback is therapeutic, serving as the basis for a longer discussion of accommodations or behavioral interventions. Health and Behavior intervention codes (i.e., 96158/96159) would also be appropriate when feedback is offered on a separate day, though some insurance agencies do not recognize those codes as reimbursable if described solely as a feedback session instead of the starting point for intervention (when applicable).

Credentialing: An important issue to address preceding provision of inpatient neuropsychological services is whether the providers (and/or psychometrists, if applicable) are correctly credentialed or have hospital-based privileges necessary for inpatient services. Direct communication with compliance officers can help confirm who has billing privileges specific to various inpatient settings. This ensures that services charged are allowable and can therefore be reimbursed. See APA's Reimbursement Guide for more information on billing facility fees (APA 2019a, b, c).

Table 2.3 Interprofessional consultation codes

Definition	Activities	Code	Time Required	wRVU Value
Interprofessional consultation, *with* discussion and both a verbal *and* written report are required to the patient's treating/requesting physician or other qualified healthcare professional	Includes telephone, internet, or electronic health record assessment and management provided by a consultative physician or other qualified healthcare professional	99446	5-10 minutes	0.35
		99447	11-20 minutes	0.70
		99448	21-30 minutes	1.05
		99449	31 minutes or more	1.40
Interprofessional consultation, *without* discussion, and *only* a written report is required to the patient's treating/requesting physician or other qualified healthcare professional	Includes telephone, internet, or electronic health record assessment and management provided by a consultative physician or other qualified healthcare professional	99451	5+ minutes	0.70

Note: wRVU totals reflect 2024 values. These codes may not be allowed due to lack of reimbursement or confusion regarding the type of provider permitted to use these codes. Do not report codes 99446, 99447, 99448, or 9949 more than once for the same patient in a 7-day period, and if greater than 50% of time is spent in data review and/or analysis. You must not have had a face-to-face visit with the patient in the last 14 days, and the consultation services should not result in a face-to-face visit with the patient in the next 14 days.

If there is not a way to ensure billing privileges for a provider and/or psychometrists, a memorandum of understanding (MOU) can be created to "lease" that individual to the hospital/facility to cover their time spent on non-billable services. MOUs are typically developed via communication between the neuropsychologist, department chair, and facility/institution administrative officials.

Reimbursement for Services

It is generally acceptable to bill a diagnostic code reflecting the content of a *referral* question and not necessarily the resulting diagnosis (Chung et al. 2018). This can be particularly important when the assessed conclusion (e.g., malingering) is not a reimbursable diagnosis. Insurance companies have unique and often confidential lists of diagnostic codes they consider reimbursable in association with certain assessments and interventions. For this reason, it is prudent to be familiar with which diagnoses are considered acceptable by various insurance companies (Barisa 2010). For example, there are regional differences in whether neuropsychological evaluations are considered behavioral *or* medical services by CMS (i.e., Medicare and Medicaid), which affects which diagnostic codes (i.e., behavioral versus medical) are deemed acceptable. This is despite advocacy efforts by psychologists for inclusion under the definition of "physician" to furnish services under Medicare in 2010 (H.R.6147; U.S. Congress, 2011). Most managed care plans treat neuropsychological evaluations as a *medical* service and require an associated medical diagnostic code, although this also varies by insurance carrier. Example diagnostic codes accepted for use with neuropsychological services include R41.3 (*Memory Loss*—a behavioral code) and R41.89 (*Signs and*

Symptoms Involving Cognition—a medical code), both of which reflect cognitive complaints as the reason for referral for a neuropsychological evaluation.

While most commercial insurances follow established federal (i.e., CMS) practices for reimbursement of medical services, state- and region-specific plans may offer different coverage of neuropsychological evaluation services (Chung et al. 2018). This appears to be particularly true for Medicaid, as some Medicaid plans routinely deny coverage of 96,138/96139 and variably cover 96132/96133. Given the disproportionate number of racially and ethnically diverse and other financially disadvantaged individuals covered by Medicaid, this unfortunately results in inequitable access—particularly among under-privileged populations— and is a source of considerable professional frustration (CMS, 2023). One possible solution to this issue, depending on state-level Medicaid Behavioral Health guidelines, is to conduct the evaluation without the assistance of a psychometrist. However, that may not be feasible from a wRVU standpoint or in terms of scheduling for clinics that operate nearly exclusively using a psychometrist model.

Pre-authorization: Pre-authorization involves proposing estimated CPT codes, expected units (or time duration), preliminary diagnostic code(s), and sometimes specific tests that may be administered to obtain advanced approval for reimbursement coverage. Failure to use accepted diagnostic codes specific to a CPT service will result in denial of coverage. Use of multiple diagnostic codes can also inadvertently result in a preemptive denial of coverage if both behavioral and medical codes are included in the pre-authorization (Barisa 2010). To avoid this, we recommend designating one code type (behavioral or medical) as the primary diagnosis. The use of pre-authorization is not universal among insurance carriers. For some, failure to submit for pre-authorization can be grounds for automatic denial, while other plans do not require any pre-authorization. There are also a few plans that do not require preauthorization but will decide after the service has been delivered if it was medically necessary. After approval is obtained, it is important to bill using the approved diagnostic and CPT codes, which is another reason to select a code reflecting the reason for referral rather than the resulting diagnosis.

It is possible for denials of pre-authorization to be appealed, even if the neuropsychologist has no prior contact with the patient. Completing a peer-to-peer insurance review can help justify the medical necessity of the evaluation. Monitoring denial patterns can minimize future error and maximize insurance coverage, which is beneficial to the patient, the provider, and the institution. Pre-authorizations can be streamlined as commercial carriers become familiar with specific neuropsychology practices and providers. However, on the clinical side, pre-authorizations can be time-consuming and require management by staff with adequate knowledge of reimbursement practices. Identifying specific contacts for each insurance carrier can facilitate a speedier process.

BUSINESS MODELS FOR INPATIENT SERVICES

Salaried Versus Non-salaried Neuropsychologists

As noted above, neuropsychologists in inpatient settings may be a salaried employee of the facility, a salaried provider for an outside entity (such as a medical school) that has an operating agreement (or affiliation) with the facility, or in private practice with a financial arrangement to cover inpatients within one or several settings. The role of the provider within the inpatient care setting is a primary determinant of the business model adopted and can significantly shape the nature of clinical care.

Salaried employees in inpatient settings often fall within one of two categories: embedded in a treatment team or serving as a consultant on an as-needed basis. For those fully employed in rehabilitation or other inpatient settings, the number of cases seen per week is often determined by wRVU productivity benchmarks that are either (1) dictated based on national published guidelines and/or (2) negotiated between the provider and their employer. It is important to take note of the source of the benchmark used, as estimates can vary widely. In academic institutions, this becomes a difficult issue because, in addition to accounting for one's salary and fringe benefits, the provider may be required to cover an equal share of their department's overhead expenses despite not equally sharing in those overhead resources (e.g., cost of rent for multiple outpatient spaces, support personnel such as physician assistants or nurse practitioners, clinic supplies).

Neuropsychologists embedded in rehabilitation settings are often assigned to a specific unit (e.g., brain injury unit) and are responsible for providing services to all or most patients on that unit at any given time. Depending on productivity benchmarks, the neuropsychologist may also be required to see a certain number of outpatients per week to "supplement" their wRVU productivity. On a positive note, patients in the hospital never fail to show up for their appointments, so there is a 0%

no-show rate unless they are occupied with another service or have been temporarily transported from the unit for a procedure that cannot be conducted in their room. Depending on the average length of stay, providers may be expected to do an initial brief assessment for all new patients prior to the initial rehabilitation team meeting (usually within the first few days after admission) and then provide serial assessments, periodic intervention, or consultative services throughout the patient's admission. This is often intermixed with evaluating outpatients who have been discharged from the inpatient unit, as well as covering for colleagues during absences. Family meetings, support groups, and educational events are commonplace and may need to occur outside of normal business hours to accommodate schedules, as there is a finite window during the day when acute and rehabilitation patients are not overly fatigued, eating, or involved in other therapy services. Of importance, many of these critical services are not billable (addressed later in this chapter).

Inpatient consultations by salaried employees can occur in institutions that have both inpatient and outpatient settings. Coverage of inpatient consults may be sporadic or routine, depending on the setting, availability of providers who may otherwise provide primarily outpatient services, and familiarity of neuropsychological services within the broader institution. In those situations, inpatient neuropsychological evaluations are often treated as being above and beyond standard clinical duties. While this offers opportunity to recoup for lost wRVU from late cancellations and no-shows, inpatient evaluations can be challenging to accommodate into an otherwise full outpatient clinic schedule. It is important to establish a triage process for inpatient referrals in this context, as doing so enables clear determination of which cases are appropriate for an inpatient neuropsychological evaluation versus which patients would be better served being seen on an outpatient basis, such as when a more comprehensive assessment is clinically indicated (e.g., differential diagnosis of dementia subtype). Any triage plan should prioritize inpatient referrals with a component of risk to self and others, or capacity for time-sensitive medical decision-making.

Non-salaried neuropsychologists, such as contract employees in private practice, must consider a wide range of factors when negotiating their contacts to ensure a reasonable financial arrangement. Billing practices associated with inpatient evaluations can greatly impact the financial prudence of inpatient coverage. For example, providers generally need to become paneled with the same insurance companies accepted by the facility/institution and will want to establish a business model that ensures financial viability. Furthermore, productivity benchmarks for contract employees are less likely to be measured in wRVUs and may instead be measured by revenue generated, which is more complex for inpatient (versus outpatient) services, in part because of greater emphasis on non-billable activities and some degree of variability in coverage across specific health insurance plans. There is a cost for providing services that account for all overhead expenses, and this amount should be fully understood before negotiating a contractual arrangement with an institution (Finnegan 2012).

Negotiating an Operating Business Agreement

When negotiating an operating business agreement, important points to raise with hospital administrators or with third-party employers include:

- Whether the neuropsychologist is expected to cover their own salary and any overhead with revenue generated from inpatient care.
- A determination if all patients within the hospital will have the same level of care regardless of their insurance status or how to address patients without insurance or who are under-insured.
- Specific parameters regarding time spent completing activities that are important to the institution but are nonetheless not reimbursable.
- A written agreement of workload proportions for inpatient/outpatient responsibilities and how to prioritize outpatient wait times versus time-sensitive inpatient consults.

Note that "money talks" when it comes to negotiation. Who you know within the institution and your ability to articulate the potential costs versus revenue or other financial benefits to your stakeholders will facilitate this process considerably.

Per Diem Services

The business side of inpatient care often operates differently than that of outpatient services. For example, many institutions charge a fixed amount for standard daily inpatient services under a per diem charge. While those services are intended to include dietary, environmental, and some routine therapy services, neuropsychological evaluations may inadvertently (and arguably inaccurately) be grouped into that charge. The result is that any charges billed by the neuropsychologist are not directly reimbursed as professional fees, so no revenue is collected by the institution. On the other hand, specialty physician charges often fall under "professional charges" that are billed separately from the per diem charge using a fee-for-service model and are more likely to be reimbursed. Some institutions circumvent this issue by generating a MOU that covers part or all the neuropsychologist's salary in return for provision of unreimbursed services considered necessary by the institution, as described above.

A similar arrangement is sometimes seen for psychologists who are part of a multidisciplinary team, where their services are considered necessary but not all services can be reimbursed (e.g., within a bariatric surgery team). In such cases, billing for multiple services may be bundled together. CMS instituted a large national bundled payment initiative in 2011, with the goal of creating a financial incentive for reducing the number and cost of combined services.

This design can help increase patient access to services which would otherwise not be reimbursable using a fee-for-service model, such as those rendered by social workers (Agency for Healthcare Research and Quality [AHRQ] 2012). Yet despite this benefit, it is our opinion that neuropsychological services should be billed separately at the professional level and not as bundled to help ensure viability of future services. If neuropsychological services must be bundled with other services at a specific institution, it is important that providers account for their time and seek appropriate financial support for their services from their institution.

If neuropsychological services do not fall under per diem charges, internal support personnel are essential to seek pre-authorizations and submit billing charges for the neuropsychologist. Even the most productive neuropsychologists cannot obtain remuneration if the hospital does not seek pre-authorizations or does not submit billing charges. Because psychology tends to be a small niche in most rehabilitation programs and other inpatient settings, infrastructure problems commonly inhibit the psychologists from obtaining payment for their services. Education is often needed for clerical staff to understand the unique process of pre-authorization of neuropsychological services.

Psychometrists and Trainees

Another issue affecting inpatient neuropsychology business models is the often-limited availability of support staff, such as psychometricians. While many neuropsychologists in outpatient settings have access to psychometrists that can administer neuropsychological measures under supervision, the use of such personnel in inpatient settings appears less common in our experience, which has direct implications for billing. While provider-administered testing codes have higher reimbursement rates (versus technician-administered testing), fewer patients can be evaluated by the provider directly and the wRVU associated with these codes (96136/96137) is especially low. Alternatively, the use of "incident-to" billing by trainees can be considered, which refers to any services considered billable by Medicare that are performed by ancillary personnel under the professional's supervision (see APA's 2023 Medicare Guidelines). While some academic medical centers prohibit or limit incident-to billing, this largely reflects a misinterpretation of standards employed for training *resident physicians* who receive federal funding for their training. This is unfortunately viewed similarly by many hospital compliance department representatives for neuropsychology trainees, even though their positions are rarely funded by the federal government.

Unreimbursed Activities

Finally, it is important to clarify the extent to which participation in non-reimbursable activities is expected. Unbillable activities include attending team meetings, bedside patient rounds, staff education services, "hallway consultations," and administrative tasks related to billing and reimbursement. Many of those activities are necessary for optimal and effective patient care. However, time spent on non-reimbursable activities does impact productivity and performance benchmarks. Unreimbursed activities also directly affect one's ability to benefit from incentive plans. Therefore, if significant time is expected for those activities, it is advisable to adopt a practice model that adequately compensates for unreimbursed time and have this documented in an employment contract or MOU. Fortunately, Medicare recently enabled coverage of some activities that were previously unbillable (Table 2.3), such as interprofessional consultation, although hospital compliance officers may have conflicting opinions regarding whether psychologists can bill those codes.

Program Development

When it comes to building a new service or expanding an existing one, the business literature is rich with methods for analysis and implementation. For example, Harvard Business Review (HBR) offers a wide range of up-to-date online resources. Establishing a business plan is a crucial part of program development, as it enables identification of not just program goals but also potential barriers and how those barriers might be mitigated (Haag 2013). Business plans are also a universal way to communicate strategy across different types of stakeholders and reflect that the presenter has looked at their proposed program from the different lenses such stakeholders perceive as important. For example, program development planning must identify what is being added to existing services and the need, review the perceived value of newly offered services, and specify how the program will function in the broader context of the institution. It is imperative to the program's success that any essential information is communicated using familiar language to increase positive reception from leadership. There are numerous ways to analyze business plan elements, as outlined by Jazayeri and Park (2019), who describe the process in the context of medicine. Figure 2.1 identifies key elements likely required of an inpatient neuropsychology service business plan. For an in-depth, accessible introduction to neuropsychology business development for inpatient practice, readers are directed to Chap. 9 of *The Business of Neuropsychology* (Barisa 2010).

In the inpatient setting, some services traditionally associated with neuropsychological practice may overlap with existing specialties (e.g., speech or occupational therapy). Thus, identifying gaps in patient care can help the neuropsychologist cultivate a unique practice niche. As stated by Boakye et al. (2019), "...neuropsychological assessment can be helpful in identifying hidden, complex and high-level impairment." Indeed, the inpatient neuropsychologist is best poised to address complex referral questions, such as whether cognitive impairment was present preceding the hospitalization or distinguishing between neurologic and psychiatric etiology of behavioral change.

While inpatient assessment traditionally encompasses the majority of services rendered, neuropsychologists in these settings may also offer supplemental activities to fill gaps in care. These can include focused assessments in the context of an ethics consultation or capacity determination, health-oriented assessment and interventions, brief inpatient psychotherapy, family-oriented education or therapy services, co-treatment with other disciplines, and/or cognitive rehabilitation/remediation. Those services can be billed using the already mentioned assessment and intervention codes (See Tables 2.1, 2.2, and 2.3 above). CPT code 90791 and the various intervention codes may be more easily reimbursed (Hart et al. 2020) and are associated with higher wRVUs than neuropsychological assessment codes. This can help offset time spent conducting non-billable services.

Marketing

Strategies

Marketing is about making connections, which can feel daunting in a large hospital setting. Communication with key physicians identified as likely referral sources, such as neurologists and hospitalists, is a good first step. As crucial allies, nurses tend to spend more time with patients and can therefore be more familiar with concerns that might otherwise go unaddressed. Social workers are another great resource, as they are often integral in addressing post-discharge care needs. Finally, coordination with complementary service providers (e.g., speech or occupational therapists) is recommended to clarify what unique services neuropsychologists can provide that would add value to the broader inpatient team.

Being physically available and seen is another straightforward way to get engaged. Attending shared educational opportunities in the institution and multidisciplinary team meetings are two ways to increase visibility in your institution. Offering to speak during educational events for medical staff provides an excellent opportunity specifically to share (1) what neuropsychologists do, (2) how we can help, and (3) how to access our services. Case presentations that engage the audience can illustrate these points as well as break the ice if you're a new provider to the institution or service. Additionally, brochures detailing inpatient neuropsychological services can be placed in high-traffic areas frequented by treatment providers and/or family members. It is important to ensure that all marketing materials specifically outline the "unique value added" by neuropsychology services, above and beyond what is already considered standard of care in inpatient settings. When implementing marketing strategies, it is critical to weigh the potential for new work generated with provider availability, as referral sources will dry up quickly if referred patients are discharged prior to being seen for inpatient neuropsychology services. Similarly, a decline in the quality of services due to competing demands placed on the provider (Barisa 2010) or delays in the completion of time-sensitive reports are likely to dissuade continued referrals and inadvertently hinder future efforts to justify workload or compensation adjustments. As such, marketing efforts may necessitate optimizing practice patterns over time to ensure high-quality, accessible, and timely care.

Communicating Value Added

As healthcare costs in the United States continue to rise, policy makers, clinical managers, and researchers have started transitioning toward a value-based care system that emphasizes the diagnostic and predictive utility of services, a return on investments in terms of cost savings and health-related outcomes, and patient satisfaction (Steel et al. 2023). In this context,

Key Element	Content Guidance
Statement of Need	• Briefly describe why creating an inpatient service makes sense. ▸ Why is it needed? ▸ Who will benefit?
Description of Services	• What services would be offered, and how they are unique or different from what is already offered at the institution? • What is the projected frequency or level of need for new inpatient services?
Financial Analysis	• How will services be billed and paid for? • Is pre-authorization required? • Do you anticipate that insurance agencies will reimburse these services? ▸ Do you anticipate a deficit or profit? ▸ Can you compare with similar services at other institutions? • Are there anticipated costs for program development? (e.g., new test materials, administrative support, etc.)
Organizational Analysis	• Who will oversee the services? • Which providers will render the services? ▸ Will psychometrist assistance be required? • What administrative support will be needed? • How will the inpatient responsibilities fit with employees' existing responsibilities? ▸ For example, how will time be split for each provider if they are expected to cover outpatient and inpatient services? • How might productivity be affected? ▸ For example, could an inpatient service or department cover a portion of the provider's time so that per diem billing does not result in a net loss for the provider's own department? • Will additional space(s) be needed for testing, scoring, interviewing, or therapeutic services, or for storage of materials?
Marketing Plan	• How will you notify potential referral sources of the new services? • Can you present data demonstrating that there will be continued need for the services? • What can you do if referrals are slow at first, or if they dry up? • What will a successful program look like?

Fig. 2.1 Key elements of an inpatient business plan

neuropsychologists must communicate the importance and value of our services to patients, their families, referring providers, and healthcare administrators—and there is a burgeoning literature to draw from.

Neuropsychological assessments yield prognostic benefit across a variety of disorders (Marcotte and Grant 2010). For example, large-scale reviews by Donders (2020; 56 studies) and Watt and Crowe (2018; 81 studies) detail the added diagnostic and predictive utility for both disease (e.g., confirmed dementia etiology, morbidity, and mortality) and functional outcomes (e.g., postsurgical decline, cognitive and communication changes, driving safety, etc.). Donders (2020) also demonstrated that neuropsychological services were associated with overall cost savings. High levels of physician satisfaction with neuropsychological assessment services across settings have also been established (Hilsabeck et al. 2014; Mahoney III et al. 2017; Temple et al. 2006), and information from neuropsychological assessments is often incorporated into physician discharge summaries (Temple et al. 2006). Satisfaction is reported for both assessment and feedback services from the patient and caregiver perspectives, as well (Foran et al. 2015; Spano et al. 2021). Further, there is a demonstrable benefit of neuropsychological assessment and feedback services on patient psychological well-being, quality of life, and social adjustment (Lanca et al. 2020; Longley et al. 2023; Rosado et al. 2018). These are important findings, particularly considering Medicare's increasing emphasis on patient care satisfaction ratings—see AMA (2024) and NAN (n.d.) for information on CMS's Merit-based Incentive Payment System (MIPS) program and neuropsychologist eligibility.

Creative Complementary Services

Neuropsychologists offer a wide range of services across diverse settings, and some activities may not always be readily identified in an inpatient context. Unique, *potentially* billable activities include treatment-related services, such as co-treatment with other rehabilitation specialties, behavioral management planning (e.g., developing and implementing a unit- or patient-specific behavior plan), individual or group-based cognitive rehabilitation, and brief inpatient psychotherapy. However, despite having corresponding CPT codes, these services may not always be reimbursed. If a psychotherapy element is added to such interventions, they can generally be billed using consistently reimbursed psychotherapy codes instead. Assessment-related services, such as cognitive evaluation during Wada testing or functional magnetic resonance imaging (fMRI) for pre-surgical brain mapping, are additional unique inpatient services. These latter services may be billed using several different CPT codes. For example, CPT code 96116 (*Mental Status Examination*) is most often used for Wada procedures, though 95958 (*Other EEG Testing Procedures*) can also be billed. Brain mapping using fMRI can similarly be billed under 96116 or 96020 (*Functional Brain Mapping*) but note that these two codes cannot be used simultaneously.

SUMMARY

Neuropsychologists working in inpatient settings provide focused evaluations and interventional services that are of demonstrable benefit to both patients and treatment teams. However, providers face unique professional challenges, including engagement in activities that are not reimbursable, leading to difficulty accurately capturing their time with available CPT codes, in part due to suboptimal associated wRVUs. Compared to our colleagues in related fields, there is still work to be done to enhance training and education in the business of neuropsychology, and to promote the recognition of the value of our work —advocacy that could lead to increased reimbursement rates and inclusion of neuropsychological services as an essential standard of care. Championing these efforts begins at the level of the individual, but national advocacy efforts are also needed. We close this chapter with a list of important key stakeholders and outline ways neuropsychologists can become more involved in business within and beyond their institutions (Fig. 2.2).

The field of neuropsychology is full of talented individuals, including many already involved in advocacy at the local and national levels. We encourage readers to join forces with these individuals. Starting with active pursuit of business education, clinicians can become empowered to help advocate for issues such as more equitable billing options and productivity targets at the local level, which can be rewarded with support for specific clinical practices and improve work-life balance. Broadening that experience to governance, neuropsychologists can address the broader benefits of specialty neuropsychology services, such as in inpatient settings, to help advance our profession on a national level. The more informed and supportive voices the field can garner, the more we are able to catalyze meaningful change.

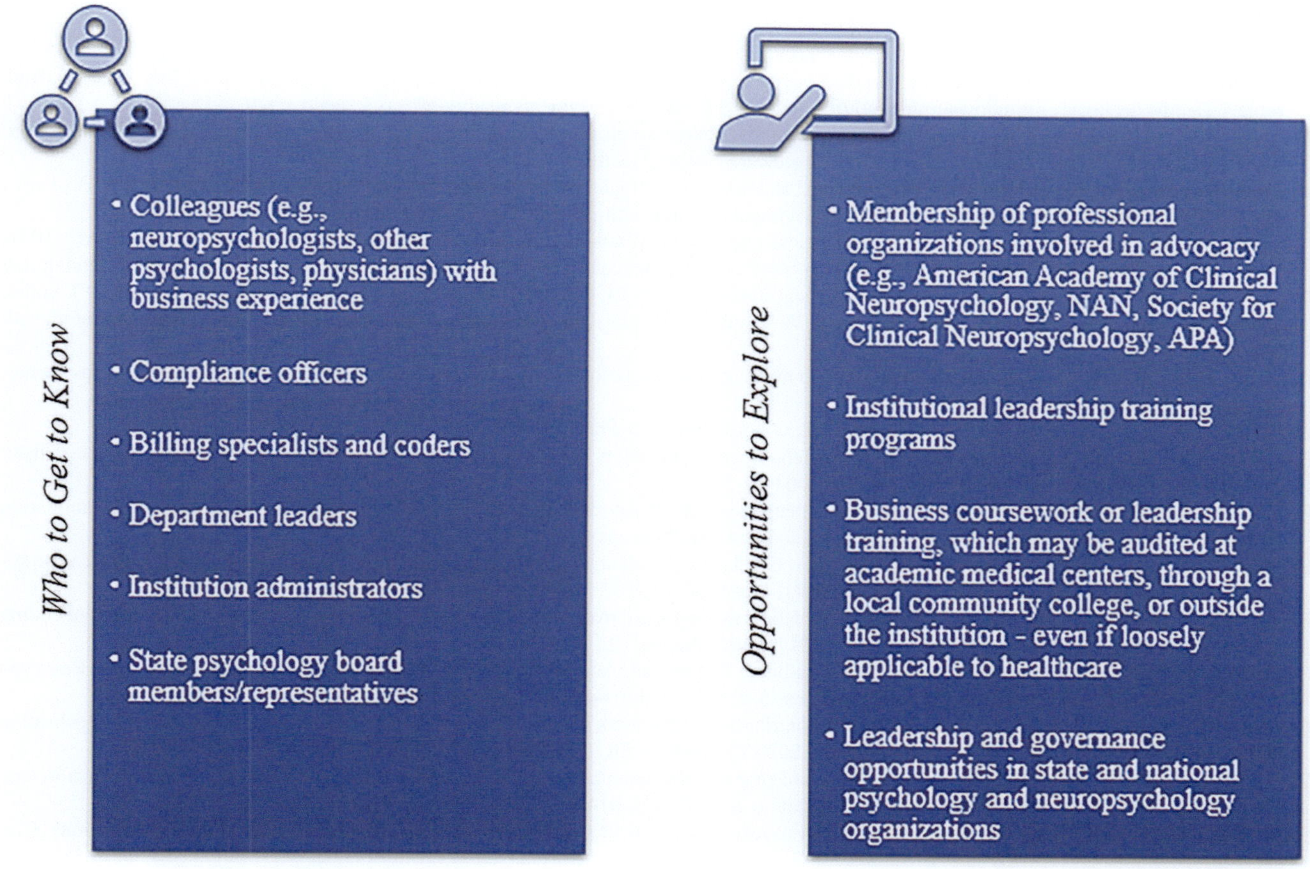

Fig. 2.2 How neuropsychologists can get more involved in business and advocacy

ADDITIONAL RESOURCES

- For the most up to date wRVU per CPT code, go to CMS's Physician Fee Lookup Tool at https://www.cms.gov/medicare/physician-fee-schedule/search
- For Medicare mandates for Telehealth Services, visit https://www.apaservices.org/practice/medicare/coverage/telehealth
- APA reimbursement information, including billing guidance can be found here: https://www.apaservices.org/practice/reimbursement
- APA guidance on practice management can be found here: https://www.apaservices.org/practice/business/management
- APA 2025 update on telehealth services can be found here:
- https://www.apaservices.org/practice/reimbursement/government/2025-medicare-changes
- For additional neuropsychology-specific business guidance, please review *The Business of Neuropsychology* by Mark T. Barisa, Ph.D., ABPP (full reference below).

REFERENCES

Agency for Healthcare Research and Quality. Bundled payment: effects of health care spending and quality. 2012. Retrieved July 22, 2024, from https://effectivehealthcare.ahrq.gov/sites/default/files/related_files/bundled-payments-quality-effects_executive.pdf

American Medical Association. Advocacy in action: leading the charge to reform Medicare pay. 2024. Retrieved on July 22, 2024, from https://www.ama-assn.org/practice-management/medicare-medicaid/advocacy-action-leading-charge-reform-medicare-pay

American Psychological Association. 2019 Psychological and neuropsychological testing billing and coding guide. 2019a. Retrieved on December 10, 2023, from https://www.apaservices.org/practice/reimbursement/health-codes/testing/billing-coding.pdf

American Psychological Association. 2019 Psychological and neuropsychological testing billing and coding guide: ADDENDUM. 2019b. Retrieved on February 27, 2025, from https://www.apaservices.org/practice/reimbursement/health-codes/testing/billing-coding-addendum.pdf

American Psychological Association. Neuropsychological testing: crosswalk for 2019 neuropsychological testing and evaluation CPT codes. 2019c. Retrieved on February 27, 2025 from https://www.apaservices.org/practice/reimbursement/health-codes/testing/neuropsychological-testing.pdf

American Psychological Association. Complete guide to medicare for psychologists. 2023. Retrieved on November 12, 2023, from https://www.apaservices.org/practice/medicare

Barisa MT. The business of neuropsychology. New York: Oxford University Press; 2010.

Barisa MT. CPT and billing codes. In: Budd MA, Hough S, Wegener ST, Stiers W, editors. Practical psychology in medical rehabilitation. Cham: Springer; 2017. https://doi.org/10.1007/978-3-319-34034-0_54.

Bercaw EL, Hanks RA, Millis SR, Gola TJ. Changes in neuropsychological performance after traumatic brain injury from inpatient rehabilitation to 1-year follow-up in predicting 2-year functional outcomes. Clin Neuropsychol. 2011;25(1):72–89.

Boakye NT, Scott R, Parsons A, Betteridge S, Smith MA, Cluckie G. All change: a stroke inpatient service's experience of a new clinical neuropsychology delivery model. BMJ Open Qual. 2019;8(1) https://doi.org/10.1136/bmjoq-2017-000184.

Borman KR. Medical coding in the United States: introduction and historical overview. In: Savarise M, Senkowski C, editors. Principles of coding and reimbursement for surgeons. Cham: Springer; 2017. https://doi.org/10.1007/978-3-319-43595-4_1.

Centers for Medicare and Medicaid. CMS releases data briefs that provide key medicaid demographic data for the first time. 2023. Retrieved 12/10/2023, from https://www.cms.gov/blog/cms-releases-data-briefs-provide-key-medicaid-demographic-data-first-time

Childers CP, Maggard-Gibbons M. Assessment of the contribution of the work relative value unit scale to differences in physician compensation across medical and surgical specialties. JAMA Surg. 2020;155(6):493–501.

Chung CY, Alson MD, Duszak R Jr, Degnan AJ. From imaging to reimbursement: what the pediatric radiologist needs to know about health care payers, documentation, coding and billing. Pediatr Radiol. 2018;48:904–014.

Dawson EL, Speelman C. Productivity measurement in psychology and neuropsychology: existing standards and alternative suggestions. Clin Neuropsychol. 2023;37(8):1569–83.

Donders J. The incremental value of neuropsychological assessment: a critical review. Clin Neuropsychol. 2020;34(1):56–87.

Finnegan J. Mitigating the risks: tips for contracting with teleradiologists. J Am Coll Radiol. 2012;9:583–5.

Foran A, Millar E, Dorstyn D. Patient satisfaction with a hospital-based neuropsychology service. Aust Health Rev. 2015;40(4):447–52.

Germine L, Reinecke K, Chaytor NS. Digital neuropsychology: challenges and opportunities at the intersection of science and software. Clin Neuropsychol. 2019;33(2):271–86.

Haag AB. Writing a successful business plan: an overview. Workplace Health Saf. 2013;61(1):19–29.

Hart T, Vaccaro M, Collier G, Chervoneva I, Fann JR. Promoting mental health in traumatic brain injury using single-session behavioural activation and SMS messaging: a randomized controlled trial. Neuropsychol Rehabil. 2020;30(8):1523–42. https://doi.org/10.1080/09602011.2019.1592761.

Hewitt KC, Block C, Bellone JA, Dawson EL, Garcia P, Gerstenecker A, et al. Diverse experiences and approaches to tele neuropsychology: Commentary and reflections over the past year of COVID-19. Clin Neuropsychol. 2022;36(4):790–805.

Hilsabeck RC, Hietpas TL, McCoy KJ. Satisfaction of referring providers with neuropsychological services within a veterans administration medical center. Arch Clin Neuropsychol. 2014;29(2):131–40.

Jazayeri A, Park KT. How to write an effective business plan in medicine. Gastroenterology. 2019;156:1243–7.

Johnson-Greene D. Clinical neuropsychology in integrated rehabilitation care teams. Arch Clin Neuropsychol. 2018;33(3):310–8. https://doi.org/10.1093/arclin/acx126.

Lanca M, Giuliano AJ, Sarapas C, Potter AI, Kim MS, West AL, Chow CM. Clinical outcomes and satisfaction following neuropsychological assessment for adults: a community hospital prospective quasi-experimental study. Arch Clin Neuropsychol. 2020;35(8):1303–11.

Longley WA, Tate RL, Brown RF. The psychological benefits of neuropsychological assessment feedback as a psychoeducational therapeutic intervention: a randomized-controlled trial with cross-over in multiple sclerosis. Neuropsychol Rehabil. 2023;33(5):764–93.

Mahoney JJ III, Bajo SD, De Marco AP, Arredondo BC, Hilsabeck RC, Broshek DK. Referring providers' preferences and satisfaction with neuropsychological services. Arch Clin Neuropsychol. 2017;32(4):427–36.

Marcotte TD, Grant I, editors. Neuropsychology of everyday functioning. New York: Guilford; 2010.

National Academy of Neuropsychology. Specialty registries can help Medicare providers increase their revenue. n.d. Retrieved July 22, 2024 from https://nanonline.org/NAN/_ProfessionalResources/_PAIC/MIPS_-_Specialty_Registries.aspx

Novack TA, Sherer M, Penna S. Neuropsychological practice in rehabilitation. In: Frank RG, Rosenthal M, Caplan B, editors. Handbook of rehabilitation psychology. Washington DC: American Psychological Association; 2010. p. 165–78. https://doi.org/10.1037/15972-010.

Rosado DL, Buehler S, Botbol-Berman E, Feigon M, León A, Luu H, et al. Neuropsychological feedback services improve quality of life and social adjustment. Clin Neuropsychol. 2018;32(3):422–35.

Spano P, Katz N, DeLuco T, Martin CO, Tam H, Montalto D, Stein CR. Parent perceptions of pediatric neuropsychological evaluations: a systematic review. Child Neuropsychol. 2021;27(7):922–48.

Steel SA, Rolin SN, Davis JJ. Relatively undervalued: comparing the work relative value units of neuropsychological evaluation to other services. Clin Neuropsychol. 2023:1–15.

Sweet, J. J., Klipfel, K. M., Nelson, N. W., & Moberg, P. J. (2021). Professional practices, beliefs, and incomes of U.S. neuropsychologists: The AACN, NAN, SCN 2020 practice and "salary survey." Clin Neuropsychol, 35(1), 7–80. https://doi.org/10.1080/13854046.2020.1849803.

Temple RO, Carvalho J, Tremont G. A national survey of physicians' use of and satisfaction with neuropsychological services. Arch Clin Neuropsychol. 2006;21(5):371–82.

United States Congress. Amendment of title XVIII of the Social Security Act to provide for treatment of clinical psychologists as physicians for purposes of furnishing clinical psychologist services under the Medicare Program. 2011. Retrieved December 10, 2023, from https://www.govinfo.gov/content/pkg/BILLS-111hr6147ih/pdf/BILLS-111hr6147ih.pdf

Watt S, Crowe SF. Examining the beneficial effect of neuropsychological assessment on adult patient outcomes: a systematic review. Clin Neuropsychol. 2018;32(3):368–90.

World Health Organization (WHO). The ICD-10 classification of mental and behavioural disorders. World Health Organization; 1993.

Medical Knowledge for Inpatient Neuropsychologists

3

Greg A. Sullivan

ABBREVIATIONS

ABG	Arterial Blood Gas
ACB	Anticholinergic Cognitive Burden (ACB)
AD	Alzheimer's Disease
ADC	Apparent Diffusion Coefficient
ADD/ADHD	Attention-Deficit/Hyperactivity Disorder
AED	Antiepileptic Drug
AGS	American Geriatrics Society
ALP	Alkaline Phosphatase
ALT	Alanine Transaminase
ART	Antiretroviral Therapy
AST	Aspartate Transaminase
BBB	Blood Brain Barrier
BMP	Basal Metabolic Panel
BUN	Blood Urea Nitrogen
Ca	Calcium
CAA	Cerebral Amyloid Angiopathy
cART	Combined Antiretroviral Therapy
CBC	Complete Blood Count
CD4	Cluster of Differentiation 4
CJD	Creutzfeldt-Jakob Disease
CMP	Comprehensive Metabolic Panel
CNS	Central Nervous System
COPD	Chronic Obstructive Pulmonary Disease
CRP	C-Reactive Protein
CSF	Cerebrospinal Fluid
CT	Computed Tomography
CVA	Cerebrovascular Accident
DAI	Diffuse Axonal Injury
DOT	Diffuse Optical Tomography
DTI	Diffusion Tensor Imaging

G. A. Sullivan (✉)
Mental Health and Behavioral Science Service (MH/BSS), James A. Haley Veterans' Hospital, Tampa, FL, USA

Department of Psychiatry and Behavioral Neurosciences, Morsani College of Medicine, University of South Florida, Tampa, FL, USA
e-mail: Gsulliv1@usf.edu

E. K. Bailey et al. (eds.), *Inpatient Neuropsychology*, https://doi.org/10.1007/978-3-032-05880-5_3

DWI	Diffusion Weighted Imaging
ECG/EKG	Electrocardiogram
EEG	Electroencephalography
ESR	Erythrocyte Sedimentation Rate
FDG-PET	Fluorodeoxyglucose-PET
FLAIR	Fluid-Attenuated Inversion Recovery
fMRI	Functional MRI
FTA-ABS	Fluorescent Treponemal-Antibody Absorption
FTD	Frontotemporal Lobar Degeneration (Frontotemporal Dementia)
GABA	Gamma (γ)-Aminobutyric Acid
GRE	Gradient Echo
H3G	Hydromorphone-3-Glucuronide
HAD	HIV-Associated Dementia
HAND	HIV-Associated Neurocognitive Disorder
HE	Herpes Encephalitis
HIV	Human Immunodeficiency Virus
HSV	Herpes Simplex Virus
LATE	Limbic-Predominant Age-Related TDP-43 Encephalopathy
LP	Lumbar Puncture
M3G	Morphine-3-Glucuronide
M6G	Morphine-6-Glucuronide
mg	Milligram
MMSE	Mini-Mental Status Examination
MRI	Magnetic Resonance Imaging
NMDA	N-methyl-D-aspartate
non-REM	Non-Rapid Eye Movement
NPH	Normal Pressure Hydrocephalus
PCT	Procalcitonin
PDD	Parkinson's Disease Dementia
PET	Positron Emission Tomography
PHPT	Primary Hyperparathyroidism
PLED	Periodic Lateralized Epileptiform Discharge
PNS	Peripheral Nervous System
PRES	Posterior Reversible Encephalopathy Syndrome
RPLS	Reversible Posterior Leukoencephalopathy Syndrome
RPR	Rapid Plasma Reagin
SPECT	Single-Photon Emission Computerized Tomography
SWI	Susceptibility-Weighted Imaging
T3	Triiodothyronine
T4	Thyroxine
TBI	Traumatic Brain Injury
TFT	Thyroid Function Test
TLE	Temporal Lobe Epilepsy
TP	Treponema Pallidum
TSH	Thyroid Stimulating Hormone
UDS	Urine Drug Screen
VCI	Vascular Cognitive Impairment
VDRL	Venereal Disease Research Laboratory
WBC	White Blood Count

INTRODUCTION

An inpatient neuropsychologist is uniquely positioned to provide expertise in neuropsychiatric symptoms and syndromes across medical, surgical, critical care, and rehabilitative settings. With a durable foundation in both neurocognitive and psychiatric pathology, the consultant integrates basic pathophysiologic processes with clinical manifestations. An integrated conceptualization such as this, that considers cognitive, emotional, and behavioral symptoms within the broader context of medical disease, provides valuable information to guide clinical assessments and treatment plan decision-making.

This approach is not without its challenges. The medical complexity of acute care and rehabilitative settings may introduce obstacles to applying these skills. These are especially common when considering unfamiliarity with standard medical nomenclature, diagnostic workups, and underlying pathophysiologic consequences of complex pharmacology. A consultant may find themselves facing a complex medical record full of arcane terms, mysterious acronyms, and peculiar jargon. Interpretation of this necessary and helpful, though complex, information frequently presents a barrier to full interdisciplinary engagement. These obstacles must be mastered to ensure accurate, efficient, and effective care. To this end, it is important to ensure a working competency in foundational medical knowledge, recognize informative patterns in laboratory diagnostics salient to neurobehavioral findings, and reflect on how clinical care itself via prescribing practices may impact cognition and function.

At the conclusion of this chapter, we provide appendices listing medical abbreviations for diagnostic studies (Appendix 1) and common medical diseases/conditions (Appendix 2), as well as common terminology used in administrative/clinical (Appendix 3), pharmacy/prescribing (Appendix 4), and neuroradiology (Appendix 5) settings.

LABORATORY DIAGNOSTICS AND INTERPRETATION

In inpatient neuropsychological assessment, diagnostic medical tests provide valuable insights into the physiologic and biochemical aspects of a patient's condition, disease, or symptom. While neuropsychologists may not have formal training in biologic or medical sciences, a fundamental understanding of common laboratory tests is essential and can easily be achieved. The significance of various laboratory diagnostic tests and their relevant interpretation may be explored, equipping neuropsychologists with the ability to decipher these elements of a patient's medical chart.

Vital Signs

Vital signs, including blood pressure, heart rate, and respiratory rate, are essential indicators of overall health. Abnormal vital signs can signify underlying medical issues contributing to neuropsychological changes, as these communicate, among many things, how effectively the cardiovascular system is supporting the central nervous system (e.g., hypotensive encephalopathy), the presence or absence of infection or inflammation (e.g., fever), or the severity of systemic or metabolic disease (e.g., rapid respiration). Critically, however, while acute vital sign abnormalities such as these may show clear chronological association with changes in behavior or cognition, vital sign measurements chronically outside the generally accepted range may be similarly associated with symptomatology (e.g., chronic subcortical microvascular changes due to long-term hypertension).

Neurological Examination

A neurological examination, often performed by Neurology or another specialty service, is used to elicit neurological pathology and inform the broader differential. Typically, this includes a thorough assessment of both central nervous system (CNS) and peripheral nervous system (PNS) functioning—some data to which may hold less relevance for the neuropsychologist. Nonetheless, as the neurological examination is designed to be efficiently performed and documented, familiarity with the common elements of a neurological exam, as well as frequently used abbreviations and terminology, will help the neuropsychologist incorporate these data into their conceptualization (Table 3.1). If a neurological examination has not been performed, or if there is reason to anticipate fluctuations in a patient's status (thus necessitating longitudinal evaluation), the inpatient neuropsychologist can incorporate a *neurobehavioral* examination (i.e., behavioral neurology/bedside examina-

Table 3.1 Common terminology in physical and neurological examinations

Common Terminology in Physical and Neurologic Exams	
Alertness & Orientation	AOx1 = Person only AOx2 = Person, plus either place or time AOx3 = Person, place, and time AOx4 = Person, place, time, and situation (*also abbreviated as AAO = Awake, alert, and oriented*)
Mental Status	Cognitive functioning (*e.g. comprehension, functional memory, attention, etc.*)
Physical/Other	Ext = Extremities BUE/BLE = Bilateral upper/lower extremity HEENT = Head, eyes, ears, nose, and throat NC/AT = Normocephalic, atraumatic NAD = No abnormalities detected NTND = Nontender, nondistended RRR = Regular rate and rhythm (*heartbeat*) CVS = Cardiovascular system
Motor	ROM = Range of motion Extension/Flexion Chorea, dystonia, ballismus, myoclonus, athetosis, tics, pronator drift, akinesia
Tremor	Essential, resting, action, intention
Tone:	
Hypertonicity	Spasticity, cogwheel rigidity, paratonia lead pipe rigidity
Hypotonicity	Flaccidity, atrophy
Medical Research Council (MRC, 1976) Muscle Strength Grading Scale:	0 = No contraction 1 = Visible twitch but no joint movement 2 = Weak contraction insufficient to overcome gravity 3 = Weak contraction able to overcome gravity but no additional resistance 4 = Weak contraction able to overcome some resistance, but not full resistance 5 = Normal; able to overcome full resistance
Coordination & Proprioception	FTN/F2N = Finger-to-nose HTS/H2S = Heel-to-shin RAM = Rapid Alternating Movements Heel-walk/Toe-walk Romberg Test (*positive test indicative of pathology*)
Gait	Normal, ataxic, spastic, asymmetric, steppage, scissored, antalgic, hemi/diplegic, hemi/paraparetic, propulsive/lurching, apraxic, circumductive, hyperkinetic, neuropathic, apraxic, magnetic

Tabel 3.1 (continued)

Cranial Nerves (CNs)		
CN II	Visual field exam, visual acuity, fundoscopic exam, afferent pupillary light reflexes	• EOMI = Extraocular movements intact • ESS/EBSS = Extinction to (bilateral) simultaneous stimulation • OD = Right eye • OS = Left eye • OU = Both eyes • VF/F = visual field (full) to confrontation • 20/20 = Normal visual acuity • Myopic = Nearsighted • Hyperopic = Farsighted • Funduscopic Examination (retinal exam)
CN III, IV, &VI	Eye lid elevation, efferent pupillary light reflexes Eye movements, saccades	• INO = Internuclear ophthalmoplegia • PERRLA = Pupils equal, round, and reactive to light and accommodation • RAPD = Relative afferent pupillary defect • VOR = Vestibular-ocular reflex • Ptosis • Saccades/smooth pursuits • Gaze palsy
CN V & VII	Facial sensation and strength	V1-V3 Intact = Ophthalmic (V1), maxillary (V2), and mandibular (V3) branches of the trigeminal nerve (CNV) intact
CN VIII	Hearing and vestibular functioning (finger rub, balance)	• Finger rub (*hearing test*) • AC = Air conduction • BC = Bone Conduction (*Rinne Test normal result = AC>BC*)
CN IX, X, XII	Palatal movements, swallowing, speech	• No uvula deviations/midline at rest • Elevation of palate • Dysphagia/Dysarthria
CN XI	Head rotation, shoulder elevation	• SCM = Sternocleidomastoid muscle
CN XII	Tongue movements	• Tongue midline during protrusion (*e.g. no deviation*) • Symmetry vs asymmetry

Tabel 3.1 (continued)

Commonly Assessed Reflexes		
Glabellar* (Myerson)	Tapping the forehead between the eyebrows resulting in involuntary blinking	• Normal: Blinking is suppressed after several taps • Abnormal: Blinking persists after multiple taps
Corneal	Touching the cornea with a clean cotton swab or fiber results in blinking	• Absence of blink is abnormal
Root* (Moro)	Stroking of cheek or mouth results in movement toward stimulus with mouth opening and tongue thrusting	• Normally present in infants • Abnormal in children and adults
Palmomental* (Marinescu)	Stroking the base of the thumb results in a chin twitch	• Normally present in infants • Abnormal in children and adults
Gag	Stimulation of the oropharynx leads to involuntary gagging	• Normally present
Palmar Grasp Reflex	Light flicking down of the middle finger leads to involuntary thumb flexion and/or index finger	• Normally present in infants • Abnormal in children and adults
Finger Flexor (Hoffman)	Light flicking down of the middle finger leads to involuntary thumb flexion and/or index finger	• Normally absent
Abdominal Reflexes	Superficial: stroking the abdomen toward the center results in umbilical movement Deep: tapping over the abdominal muscles results in contraction	• Absence of either is abnormal
Cremasteric	Stroking the inner thigh leads to testicular elevation	• Normally present in males
Geigel	Stroking the inner thigh leads to muscle contraction at the inguinal ligament	• Normally present in females

Tabel 3.1 (continued)

Anal	Stroking the perianal area or inserting finger in the rectum leads to anal sphincter contraction	• Normally present
Plantar (Babinski)	Blunt stimulation of the lateral sole of the foot up from the heel and across the ball of the foot	• Normal: Toes flex inward toward stimulus • Abnormal: Toes fan outward (normal finding in infants) • No response: equivocal finding
Deep Tendon Reflexes (DTRs)		
Scale: 0 = Absent 1+ = Hypoactive 2+ = Normal 3+ = Hyperactive 4+ = Hyperactive with clonus	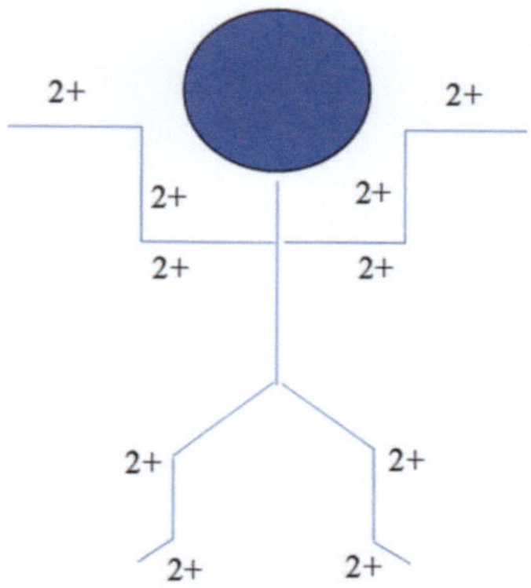	*(Ascending to descending)*: Brachioradialis (forearm), biceps, triceps, abdominal, patellar (knee), and Achilles tendon reflexes

** Indicates reflexes (glabellar, root, palmomental, palmar, babinski) associated with frontal release signs, all generally present in infancy and disappear in adulthood. While normal population variation exists (e.g. abnormal palmomental reflex seen in ~10% of normal adults), frontal release signs may present with neurodegeneration or other neurologic pathology.*

tion) into the neuropsychological assessment. Like a neurological examination, a neurobehavioral examination is a targeted assessment used to identify clinical signs and symptoms of CNS dysfunction that may have neuropsychiatric implications. The presence of frontal release signs, for instance, may suggest the presence of frontotemporal degeneration or a predisposition to postoperative delirium, while motor and sensory deficits can localize pathology within the central nervous system.

Complete Blood Count

A complete blood count (CBC) panel is a general blood test providing information about a patient's overall health status. It assesses red blood cells, white blood cells, and platelets. Elevated or decreased values on the CBC may be reflective of infection, inflammatory reactions, blood cancers, autoimmune processes, or nutritional deficiencies. Neuropsychologists can utilize CBC results to identify potential contributors to cognitive changes, such as poor motivation and fatigue associated with low red blood cells (anemia), or delirium associated with underlying infection leading to high white blood cells (leukocytosis) (Table 3.2).

Comprehensive Metabolic Panel

The comprehensive metabolic panel (CMP) is a blood test that evaluates various biochemical markers, including glucose, electrolytes, and both liver and kidney function (Table 3.3). When combined with the CBC, these broad tests provide an

Table 3.2 Complete blood count findings across broad disease classes

	Elevated	Decreased
White blood cells	Infection Inflammation Blood cancer	Malnutrition Immune deficiency Bone marrow disease
Red blood cells	Smoking Chronically low oxygen Dehydration	Iron deficiency Blood cancer Chronic kidney disease
Platelets	Infection Inflammation Surgery or trauma	Alcohol or other toxins Autoimmune disease Bone marrow disease

overall picture of general physical health and will frequently be used to guide further medical evaluation or diagnostic workup. A similar, but less comprehensive test is the basal metabolic panel (BMP), which includes electrolytes (sodium, potassium, chloride), calcium, creatinine, blood urea nitrogen (BUN), and glucose. A CMP is typically ordered over a BMP when clinical indication warrants more in-depth analysis of liver function. Determinations of elevated or decreased lab values are made based on reference ranges which vary depending on the individual lab and the patient's age, sex, and in some cases—even ethnicity. The ranges should always be reported in the lab results according to those parameters. In medical documentation, lab results are often displayed in common templated diagrams for ease of identification. Examples of these are found in Fig. 3.1.

Understanding CMP results enables neuropsychologists to identify metabolic imbalances that may underlie neuropsychological symptoms. Examples might include elevated sodium and creatinine, suggesting dehydration as a cause for ongoing disorientation, or altered liver enzymes suggesting impaired hepatic function reducing medication metabolism. Abnormal calcium (Ca) levels can also impact cognitive and behavioral functioning in unique ways relative to other measures included in the CMP. Understanding calcium results allows neuropsychologists to recognize potential associations with delirium (high calcium) or changes in muscle tone (low calcium). Calcium disruptions can cause psychiatric changes, such as depression, anxiety, and in rare cases, psychosis. For example, symptoms of primary hyperparathyroidism (PHPT), an endocrine disorder associated with hypercalcemia, are colloquially described in the mnemonic "stones, bones, groans, and psychiatric moans"—referring to kidney stones, bone pain, gastrointestinal symptoms, and changes in mood/mental status. Seizures, coma, and death may also occur with severe calcium disruption.

Urinalysis and Culture

Urinalysis provides information about kidney function by measuring byproducts of metabolism. This test may also alert to the presence of urinary tract infections through evidence of inflammation or immune system reactions. The term "culture" refers to the use of a suitable laboratory material to encourage further microbial growth. If there is evidence of infection, the sample will be cultured to identify what microorganisms may be present and which antibiotics may be most effective. Urinalysis and culture are most important in the assessment of acute changes in neurocognitive status, as these tests may suggest underlying kidney dysfunction or infection as a primary contributor to delirium.

Urine Drug Screen

The urine drug screen (UDS) is a tool to detect the presence of specific drugs or their metabolites in the urine. It aids in assessing substance use or misuse, which can have significant implications for intoxication, withdrawal, or drug-related neuropsychiatric symptoms. It is important to note that the UDS will typically measure a specific subset of substances and will not assess for every substance of misuse. This is particularly true regarding newer synthetic substances, which may require additional complex testing to screen or identify.

Table 3.3 Complete metabolic panel findings across broad disease classes

	Elevated	Decreased
Sodium (Na)	*Hypernatremia* Dehydration Kidney dysfunction Endocrine disease	*Hyponatremia* Excessive water intake Medication effects Endocrine disease
Potassium (K)	*Hyperkalemia* Excessive intake Kidney dysfunction Endocrine disease	*Hypokalemia* Diarrhea or vomiting Medication effects Kidney or endocrine disease
Chloride (Cl)	*Hyperchloremia* Toxic exposures Kidney or endocrine disease	*Hypochloremia* Excessive sweating Excessive water retention Endocrine disease
Carbon Dioxide (CO2)	*Hypercarbia* Excessive sweating Excessive water retention Endocrine disease	*Hypocarbia* Toxic exposures Kidney or endocrine disease
Blood urea nitrogen (BUN)	*Azotemia* * High protein diet Dehydration Kidney or urinary disease	*Low BUN* Low protein diet Excessive water intake Low muscle mass
Creatinine (Cr)	*High CR* Dehydration Kidney or urinary disease Muscle breakdown	*Low CR* Malnutrition Low muscle mass
Glucose (Gl)	*Hyperglycemia* Diabetes Thyroid or pancreas disease Physiologic stress reaction	*Hypoglycemia* Malnutrition Medication effects Thyroid, pancreas, or pituitary disease
Calcium (Ca)	*Hypercalcemia* Thyroid or parathyroid disease Excessive intake or absorption Bone disease	*Hypocalcemia* Thyroid or parathyroid disease Malnutrition Kidney disease
Albumin	*Hyperalbuminemia* Kidney or liver disease Inflammatory bowel disease Low protein diet or malnutrition	*Hypoalbuminemia* Dehydration High protein diet

Tabel 3.3 (continued)

Total Protein	*Hyperproteinemia* Chronic inflammation Blood cancer	*Hypoproteinemia* Malnutrition Hemorrhage or burns Kidney or liver disease
Alanine transaminase (ALT)	*Transaminitis* + Mild or moderate liver disease Medication or toxin effects Infection or inflammation	*Low ALT* End stage liver disease Malnutrition
Aspartate transaminase (AST)	*Transaminitis* + Mild or moderate liver disease Medication or toxin effects Infection or inflammation	*Low AST* End stage liver disease
Bilirubin	*Hyperbilirubinemia* Liver or gallbladder disease Gallstones Red blood cell breakdown	*Hypobilirubinemia* Medication effects Inflammatory disease
Alkaline phosphatase (ALP)	*High ALP* Liver disease Bone disease Rapid bone healing or growth	*Low ALP* Thyroid disease Malnutrition

* *Typically diagnosed in combination with elevated Cr*
+*Typically diagnosed in combination with elevated AST and/or ALT*

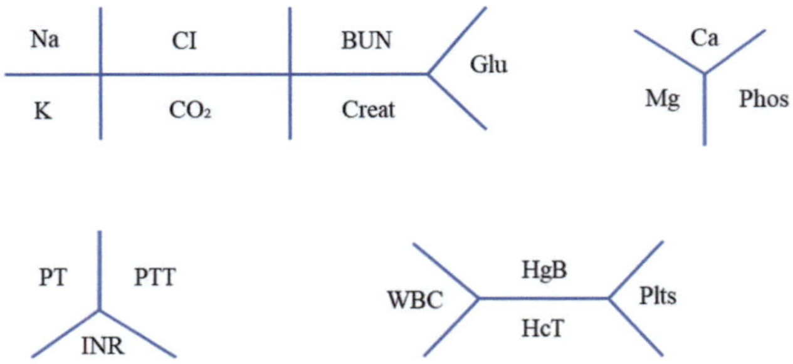

Fig. 3.1 Common lab result diagrams for sodium (Na), potassium (K), chloride (Cl), bicarbonate (CO_2), blood urea nitrogen (BUN), creatinine (Creat), glucose (Glu), calcium (Ca), magnesium (Mg), phosphate (Phos), prothrombin (PT), partial thromboplastin time (PTT), international normalized ratio (INR), white blood cell count (WBC), hemoglobin (HgB), hematocrit (HcT), and platelets (Plts)

Nutritional Deficiencies

Deficiencies in essential vitamins like B12, Vitamin D, niacin, and folate can affect cognitive function, in addition to variable changes in neuromotor and psychiatric findings. This is particularly true when presented alongside other indicators of nutritional deficiency, such as low protein production or a physical examination revealing an underweight patient. A panel evaluating these vitamins, along with other lab values, is often included as part of the diagnostic work-up to evaluate reversible causes of cognitive change (i.e., frequently termed "cognitive" or "dementia" labs).

Infectious Markers

Infectious markers like human immunodeficiency virus (HIV) antibodies, rapid plasma reagin (RPR) or treponemal antibody (anti-*treponema pallidum* or anti-TP) testing for syphilis, and respiratory viral panels can provide insights into infectious etiologies of neuropsychological symptoms. Through extensive study, the potential neurobehavioral effects of infections such as HIV and syphilis have been well-established. Other infections, however, may have more variable effects, particularly during the acute phase of illness. This includes such common infections as influenza or other respiratory viruses. It is generally held that as a downstream result of infection, for instance, neurophysiologic changes resulting from inflammation may manifest in a variety of ways, such as delirium (Maldonado 2018) or depression (Collins et al. 2021). As more recent research has extensively examined the impact of chronic infection and resulting inflammation, it is especially important to understand variances which exist in literature prior to—and following—the development and widened availability of gold standard therapies such as antibiotics for syphilis or combined antiretroviral therapy (cART) for HIV.

Human Immunodeficiency Virus

Acute HIV infection is associated with inflammation, rapid increases in systemic viral replication, and may lead to direct neuronal toxicity. This may result in a highly variable clinical picture more akin to delirium. The most common initial cognitive manifestation of chronic infection is bradyphrenia, sometimes attributed to apathetic or depressed presentations. Similarly noted are inattention, broad executive dysfunction, and deficits of fine motor function. Later in the clinical course impairments in memory and visuospatial function may exist alongside mood changes, with more pronounced motor impairments and gait disturbances occurring over time (Rumbaugh and Tyor 2015). Historical diagnoses of HIV-associated dementia (HAD) in the pre-cART era referred to a cognitive syndrome arising as a consequence of untreated HIV infection. However, contemporary clinical and research nomenclature in the post-cART era allows for the designation of a cognitive disorder despite optimal treatment (Antinori et al. 2007).

Unfortunately, objective laboratory findings such as high HIV viral load in the blood or cerebrospinal fluid (CSF) correlate poorly with the degree of chronic cognitive impairment. The diagnosis of HIV-associated neurocognitive disorder (HAND) will generally depend on course, quality, and degree of impairment noted on a clinical examination relative to the duration of infection and optimization of cART therapy (Fig. 3.2). A subset of immune system cells, CD4 lymphocytes, may be measured to monitor for additional indirect effects associated with HIV infection. A range of 500–1500 CD4 cells per cubic millimeter of blood generally represents the normal range, while lower levels (<500) represent immune system impairment. Severe reductions, particularly when less than 200 cells per cubic millimeter, confer increased risk of opportunistic infections and other complications such as abscesses, encephalitis, or lymphoma which may lead to an additional broad constellation of cognitive impairments.

Syphilis

Neurosyphilis, the term for central nervous system involvement of untreated infection by the *T. pallidum* bacteria, is less common following development of effective treatments such as penicillin. Screening for infection, however, remains a common element in the assessment of neurocognitive symptoms, particularly when suspicious symptoms are present. Early infection may be identified by an asymptomatic meningitis serendipitously observed on imaging studies, though headache, photophobia, confusion, and cranial nerve deficits may arise. Late infection historically has been associated with two primary categories of symptomatology: *general paresis* (dementia, dysarthria, mood/personality changes) and *tabes dorsalis*

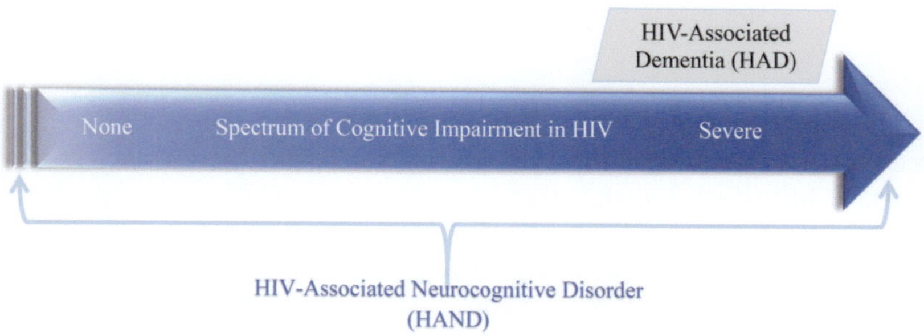

Fig. 3.2 HAND and HAD

(ataxia, paraparesis, loss of deep and proprioceptive sensation). A bilateral pupillary exam revealing pinpoint pupils that constrict with accommodation but not light (Argyll Robertson pupils) is highly specific for neurosyphilis (Ropper 2019).

Screening for syphilis infection is generally based on blood and CSF assays. Treponemal tests such as fluorescent treponemal-antibody absorption (FTA-ABS) detect antibodies directly against *T. pallidum* bacteria, while nontreponemal tests such as Venereal Disease Research Laboratory (VDRL) or RPR detect antibodies against another substance associated with infection. Treponemal tests are generally positive earlier in the course of an infection and remain detectable for life, making them somewhat less specific for identifying active infection. Non-treponemal tests, however, will generally return to a nonreactive state following treatment. As a result, accurate diagnosis will almost always require a screening and confirmatory test phase to differentiate between historic and active infections.

Procalcitonin

Procalcitonin (PCT) is produced in certain cells in the body, especially in response to bacterial infections. Elevations in procalcitonin levels may indicate the presence of an inflammatory response to an infectious process. This inflammation may impact neurologic and neurocognitive function in a variety of ways, but in the acute setting presents most frequently as delirium.

Inflammatory Markers

Commonly measured markers for inflammation include the erythrocyte sedimentation rate (ESR) and C-reactive protein (CRP). Elevated levels of these biomarkers may suggest underlying infection, autoimmune disease, or chronic inflammatory conditions. Unfortunately, these markers are generally nonspecific. In isolation these markers can neither identify the underlying causes of inflammation, nor can their levels be used to extrapolate the severity of neurocognitive disturbance. Generally speaking, acute inflammation tends to cause delirium (particularly in patients with preexisting cognitive impairment or vulnerability), while chronic inflammation commonly causes mood disturbances, amotivation, fatigue, and broad (but more subtle) cognitive symptoms.

Thyroid Function Tests

Thyroid function tests (TFTs) evaluate the thyroid gland's activity. Specific markers for this evaluation may include thyroid-stimulating hormone (TSH), which triggers the thyroid to release additional hormones such as thyroxine (T4), which is then converted to triiodothyronine (T3). Broadly, hypothyroidism is associated with elevations of TSH and low or normal T3 and T4. Contrastingly, hyperthyroidism is associated with low TSH and normal to high T3 and T4 levels. There are several specific clinical syndromes associated with particular combinations of complex lab findings that may warrant further testing. By evaluating the various steps in the thyroid function pathway, these laboratory tests may indicate the point along the pathway in which a disease process is occurring. Abnormalities in thyroid function can lead to reversible global cognitive dysfunction or more specific changes in attention, visuospatial processing, working memory, or motor speed. Hypothyroidism (high TSH) can also lead to depression, fatigue, and lethargy, while hyperthyroidism (low TSH) can cause anxiety and hyperactivity.

Arterial Blood Gas

Blood may be analyzed to assess the levels of important arterial blood gases (ABGs) such as oxygen and carbon dioxide. This test may also show how acidic or alkaline the blood is, which is critical in cases of respiratory distress or metabolic disturbances that can affect brain function. Poor ventilation and acute reductions in blood oxygen (anoxia), for example, will lead to acute global transient cognitive dysfunction. More severe and prolonged periods of anoxia are associated with greater severity of global symptomatology, but regionally associated selective vulnerabilities do exist. Cerebral cortical, hippocampal, cerebellar, and basal ganglia regions are generally most affected by anoxia and show structural changes earlier. The most common chronic symptom of anoxia is memory impairment, though considerable variability exists (Garcia-Molina et al. 2006).

Lumbar Puncture

The CSF surrounding the brain and spinal cord may provide valuable information regarding the condition of the central nervous system and many disease processes that can impact it. Lumbar puncture (LP), or spinal tap, is a technique in which a needle is passed between the bones of the spinal column to extract a sample of this fluid. A specific type of LP that removes a larger volume of fluid is referred to as the Miller Fisher Test or high-volume tap. This test supports the clinical examination of a patient in whom normal pressure hydrocephalus (NPH) is suspected. In acute NPH, the characteristic findings of cognitive, gait, and urinary dysfunction may show rapid improvement with removal of a large volume of CSF. In chronic disease, however, the recovery may be less rapid and less robust.

Regardless of the technique, an LP allows the analysis of CSF and aids in diagnosing conditions like infection, bleeding, or increased intracranial pressure that may lead to global cognitive dysfunction, focal neurologic deficits, or delirium. Basic analysis of CSF includes the pressure of the fluid, the total number and types of cells contained, and respective levels of protein and glucose. Elevated white blood cells associated with high protein and low glucose, for instance, may suggest bacterial CNS infection, while normal protein and glucose may be more suggestive of viral infection. Additional more specific assays may be requested, such as a Gram stain to differentiate bacterial subtype, or antibody titers to rule out inflammatory, autoimmune, or paraneoplastic conditions. For instance, in a case of suspected encephalitis secondary to autoimmune N-methyl-D-aspartate (NMDA) receptor antibodies, CSF testing will yield a much more sensitive diagnostic screen than serum studies.

Noninvasive Diagnostic Studies

Electroencephalography

Electroencephalography (EEG) records brain electrical activity as a signal for underlying neurological functioning. Based on the patterns identified, EEG may help diagnose various neurological conditions, such as seizures and encephalopathies, while also assessing normal changes occurring across a variety of brain states, such as sleep and wakefulness. Initial interpretation assesses the general background level of activity and the degree of symmetry which is present.

Background activity will generally fall into one of five categories. High frequency waveforms are referred to as Alpha (8–12 Hz), associated with resting wakefulness; Beta (12–30 Hz), associated with generally active thought processes; or Gamma (30 Hz and above), associated with heightened concentration and active problem-solving. Lower frequency waveforms are referred to as Theta (4–7 Hz), associated with meditative states and light sleep, or Delta (0.1–4 Hz), associated with deep non-rapid-eye-movement (non-REM) sleep or unconsciousness.

Asymmetry on an EEG will generally be indicative of focal lesions following surgery, trauma, or other intracranial pathology. An example of this would be periodic lateralized epileptiform discharges (PLEDS), a 1–2 Hz waveform occurring only in one hemisphere, and generally associated with cerebrovascular accidents, intracranial bleeds, or certain transmissible diseases such as herpes encephalitis (HE) or Creutzfeld-Jakob disease (CJD).

Certain clinical conditions with specific neurobehavioral presentations may demonstrate stereotyped waveform findings on EEG. Spike, sharp, and slow waves, for instance, are each a distinct finding that may occur on a normal EEG and stand out from the background activity. However, when these occur together in close proximity or chronology, they may suggest underlying seizure activity. Generalized slowing that is inconsistent with the clinical wakefulness of the patient is suggestive of an underlying delirium or encephalopathy. Triphasic waves are another well-defined but nonspecific waveform and are generally suggestive of underlying metabolic or toxic encephalopathy.

Electrocardiogram

Electrocardiograms (ECGs or EKGs) record the heart's electrical activity and help detect problematic rhythms or other cardiac conditions affecting brain perfusion and function. An acute problem with the heart, such as a rhythm disturbance or a failure to provide sufficient oxygen to the cardiac muscle, will generally also be associated with physical symptoms such as chest pain or numbness, in which case an EKG may serve as a confirmatory test. Additional problems, however, may be incidentally diagnosed based on the EKG report, even without other symptoms. In such cases, the presence of cardiac disease such as coronary artery disease or atrial fibrillation may indicate global vascular disease; these are particularly important in neurocognitive profiles suggestive of chronic subcortical microvascular disease.

NEUROIMAGING TECHNIQUES AND CLINICAL APPLICATIONS

In neuropsychological assessment, the ability to visualize and understand the brain's structure and function is paramount. Neuroimaging techniques provide the necessary window into the central nervous system, aiding in the diagnosis and evaluation of certain conditions. A broad overview of the science, techniques, and applications of common neuroimaging modalities may provide neuropsychologists with the foundation to interpret and incorporate these tools into practice. A list of radiologic terminology commonly used in neuroimaging reports can be found in Appendix 5 at the end of this chapter.

Computed Tomography

Computed tomography, or CT scan, uses a narrow beam of x-ray radiation, which is rapidly rotated around the body, to generate detailed cross-sectional images of the brain. A CT of the brain will take approximately 15–30 minutes; studies may require additional time if the scan is duplicated after the administration of an intravenous agent to improve contrast of neuroanatomic structures. Higher density tissues, such as bone or brain, will block more of the beam than lower density regions, such as the ventricles. This neuroimaging modality is useful in quickly detecting gross neuroanatomic conditions such as hemorrhages, tumors, subacute infarct, or other structural abnormalities that may contribute to cognitive changes.

Structural Magnetic Resonance Imaging

Magnetic resonance imaging (MRI) employs strong magnetic fields and radio waves to create highly detailed images of the brain's structure. This is accomplished by analyzing movement of subatomic particles (e.g., protons) contained in living tissue. Whereas a CT scan relies upon the density of tissue itself to absorb x-ray energy in constructing an image, MRI is better able to distinguish between tissues which are of a similar density. There are several ways in which the image information may be recorded and processed, allowing the same tissue to be presented in ways that might better inform a clinical or diagnostic assessment. See Table 3.4 for general strategies of interpretation for specific imaging series. Structural MRI is instrumental in visualizing gross lesions, subtle abnormalities, and structural changes associated with neuropsychological disorders, such as dementia and multiple sclerosis. Additional specialized MRI studies, particularly apparent diffusion coefficient (ADC) and diffusion weighted imaging (DWI) sequences, can be extremely valuable in identifying acute cerebrovascular accidents and show tissue changes within minutes of an ischemic infarct.

Diffusion Tensor Imaging

Diffusion tensor imaging (DTI) is an advanced MRI technique; rather than showing neuroanatomic structures, DTI may be used to map brain connectivity. By calculating how the random movement of water in brain tissue is affected by cellular structures, DTI is especially useful in illustrating the arrangement and microstructural integrity of long white matter tracts. This modality aids in the assessment of structural connectivity and is essential for understanding conditions involving disruptions in neural pathways, such as traumatic brain injury (TBI) with diffuse axonal injury (DAI) or demyelinating diseases.

Table 3.4 Signal intensity of various tissues across MRI sequences

	T1	T2	FLAIR
White matter	◔	●	●
Gray matter	◕	◔	◔
CSF or Fluid	●	○	●
Fat	○	◔	◔

Appearance: dark ●, *dark gray* ◕, *neutral* ◑, *light* ◔, *bright* ○

Positron Emission Tomography

Positron emission tomography (PET) scans are valuable in detecting functional changes related to conditions such as Alzheimer's disease (AD), epilepsy, and brain tumors. PET uses radioactive tracer compounds to measure specific characteristics of function or structure in the brain. There are a variety of PET modalities used across clinical settings (e.g., cancer staging), with the following imaging studies used more frequently in examining neurocognitive disorders. Fluorodeoxyglucose (FDG) PET shows the levels of metabolic activity in the brain, by measuring how much of a special radioactive glucose neuronal tissue uses; by comparing variable levels of metabolic activity across brain regions, distinct disease processes such as frontotemporal lobar degeneration (FTD) and AD may be differentiated.

Other types of PET use radioactive tracer compounds which attach to specific proteins, such as amyloid or tau; rather than measuring metabolic activity, these show the distribution and quantity of specific proteins throughout the brain and may be used to confirm neurodegenerative diagnoses. These amyloid tracer imaging studies are increasingly incorporated into structured clinical care protocols involving antibody-mediated removal of amyloid plaques in AD. However, there is limited evidence to suggest that these should be used as general screening, as amyloid may be a nonspecific finding even in individuals without a neurodegenerative process.

Other Functional Neuroimaging Techniques

There are a variety of additional imaging modalities, such as functional MRI (fMRI) or single-photon emission computerized tomography (SPECT), which may be used in research settings or particularly advanced assessment of cognitive conditions. These will generally not be used in routine clinical care. Some considerations for the future practice of functional imaging include advanced techniques that allow real-time comparison of cortical activity across brain regions, without the invasiveness of PET or unwieldy nature of fMRI; the most recent example of this includes diffuse optical tomography (DOT), which uses near-infrared light through the skull to measure brain structure and activity near-instantaneously.

Understanding these neuroimaging modalities and their roles in clinical practice empowers neuropsychologists to collaborate effectively with radiologists and other healthcare professionals. It enables them to recommend the most appropriate imaging method for a given clinical scenario and interpret the findings in the context of their neuropsychological evaluations, ultimately enhancing the precision of their assessments.

APPLYING MEDICAL DIAGNOSTICS: DELIRIUM

In neuropsychological assessment, the ability to decipher complex neuropsychiatric syndromes within the broader context of medical diagnostic findings is of paramount importance. This is especially true when faced with acute changes in mental status observed in conditions like delirium. To aid neuropsychologists in this critical endeavor, the acronym "IWATCHDEATH" holds value as a mnemonic for considering possible contributions to acutely altered mentation, while also serving as a clinical reminder underscoring the urgency and significance of acute changes in mental status. Both the severity and duration of diagnoses such as delirium, after all, are highly positively associated with increased morbidity and mortality, making rapid and complete resolution a primary goal of clinical care.

The components of the IWATCHDEATH acronym are outlined below, as an example for how a neuropsychologist may garner insights from the laboratory findings encountered during assessments.

I—Infection:

Laboratory Findings:

- Elevated WBC or platelets on CBC.
- Elevated infectious or inflammatory markers, such as CRP or procalcitonin.

Vital Signs and Physical Exam Findings:

- Fever and tachycardia may be present during acute infection.
- Severe infections (sepsis) may be associated with low blood pressure.

W—Withdrawal:

Laboratory Findings:

- Positive urine drug screen indicating the presence of withdrawal-inducing substances or their metabolites.
- Electrolyte imbalances, such as low potassium or sodium levels, often seen in alcohol intoxication or withdrawal.
- Blood alcohol levels can be helpful to assess level of intoxication or, if absent or low in an individual with chronic alcohol ingestion, may suggest presence of withdrawal.

A—Acute Metabolic (acidosis, alkalosis, electrolyte):

Laboratory Findings:

- CMP revealing abnormal electrolyte levels, such as elevated sodium and creatinine in dehydration or high glucose with low sodium and chloride levels in uncontrolled diabetes.
- ABG showing acid-base disturbances (e.g., metabolic acidosis) that may arise from metabolic disease (e.g., diabetes), toxic exposure (e.g., methanol, ethylene glycol), or respiratory disease (e.g., COPD).

T—Trauma (closed head injury, hemorrhage, fat embolus):

Laboratory Findings:

- Evidence of traumatic brain injury on neuroimaging, like CT or MRI.
- Elevated levels of D-dimer, a protein fragment that is made when a blood clot dissolves in the body, indicating possible thromboembolic events.

C—Central Nervous System (CNS) pathology (seizure, stroke, encephalitis):

Laboratory Findings:

- Abnormal EEG patterns indicative of varying pathology seizures (rapid spiking waves), encephalopathy (generalized slowing), or encephalitis (slow wave with spike discharges).
- Elevated WBC count in cerebrospinal fluid (CSF) for encephalitis.
- Lower Aβ-42/Aβ-40 ratio in CSF testing for Alzheimer's disease.

H—Hypoxia/Hypercarbia:

Laboratory Findings:

- ABG showing hypoxemia (low oxygen levels) or hypercapnia (high carbon dioxide levels).
- Elevated lactate levels in cases of low tissue oxygen, indicating cellular use of an alternative energy source.

D—Deficiencies (thiamine, niacin, B12, folate):

Laboratory Findings:

- Low serum levels of B12 or folate leading to global cognitive dysfunction in older adults.
- Low serum levels of thiamine or niacin in chronic alcohol use, leading to encephalopathy and dementia.
- Abnormally small or abnormally large red blood cells on CBC, suggesting nutritional deficiencies such as iron or cobalamin deficiencies.

E—Endocrinopathies (thyroid, glucose, adrenal):

Laboratory Findings:

- Abnormal thyroid function tests indicating hypo- or hyperthyroidism.
- Presence of antibodies against thyroid cells, indicating a maladaptive immune reaction to thyroid tissue.
- Dysregulated blood glucose levels, which may signify acute or chronic diabetes-related mental status changes.

A—Acute Cardiovascular (hypertensive crisis, arrhythmia):

Laboratory Findings:

- Elevated blood pressure measurements or evidence of blood pressure crisis, with hypertensive emergency affecting brain perfusion.
- Distinctive imaging findings of occipital and parietal lobe swelling and white matter changes suggestive of reversible posterior leukoencephalopathy syndrome (RPLS, formerly PRES or posterior reversible encephalopathy syndrome) due to uncontrolled hypertension.
- EKG abnormalities indicating arrhythmias or cardiac stress which may lead to central nervous system perfusion problems or vascular disease.

T—Toxins/Drugs:

Laboratory Findings:

- Blood or urine drug screens positive for specific substances of abuse.
- Elevated levels of toxic substances in blood or urine samples, such as through organophosphate exposure.
- Supratherapeutic levels of prescribed medications, such as lithium leading to confusion or cerebellar dysfunction.

H—Heavy Metals:

Laboratory Findings:

- Elevated levels of heavy metals like lead or mercury in blood or urine tests.
- Abnormal metal profiles suggesting metal poisoning.

While these serve as examples of how a broad comprehensive medical examination might inform the assessment of a patient with acute changes in mental status, there are a variety of additional subspecialty tests which may further refine the differential diagnosis based on this essential framework.

COMMON MEDICATION CLASSES AND COGNITIVE EFFECTS

Across the various domains of cognitive function, medications wield considerable influence, both as therapeutic tools and potential contributors to cognitive changes. While the complex nature of psychopharmacology and pharmaceutical interventions may be challenging, understanding their cognitive effects and potential indirect implications remains accessible. This section explores the role of key medications in neuropsychological assessment, with focus on medications showing especially relevant cognitive effects.

AGS Beers Criteria

Originally developed by the American Geriatrics Society (AGS) in 2011 and most recently updated in 2023, The Beers Criteria was developed as a clinical resource aid in identification of potentially inappropriate medications in older adults (AGS 2023). The problems associated with these medications represent a broad spectrum of potential adverse outcomes. Based on these recommendations, some of the most recognized medications with adverse cognitive effect profiles will be highlighted.

Anticholinergic Medications

Anticholinergic medications have direct effects on the central nervous system itself. The neurotransmitter acetylcholine has long been recognized as a prominent factor in frontal cognitive function, particularly in theories of dementia states such as Alzheimer's disease, and the central cholinergic system is now recognized to be made up of neuron clusters across the brain. These are most highly organized in the medial and lateral cholinergic pathways linking the nucleus basalis to the cerebral cortex and amygdala (Mufson et al. 2003). With the introduction of cholinergic blocking agents, these pathways will be less efficient and lead to variable, but potentially global, cognitive decrements. Many studies have highlighted additional potential risk associated with these central effects, which exceeds the short-term cognitive changes associated with anticholinergic agents. For example, some studies suggest an association between the magnitude of anticholinergic burden, duration of exposure, and future development of dementia (Coupland et al. 2019; Risacher et al. 2016). However, this is recognizably a complex phenomenon, and other research including meta-analyses have found no such link (Moriarty et al. 2021; Taylor-Rowan et al. 2021).

Even when not directly classified as an anticholinergic, systemic potential for acetylcholine blockade is present across a variety of medication classes, and in older adults may result in whole-body adverse effects with both direct and indirect cognitive effects (AGS 2023). Dry mouth, for example, may be common with these medications, but is less likely to lead to cognitive deficits. Constipation and urinary retention, however, may have potent indirect effects on cognitive function. One such example, informally referred to as "Cystocerebral Syndrome," speaks to a cognitive manifestation of extreme bowel or bladder distension. Normally a distended bladder will, upon reaching a critical capacity of urine, receive sympathetic activation that allows bladder relaxation to accommodate additional urine. As further expansion occurs, additional sympathetic upregulation leads to central catecholamine surges (Thelmo et al. 2020). An anticholinergic drug may enhance bladder outlet muscle tone, making voiding more difficult, leading to excessive urinary retention, and secondarily leading to excessive central noradrenergic effects. These effects may manifest as anxiety, restlessness, agitation, or confusion in older adults.

The most recognized types of anticholinergic agents are antihistamines used for allergies or as antiemetics. Historically, such agents are also prescribed off-label for anxiety or insomnia. These agents are available over the counter, in addition to prescription formulations, and include diphenhydramine, doxylamine, meclizine, scopolamine, and promethazine (AGS 2023). Certain antihistamines such as hydroxyzine are, by virtue of their inclusion in the class, misrepresented in their anticholinergic potential (Orzechowski et al. 2005); therefore, it remains important to assess the level of anticholinergic potential across the spectrum. Additional anticholinergic agents include but are not limited to some anti-Parkinsonian agents (e.g., benztropine, trihexyphenidyl), antidepressants (e.g., amitriptyline, desipramine, doxepin, paroxetine), and urinary antispasmodics (e.g., oxybutynin, tolterodine).

Determining Anticholinergic Burden

There is no definitive way to estimate the degree to which an anticholinergic medication may lead to cognitive dysfunction, and even direct biochemical testing may not have correspondingly direct clinical application. This varies according to agent, dose, pharmacokinetic properties at the acetylcholine receptor, individual metabolic processes, individual cholinergic pathway resilience, and a variety of other factors. Clinical providers will generally seek to review overall cumulative anticholinergic load when faced with obvious adverse effects, or when concerned about potential sensitivities. One strategy for this review would include the Anticholinergic Cognitive Burden (ACB) Scale, which stratifies several medications according to broad anticholinergic potential to estimate the overall anticholinergic load (Boustani et al. 2008). See Table 3.5 for details of the ACB. Agents with a score of 2 or 3 are considered highest priority for deprescribing, however the needs of individual clinical cases may vary.

Table 3.5 Anticholinergic Burden (ACB) Scale

Score 1	Score 2	Score 3
Alimemazine	Amantadine*	Amitriptyline* †
Alverine	Belladonna alkaloids	Amoxapine
Alprazolam †	Carbamazepine	Atropine*
Atenolol	Cyclobenzaprine* †	Benztropine*
Brompheniramine maleate	Cyproheptadine	Brompheniramine
Bupropion hydrochloride	Empracet	Carbinoxamine
Captopril	Loxapine	Chlorpheniramine*
Chlorthalidone	Meperidine	Chlorpromazine*
Cimetidine hydrochloride	Methotrimeprazine	Clemastine
Ranitidine*	Molindone*	Clomipramine
Clorazepate	Oxcarbazepine	Clozapine †
Codeine	Pethidine hydrochloride	Darifenacin †
Colchicine	Pimozide*	Desipramine
Coumadin		Dicyclomine* †
Diazepam †		Dimenhydrinate
Digoxin		Diphenhydramine* †
Dipyridamole		Doxepin
Disopyramide phosphate		Flavoxate*
Fentanyl		Hydroxyzine*
Furosemide*		Hyoscyamine*
Ruvoxamine		Imipramine*
Haloperidol*		Meclizine*
Hydralazine		Nortriptyline
Hydrocortisone		Olanzapine
Isosorbide*		Orphenadrine
Loperamide		Oxybutynin*
Metoprolol		Paroxetine
Morphine		Perphenazine*
Nifedipine		Procyclidine
Prednisone		Promazine
Quinidine		Promethazine*
Risperidone*		Propantheline
Theophylline		Pyrilamine
Trazodone*		Quetiapine
Triamterene		Scopolamine
		Solifenacin
		Thioridazine* †
		Tolterodine †
		Trifluoperazine*
		Trihexyphenidyl*
		Trimipramine

Adapted from with publisher permission from Boustani et al., 2008.
Commonly prescribed medications with anticholinergic effects may be stratified by relative anticholinergic activity, with a total score calculated by summation of each relevant agent. A total score ≥3 is considered clinically relevant.
** The same score given by Anticholinergic Risk Scale; † The same score given by Anticholinergic Drug Scale*

Opiates and Opioids

The earliest known opioids were derived directly from the sap of the poppy flower, *Papaver somniferum*, and have been used medicinally and recreationally by humans for thousands of years. The terms opiate and opioid are often used interchangeably; however, these terms describe different spectrums of agents with the potential for sedation and pain relief (Pathan and Williams 2012). *Opiates* are naturally occurring substances extracted from the milky sap of the opium poppy. These include opium, heroin, codeine, and morphine, along with thebaine, used in the industrial manufacture of other opioid receptor activators (Pathan and Williams 2012). In contrast, *opioids* refer to a broad classification of substances acting on opioid receptors in the brain, associated with reduction in intensity of pain signals and subjective experience of pain. This category includes the naturally occurring opiates, in addition to semi-synthetic agents (e.g., hydrocodone, hydromorphone, oxycodone, and oxymorphone), as well as fully synthetic agents (e.g., methadone, fentanyl). In short, all opiates are opioids, but not all opioids are opiates. The pharmacologic relationships and metabolic conversions among the opioids are quite complex but are summarized in Fig. 3.3.

Opioid medications, while effective for pain management, can cause sedation in addition to a spectrum of cognitive and mental status changes. To better understand the nature and breadth of opioid-associated cognitive dysfunction, literature reviews have sought to objectively describe these changes among older adults. Generally, these reviews find that overall cognition as measured by global screening instruments—the Mini-Mental Status Examination (MMSE; Folstein et al. 1975) being the most reported—is unchanged by opioid medications (Allen et al. 2003; Pask et al. 2020; Swart et al. 2017). However, the imprecision of such instruments leads to limited ecological validity in practice. With consideration given to specific domain-associated changes, decrements in attention and memory function are commonly described. However, several reviews have noted that similar deficits may be present in individuals with chronic pain who have not received opioids, making distinction of the isolated cognitive effects in opioid use challenging (Moriarty et al. 2011). Unfortunately, these data are affected by methodological challenges that impact reporting, including a need for consistency of testing to facilitate cross-study comparisons, necessity of sample sizes sufficient to draw conclusions regarding age- or comorbidity-based factors, and importance of clear distinctions between short- and long-term use of opioid agents. Furthermore, the range of individual characteristics among opioid medications makes direct comparison particularly challenging. Considering this, attention should be paid to these specific effects of the most used individual agents, including their theoretical effects on cognition.

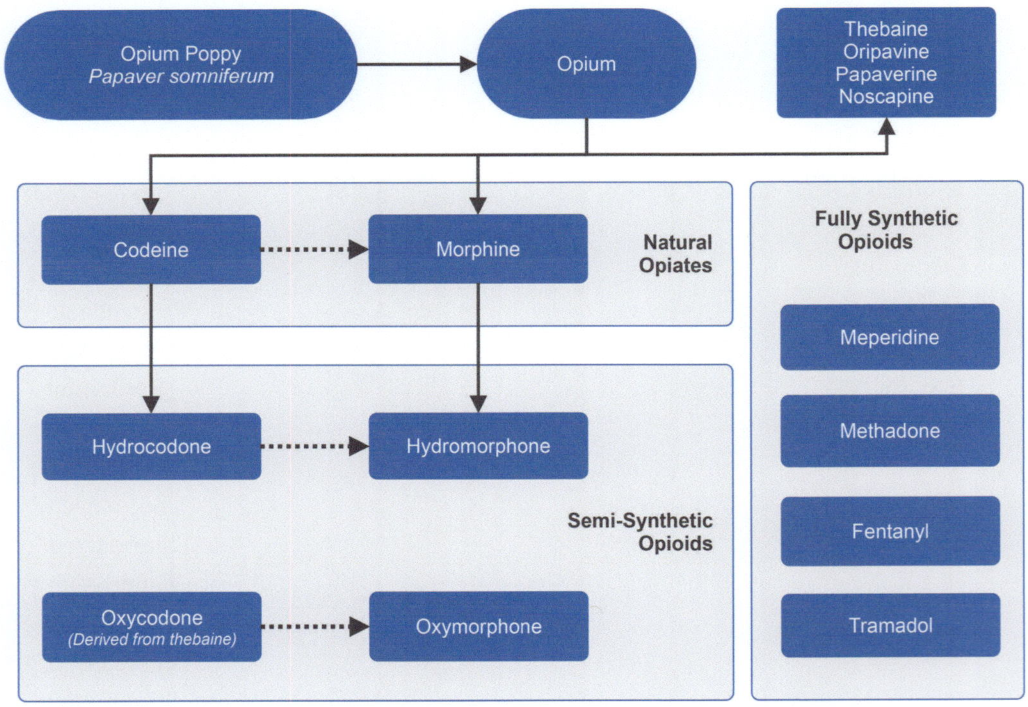

Fig. 3.3 Pharmacologic and metabolic derivatives of natural and synthetic opioids. *Note*: All natural opiates or semi-synthetic opioids are derived from the poppy flower (Papaver somniferum), while fully synthetic opioids are created in a laboratory. Solid lines denote direct extraction or derivation, while dashed lines denote metabolic conversion within the body

Codeine and Hydrocodone

Historically used to target mild pain, particularly in combination with acetaminophen or nonsteroidal anti-inflammatory agents, codeine has additionally entered medical practice for the management of coughing and diarrhea. Newer evidence, however, neither supports its use as an antitussive nor recommends it for the management of gastrointestinal symptoms. Furthermore, in spite of its reputation as a relatively benign narcotic, anecdotal patterns among experienced clinicians suggest that its use should be avoided for management of pain in older adults in favor of agents with fewer adverse effects; one strong consideration is the fact that a high proportion of the population lacks the metabolic ability to convert codeine to its active metabolites, leading to the adverse buildup of metabolic byproducts which do not alleviate pain (Gazelka et al. 2020; Jassal et al. 2020). Controversy exists regarding its anticholinergic potential, though this buildup of ineffective metabolic byproducts may explain its anecdotal association with delirium.

Hydrocodone is a semi-synthetic opioid derived from codeine. When compared directly to codeine, hydrocodone is a higher potency analgesic and effective at lower doses for pain. Studies have suggested hydrocodone has negligible impact on complex cognitive domains, with predominant effects noted on simple tracking and simple reaction-time tasks (Allen et al. 2003).

Morphine and Hydromorphone

Perhaps the most well-recognized prescription opioid, morphine is exceptionally common in clinical practice owing to its dosing form availability and versatility for management of pain across clinical settings. However, despite its versatility, there is clear potential for adverse cognitive effects. Morphine is predominantly metabolized in the body through a process called glucuronidation, which makes the compound water-soluble for excretion by adding a specific molecule at one of two specific areas of its structure; the site of this addition determines which of two substances are created, morphine-3-glucuronide (M3G) or morphine-6-glucuronide (M6G). While M6G is thought to be responsible for morphine's pain alleviation effects, M3G exerts a potentiating effect on glutamate, a potent excitatory neurotransmitter. At elevated levels, this glutamate activation may be associated with anxiety, restlessness, agitation, disorientation, and hallucinations, with prolonged activation potentially associated with neurotoxicity (Gabel et al. 2022). These effects may be particularly prominent in individuals with acute or chronic kidney disease, owing to the excessive buildup of M3G due to insufficient excretion. A semi-synthetic opioid derived from morphine, hydromorphone has similarly theorized potential for neuroexcitatory effects and delirium due to its equivalent metabolite, hydromorphone-3-glucuronide (H3G; Kullgren et al. 2013). However, there is no definitive data demonstrating clinically meaningful deficits associated with hydromorphone, and anecdotal experience generally supports its preferred use when intravenous opioid therapy is required.

Oxycodone

Developed from thebaine, a naturally occurring compound found in opium, oxycodone is considered a semi-synthetic opioid. One advantage of oxycodone relates to its stable pharmacokinetic profile, even among older adults in which medication metabolism will generally be reduced. With comparison of middle-aged and older adults there are deficits noted on simple, complex, and sustained attention tasks, which are consistent among most other opioid agents. These effects are generally noted immediately following a medication dose, following a dose increase, or addition of an additional dose; few clinically relevant cognitive deficits are identified with long-term therapy (Cherrier et al. 2009).

Fully Synthetic Opioids

Fentanyl: The multitude of available dosing formulations (i.e., oral, intravenous, transdermal, etc.) makes fentanyl a particularly common synthetic opioid in the management of both acute and chronic pain. Theoretical models have suggested that the agent may play a role in potentiating anticholinergic burden with a subsequent worsening of cognitive function (Hustveit 1996). However, this has not been demonstrated clinically and may be relevant only to seriously clinically ill patients receiving high-dose fentanyl for sedation. At normal doses for management of both acute postoperative pain and chronic pain management, there is little evidence of cognitive dysfunction (Menefee et al. 2004; Silbert et al. 2006).

Methadone: Historically unique among the opioids, methadone is most used as a maintenance therapy within broader treatment programs for opioid use disorder, owing to its distinctive pharmacokinetic profile. As a consequence of this common use, there is an extensive literature base seeking to determine its potential for long-term cognitive changes; systematic reviews have highlighted challenges with the existing data and provide little opportunity for definitive findings (Baldacchino et al. 2017). The pre-existing substance use, with or without chronic pain, of individuals receiving methadone maintenance treatment is thought to play a significant role in confounding these studies. It is, however, definitively accepted that the benefits of managing opioid misuse outweigh the potential adverse effects.

Meperidine: Clinical practice has historically called for the use of meperidine for specific clinical cases given its uniquely potent serotonergic and anticholinergic effects relative to other opioids (e.g., acute diverticulitis given its potential to decrease pressure in the bowel). However, there is rarely any advantage to using the agent for general practice, and the potential to precipitate serotonin syndrome or anticholinergic delirium generally precludes its use. There is an extensive literature base recommending that it be avoided in older adults due to its considerable potential for adverse cognitive effects (Adunsky et al. 2002; Fong et al. 2006).

Tramadol: Despite its similarity in serotonergic and anticholinergic effects to that of meperidine, tramadol has less evidence supporting or guiding its use. It is perceived in clinical practice to be a less potent and potentially safer opioid for management of moderate chronic pain, owing to its mechanism as a partial opioid receptor activator. Efforts to quantify its potential cognitive effects have ranged from showing no statistical significance (Ahmad et al. 2019), presence of visuospatial memory and executive function deficits (Shalaby et al. 2022), significantly increased risk of postoperative delirium (Brouquet et al. 2010), and increases in all-cause mortality in older adults (Zeng et al. 2019). Unfortunately, there is little consistency across these findings to guide clinical considerations. It is generally thought that low-dose tramadol for management of chronic pain may be safe, with careful monitoring for mental status changes associated with serotonergic or anticholinergic burden. If requiring high doses, in settings of uncontrolled pain, or for postoperative management of symptoms, alternative opioids should be considered.

Ultimately there is compelling evidence to suggest that, despite much argument to the contrary, opioids may be relatively safer than has been traditionally theorized regarding cognitive outcomes. There are likely to be transient decrements in attentional and memory tasks, attributable to simple cognitive "clouding" due to sedation, however the effective management of pain is a critical component of optimal cognitive preservation. Notably, additional studies suggest that rather than exclusively worsening cognitive function, opioids may improve specific measures (Jamison et al. 2003). This appears to be true in individuals with pre-existing deficits such as inattention and poor memory recall due to the known effects of acute and chronic pain. In such cases, the short-term and transient effects of opioid agents likely do not worsen the pre-existing pain-associated deficits and may improve cognition long-term.

With these considerations, a few pragmatics may be considered. Codeine, morphine, and meperidine should be avoided in older adults where possible, particularly among those with pre-existing cognitive disorders, those at risk of delirium due to other comorbidities, or those with metabolic dysfunction. Oxycodone, hydromorphone, and fentanyl are likely to be the safest agents with the lowest theorized risk of cognitive dysfunction.

Benzodiazepines

Commonly prescribed for both psychiatric and medical purposes, benzodiazepines are among the most cited medications with adverse impacts on cognition. The effect of these agents as anxiolytics or antispasmodics is attributable to inhibition of gamma-aminobutyric acid (GABA), like a variety of other sedating compounds, including ethanol. Like opioids, short-term cognitive effects of these agents have long been recognized and suggest similar deficits of attention and processing speed as a consequence of sedating effects. Delirium is the most serious neurocognitive concern, particularly in older adults receiving benzodiazepines acutely, and benzodiazepines are featured prominently in the AGS Beers Criteria for this reason (AGS 2023). Some data do suggest specific anterograde amnestic effects and executive function changes, though historically this has been attributed to response time slowing and secondary effects on more complex executive function and memory processing tasks (Curran et al. 1991; Curran 1999). Systematic reviews of current users have suggested statistically significant decrements in working memory, processing speed, divided attention, recent memory, and expressive language. Individuals recently discontinuing benzodiazepines have shown decrements in all domains except for executive function.

Medium- and long-term effects of benzodiazepines have proven controversial, though generally there is robust evidence to suggest that these nearly pan-domain deficits persist (Barker et al. 2004; Crowe and Stranks 2018; Zetsen et al. 2022). These data tend to be confounded by the multitude of variables which exist, not the least of which are the comorbidity of other neurologic and psychiatric processes for which benzodiazepines are commonly prescribed, and their independent potential consequences on cognition. It is generally accepted that short-term benzodiazepine use is likely benign when used judiciously and briefly, but long-term chronic benzodiazepine use should be avoided. The exceptions to this include individuals at substantial risk of delirium, in which case benzodiazepines should always be avoided when possible.

Antiepileptic Drugs

Traditional antiepileptic drugs (AEDs) have demonstrable efficacy in seizure control but are associated with negative side-effects that have historically limited tolerance and compliance. To address this issue, newer AEDs have been developed that offer fewer adverse effects and more consistent pharmacokinetics (e.g., metabolism, drug interactions). Understanding the differences in side effect profiles across traditional vs newer AEDs can help the neuropsychologist understand how a patient's seizure disorder is being managed and the possible impact of AEDs on cognitive functioning. Older antiepileptic drugs (AEDs), such as phenobarbital, phenytoin, carbamazepine, and valproate, show variable effects on cognition, most commonly specific deficits in attention and processing speed. Among these, carbamazepine and valproate may have fewer adverse impacts (Mula and Trimble 2009). Several newer AEDs, including oxcarbazepine, vigabatrin, lamotrigine, tiagabine, levetiracetam, and zonisamide, are thought to be relatively benign regarding specific cognitive domain impairments (Brunbech and Sabers 2002), with lamotrigine and oxcarbazepine likely showing the fewest adverse cognitive effects (Gualtieri and Johnson 2006). One exception to this is topiramate, which has historically been associated with an idiosyncratic verbal fluency deficit and word-finding difficulty (Marino et al. 2012; Mula 2012; Mula et al. 2003), though some suggestion of global cognitive impacts in at least short-term use has been made (Sommer et al. 2013). These adverse changes appear to be exacerbated by premorbid cognitive impairment, with early cognitive changes predicting worse long-term outcomes.

Dose-dependent effects may act upon premorbid vulnerability factors, with dosages below 100 mg daily showing limited effect, 100–200 mg daily dosages showing modest impairment which improved over time, and dosages at 200 mg and greater showing more pronounced deficits across multiple domains (Callisto et al. 2020; Loring et al. 2011; Shorvon 1996). Importantly, treatment-refractory epilepsy will frequently necessitate the use of multiple AEDs. Even when selected from among the most cognitively benign agents, overlaps in mechanisms of action and downstream adverse effects may confer greater overall cognitive burden than any individual agent, with a greater overall impact on global executive function (Witt et al. 2015). These effects are likely mediated by several additional factors, including age, type of epilepsy, and history of neurologic or psychiatric disease.

In addition to seizure control, many AEDs hold clinical utility for management of psychiatric syndromes. Given clear efficacy of some of these agents in psychiatric symptoms (Grunze 2008), it stands to reason that they additionally hold the potential to induce new or exacerbate existing symptomatology. Given the considerable overlap between neurobehavioral and epileptic syndromes, this is challenging to definitively prove. However, there is theory to suggest that a broad classification according to mechanisms of action may be informative. Agents augmenting GABA, for instance, may simultaneously be more effective in targeting psychiatric symptoms such as mania or panic, while conferring a higher overall risk of fatigue, psychomotor slowing, or overall depressive effects. These include phenobarbital, valproate, gabapentin, and vigabatrin. Other agents likely modulate the effects of glutamate and may show higher rates of activating neurobehavioral symptoms such as anxiogenesis (substance-induced anxiety), irritability, agitation, and racing thoughts. These include levetiracetam, felbamate, lamotrigine, and oxcarbazepine (Eddy et al. 2012; Ketter et al. 1999; Loring et al. 2007).

Antipsychotics

Antipsychotics are frequently prescribed across acute care and rehabilitative settings to manage both primary psychiatric disorders and neurobehavioral syndromes secondary to medical or neurologic disease. Despite frequency of use, the potential cognitive effects are not well understood. It is well documented that typical antipsychotics show a greater degree of dopamine receptor occupancy and blockade, which may lead to both parkinsonistic motor findings in addition to impairment of cognitive tasks relying on either fine motor skills or psychomotor speed (Kasper and Resinger 2003). Some of these agents may additionally demonstrate potent anticholinergic potential, the effects of which have been previously reviewed in this chapter. Others may have little inherent anticholinergic potential but may require coadministration with an anticholinergic agent to target the extrapyramidal motor changes associated with dopamine blockade in the nigrostriatal pathway. Atypical, or second-generation, antipsychotics are generally thought to demonstrate lower overall cognitive burden as a consequence of their alternative mechanisms of action. Through their effects at serotonin receptors there may be selective modulation of frontal speed and motor control systems, coupled with sparing or even enhancement of learning and memory functions (Kasper and Resinger 2003). This may be particularly true in treating a primary psychotic disorder, in which a known association with cognitive impairment exists independently of treatment.

Outside of a primary psychotic disorder, antipsychotics may commonly be prescribed to manage behavior secondary to an underlying neurodegenerative disease. Unfortunately, the heterogeneity of this population makes global determination of cognitive adverse effects challenging. Similar patterns to those demonstrated in schizophrenia have been shown, such that atypical antipsychotics may have an overall more benign cognitive profile due to more favorable modulation of dopaminergic, cholinergic, and histaminergic systems relative to typical antipsychotics. However, atypical antipsychotics do show clear global decrements in cognition relative to baseline, and in AD may be associated with a 1-year magnitude of decline (Vigen et al. 2011). Furthermore, more consistent adverse effects have been demonstrated within clinical subgroups. For example, individuals with dementia associated with Parkinson's disease (PDD) demonstrate greater adverse response to agents with more pronounced dopaminergic blockade, while individuals with AD have greater sensitivity to those with anticholinergic effects (Byerly et al. 2001). Ultimately, antipsychotics are associated with statistically significant decrements of cognition that show varying clinical significance according to other factors associated with the clinical condition.

Stimulants

Psychostimulant drugs represent a broad class of medications affecting various cognitive domains, including attention, arousal, motivation, memory, and locomotion. The literature is vast regarding individual cognitive effects of these agents, particularly in treating attention-deficit/hyperactivity disorder (ADD/ADHD). These findings suggest a broad role for stimulants in enhancing prefrontal regulation of behavior and attention (Arnsten 2006) along with broader reinforcement of frontoparietal functional connectivity (Wong and Stevens 2012).

Stimulating agents are commonly held during hospitalization, depending upon medical comorbidities such as cardiovascular disease, thus potentially impacting an individual's ability to participate in clinical treatment planning or formal neurobehavioral cognitive assessment. In rehabilitative settings, it may be prudent to consider whether inattentiveness, amotivation, or psychomotor slowing may adversely impact recovery, thus warranting review for potential reinitiation of the agent in a medically safe manner. If an individual has been prescribed stimulant medications for a preexisting ADHD diagnosis, it is generally prudent to resume this agent when medically appropriate (Coris et al. 2022).

It is common for stimulants to be started de novo during rehabilitation and recovery following concussion and TBI, however the evidence underlying their use in these scenarios remains variable. Methylphenidate and modafinil may have a favorable benefit-to-risk ratio in this group, with important risk caveats to be considered such as cardiac disease or seizure disorders. Furthermore, there is rarely a rehabilitative scenario in which a stimulant alone will improve post-injury cognition, and typically these agents must be prescribed alongside robust physical and cognitive rehabilitation (Coris et al. 2022; Kakehi and Tompkins 2021). The agent with the most robust literature supporting its role in cognitive recovery following TBI is, rather than a conventional stimulant, amantadine by virtue of various mechanisms of action (Giacino et al. 2012).

SUMMARY

Through a comprehensive review of the medical record, inpatient neuropsychologists will gain a rich perspective of both the medical history and potential contributors to the neurobehavioral presentation. By understanding the potential impact of findings within the laboratory workup, a framework may be established upon which further assessment can be placed. This framework then creates opportunities for additional investigation that inform not only the diagnosis, but treatment planning and rehabilitative trajectories. Finally, these clinical trajectories will always be subject to iatrogenic effects, whether positively or negatively, and can be optimized by applying principles of basic pharmacologic understanding. Having a foundational knowledge of the cognitive effects of medications, neuropsychologists can make informed decisions in their practice, balancing the benefits of pharmaceutical interventions with potential cognitive risks. This knowledge empowers neuropsychologists to optimize patient care and ensure the highest quality of clinical services in inpatient settings. Ultimately, it is the unique role of an inpatient neuropsychologist that allows integration across all such domains, synthesizing the available medical information in an accurate, timely, and effective manner.

Appendices

Appendix 1: Abbreviations of Common Diagnostic Studies

Abbreviation	Diagnostic Study
ABG	Arterial Blood Gas
ACTH	Adrenocorticotropic Hormone
Alb	Albumin
ALK/ALP	Alkaline Phosphatase
ALT	Alanine Aminotransferase
AST	Aspartate Aminotransferase
B1	Thiamine
B12	Cobalamin
B3	Niacin
B9	Folate or Folic Acid
Bili	Bilirubin
BMP	Basal Metabolic Panel
BP	Blood Pressure
BUN	Blood Urea Nitrogen
CBC	Complete Blood Count
CDT	Carbohydrate-deficient Transferrin
Cl	Chloride
CMP	Complete Metabolic Panel
CO2	Carbon Dioxide
CO2	Carbon Monoxide
Cr	Creatinine
CSF	Cerebrospinal Fluid
CT/CAT	Computed (Axial) Tomography
CTA	Computed Tomography Angiography
CXR	Chest X-ray
DTI	Diffusion Tensor Imaging
ECG/EKG	Electrocardiogram
Echo	Echocardiogram
EEG	Electroencephalogram
EMG	Electromyography
ESR	Erythrocyte Sedimentation Rate
GFR	Glomerular Filtration Rate
GGT	Gamma-glutamyl transpeptidase
Glob	Globulin
Hb/Hgb	Hemoglobin
HCO3	Bicarbonate
Hct	Hematocrit
HDL	High Density Lipoprotein
HgbA1c	Hemoglobin A1C
HR	Heartrate
INR	International Normalized Ratio
K	Potassium

LDL	Low Density Lipoprotein
LP	Lumbar Puncture
MRA	Magnetic Resonance Angiography
MRI	Magnetic Resonance Imaging
Na	Sodium
O2	Oxygen
PaCO2	Partial Pressure of Carbon Dioxide
PaO2	Partial Pressure of Oxygen
PET	Positron Emission Tomography
PFT	Pulmonary Function Test
pH	Potential of Hydrogen
Plt	Platelets
PO4	Phosphorus
PSA	Prostate Specific Antigen
PT	Prothrombin Time
PTH	Parathyroid Hormone
PTT	Partial Thromboplastin Time
Pulse Ox	Pulse Oximetry Saturation
RBC	Red Blood Cell
RPR	Rapid Plasma Reagin
SPECT	Single-photon Emission Computed Tomography
T3	Triiodothyronine
T4	Thyroxine
TFT	Thyroid Function Test
TIBC	Total Iron Binding Capacity
TProt	Total Protein
TSH	Thyroid Stimulating Hormone
UA	Urinalysis
UCx/UC/S	Urine Culture & Sensitivities
UDS	Urine Drug Screen
WBC	White Blood Cell

Appendix 2: Abbreviations of Common Medical Conditions or Diseases

Abbreviation	Medical Condition or Disease
AD/DAT	Alzheimer's Disease or Alzheimer's Dementia/ Dementia of the Alzheimer's Type
ADEM	Acute Disseminated Encephalopathy
AF/AFib	Atrial Fibrillation
AIDS	Acquired Immunodeficiency Syndrome
AKA/BKA	Above (Below) the Knee Amputation
ALS	Amyotrophic Lateral Sclerosis
AMS	Altered Mental Status
AVM	Arteriovenous Malformation
BPH	Benign Prostatic Hyperplasia
BSE	Bovine Spongiform Encephalopathy
CA	Cancer
CABG	Coronary Artery Bypass Graft
CAD	Coronary Artery Disease
CADASIL	Cerebral Autosomal Dominant Arteriopathy with Subcortical Infarcts and Leukoencephalopathy
CF	Cystic Fibrosis
CHF	Congestive Heart Failure
CIDP	Chronic Inflammatory Demyelinating Polyneuropathy
CJD	Creutzfeldt-Jakob Disease
CKD	Chronic Kidney Disease
CNS	Central Nervous System
COPD	Chronic Obstructive Pulmonary Disease
CP	Cerebral Palsy
CRF	Chronic Renal Failure
CRPS	Complex Regional Pain Syndrome
CSA	Central Sleep Apnea
CTE	Chronic Traumatic Encephalopathy
CVA	Cerebrovascular Accident
CVD	Cardiovascular Disease
DAI	Diffuse Axonal Injury
DJD	Degenerative Joint Disease
DLB/LBD	Dementia with Lewy Bodies or Lewy Body Dementia
DM	Diabetes Mellitus
DOE	Dyspnea on Exertion
DTs	Delirium Tremens
DVT	Deep Vein Thrombosis
EPS	Extrapyramidal Symptoms
ESRD	End-stage Renal Disease
FAS	Fetal Alcohol Syndrome
FTD/FTLD	Frontotemporal Dementia or Frontotemporal Degeneration/ Frontotemporal Lobar Degeneration
GBS	Group B streptococcal disease or Guillain-Barre Syndrome

GD	Gestational Diabetes
GERD	Gastroesophageal Reflux Disease
HA	Headache
HAV/HBV/HCV	Hepatitis A/B/C Virus
HD	Huntington's Disease or Hemodialysis
HIV	Human Immunodeficiency Virus
HPA	Hypothalamic-Pituitary-Adrenal
HPV	Human Papillomavirus
HSV	Herpes Simplex Virus
HTN	Hypertension
IBD	Inflammatory Bowel Disease
IBS	Irritable Bowel Syndrome
ICD/AICD	(Automatic) Implantable Cardioverter-Defibrillator
MI/AMI	(Acute) Myocardial Infarction
MMR	Measles, Mumps, Rubella (Vaccine)
MND	Motor Neuron Disease
MRSA	Methicillin-Resistant Staphylococcus Aureus
MS	Multiple Sclerosis
MSA	Multiple Systems Atrophy
NF	Neurofibromatosis
NP	Niemann-pick Disease
NPH	Normal Pressure Hydrocephalus
OSA	Obstructive Sleep Apnea
PAD	Peripheral Artery Disease
PD/PDD	Parkinson's Disease/ Parkinson's Disease Dementia
PE	Pulmonary Embolism
PID	Pelvic Inflammatory Disease
PM	Pacemaker
PML	Progressive Multifocal Leukoencephalopathy
PNA	Pneumonia
POTS	Postural Orthostatic Tachycardia Syndrome
PRES	Posterior Reversible Encephalopathy Syndrome
PSP	Progressive Supranuclear Palsy
PUD	Peptic Ulcer Disease
RA	Rheumatoid Arthritis
RLS	Restless Legs Syndrome
RPLS	Reversible Posterior Leukoencephalopathy Syndrome
RTI	Respiratory Tract Infection
SARS	Severe Acute Respiratory Syndrome
SDH	Subdural Hematoma
SLE	Systemic Lupus Erythematosus
STD/STI	Sexually Transmitted Disease (Infection)
TAH	Total Abdominal Hysterectomy
TdP	Torsades de pointes
TB	Tuberculosis
TBI	Traumatic Brain Injury
TIA	Transient Ischemic Attack
UTI	Urinary Tract Infection
VFib	Ventricular Fibrillation
VTAch	Ventricular Tachycardia
WD	Wilson's Disease

Appendix 3: Abbreviations of Common Terms in Administrative and Clinical Settings

Abbreviation	Clinical Setting & Administrative Terms
ADL	Activities of Daily Living
AMA	Against Medical Advice
ASAP	As Soon As Possible
BP	Blood Pressure
Bx	Biopsy
CC	Chief Complaint
CCU	Critical Care Unit or Cardiac Care Unit
CICU	Coronary Intensive Care Unit
CMO	Comfort Measures Only
CPR	Cardiopulmonary Resuscitation
DNI/DNR	Do Not Intubate/Resuscitate
Dx	Diagnosis
ED/ER	Emergency Department (Room)
EMR	Electronic Medical Record
FMLA	Family Medical Leave Act
HPI	History of Presenting Illness
Hx	History
IADL	Independent Activities of Daily Living
LTAC	Long-Term Acute Care
LTC	Long-Term Care
MAR	Medication Administration Record
MICU	Medical Intensive Care Unit
NG	Nasogastric
NICU	Neonatal Intensive Care Unit or Neurologic Intensive Care Unit
NKA/NKDA	No Known (Drug) Allergies
OR	Operating Room
P	Pulse
PACU	Post-Anesthesia Care Unit
PICU	Pediatric Intensive Care Unit
ROS	Review of Systems
SICU	Surgical Intensive Care Unit
SNF	Skilled Nursing Facility
STAT	Immediately *(Latin: statim)*
T	Temperature
Tx	Treatment or Transfer

Appendix 4: Abbreviations of Common Medication-Related Terms

Abbreviation	Medication-Related (dosing, route, frequency)
AC	Before meals
BID	Twice daily (Latin: *bis in die*)
G	Grams
IM	Intramuscular
IR	Immediate Release
IV	Intravenous
LAI	Long-Acting Injectable
mcg	Micrograms
mg	Milligrams
NPO	Nothing by mouth (Latin: *nihil per os*)
PO	By mouth (Latin: *per os*)
PR	Per rectum
PRN	As needed (Latin: *pro re nata*)
QAM	Each morning (Latin: *quaque ante meridiem*)
QD	Once daily (Latin: *quaque die*)
QHS	Each night (Latin: *quaque hora somni*)
QID	Four times daily (Latin: *quater in die*)
Rx	Prescription
SA	Sustained Action
SQ/SubQ	Subcutaneous
TID	Three times daily (Latin: *ter in die*)
XL	Extended
XR	Extended Release

Appendix 5: Common Terms in Structural Neuroimaging Reports

Term	Description	Appearance	Cause/Pathophysiology
Arachnoid Cyst	Common, benign, CSF-filled cyst between the brain and meninges.	Typically seen in the middle cranial fossa. **CT**: Well-defined, displace adjacent tissue. **MRI**: Follows CSF on all sequences, well circumscribed with very thin wall.	Typically, sporadic and congenital; arachnoid layer splits during development and creates a pocket that fills with CSF.
Arterial Stenosis	Narrowing of an artery, commonly caused by ischemic strokes and TIAs.	Most common in the internal carotid arteries but can affect any vessel. Visualized with angiography (CTA or MRA)	Typically, due to atherosclerosis (similar to causes of heart disease); most common in elderly males.
Callosal Angle	Angle of the corpus callosum. Typically calculated at the level of the posterior commissure.	Normal values are ~100 - 120° (greater than 90°) Values less than 90° (typically between 50-80°) are compatible with NPH.	Can be useful in distinguishing NPH from atrophy.
Clinical Correlation	Standard term indicating a need for symptoms or other clinical findings to confirm a diagnosis (e.g., "callosal angle is <90°, clinical correlation recommended")		
Cortical Ribboning	Area of restricted diffusion in the cerebral cortex, often longitudinal (i.e. ribbon-like).	MRI: Hyperintense on DWI with corresponding hypointense ADC.	Classically seen in CJD, but is more commonly seen with ischemic stroke, post-ictal changes including status epilepticus, HSV encephalitis, PRES/RPLS.
Diffuse Axonal Injury (DAI)	Shearing of axonal tracts from rotational forces.	**CT**: Very difficult to appreciate. **MRI**: Best seen on GRE/SWI sequences which are very sensitive to DAI-related microbleeds. Hyperintense FLAIR signal can be seen subacutely. Tractography (e.g. DTI) may be useful.	DAI is essentially pathognomonic for TBI or other mechanical injury to the brain.

Downward Herniation of Cerebellar Tonsils	Extension of the inferior cerebellum through the foramen magnum, can be congenital or acquired.	Can be visualized on CT or MRI, best seen on a sagittal slice.	Typically caused by Chiari Malformation but can also be due to increased intracranial pressure (a medical emergency).
Edema	Water accumulation in the brain, which leads to swelling of cells.	**CT**: Loss of grey-white junction; uniform areas of hypodensity. **MRI**: Restricted diffusion (hyperintense on DWI and hypointense on ADC).	Can be caused by most intracranial pathology; large areas are commonly reflective of acute ischemia or TBI, but smaller areas can be seen with inflammation, BBB breakdown, surrounding tumors, etc.
Effacement	Closing of a CSF-filled cavity (e.g., ventricles, sulci) due to mass effect from some intracranial pathology.	Lack of normal anatomy where such is expected, may be accompanied by midline shift or other gross evidence of mass effect.	Can be seen in NPH (more general sulcal effacement), tumors, abscesses, hemorrhagic stroke, edema from acute TBI, or any other space-occupying lesion.
Empty Sella	Term used when the pituitary gland is shrunken or absent.	CSF in the Sella Turcica (pouch in the sphenoid bone), which is normally filled by the pituitary gland.	May be incidental but can be caused by intracranial hypertension or following removal of a pituitary adenoma.
Encephalomalacia	Loss of cerebral tissue as the result of liquefactive necrosis, may serve as a subsequent seizure focus.	**CT**: Mildly hypodense compared with surrounding tissue, often accompanied by volume loss. **MRI**: Follows CSF on all sequences, often surrounded by a hyperintense area of gliosis.	Causes are broad but is classically seen with stroke (both ischemic and hemorrhagic), TBI, results of surgery (e.g., encephalomalacia following the surgical approach).
Enhancement	Appears brighter than surrounding tissues after administration of contrast		

Extra-axial Mass	A tumor or other growth outside of the brain parenchyma, typically between the brain and meninges; often synonymous with meningioma.	Highly dependent on the cause; meningiomas typically appear as: **CT**: Usually slightly hyperdense, may have some calcification or surrounding edema. **MRI**: Typically, similar to brain parenchyma but may be hypointense (T1) or hyperintense (T2); homogeneously enhances with contrast.	Typically refers to meningioma, but may also reference pituitary tumors, cranial nerve tumors (schwannomas), other neoplasms, cysts, subdural/epidural/subarachnoid hemorrhage, vascular growths or malformations.
Gliosis/Gliotic Scarring	Subacute/chronic proliferation of glial cells, generally astrocytes, following CNS insults or injuries; may be termed "reactive gliosis."	**CT**: Difficult to visualize but may appear as subtle hypoattenuation. **MRI**: Increased area of T2/FLAIR surrounding or adjacent to an area of encephalomalacia; may also evidence mild hypotensity on T1.	Glial cells are activated and proliferate as a response to a brain injury of any type, but most often from stroke or TBI; astrocytes in particular increase to maintain the BBB, reduce oxidative stress or inflammation, and attenuate excitotoxicity.
Hemosiderin	Metalloprotein complex of iron, which is usually the result of hemorrhage. When blood cells die their iron gets scavenged by macrophages; as most tissues turn over the iron will get recycled; but this doesn't happen in the brain, leading to often permanent iron staining.	**CT**: Not visible. **MRI**: Typically, not visible on standard T1, T2, FLAIR, ADC, or DWI sequences; as hemosiderin is paramagnetic it is quite visible on GRE scans (e.g. T2*, SWI), where it is markedly hypointense and often exhibits blooming artifact (appearing larger than it actually is due to distortion of the magnetic field).	Hemosiderin is generally pathognomonic for prior hemorrhage or extravascular blood products; causes may include DAI from TBI, hemorrhagic strokes, CAA, etc. Can also be seen in cases of altered iron metabolism in neurodegenerative diseases like PD, which may not always signify distal hemorrhage.

Hippocampal Atrophy or Sclerosis	Degeneration of the hippocampus, hippocampal formation, and/or mesial temporal lobe. Should be notated as either commiserate with, or greater than, the overall degree of cerebral atrophy.	Can be visualized on CT or MRI, but MRI is more sensitive when atrophy is due to seizures. Easiest to see on T2 MRI with coronal slices; may be unilateral or bilateral.	In older adults, is most likely due to Alzheimer's disease, LATE, or other idiopathic neurodegenerative disease. Can also be seen in cases of TLE (more likely to be unilateral), which would increase the likelihood of cognitive impairments
Hydrocephalus ex Vacuo	Enlargement of the ventricles and subarachnoid spaces, which can mimic NPH.	Easiest to see on axial CT or MRI. Ventricles will appear large and NPH-like, but without sulcal effacement or acute callosal angle.	Typically, due to generalized or central-predominant cerebral atrophy; may be non-pathological, due to remote stroke or TBI, or due to neurodegeneration.
Hyperdense / Hypointense	Appears brighter than surrounding tissues on CT (hyperdense) or MRI (hyperintense).		
Hypodense / Hypointense	Appears darker than surrounding tissues on CT (hypodense) or MRI (hypointense).		
Isodense / Isointense	Appears the same as surrounding tissues on CT (isodense) or MRI (isointense, not commonly used).		
Lacunar Infarct	Small ischemic infarct, typically in a lenticulostriate or other small penetrating artery, commonly seen in the basal ganglia or cerebellum.	**CT**: Small, often rounded hypodensity. **MRI**: Hypointense (T1) or hyperintense (T2) small, often rounded focus.	Lenticulostriate and other penetrating arteries are small and branch off larger vessels at right angles. For these reasons they are more likely to occlude or be subject to ischemic disease.
Laminar Necrosis	The cortex is arranged in six layers (laminae). Laminar necrosis refers to death of cortex in a sheet-like fashion along one of the laminae (i.e., vs. a more typical infarct that affects multiple layers and white matter).	**CT**: Subtle, hypo- or hyperdense gyriform changes (depending on timing). **MRI**: Initial high DWI/low ADC like other ischemic injuries. T1 hyperintensities in a sheet-like or curvilinear fashion can be seen subacutely.	Neurons have higher metabolisms than glial cells, and some layers of the cortex are specifically vulnerable to hypoxic or ischemic injuries. Laminar necrosis is typically a worse prognostic sign and can signal more significant loss of brain tissue.

Leptomeningeal Enhancement	The leptomeninges consist of the arachnoid and pia maters. Enhancement of these meninges with contrast, which can be diffuse or focal, typically infers pathology distinct from that which affects the dura mater. May be used to refer to meningitis or encephalomeningitis		
Leptomeningeal Metastases / Leptomeningeal Carcinomatosis	The spread of malignancy into and through CSF and the extra-parenchymal space. The meninges are not protected by the BBB. Relatively common with late-stage cancers and a poor prognostic sign.	**CT**: Difficult to appreciate **MRI**: Often sheet-like, "sugar coated" enhancement of the leptomeninges on T1 with contrast. Can show hyperintensities on non-contrast T2/FLAIR.	Virtually any cancer can lead to intracranial spread if metastatic disease is widespread. Can be seen with primary intracranial solid tumors; other common primary sites include breast, lung, skin (melanoma), GI, and hematological.
Leptomeningeal Siderosis	Also known as Superficial Siderosis; refers to hemosiderin deposition along the leptomeninges.	**CT**: Typically, unremarkable **MRI**: Hyperintensities (both T1 and T2), will show low signal with blooming on GRE/SWI, can be widespread or 'worm-like'	Can be idiopathic, due to prior intracranial bleeds (major or chronic/minor), or intracranial neoplasm.
Mass Effect	Intracranial volume is fixed and a balance of CSF, intracranial blood, and brain parenchyma. Any increase in one of these three leads to decreases in the others (the Monro-Kellie Doctrine). Mass effect refers to displacement of brain tissue, compression of the ventricles possibly leading to obstruction, etc. Displacement can be lateral or vertical, with the latter leading to cerebellar herniation, a medical emergency.		
Midline Shift	Lateral mass effect leading to displacement of the midline to one side, often accompanied by ventricular effacement. A helpful tip is to compare the right and left sides of an axial slice for symmetry, noting if structures do not appear similarly placed or evident.		
No Abnormal Enhancement	Standard rule-out, essentially synonymous with "negative" or "normal," but more precisely indicating a lack of unusual or atypical findings.		
No Acute Ischemia	Standard rule-out indicating a lack of acute stroke; can sometimes imply a more general lack of acute findings or a "negative" or "normal" scan.		
Patent	Means "open" or "unobstructed;" usually standard language used in conjunction with a negative scan (e.g., "Ventricles were patient").		

Ring Enhancing	Rim or ring of hyperdense (CT) or hyperintense (MRI) tissue surrounding a hypodense/hypointense center when contrast agents are administered.	Visible on either contrast CT or MRI. May be surrounded by edema. Different characteristics (e.g. thickness, degree of surrounding edema, centrally restrictive diffusion) will favor specific differentials.	Contrast accumulates due to active inflammation or BBB breakdown. While nonspecific, this sign can be seen with primary and metastatic intracranial neoplasms, abscesses, demyelinating lesions, radiation necrosis, etc.
Volume Loss	Typically synonymous with atrophy, refers to loss of brain parenchyma in a typically gradual or involutional fashion (i.e., less frequently used with strokes or other acute pathologies). Can be diffuse, central, predominant in a specific lobe, or a combination. A degree of cerebral volume loss is normal for aging, which may be termed "volume loss appropriate for age."		
Wallerian Degeneration	Distal axonal degeneration due to damage or loss of neuronal cell bodies, often evolves over time	CT or MRI typically show volume loss apart from the initial lesion; restricted diffusion (high DWI/low ADC) may be seen subacutely.	Nonspecific but is typically seen in TBI and stroke, with the area of distal degeneration commensurate with proximal areas of injury (e.g., damage to the motor strip can lead to atrophy in the brainstem from loss of the corticospinal tracts).
White Matter Disease, White Matter Hyperintensities, Microvascular Disease, Leukoaraiosis	Extremely common, increases with aging. Can be normal/age-related but causes VCI if present to a larger degree.	**CT**: Hypodensities in the white matter, can be somewhat subtle/difficult to visualize. **MRI**: Visible on T1 (hypointense) and T2 (hyperintense), but FLAIR is sequence of choice, particularly for periventricular lesions. Typically, easiest to visualize on axial slices, but MS lesions are often perpendicular to the corpus callosum (follow small penetrating vessels).	White matter is marginally perfused, with less redundancy than cortical areas; this is particularly true for periventricular white matter. White matter is also more vulnerable to oxidative stress and inflammation. Lipohyalinosis and microatheromas preferentially affect the small blood vessels that distally perfuse white matter.

ADC = Apparent Diffusion Coefficient; BBB = Blood Brain Barrier; CAA = Cerebral Amyloid Angiopathy; CJD = Creutzfeldt-Jakob Disease; CSF = Cerebrospinal Fluid; DAI = Diffuse Axonal Injury; DWI = Diffusion-Weighted Imaging; FLAIR = Fluid-Attenuated Inversion Recovery; GRE = Gradient Echo; NPH = Normal Pressure Hydrocephalus; HSV = Herpes Simplex Virus LATE Limbic-Predominant Age-Related TDP-43 Encephalopathy; PRES = Posterior Reversible Encephalopathy Syndrome; RPLS = Reversible Posterior Leukoencephalopthy Syndrome; SWI = Susceptibility-Weighted Imaging; TBI = Traumatic Brain Injury; TLE = Temporal Lobe Epilepsy; VCI = Vascular Cognitive Impairment

REFERENCES

Adunsky A, Levy R, Heim M, Mizrahi E, Arad M. Meperidine analgesia and delirium in aged hip fracture patients. Arch Gerontol Geriatr. 2002;35(3):253–9.

Ahmad HEK, Darweesh AEM, Hassaan SHM, Nooman M, Shaaban I, Sayed SM. The effect of duration of dependence and daily dose of tramadol in tramadol dependent patients on cognitive performance. Middle East Curr Psychiatry. 2019;26:1–5.

Allen GJ, Hartl TL, Duffany S, Smith SF, VanHeest JL, Anderson JM, et al. Cognitive and motor function after administration of hydrocodone bitartrate plus ibuprofen, ibuprofen alone, or placebo in healthy subjects with exercise-induced muscle damage: a randomized, repeated-dose, placebo-controlled study. Psychopharmacology. 2003;166:228–33.

American Geriatrics Society Beers Criteria® Update Expert Panel. American Geriatrics Society 2023 updated AGS Beers Criteria® for potentially inappropriate medication use in older adults. J Am Geriatr Soc. 2023;71(7):2052–81.

Antinori A, Arendt G, Becker JT, Brew BJ, Byrd DA, Cherner M, et al. Updated research nosology for HIV-associated neurocognitive disorders. Neurology. 2007;69(18):1789–99.

Arnsten AF. Stimulants: therapeutic actions in ADHD. Neuropsychopharmacology. 2006;31(11):2376–83.

Baldacchino A, Armanyous M, Balfour DJK, Humphris G, Matthews K. Neuropsychological functioning and chronic methadone use: a systematic review and meta-analysis. Neurosci Biobehav Rev. 2017;73:23–38.

Barker MJ, Greenwood KM, Jackson M, Crowe SF. Persistence of cognitive effects after withdrawal from long-term benzodiazepine use: a meta-analysis. Arch Clin Neuropsychol. 2004;19(3):437–54.

Boustani M, Campbell N, Munger S, Maidment I, Fox C. Impact of anticholinergics on the aging brain: a review and practical application. Aging Health. 2008;4(3):311–20.

Brouquet A, Cudennec T, Benoist S, Moulias S, Beauchet A, Penna C, et al. Impaired mobility, ASA status and administration of tramadol are risk factors for postoperative delirium in patients aged 75 years or more after major abdominal surgery. Ann Surg. 2010;251(4):759–65.

Brunbech L, Sabers A. Effect of antiepileptic drugs on cognitive function in individuals with epilepsy: a comparative review of newer versus older agents. Drugs. 2002;62:593–604.

Byerly MJ, Weber MT, Brooks DL, Snow LR, Worley MA, Lescouflair E. Antipsychotic medications and the elderly: effects on cognition and implications for use. Drugs Aging. 2001;18:45–61.

Callisto SP, Illamola SM, Birnbaum AK, Barkley CM, Bathena SPR, Leppik IE, Marino SE. Severity of topiramate-related working memory impairment is modulated by plasma concentration and working memory capacity. J Clin Pharmacol. 2020;60(9):1166–76.

Cherrier MM, Amory JK, Ersek M, Risler L, Shen DD. Comparative cognitive and subjective side effects of immediate-release oxycodone in healthy middle-aged and older adults. J Pain. 2009;10(10):1038–50.

Collins ML, Cheney GA, Yehl JL, Sullivan GA, Stewart JT, Catalano G. Postviral depression. J Psychiatr Pract. 2021;27(2):126–30.

Coris EE, Moran B, Sneed K, Del Rossi G, Bindas B, Mehta S, Narducci D. Stimulant therapy utilization for neurocognitive deficits in mild traumatic brain injury. Sports Health. 2022;14(4):538–48.

Coupland CA, Hill T, Dening T, Morriss R, Moore M, Hippisley-Cox J. Anticholinergic drug exposure and the risk of dementia: a nested case-control study. JAMA Intern Med. 2019;179(8):1084–93.

Crowe SF, Stranks EK. The residual medium and long-term cognitive effects of benzodiazepine use: an updated meta-analysis. Arch Clin Neuropsychol. 2018;33(7):901–11.

Curran HV. Effects of anxiolytics on memory. Hum Psychopharmacol Clin Exp. 1999;14(S1):S72–9.

Curran HV, Schifano F, Lader M. Models of memory dysfunction? A comparison of the effects of scopolamine and lorazepam on memory, psychomotor performance and mood. Psychopharmacology. 1991;103:83–90.

Eddy CM, Rickards HE, Cavanna AE. Behavioral adverse effects of antiepileptic drugs in epilepsy. J Clin Psychopharmacol. 2012;32(3):362–75.

Folstein MF, Folstein SE, McHugh PR. "Mini-mental state": a practical method for grading the cognitive state of patients for the clinician. J Psychiatr Res. 1975;12(3):189–98.

Fong HK, Sands LP, Leung JM. The role of postoperative analgesia in delirium and cognitive decline in elderly patients: a systematic review. Anesth Analg. 2006;102(4):1255–66.

Gabel F, Hovhannisyan V, Berkati AK, Goumon Y. Morphine-3-glucuronide, physiology and behavior. Front Mol Neurosci. 2022;15:882443.

Garcia-Molina A, Roig-Rovira T, Ensenat-Cantallops A, Sanchez-Carrion R, Pico-Azanza N, Bernabeu M, Tormos JM. Neuropsychological profile of persons with anoxic brain injury: differences regarding physiopathological mechanism. Brain Inj. 2006;20(11):1139–45.

Gazelka HM, Leal JC, Lapid MI, Rummans TA. Opioids in older adults: indications, prescribing, complications, and alternative therapies for primary care. Mayo Clin Proc. 2020;95(4):793–800.

Giacino JT, Whyte J, Bagiella E, Kalmar K, Childs N, Khademi A, et al. Placebo-controlled trial of amantadine for severe traumatic brain injury. N Engl J Med. 2012;366(9):819–26.

Grunze HC. The effectiveness of anticonvulsants in psychiatric disorders. Dialogues Clin Neurosci. 2008;10(1):77–89.

Gualtieri CT, Johnson LG. Comparative neurocognitive effects of 5 psychotropic anticonvulsants and lithium. Medscape Gen Med. 2006;8(3):46.

Hustveit O. Interaction between opioid and muscarinic receptors in the guinea-pig ileum preparation: a mathematical model. Pharmacol Toxicol. 1996;78(3):167–73.

Jamison RN, Schein JR, Vallow S, Ascher S, Vorsanger GJ, Katz NP. Neuropsychological effects of long-term opioid use in chronic pain patients. J Pain Symptom Manag. 2003;26(4):913–21.

Jassal M, Egan G, Dahri K. Opioid prescribing in the elderly: a systematic review. J Pharm Technol. 2020;36(1):28–40.

Kakehi S, Tompkins DM. A review of pharmacologic neurostimulant use during rehabilitation and recovery after brain injury. Ann Pharmacother. 2021;55(10):1254–66.

Kasper S, Resinger E. Cognitive effects and antipsychotic treatment. Psychoneuroendocrinology. 2003;28:27–38.

Ketter TA, Post RM, Theodore WH. Positive and negative psychiatric effects of antiepileptic drugs in patients with seizure disorders. Neurology. 1999;53(5 Suppl 2):S53–67.

Kullgren J, Le V, Wheeler W. Incidence of hydromorphone-induced neuroexcitation in hospice patients. J Palliat Med. 2013;16(10):1205–9.

Loring DW, Marino S, Meador KJ. Neuropsychological and behavioral effects of antiepilepsy drugs. Neuropsychol Rev. 2007;17:413–25.

Loring DW, Williamson DJ, Meador KJ, Wiegand F, Hulihan J. Topiramate dose effects on cognition: a randomized double-blind study. Neurology. 2011;76(2):131–7.

Maldonado JR. Delirium pathophysiology: an updated hypothesis of the etiology of acute brain failure. Int J Geriatr Psychiatry. 2018;33(11):1428–57.

Marino SE, Pakhomov SVS, Han S, Anderson KL, Ding M, Eberly LE, et al. The effect of topiramate plasma concentration on linguistic behavior, verbal recall and working memory. Epilepsy Behav. 2012;24(3):365–72.

Medical Research Council. Aids to the examination of the peripheral nervous system. London: Her Majesty's Stationary Office; 1976.

Menefee LA, Frank ED, Crerand C, Jalali S, Park J, Sanschagrin K, Besser M. The effects of transdermal fentanyl on driving, cognitive performance, and balance in patients with chronic nonmalignant pain conditions. Pain Med. 2004;5(1):42–9.

Moriarty O, McGuire BE, Finn DP. The effect of pain on cognitive function: a review of clinical and preclinical research. Prog Neurobiol. 2011;93(3):385–404.

Moriarty F, Savva GM, Grossi CM, Bennett K, Fox C, Maidment I, et al. Cognitive decline associated with anticholinergics, benzodiazepines and Z-drugs: findings from the Irish Longitudinal Study on Ageing (TILDA). Br J Clin Pharmacol. 2021;87(7):2818–29.

Mufson EJ, Ginsberg SD, Ikonomovic MD, DeKosky ST. Human cholinergic basal forebrain: chemoanatomy and neurologic dysfunction. J Chem Neuroanat. 2003;26(4):233–42.

Mula M. Topiramate and cognitive impairment: evidence and clinical implications. Ther Adv Drug Saf. 2012;3(6):279–89.

Mula M, Trimble MR. Antiepileptic drug-induced cognitive adverse effects: potential mechanisms and contributing factors. CNS Drugs. 2009;23:121–37.

Mula M, Trimble MR, Thompson P, Sander JW. Topiramate and word-finding difficulties in patients with epilepsy. Neurology. 2003;60(7):1104–7.

Orzechowski RF, Currie DS, Valancius CA. Comparative anticholinergic activities of 10 histamine H1 receptor antagonists in two functional models. Eur J Pharmacol. 2005;506(3):257–64.

Pask S, Dell'Olio M, Murtagh FE, Boland JW. The effects of opioids on cognition in older adults with cancer and chronic noncancer pain: a systematic review. J Pain Symptom Manag. 2020;59(4):871–93.

Pathan H, Williams J. Basic opioid pharmacology: an update. Br J Pain. 2012;6(1):11–6.

Risacher SL, McDonald BC, Tallman EF, West JD, Farlow MR, Unverzagt FW, et al. Association between anticholinergic medication use and cognition, brain metabolism, and brain atrophy in cognitively normal older adults. JAMA Neurol. 2016;73(6):721–32.

Ropper AH. Neurosyphilis. N Engl J Med. 2019;381(14):1358–63.

Rumbaugh JA, Tyor W. HIV-associated neurocognitive disorders: five new things. Neurol Clin Pract. 2015;5(3):224–31.

Shalaby AS, El-Seidy AM, Zayed MA, Allam AR. Does tramadol dependence impair cognitive functions? Int Clin Psychopharmacol. 2022;37(2):67–71.

Shorvon SD. Safety of topiramate: adverse events and relationships to dosing. Epilepsia. 1996;37:S18–22.

Silbert BS, Scott DA, Evered LA, Lewis MS, Kalpokas M, Maruff P, et al. A comparison of the effect of high-and low-dose fentanyl on the incidence of postoperative cognitive dysfunction after coronary artery bypass surgery in the elderly. Anesthesiology. 2006;104(6):1137–45.

Sommer BR, Mitchell EL, Wroolie TE. Topiramate: effects on cognition in patients with epilepsy, migraine headache and obesity. Ther Adv Neurol Disord. 2013;6(4):211–27.

Swart LM, van der Zanden V, Spies PE, de Rooij SE, van Munster BC. The comparative risk of delirium with different opioids: a systematic review. Drugs Aging. 2017;34:437–43.

Taylor-Rowan M, Edwards S, Noel-Storr AH, McCleery J, Myint PK, Soiza R, et al. Anticholinergic burden (prognostic factor) for prediction of dementia or cognitive decline in older adults with no known cognitive syndrome. Cochrane Database Syst Rev. 2021;5:CD013540.

Thelmo FL Jr, Tzarnas S, Rosal NR, Kramer M, Walters L, Rosal N. Cystocerebral syndrome: an updated review and a new proposed mechanism for an often-forgotten cause of delirium. Cureus. 2020;12(10):e11034.

Vigen CL, Mack WJ, Keefe RS, Sano M, Sultzer DL, Stroup TS, et al. Cognitive effects of atypical antipsychotic medications in patients with Alzheimer's disease: outcomes from CATIE-AD. Am J Psychiatry. 2011;168(8):831–9.

Witt JA, Elger CE, Helmstaedter C. Adverse cognitive effects of antiepileptic pharmacotherapy: each additional drug matters. Eur Neuropsychopharmacol. 2015;25(11):1954–9.

Wong CG, Stevens MC. The effects of stimulant medication on working memory functional connectivity in attention-deficit/hyperactivity disorder. Biol Psychiatry. 2012;71(5):458–66.

Zeng C, Dubreuil M, LaRochelle MR, Lu N, Wei J, Choi HK, et al. Association of tramadol with all-cause mortality among patients with osteoarthritis. JAMA. 2019;321(10):969–82.

Zetsen SP, Schellekens AF, Paling EP, Kan CC, Kessels RP. Cognitive functioning in long-term benzodiazepine users. Eur Addict Res. 2022;28(5):377–81.

Bedside Examination and Inpatient Report Writing

4

Erin K. Bailey, Andrea Kurasz, and Emanuel M. Boutzoukas

ABBREVIATIONS

2/2	Secondary to
3MS	Modified Mini-Mental Status Examination
A/V	Auditory/Visual
A1C	Hemoglobin A1C (Glycohemoglobin)
ABS	Agitated Behavior Scale
AChE	Acetylcholinesterase
AD	Alzheimer's Disease
AGS	American Geriatric Society
ALF	Assisted Living Facility
BDI-2	Beck Depression Inventory-II
BP	Blood Pressure
C.Diff	Clostriodiodes Difficile
CAM	Confusion Assessment Method
CAM-ICU	Confusion Assessment Method for the Intensive Care Unit
CBC	Complete Blood Count
CDC	Centers for Disease Control
CDT	Clock Drawing Test
COG-LOG	Cognitive Log
COVID-19	Coronavirus Disease-2019
CPAP	Continuous Positive Airway Pressure
CT	Computed Tomography
CVA	Cerebrovascular Accident
d/t	Due to
DM2/DMII	Diabetes Mellitus, Type II
DOC	Disorder of Consciousness

E. K. Bailey (✉)
James A. Haley Veterans' Hospital, Mental Health and Behavioral Science Service (MS/BSS), University of South Florida, Morsani College of Medicine, Department of Psychiatry and Behavioral Sciences, Tampa, FL, USA
e-mail: Erin.Bailey@va.gov

A. Kurasz · E. M. Boutzoukas
Mental Health and Behavioral Science Service (MH/BSS), James A. Haley Veterans' Hospital, Tampa, FL, USA
e-mail: Emanuel.Boutzoukas@va.gov

E. K. Bailey et al. (eds.), *Inpatient Neuropsychology*, https://doi.org/10.1007/978-3-032-05880-5_4

ECS	Electrocortical Stimulation
ED	Emergency Department
EMR	Electronic Medical Record
EMS	Emergency Medical Services
FAB	Frontal Assessment Battery
GCS	Glasgow Coma Scale
GOAT	Galveston Orientation & Amnesia Test
h/o	History of
HADS	Hospital Anxiety and Depression Scale
HCS	Healthcare Surrogate
HDL	High-Density Lipoprotein
HLD	Hyperlipidemia
HPI	History of Presenting Illness
HTN	Hypertension
IADLs	Instrumental Activities of Daily Living
ICDSC	Intensive Care Delirium Screening Checklist
IV	Intravenous
Lb	Pounds
LDL	Low-Density Lipoprotein
LOC	Loss of Consciousness
MAST	Mississippi Aphasia Screening Test
MMSE	Mini-mental Status Examination
MOAT	Modified Galveston Orientation & Amnesia Test
MoCA	Montreal Cognitive Assessment
MRI	Magnetic Resonance Imaging
MRSA	Methicillin-Resistant Staphylococcus aureus
O-LOG	Orientation Log
OSA	Obstructive Sleep Apnea
OT	Occupational Therapy
OTC	Over the Counter
PET	Positron Emission Tomography
PHQ-9	Patient Health Questionnaire-9 Item
PNEA	Psychogenic Non-epileptic Attacks
POA	Power of Attorney
PPE	Personal Protective Equipment
PT	Physical Therapy
PTC	Posttraumatic Confusion
s/p	Status Post
SDH	Subdural Hematoma
SI/HI	Suicidal Ideation/Homicidal Ideation
SLP/SPT	Speech-Language Pathology/Speech Pathology Therapy
SLUMS	Saint Louis University Mental Status Exam
SNF	Skilled Nursing Facility
TBI	Traumatic Brain Injury
TME	Toxic Metabolic Encephalopathy
TNT	Tampa Neurobehavioral Test
UDS	Urine Drug Screen
VA	Veterans Affairs
w/o	Without
WBC	White Blood Cells

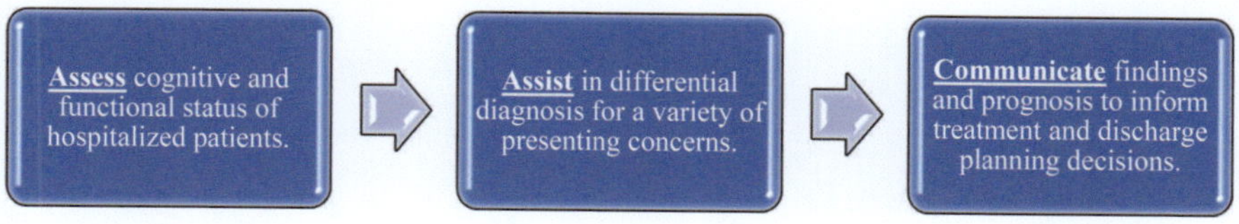

Fig. 4.1 Goals of the bedside examination

INTRODUCTION

Inpatient neuropsychologists are frequently consulted to assess patients with acute and/or severe medical conditions impacting cognition, emotions, and behavior. Unlike comprehensive outpatient neuropsychological evaluations, bedside examinations are typically brief and designed to provide rapid, clinically useful information for the purposes of differential diagnosis, treatment planning, and coordination of care (Pavol 2019; Schnider 2008). As many hospitalized patients may be too physically or cognitively impaired to undergo robust cognitive testing, the bedside examination should be tailored to the unique needs of the patient and designed to address the specific referral question at hand.

Goals of the Bedside Examination

The goals of bedside examinations are varied and include establishing an acute baseline, monitoring cognitive changes throughout an admission, facilitating treatment and discharge planning, and confirming a suspected diagnosis (Fig. 4.1). Goals may differ depending on the neuropsychologist's role within the larger treatment team. In some settings, neuropsychologists serve as consultants, with primary aims of providing diagnostic opinions to the medical team and assisting with treatment planning. In other settings, neuropsychologists are embedded in interdisciplinary systems of care and make recommendations to optimize patient engagement or conduct serial bedside examinations to inform prognosis. In either circumstance, the neuropsychologist should generally aim to obtain clinically meaningful information that will assist in characterizing the cognitive and functional status of hospitalized patients.

HOW TO PREPARE

There are several unique aspects of inpatient consultation that warrant consideration when planning a bedside evaluation. Limitations in test selection or time availability, for example, may restrict what is possible at bedside. For this reason, we recommend providers prioritize information that is *sufficient* and *necessary* to address the referral question, rather than aiming to be comprehensive. Below, we highlight practice considerations for efficient bedside neuropsychological examinations.

Chart Review

In general, the purpose of the chart review is to gain greater contextual information about the patient. Gathering all pertinent information from the medical record prior to seeing the patient can assist in developing a preliminary case conceptualization, shape expectations for the evaluation, and narrow the scope of the clinical interview—thus saving time. The volume of available records is often contingent on the frequency and extent of a patient's interaction within the healthcare system(s). However, at minimum, hospital admission requires an initial medical work-up, which is usually the first datapoint during chart review in developing a working conceptualization of the patient's present circumstance. The initial admission note may be titled differently depending on your setting (*History of the Problem, Admission Note, Emergency Department [ED] Note,* etc.), but typically includes the patient's admitting complaint, history of present illness, past medical history, medications, available social history, review of systems, physical examination, and assessment/plan. If the patient has received therapy services (physical therapy [PT], occupational therapy [OT], speech-language pathology/therapy [SLP/SPT]) or specialty medical care (neurology, psychiatry, etc.) during their admission, reviewing these notes can give insight into a patient's functional status, their ability to participate and engage with evaluators, and a preliminary sense of their mental status across time and circumstance (Fig. 4.2).

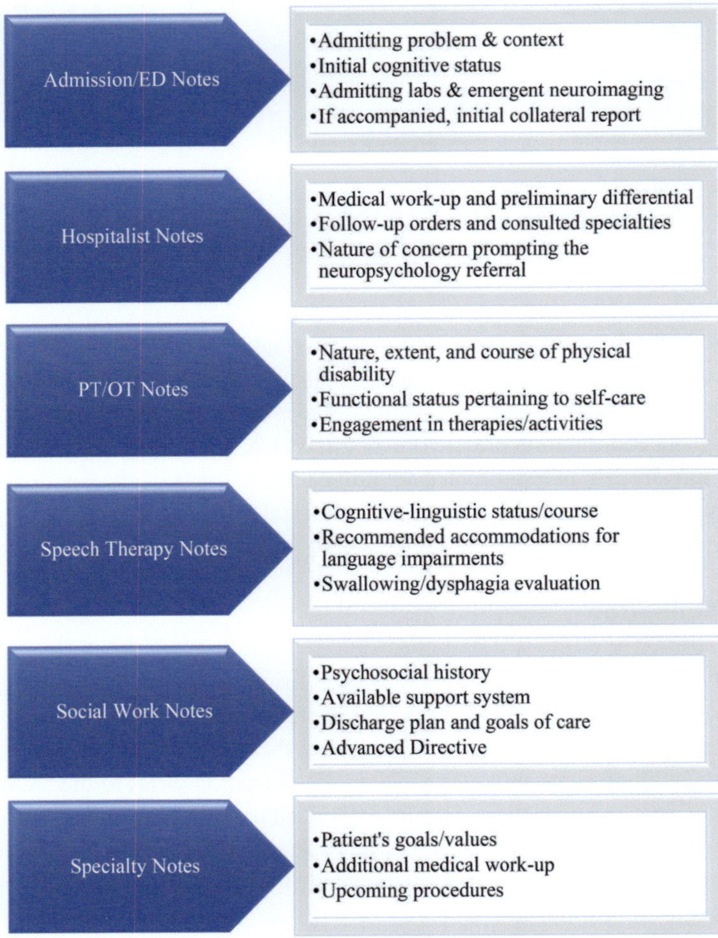

Fig. 4.2 Chart review guide

Labs, neuroimaging, behavioral management notes, and other inpatient consultation service notes (e.g., social work, infectious disease) should also be reviewed. Similarly, reviewing available outpatient records that predate a patient's current hospital admission can help characterize the onset and course of chronic health conditions and identify longitudinal variations in symptoms, lab values, or imaging results. When available, outpatient mental health notes can offer important insights about a patient's baseline psychiatric and cognitive status, as well as whether there is any history of psychotherapeutic treatment. Chart review is also helpful in identifying what information still needs to be ascertained. While reviewing records, we recommend keeping a list of important questions to clarify with the patient. Having a rough outline of missing/unclear information revealed from chart review will ensure you have the information you need at the conclusion of the clinical interview.

Lastly, it should be noted that working in an inpatient setting—where acute medical illness and treatment are often intertwined with neuropsychological practice—requires a higher degree of medical literacy on the part of the provider as compared to traditional outpatient settings. It is not uncommon to come across medical terminology, medications, procedures, or other pieces of data that necessitate further investigation to understand how this data may (or may not) be germane to the neuropsychologist's evaluation. Having access to easily referenced resources, such as a list of medical abbreviations or the American Geriatric Society (AGS) Beers List Criteria (2023)—a list of medications with possible contraindications in older adults—can improve the fluidity of chart review.

Identifying the Referral Question

Inpatient neuropsychology provides clinical utility to a variety of stakeholders (e.g., treating physicians, hospital administrators, nursing staff, allied health professionals, patients, and families), some of whom may be unfamiliar with the scope of our services. After conducting an initial chart review, the neuropsychologist should carefully re-examine the referral question. While consultation processes vary across facilities, referrals for bedside examinations may be poised with little to no context (e.g., 'assess cognition') or may be contraindicated at first glance (e.g., request to 'evaluate for dementia' in a patient who is acutely delirious), requiring further clarification to guide a meaningful and tailored evaluation (Fig. 4.3). Therefore, effective communication with physicians, therapists, nurses, and social workers within your treatment setting is vital to foster consistent goals for patient care.

Discussions with team members can also elucidate providers' own perceptions or expectations for the role of the neuropsychologist. This is helpful for two reasons. First, this can give context to an otherwise ambiguous referral question. For example, is your expected role on the team that of a consultant who is being asked to provide a one-time diagnostic opinion? Alternatively, are you perceived as embedded in the interdisciplinary team with expectations to provide serial examinations, and possibly intervention, throughout the duration of a patient's admission? Team consultation also provides opportunities to educate referring providers on the numerous ways neuropsychologists can be helpful in inpatient settings and clarify any

Original referral question: *"Assess for dementia"*	
Chart review	Chart review delineates progressive cognitive and functional decline over a two-year period. Strong concern for dementia has been documented by multiple providers, although he has never been referred for an outpatient neuropsychological evaluation. The patient is currently hospitalized for a fall. During this admission, PT has recommended he discharge to a skilled nursing facility (SNF). However, the local SNF will not accept the patient until a formal neurocognitive (dementia) diagnosis is rendered.
Decoded referral question: *"Stage cognitive and functional deficits to assist discharge planning"*	
How to be helpful	On exam, the patient presents with prominent anosognosia and executive difficulties. Formal testing via a cognitive screening measure confirms deficits in organization, abstract reasoning, and memory. Data gathered during the evaluation is consistent with records documenting a clear decline in cognitive and functional status. Based on this bedside evaluation, the patient is diagnosed with dementia. The neuropsychologist provides prognostic impressions and gives recommendations regarding the anticipated level of supervision and assistance he will require following discharge (e.g., SNF-level or home equivalent). In this case, clear documentation of the patient's functional limitations is particularly relevant to the team's goal (i.e., documenting care needs after discharge).
Why it matters	Without determining the 'question behind the question', you might have conducted an evaluation focused on differential diagnosis and missed the opportunity to provide clear recommendations regarding this patient's anticipated level of care needs following discharge. A clear understanding of the referral question enhances treatment and discharge planning through more meaningful conclusions and tailored recommendations that align with the broader goals of the patient and treatment team.

Fig. 4.3 Decoding the referral question

misunderstanding of what is (and is not) within our scope of practice. Sometimes, thorough chart review and familiarity with the expectations of the treatment setting is sufficient to identify the true intention behind a vaguely worded consult (what we call the 'question behind the question'). However, if the referral question remains unclear after completing the chart review, we recommend reaching out directly to the referring provider for additional context.

Time Management

Acute care settings necessitate prompt assessment as circumstances may rapidly change. Inpatient neuropsychologists must be efficient, while also gathering sufficient information to support conclusions. For example, while a thorough chart review can clarify the referral question and goals of the assessment, incorporating every single lab value, progress note, or data point into the background section of a report is likely to delay feedback to the team while not adding much unique value to your conclusions or recommendations. This again highlights the importance of brevity and paring down elements of the evaluation for a targeted purpose.

As part of time management, inpatient neuropsychologists may be required to adjust their own expectations for the nature of face-to-face evaluation. Many neuropsychologists primarily practice in outpatient settings where long-format interviews and comprehensive test batteries are the norm. In fact, the opportunity to really 'get to know' patients afforded by extended appointment durations is often highlighted as a unique and positive characteristic of neuropsychology relative to other health professions. It stands to reason that more information is better, and that greater data helps to converge on a more accurate differential diagnosis. However, within inpatient settings, it is generally neither practical nor feasible to conduct a comprehensive neuropsychological battery at patients' bedside. The typical bedside examination can last anywhere from few minutes to an hour, but rarely extends longer than 1 h per session. Evaluations may be split across two or more visits to accommodate patient fatigue, interruptions for medical or nursing care, or visiting friends and family. Accepting that a brief and often time-limited evaluation is the 'rule' and not the exception can help normalize what may otherwise feel inadequate.

Familiarity in navigating the medical environment can improve efficiency, increase effectiveness, and help avoid common pitfalls in the provision of inpatient neuropsychological services. Given the medical complexity of patients and the residential nature of a hospital admission, there is no guarantee that a patient will be easily located in their designated room at the time convenient to the neuropsychologist. Patients may walk outside, venture to other areas of the hospital, or be 'off unit' for medical tests. If your designated workspace or office is not located on or near the unit with easy access to patients' comings-and-goings, we recommend calling the nurse's station—or in some cases, the patient directly—in advance to coordinate the ideal time for evaluation.

Similarly, we encourage keeping apprised of regularly scheduled activities on the unit, such as mealtime deliveries or medical rounds, which is useful for identifying windows of time when patients are (or are not) likely to be available. Again, liaising with medical and nursing staff can provide valuable information about a particular patient's schedule and optimal times for evaluation (when they receive their medications, if they are more cogent in the mornings, etc.). This can also be helpful should you desire to co-evaluate patients with another service, or to participate in rounds or other unit activities.

The Inpatient Environment

Each inpatient unit is unique and the neuropsychologist must be prepared to adapt and optimize their approach within the parameters of the given setting. Below, we outline important practice considerations for conducting evaluations in inpatient settings, including selection of testing materials, navigating inpatient units, handling interruptions, and protecting patient rights and confidentiality.

Testing Materials

The inpatient neuropsychological test battery should be designed with the referral question, medical history, and preliminary hypotheses in mind. To this end, your evaluative approach and selection of test materials should be based on two questions:

- *"What will help me reliably and validly answer the true referral question?"*
- *"What is the patient able to tolerate?"*

We recommend beginning with a minimalistic approach, as you may have to carry items between buildings or across campuses. Even if you have an office in (or near) the unit, you should not bring extensive materials into the patient's room given the potential for space and time constraints. In our experience, we find it most beneficial to bring cognitive, neurobehavioral, and emotional screening measures that allow flexibility in capturing important qualitative information, rather than comprehensive test systems. For example, the Montreal Cognitive Assessment (MoCA; Nasreddine et al. 2005) and the Mini-Mental Status Examination (MMSE; Folstein et al. 1975) have demonstrated utility for detecting cognitive impairment and predicting future outcomes in several medical populations including those with heart failure, traumatic brain injury (TBI), and stroke (Harmon and Gillen 2023; Hawkins et al. 2014; Toglia et al. 2011). If chart review or preliminary conversation with the treatment team suggests a patient is prone to fluctuations in mental status, we also encourage including testing options that span complexity and sensitivity as this provides flexibility once a patient's cognitive status can be appreciated in vivo at bedside. A 'starter kit' for bedside evaluation may be as simple as a clipboard, 1–2 screening measures, a stopwatch, extra paper, and a pen or pencil. If available, you may also consider bringing adaptive technologies such as spare glasses, large print instructions, or personal amplification devices ('pocket talkers') to improve accessibility in cases of visual or hearing impairment. Similarly, if chart review suggests the presence of a communication disorder such as aphasia, we recommend preemptively finding creative ways to adapt the interview and testing material. This can include using an alphabet communication board, yes/no cards, written multiple choice options, or any other accommodations that may enhance patient engagement.

Navigating the Medical Environment

Several factors can significantly influence the efficacy and validity of cognitive testing in the inpatient setting. Below we discuss some considerations for improving proficiency of inpatient neuropsychological care.

Space and Mobility Constraints: Testing space in the inpatient setting is sparse. In most instances, patients are evaluated at bedside which often limits the usable desktop surface to a bedside tray table. Available seating options for the evaluator are also scant and another reason to make bedside examinations brief. Chances are, you will stand for the duration of the evaluation so dress in comfortable closed-toed shoes. Wearing pants instead of skirts or dresses is also recommended in case squatting or kneeling closer to the patient is required. If seating is desired, we recommend requesting a chair from a nearby nurse's station—never sit on a patient's bed! We also recommend asking permission from the patient to move any personal items, including clearing space on a bedside tray table. If you do relocate a patient's personal belongings, make sure to clearly state the new location—"*Mr. Jones, I am moving your wallet to your chair right here.*"—and return the items to their original location when the evaluation is completed.

Relocation from the patient's hospital room to a designated testing room on, or near, the unit is often unfeasible for several reasons. First, various medical conditions, treatments, or interventions (e.g., intravenous ([IV] lines) can restrict patients' ability to move, engage with test materials, and/or relocate to a designated testing space—complicating the logistics of the assessment. Second, many inpatient settings will not have any unused, designated space that would otherwise be appropriate for patient care. Instead, inpatient neuropsychologists must manage the limited space allocation and be mindful of how mobility restrictions may impact test performance.

Transmission-Based Precautions: In inpatient medical settings, there is an increased risk of exposure to bodily fluid and/or infectious agents. Infection prevention and control (IPC) policies, defined by the Centers for Disease Control and Prevention (CDC 2024), are standard healthcare precautions implemented by hospital staff to prevent the spread of infectious agents (Siegel, Rhinehart, Jackson, Chiarello 2019). These include proper hand hygiene before and after patient contact (e.g., handwashing with soap and water, using alcohol-based hand sanitizer) and the use of gloves or other personal protective equipment (PPE) when exposure is possible. Inpatient providers should know where to obtain necessary PPE for each type of precaution. IPC policy and environmental cleaning procedures are mandated at a systems-level by federal (CDC) and nationally accrediting agencies (e.g., The Joint Commission [TJC]).

In addition, individual transmission-based precautions may be designated according to national, local, or institution policy when a patient has a known or suspected communicable disease. The specific type of transmission-based precaution is usually listed outside the patient's room (and/or included in their medical chart) and should be reviewed prior to entering the patient's room. Common transmission-based precautions include contact, droplet, and airborne precautions. Depending on policy, specific types and levels of PPE may be designated depending on the type of infection and route of transmission. Contact precautions are implemented for infections spread by direct or indirect contact, such as methicillin-resistant staphylococcus aureus (MRSA) or clostridioides difficile (C.diff), and require gloves and gowns when entering a patient's room.

Droplet precautions are used for patients with known or suspected infections transmitted through respiratory droplets, such as influenza, and may require surgical masks and/or eye protection. Airborne precautions are required for patients with pathogens dispersed via airborne droplet nuclei (e.g., tuberculosis, measles, chickenpox, disseminated herpes zoster) and often necessitate the use of N95 other higher-level respirators.

Transmission-based precautions are also used when working with patients who are immunocompromised, necessitating the use of PPE by all who enter the patient's room. While the neuropsychologist must adhere to transmission protocols of their facility to reduce the spread of disease, donning PPE can introduce challenges for bedside neuropsychological testing. For example, face masks can obscure facial expressions, reduce the volume or articulation of speech, and impede lip movement making it difficult for patients with hearing impairments to understand verbal instructions. Masks can also interfere with the examiner's ability to accurately capture the nuanced qualities of a patient's speech. Gloves can impede motor speed and dexterity, potentially affecting the ability to manipulate test materials.

To adhere to transmission precaution policy while ensuring provision of necessary patient care, it is essential to use materials that can be easily disinfected. We recommend the use of plastic and/or laminated materials. This includes clipboards, pens, and stimulus cards that can be wiped down with hospital-grade disinfectant wipes. Technological devices, such as tablets, used for test administration (when available) can also be disinfected easily. Alternatively, consider using disposable materials including paper-based stimuli or single-use gloves, to minimize the risk of cross-contamination. By considering these factors, neuropsychologists can ensure effective bedside evaluations while maintaining safety and adhering to important infection control protocols in inpatient settings.

Fatigue and Engagement: While many neuropsychologists may be accustomed to conducting comprehensive clinical assessments, we strongly encourage a targeted approach when working with acutely ill patients (Kessler 2006). Interviews should focus on gathering pertinent background data, rather than being exhaustive of all clinical history. By nature of a hospital admission, patients will have likely been asked similar questions at each new intake and may be frustrated with 'retelling' their story. Further, it not uncommon for patients to experience anxiety, anger, remorse, and/or hopelessness during hospital admission or overwhelmed by their current situation. The neuropsychologist can preemptively ameliorate patients' apprehensiveness by first expressing empathy for the patient's current situation and acknowledging the understandable annoyance of (often redundant) questioning by a continuous flow of providers prior to initiating the interview.

Similarly, the clinical utility of formal assessment data may be limited by diminished stamina or poor cooperation on the part of the patient arising from several possible causes, including fatigue, depression, medical/postsurgical recovery, disrupted sleep, delirium, and medication side effects. If there is poor engagement, the neuropsychologist can split the evaluation into multiple shorter visits, using the initial encounter(s) to solely provide compassionate support and build therapeutic rapport. This can later inform recommendations to the family and treatment team on how to best engage with the patient. We encourage taking advantage of the ability to collect longitudinal data across time points, which can help characterize any changes in cognition. If an abbreviated patient interview is anticipated and the patient is unlikely to participate in future visits, gathering data most pertinent to the referral question should be prioritized.

Interruptions and Visitors: Semi-private rooms, the presence of visitors, proximity of electronic devices (televisions, cell phones, etc.), and the routine necessity of nursing care can create unwelcome distractions during bedside examinations. Effective management of these factors is crucial for increasing the validity of one's assessment. While it is commonplace to exclude third-party observation of outpatient evaluations, this may not always be feasible in inpatient environments.

Distractions and the presence of visitors can sometimes provide valuable data. For example, the neuropsychologist may be able to comment on the patient's ability to refocus after a hallway interruption. Similarly, family members are often a valuable source of background information, especially for patients with altered mental status, and in some cases, having a family member present can help calm the patient and facilitate a better assessment.

However, it is essential to prioritize a focused and confidential evaluation whenever possible. Family members may inadvertently provide cues or interfere with test instructions and as such, setting clear expectations with family members at the start of the session can help ensure their involvement is constructive, or at least, as neutral as possible. Attempts to minimize distractions are also recommended, and may include simply closing the door, asking other staff and visitors to wait a moment if interrupted, or placing an 'In Session' sign on the door. Use of professional attire, including lab coats, may assist with minimizing interruptions, as you are less likely to be confused for a visitor. In busy hospital settings, it may be helpful to coordinate with other services (e.g., PT, OT, SLP) to reduce the likelihood of being interrupted. Please note that nursing and other acute care providers may need access to patients regardless. If distractions are present, we recommend modifying your evaluation as best as possible and document any testing caveats (e.g., "*Extent of memory difficulties unclear given the impact of environmental distractions during learning trials*").

Confidentiality: Expectations of confidentiality are limited within the hospital setting, particularly in facilities with "semi-private" rooms where two or more patients share a room. While healthcare providers must adhere to ethical and legal standards, third parties (such as roommates or their visitors) may overhear portions of the evaluation despite best efforts to maintain privacy. The neuropsychologist must navigate potential breaches of confidentiality carefully and take all necessary steps to maximize privacy. Never assume that the patient is comfortable with other parties being present when discussing their health information. As a rule, the neuropsychologist is expected to ensure that personal health information is shared discretely, on a need-to-know basis, and to minimize unintended disclosures. Some recommended strategies to maintain as much confidentiality as possible in shared hospital rooms are listed below.

- *Inform the patient:* Educate the patient about the importance of confidentiality and their right to request a more private setting (if possible). Remind the patient that consent is an ongoing process and that they can voice concerns at any point during the evaluation.
- *Use of privacy measures:* Draw curtains or room dividers to create a visual barrier during testing. Lower your voice and use discrete language when discussing sensitive topics.
- *Timing of evaluations:* Schedule testing sessions during times when the roommate is away for therapy or medical procedures. Coordinate with nursing staff to identify the best time for a more private evaluation. If necessary, conduct an evaluation across multiple visits.
- *Modified communication techniques:* Use written communication (e.g., handing the patient a notepad for responses), non-verbal cues, or visual rating scales to reduce verbal disclosure.

By employing these strategies, inpatient neuropsychologists can optimize the assessment process while respecting the patient's right to confidentiality and privacy.

Professional Role: As opposed to outpatient practice, where patients are expected to adhere to social norms that dictate professional behavior in an office or clinic environment, the inpatient setting requires flexibility on the part of the provider to adapt to the needs of the patient while still maintaining professionalism. Inpatient providers may find themselves assisting patients with changing the channel on their television after testing is completed or refilling the patient's water cup midway through the evaluation. The hospital room is a multipurpose environment, simultaneously functioning as a hospital, rehabilitation center, office, dining room, locker room, living room, and bedroom. In some ways, conducting an exam at bedside is akin to evaluating a patient in all these different settings and as expected, interpersonal dynamics may shift accordingly in these contexts.

Remember that patients are being evaluated in their temporary living quarters. As such, you should always be respectful when entering the patient's room. Knocking on the door before entering, slowly turning on lights (as appropriate), and the judicious use of gentle touch, if needed, to wake a sleeping patient is encouraged. Providers should also use a soft tone of voice to introduce yourself, the purpose of the evaluation, and ask the patient's permission to proceed. Interviews may need to be conducted while the patient is eating lunch or be interrupted briefly for bowel or wound care. Be mindful of factors that may undermine the patient's confidence and self-esteem; for example, nonverbal cues or reactions to normal body functions that may elicit embarrassment or shame on the part of the patient. This is especially important for patients who are unable to advocate for themselves, communicate their preferences, or who are suspicious of healthcare providers. A not-uncommon occurrence is an accidental view of nudity due to a loosely tied hospital gown or an askew blanket. In all these instances, we strongly encourage prioritizing patients' dignity.

Similarly, inpatient neuropsychologists may be tasked with assisting nursing staff in an emergency or to bear witness to provision of acute medical care. Although practicing within the scope of competency is always encouraged, simple actions such as offering an arm to an unstable patient or adjusting a patient's pillow at their request can go a long way in building rapport and providing comfort. As these types of situations are innumerable and typically unpredictable, neuropsychological practice in inpatient settings requires good clinical judgement to problem solve in the moment (Fig. 4.4).

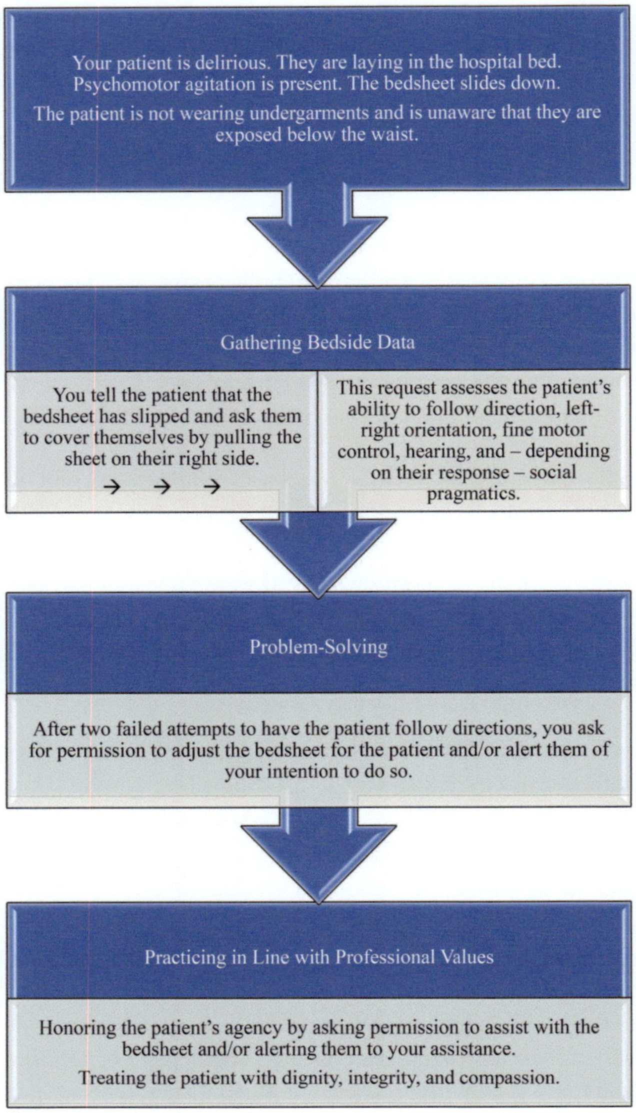

Fig. 4.4 Protecting patient dignity

Obtaining Informed Consent

Considering the above factors, neuropsychologists must adapt their approach to optimize patient cooperation. Patients may be taken by surprise when informed of the purpose of the evaluation. There may be feelings of anxiety, suspicion, or even resentment toward the notion of being evaluated. To address this, provide the patient with a clear explanation of the assessment. For instance, stating, "*Your medical team wants to check for any changes in your thinking related to your (current health) condition*," can help demystify the process and alleviate concerns (Fig. 4.5). Limitations of confidentiality, including sharing of results with the medical team, should be identified, and discussed as part of the informed consent process. If there are concerns about a patient's capacity to give informed consent, we recommend reviewing the chart or asking nursing staff whether there is a designated Power of Attorney (POA) or Healthcare Surrogate (HCS) on file. For patients who do not have capacity to give informed consent, regardless of whether (or not) they have a consenting surrogate, the purpose of the evaluation should still be explained to the patient in layman's terms and assent should be obtained, whenever possible.

For patients who remain hesitant to engage, offering the option to refuse the exam may bolster empowerment, as the ability to exercise control can enhance willingness to engage. If a patient continues to decline bedside examination after a thorough discussion, including possible outcomes of non-participation (e.g., less information to the treatment team), the

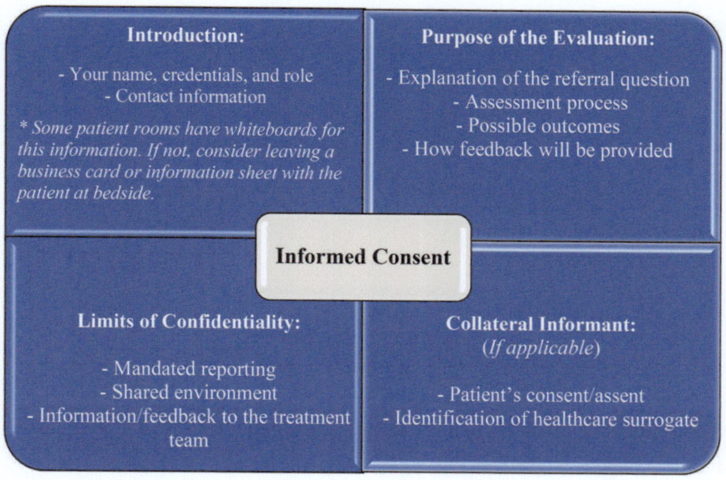

Fig. 4.5 Obtaining informed consent

neuropsychologist should not attempt testing. Trying to complete an assessment with an unwilling patient would be unethical, counterproductive, and likely to produce unreliable or invalid responses. In cases of continued resistance, it is important to be non-defensive and reiterate the patient's right to refusal as it relates to all aspects of their health care. Patients' right to other therapies, services, and privileges should never be used as contingent reinforcement, withheld, or bargained in exchange for participation in testing (or any other procedure).

DATA GATHERING

One of the primary responsibilities of a neuropsychologist when conducting bedside evaluations is to provide diagnostic and prognostic clarity. While many neuropsychologists are accustomed to conducting comprehensive neuropsychological evaluations for differential diagnosis, this will often not be the best approach for inpatient bedside examination as described above. Rarely, there may be some settings where it would be appropriate for the neuropsychologist to conduct a larger battery, such as on epilepsy monitoring units or post-acute traumatic brain injury (TBI)/stroke rehabilitation units, particularly if the patient is stable and able to tolerate the demands of an extended evaluation. However, a brief exam encompassing behavioral observations, a neurobehavioral examination, and/or brief cognitive or psychiatric screening measures is often the more pragmatic and sufficient approach in acute inpatient settings.

Behavioral Observations

Behavioral observations provide valuable data for any neuropsychological evaluation, though are of particular importance when assessing inpatients—a population prone to fatigue and limited tolerance for lengthy examinations. The inpatient setting is a unique opportunity for rich behavioral observations and, as a rule of thumb, your direct evaluation with the patient begins the moment you enter the patient's room. Standard behavioral observations include, but are not limited to, assessing appearance, orientation, attention, communication/language, mood/affect, insight, thought processes, and sensory/motor abilities. Assessing a patient's arousal and alertness is also essential in acute settings, where delirium or sedation due to medications is common. Observations such as whether a sleeping patient is responsive to verbal, tactile, or light stimulation are helpful in quantifying somnolence and sustained arousal.

Behavioral observations are paramount when formal testing is precluded, and we encourage providers to consider all environmental stimuli at their disposal to drive hypothesis testing across cognitive domains (Table 4.1). For example, is the patient able to successfully use the call buttons or the TV remote? Observing how a patient interacts with technology may reveal whether they have impairments in learning, praxis, or visuospatial abilities. Temporal observation of a patient across time points (see *Serial Monitoring* discussed later in this chapter) also yields information pertaining to the consistency of a patient's functional abilities, such as their ability to self-care. Similarly, building redundancy into the clinical interview can help the evaluator gauge the reliability of a patient's self-report, thus giving insight into their functional memory. Providing

Table 4.1 Behavioral observations and cognition

Example Behavioral Observation	Example Impacted Domain
Positional (e.g. lying in bed, sitting up)	Arousal, mobility, accommodation of vison or hearing loss
Alertness (e.g. responsive to stimulation)	Arousal
Sustained eye-contact, visual tracking, visual search	Attention, saccades, visual fields, visual neglect
Distractibility and response to redirection	Attention, executive control
Response latencies	Verbal/motor Initiation, processing speed
Motor movements	Motor coordination, dysmetria, functional capabilities
Repetition	Aphasia, echolalia
Affect	Mood, pseudobulbar affect
Yes/No accuracy	Comprehension, memory recognition, aphasia
Command following	Aphasia, comprehension, working memory
Error Monitoring	Attention, executive control, reliability of self-report, visual neglect (if visuospatial)
Recounted information from recent events	Extent of learning, functional memory
Utilization, stimulus-bound behavior	Executive control

Table 4.1 (continued)

Impulsive, perseverative behavior	Executive control, insight
Self-reflective statements, characterization of current circumstance	Insight, orientation, mood
Interactions with others (e.g. staff, visitors)	Orientation, insight, automaticity of speech and motor movements, reliability of presentation across circumstance, mood, social pragmatics
Interactions with environmental stimuli	Automaticity of speech and motor movements, apraxia, functional capabilities, executive functioning/problem-solving

cues/prompts or modifying the environment can help elucidate the nature of observed difficulties and shape recommendations for compensatory strategies.

The dynamic nature of the inpatient setting also allows for observing social interactions between the patient, healthcare providers, and visitors. The ability to observe various social interactions and dynamics allows the neuropsychologist to learn whether social comportment is appropriate, and if present, determine the underpinnings for concerning behavior. For instance, is emotional dysregulation or poor engagement secondary to a pre-existing psychiatric or cognitive disorder, or in response to an environmental antecedent? Being able to answer this question can be of extreme value to the medical team, particularly when disruptive behavior is interfering with the patient's care. When there is concern for poor engagement, we strongly recommend the neuropsychologist assess for volitional versus unintentional disengagement and identify underlying contributors to disengagement, with the goal of assisting the treatment team in adapting to the patient's needs. Strategies might include increasing rapport, spending more time providing education, or encouraging the patient to collaborate in their medical care.

Bedside Psychiatric Evaluations

The inpatient neuropsychologist will often encounter patients who are experiencing psychiatric symptoms. These can be neurogenic in origin—affectual lability seen in delirium, for example—or due to a primary psychiatric disorder. Acute or chronic medical illness and associated disability can also increase susceptibility to adjustment-related emotional reactions. Depending on the referral question, the type of setting, time constraints, and the availability of other mental health providers, the inpatient neuropsychologist may serve in a dual role by providing brief interventions in addition to assessment. Consequently, evaluating symptoms and possible contributors to psychiatric pathology can improve patients' recovery during their admission and guide treatment planning. For example, the quantity and quality of sleep is often reduced in the inpatient setting due to discomfort, interruptions (e.g., light, noise, care delivery), or other situational stressors. Poor sleep has been linked with protracted wound healing, worsened perceptions of pain, detrimental effects to cardiovascular, immunological, and metabolic systems, reduced cognitive efficiency, and low mood among hospitalized patients (Delaney et al. 2015; Doğan et al. 2005; Doraciak 1990; Raymond et al. 2001; Reid 2001).

The complex interplay between mood and behavioral health can significantly impact a patient's response to medical treatment, including immediate and long-term outcomes. For example, a patient with depressed mood can experience an exacerbation in pain perception, sleep disturbance, executive dysfunction, and decreased motivation to engage with their treatment team or rehabilitation efforts (Duggan et al. 2017; Gillen et al. 2001; IsHak et al. 2018; Lenze et al. 2004). Given the synergistic nature of behavioral health factors, depressive symptoms can be exacerbated in the context of decreased behavioral activation (refusing to participate in therapies, etc.), poorly managed pain, and sleep disturbance. Neuropsychologists in the inpatient setting are well poised to identify interconnected physical, somatic, and psychological factors to enhance patients' emotional and neurocognitive well-being through psychoeducation (e.g., gate theory of pain) or targeted interventions (sleep hygiene, brief cognitive behavioral therapy, etc.).

Unfortunately, psychological problems are significantly less likely to be measured in a standardized fashion or documented in a patient's medical record (Bijnsdorp et al. 2023). This is concerning given that nearly 50% of patients with comorbid depression do not have the condition detected when hospitalized for non-psychiatric reasons (Rentsch et al. 2007). An inpatient neuropsychologist's toolkit should include assessing psychological/psychiatric status in interview and behavioral observation, as well as the use of validated self-report screening tools that are easy and quick to administer. A commonly used self-report depressive symptom screener used with a wide range of medical presentations is the Patient Health Questionnaire-9 item (PHQ-9; Kroenke et al. 2001). Although the PHQ-9 was developed in the primary care setting, it has been identified as a generally reliable measure with medically hospitalized patients (Manea et al. 2015; Daray et al. 2019). Another available measure is the Hospital Anxiety and Depression Scale (HADS; Zigmond and Snaith 1983). The HADS measures both anxiety and depressive symptoms and does not include somatic symptoms, unlike the PHQ-9 and other commonly used measures (e.g., Beck Depression Inventory). Although somatic symptoms may very well be reflective of psychological distress, it is important to keep in mind that somatic symptoms may also overlap with medical illness.

Inpatient assessment of psychiatric symptoms can also guide discharge planning. While outpatient medical follow-up is usually standard practice in most settings, outpatient mental health treatment is less often recommended in patients discharged from a medical admission. This is concerning as research shows high rates of mood symptoms in individuals presenting to Emergency Departments, particularly among those who have a chronic illness (Boudreaux et al. 2008). Although beyond the scope of this chapter, several studies highlight the psychiatric toll of acute hospitalization and the risk of mood symptoms following medical admission (Garrigues et al. 2020; Jiang et al. 2013; Kim et al. 2018; Worsham et al. 2021).

Bedside Neurocognitive Examinations

Conducting bedside neurocognitive examinations is a crucial skill for neuropsychologists working in inpatient settings, especially when more comprehensive neuropsychological evaluation is not feasible. Bedside neurocognitive examination offers an efficient and practical way to evaluate a wide range of brain-behavior relations, including cognitive, sensory, motor, and behavioral functions. Examinations aim to detect acute changes (e.g., post-stroke), rule out neurodegenerative diseases, and monitor change throughout hospitalization. Advantages of brief bedside evaluations include little or no need for stimuli/testing material and flexible administration, with options to break standardization when necessary. While not replacing the breadth and depth of more comprehensive neuropsychological testing, these examinations provide valuable insight into the patient's current neurocognitive status and can quickly identify areas for further evaluation and monitoring.

The bedside examination involves several key domains, including orientation, neurobehavioral (e.g., sensorimotor) and cognitive (language, visuospatial abilities, memory, executive functioning, etc.) functioning (Losier et al. 2020). If the patient's baseline functioning is largely unknown and significant deficits are evident, you may consider administering all components of the bedside examination. In contrast, you may only administer select components if localized deficits are apparent and you are interested in ruling out a priori diagnoses.

It is important to adapt your examination to the patient's condition. Consider the patient's medical status, level of alertness, and ability to participate. The neuropsychologist must also be prepared to modify testing approaches based on fatigue, pain, physical/sensory deficits, or medical interventions. As there is no single best screening tool, neuropsychologists have the option to choose from several well-validated cognitive screening measures that can be administered at bedside within 5 to 25 minutes (see Appendix 1). Popular global cognitive screening measures include the Confusion Assessment Method (CAM; Inouye et al. 1990; Waszynski 2003), the MoCA (Nasreddine et al. 2005), the MMSE (Folstein et al. 1975), the Modified Mini-Mental Status Examination (3MS, Teng and Chui 1987), and the Mini-Cog (Borson et al. 2000). Many of these instruments have been adapted for certain populations, such as patients who are visually impaired (e.g., Blind MoCA) or on a mechanical ventilator (e.g., CAM-ICU; Tate and Happ 2018). Of note, some of these measures are tailored for use in specific circumstances (e.g., CAM-ICU for assessing delirium) and may not be appropriate for all patients (Gusmao-Flores et. al., 2012).

Within traumatic brain injury, practice standards suggest using the Orientation Log (O-Log; Jackson et al. 1998), the Cognitive Log (Cog-Log; Alderson and Novack 2003), the Agitated Behavior Scale (ABS; Corrigan 1989), and the Galveston Orientation & Amnesia Test (GOAT; Levin et al. 1979). These are popular, brief assessments for capturing the recovery of post-traumatic confusional state (Jain et al., 2000). Additional bedside neurocognitive tests that assess specific domains of cognitive function include the Clock Drawing Test (CDT; Goodglass and Kaplan 1983), the Mississippi Aphasia Screening Test (MAST; Nakase-Thompson et al. 2005), and the Frontal Assessment Battery (FAB; Dubois et al. 2000). We have also included a previously unpublished measure that comprises several bedside tests, the Tampa Neurobehavioral Test (TNT, Appendix 2). While this measure does not presently have normative data, it is designed to be used flexibly and creatively to capture a wide range of rich, qualitative neurobehavioral data, much of which may be otherwise under-assessed.

While standard practice for any neuropsychologist, consideration of cultural and linguistic factors is particularly important in the inpatient setting due to the rapid exchange of information and the impact language barriers may have on the informed decision-making process. When available, ensure that language barriers are mitigated as much as possible using existing resources (e.g., interpreter services) and culturally appropriate measures. For example, the Saint Louis University Mental Status Exam (SLUMS), a cognitive screen commonly used in Veterans Affairs (VA) settings, has been translated into over 15 languages for ease of use with a diverse array of patient populations (Morley and Tumosa 2002). By considering these factors, neuropsychologists can enhance the effectiveness of bedside neurocognitive examinations and provide valuable clinical insights in inpatient settings.

THE INPATIENT REPORT

Documentation in inpatient settings requires adaptation of a typical outpatient neuropsychological report. Adjustments to structure, organization, and function of these time-sensitive evaluations are discussed below. The target audience of an inpatient neuropsychological report is largely the medical team. For this reason, it is important to remain succinct and consistent in the use of medical acronyms and terminology. As a general guideline, we strongly advise against including non-pertinent information in the patient's medical record. For example, a pending civil lawsuit may be relevant to an outpatient mental health provider when conducting an initial assessment for depression, but this should not be included in an inpatient report for a patient hospitalized for hepatic encephalopathy unless relevant to the cognitive evaluation or referral question. Eliminating non-essential information will reduce the time-burden for the reader and spotlight conclusions more effectively. Before including information in your report, take note of the considerations in Fig. 4.6.

Heading

Report headings will vary depending on style and practice location. In general, a heading should include the facility, identifying information of the patient, the name and credentials of the examiner, the date of the evaluation(s), and billing/services. Providers may also wish to include a statement of informed consent, if applicable (Fig. 4.7).

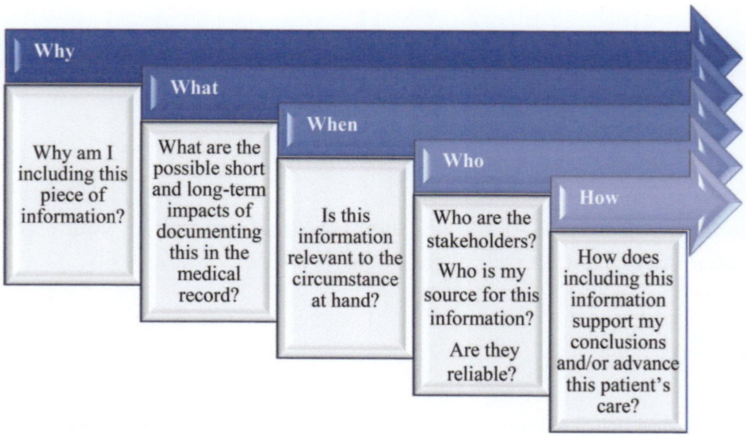

Fig. 4.6 Decision-making for data inclusion

===

INPATIENT NEUROPSYCHOLOGY EVALUATION
Local Hospital, City, State

===

Patient Name: Last, First — Handedness: Right
DOB: 00/00/1924 — Education: 12 Years
Age: 76 Years

Evaluation Date: 00/00/2000

Service Provider: Dr. XX, Ph.D., Staff Neuropsychologist

Services Provided:
- Neuropsychological evaluation (includes evaluative services, record review, clinical interpretation/decision making, report preparation, feedback to patient and other stakeholders, and liaising with other healthcare professionals):
96132 - 1 unit & 96133 - 1 units - Total Time spent in Evaluation: 2 hours

Fig. 4.7 Report heading

===

REASON FOR REFERRAL:
Ms. G is a 76-year-old, widowed, white non-Hispanic female referred by the inpatient medical team in the context of left frontal subdural hematoma s/p unwitnessed fall (presumed) for cognitive evaluation to determine current functioning and make treatment recommendations.

===

s/p = status post

Fig. 4.8 Reason for referral

=====================================

TBI Severity: Complicated Mild
- Initial GCS = 13 (at scene); 14 (in ED)
- Duration of LOC = Estimated 20 min
- Duration of PTC = approximately 2 hours
- Neuroimaging: + L Frontal SDH

=====================================

GCS = Glasgow Coma Scale; ED = emergency department;
PTC = posttraumatic confusion; L = left;
LOC = loss of consciousness; SDH – subdural hematoma; + = positive for

Fig. 4.9 TBI severity metrics

Reason for Referral

The referral source and referral question(s) should be listed near the beginning of your report (Fig. 4.8). This helps the reader understand the context of the consult, as well as the goals of your evaluation.

History of the Problem

Referrals for a cognitive evaluation in acute care or rehabilitation settings tend to be based on the patient's presenting reason for admission. As such, we recommend a concise, but throughout overview of the pertinent History of the Problem (also called History of Presenting Illness; HPI). Specific details of a patient's condition, such as severity metrics for TBI, can be listed at the top of the report for ease of identification (Fig. 4.9) along with a chronological account of relevant symptoms, medical documentation, reports, and interventions leading up to the neuropsychology referral (Fig. 4.10). We advise against including all

==

HPI:
Ms. G is a 76-year-old woman PMH of DMII, OSA, HTN, HLD, alcohol abuse (in sustained remission), and depression. Medical records note a history of poor compliance with DMII treatment. Ms. G has had one prior hospitalizations in the past 5 years for left-sided leg weakness though stroke w/u was negative at that time.

Recently, Ms. G was brought to the ED by EMS on 00/00/2000 after being found down by her friend in her home at approximately 8pm. Patient lives alone and event was unwitnessed. Patient was awake at the scene. Initial GCS per EMS was 13 at the scene and patient appeared disoriented. Patient's friend presented to ED with EMS and expressed concern for her living alone, noting that the patient is prone to bouts of confusion. ED work-up was notable for severe hypoglycemia. UDS negative. CBC notable for mildly elevated WBC though otherwise normal. Stroke protocol was initiated. Head CT showed left frontal SDH s/p presumed fall vs syncope. GCS improved to a 14 in the ED. Neurosurgery was consulted and recommended monitoring at this time. Patient did not sustain any orthopedic injury though contusions were noted on her left arm and hip. She was admitted to inpatient medicine on 00/00/2000 for continued management.

Hospital course has been notable for improved mental status, though patient remains intermittently disoriented to circumstance. Recent PT note documents patient inquiring about "checking out" of this "hotel." Neuropsychology was consulted to evaluate current cognitive functioning, aid in differential diagnosis, and guide discharge planning. Current exam conducted five days post-admission.

==

HPI = history of presenting illness; PMH = personal medical history; DMII = diabetes mellitus type II; OSA = obstructive sleep apnea; HTN = hypertension; HLD = hyperlipidemia; w/u = workup; ED = emergency department; EMS = emergency medical services; GCS = Glasgow Coma Scale; UDS = urine drug screen; CT = computerized tomography; CBC = complete blood count; WBC = white blood cell [count] SDH – subdural hematoma; s/p = status post; PT = physical therapy

Fig. 4.10 History of presenting illness

historical information in an HPI section, as this can be redundant when later listed in other sections (e.g., medical history, psychiatric history) and cumbersome to read. If a patient has an extensive or complicated history of hospital admissions or medical complications, bullet points may be useful in conveying a clear timeline.

Though dependent on record availability and applicability, the HPI section should start by describing the context of the admission and acute work-up, such as labs or imaging done in the Emergency Department (ED). This can be followed by the course of admission, including any complications and responses to treatments and interventions. The course of any inpatient therapies, such as PT, OT, or Speech therapy, can also be summarized. Psychosocial information, such as whether family members have visited the patient or been in contact with the treatment team, can also be helpful in characterizing the patient's support network and providing recommendations for outpatient care. Any inclusion of second or third-hand information should be clearly caveated to avoid perpetuating bias/opinion or inaccurate information. When information from two sources is in direct conflict, we recommend mentioning the discrepancy, or—in certain instances—only including the most reliable source.

Direct Report

When interviewing the patient and collateral informants, we recommend being as specific as possible about *who* is providing the information. Using first and/or last names to identify the reporter can help maintain consistency throughout your report. A good example of how this can be helpful is when working with multiple family members. A patient may have several sons or daughters and identifying these individuals by name can avoid confusion to the reader.

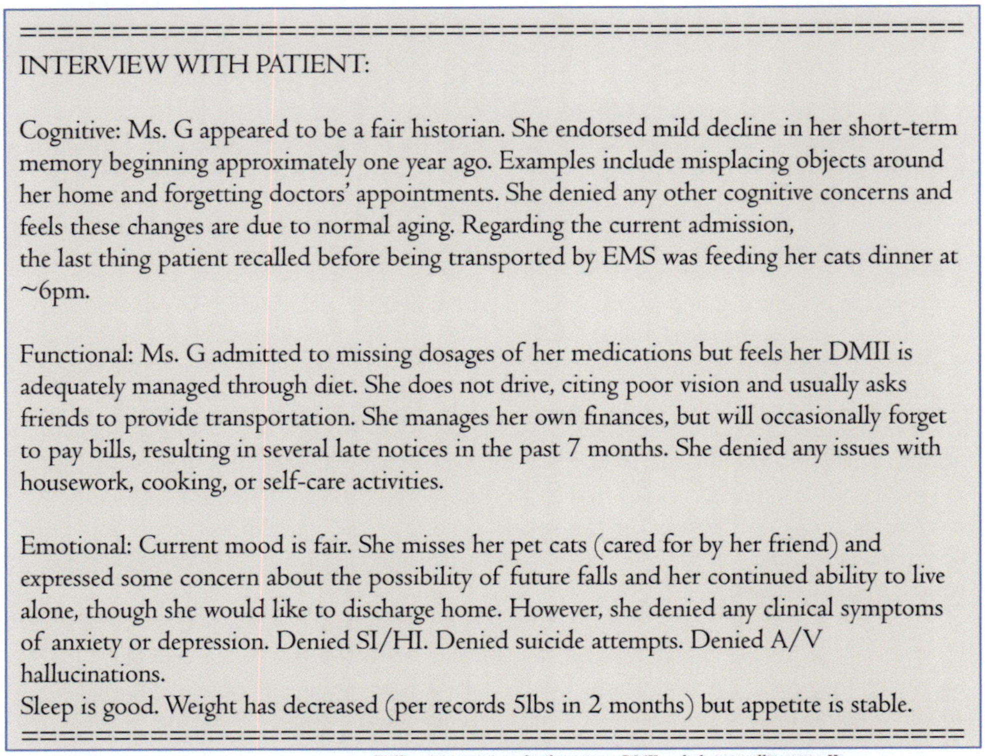
==
INTERVIEW WITH PATIENT:

Cognitive: Ms. G appeared to be a fair historian. She endorsed mild decline in her short-term memory beginning approximately one year ago. Examples include misplacing objects around her home and forgetting doctors' appointments. She denied any other cognitive concerns and feels these changes are due to normal aging. Regarding the current admission, the last thing patient recalled before being transported by EMS was feeding her cats dinner at ~6pm.

Functional: Ms. G admitted to missing dosages of her medications but feels her DMII is adequately managed through diet. She does not drive, citing poor vision and usually asks friends to provide transportation. She manages her own finances, but will occasionally forget to pay bills, resulting in several late notices in the past 7 months. She denied any issues with housework, cooking, or self-care activities.

Emotional: Current mood is fair. She misses her pet cats (cared for by her friend) and expressed some concern about the possibility of future falls and her continued ability to live alone, though she would like to discharge home. However, she denied any clinical symptoms of anxiety or depression. Denied SI/HI. Denied suicide attempts. Denied A/V hallucinations.
Sleep is good. Weight has decreased (per records 5lbs in 2 months) but appetite is stable.
==

EMS = emergency medical services; DMII = diabetes mellitus type II;
SI/HI – suicidal ideation/homocidal ideation; A/V = auditory/visual; lb = pounds

Fig. 4.11 Patient interview

Patient Interview

Sometimes titled 'Patient Complaints' or 'Patient Report,' this section documents information obtained from your direct interview with the patient (Fig. 4.11). Although most behavioral caveats should be reserved for the Behavioral Observations section, we do encourage providers to indicate if the patient is believed to be an unreliable historian at the beginning of this section. Content should include cognitive/physical/emotional/functional complaints, patient's self-report of the onset and course of these symptoms, as well as any relevant opinions, goals, values expressed regarding their current admission or general health status.

Collateral Interview

Any information obtained directly from collateral report, such as friends or family, should be included in the report. We encourage documentation of patient consent, the name of the collateral informant, the mechanism of contact, and the relation to the patient. Like the patient interview, collateral perceptions of the nature, onset, and course of the patient's symptoms should be clearly documented, noting any discrepancies (Fig. 4.12).

==

INTERVIEW WITH COLLATERAL:

Patient expressed verbal consent for this provider to contact her friend, Ms. L (555-222-1234) regarding patient's current hospitalization. Ms. G expressed consent to share results and recommendations of the current evaluation with Ms. L.

Ms. L was interviewed via telephone separately. Per Ms. L, Ms. G has demonstrated a slow decline in her memory and problem-solving for the past two years, with more rapid changes noticed over the past few months. Ms. L lives next door and assists the patient with transportation as needed. Due to the patient's increasing difficulty managing her IADLs, Ms. L recently began assisting the patient with setting up her pill box on a weekly basis and writing out checks for utility bills. The patient does not use the stove and limits cooking to the microwave only. Ms. L states that the patient continues to do household chores and self-care tasks independently, without issue. The two parties regularly play cards, although this has become more confusing for the patient. Three weeks ago, Ms. L contacted the patient's son, who lives out of state, and discussed the possibility of Ms. G moving closer to family.

On the day of the patient's admission, Ms. L had spoken with Ms. G on the phone ~25 minutes prior to finding the patient on the floor of her home, unconscious. Ms. L later stated that Ms. G "did not seem like herself" on the phone and wanted to check on her. Per Ms. L, the patient regained consciousness ~2 min after her arrival and was awake when EMS arrived, though Ms. L stated that the patient remained confused after regaining consciousness.

==

IADLs = instrumental activities of daily living; EMS = emergency medical services

Fig. 4.12 Collateral report

Background Sections

Like the HPI, which summarizes the current context for the neuropsychology referral, the background sections of an inpatient report should include historical information that is pertinent to the clinical question at hand and will aid in supporting your conclusions. Due to the often time-sensitive nature of inpatient evaluations, maintaining brevity will help reduce the turnaround time for reports and communicate your results to the referring provider as efficiently as possible. We do not advise documenting an exhaustive narrative of all history. Again, we recommend using bullet points or other design elements, to structure background sections, such as pertinent medical, psychiatric, substance use, psychosocial, and developmental history.

Medical History

The medical history section should give a list of all known diagnoses and relevant surgical procedures or other interventions (Fig. 4.13). If available, it is also helpful to document the status of how conditions are currently being managed and relevant dates of diagnosis or treatment. If possible, templates or auto-populated information obtained through electronic health records software should be used to maximize efficiency of report writing. Family medical history, including family psychiatric history can also be reported here.

The patient's psychiatric and/or substance use history can be included in this section or broke into a separate section, depending on the case and hand (Fig. 4.14).

======================================

PATIENT MEDICAL HISTORY:

- Chronic alcoholism
  - In sustained remission
- Depression
  - No current mental health treatment
- Obstructive Sleep Apnea of Adult
  - Noncompliant with CPAP
- Diabetes Mellitus II
  - A1C in ED: 11.6
- Hyperlipidemia
  - HDL in ED: 20
  - LDL in ED: 215
- Hypertension
  - BP in ED: 106/85

SURGICAL HISTORY:

- Cholecystectomy
  - 6 years ago, no reported complications

FAMILY HISTORY:

- Father: Depression, deceased at 82 d/t CVA
- Mother: Breast Cancer, deceased at 79 d/t breast cancer
- Paternal Grandfather: h/o alcohol abuse
- Maternal Cousin: Completed suicide at age 25

======================================

HDL = high density lipoprotein; LDL = low density lipoprotein; ED = emergency department; BP = blood pressure; CVA = cerebrovascular accident; CPAP = continuous positive airway pressure A1C = hemoglobin A1C (glycohemoglobin); h/o = history of; d/t = due to

Fig. 4.13 Medical history

======================================

PSYCHIATRIC AND SUBSTANCE USE HISTORY:

- H/o depression (per chart). No current psychotropic medications.
  - No documented h/o inpatient psychiatric hospitalizations
- H/o alcohol abuse in sustained remission. Last use 15+ years ago
- No current substance use

======================================

h/o = history of

Fig. 4.14 Psychiatric and substance use history

Medications

A report should include a list of all known prescribed and over-the-counter (OTC) medications and supplements, as well as dosage schedules. Take note if a patient's report of their current medications (or regimen) differs from their medical chart, although bear in mind that outpatient medications are often adjusted or changed once a patient is admitted. If available and relevant to the case, you can also include information about when medications were last filled. Identifying barriers to medication compliance has important implications for treatment planning and neuropsychologists can assist the medical team by addressing medication non-compliance from a biopsychosocial perspective (see examples in Fig. 4.15).

Some settings may grant access to patient-provider communication records (e.g., secure messaging), which can speak to a patient's autonomy and diligence in managing their medications. For example, does the patient have a history of messaging their provider in crisis and/or presenting to the ED to fill medication emergently? Do questions posed to their provider reflect someone with a high degree of health literacy, or one who is diligent about making sure they understand their medications? Similarly, analyzing longitudinal patterns of communication about medications in the patient's chart can give insight into whether there has been a change or decline in the patient's ability to manage their medical treatment. Lastly, any documented involvement in medication coordination from third parties, such as a spouse or caregiver, may suggest reduced functional abilities on the part of the patient in this area and/or help identify an appropriate source of collateral information.

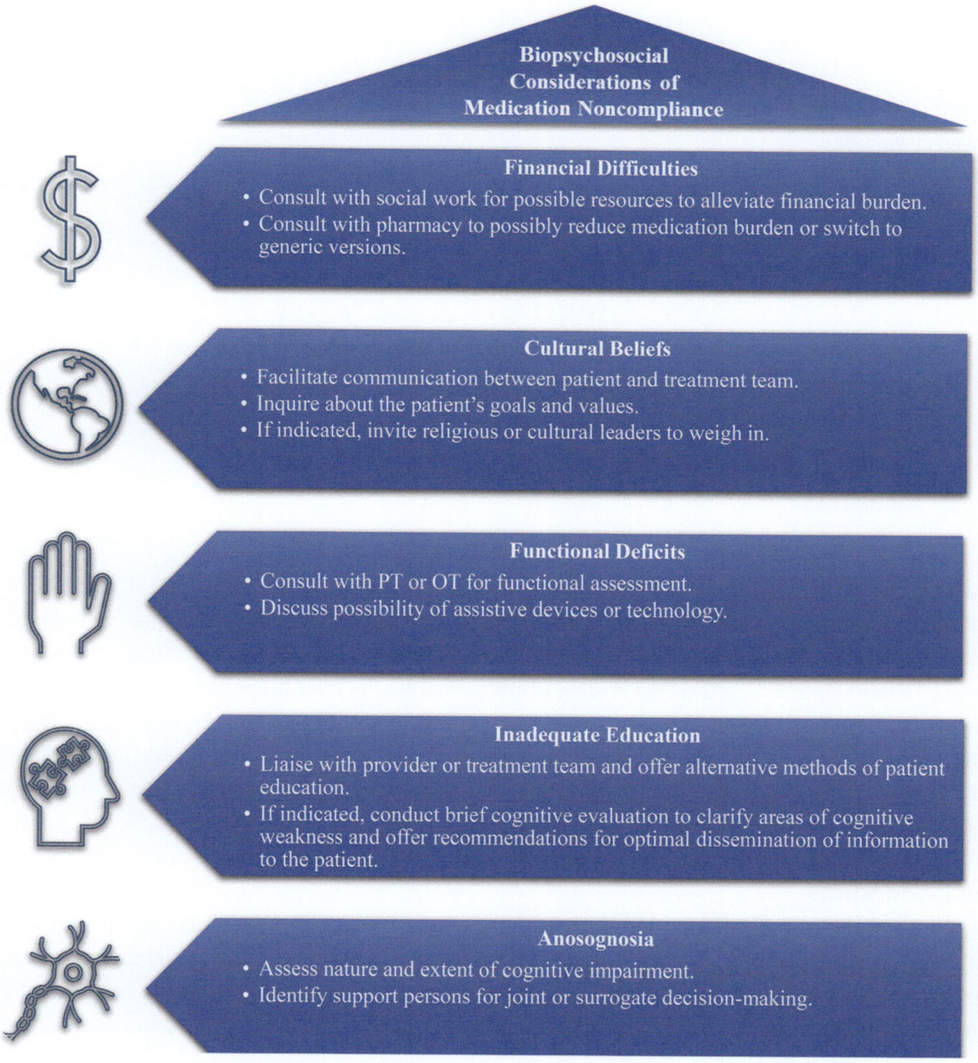

Fig. 4.15 Biopsychosocial considerations of medication noncompliance

==

NEUROIMAGING:

- 00/00/00: Head CT w/o Contrast
  - Indication: Fall versus syncope
  - Impressions: 1 cm crescent-shaped hyperdensity in left frontal area consistent with subdural hematoma. There is no evidence of midline shift, mass effect, or hydrocephalus. Cerebral volume loss.
- 00/00/99: Head CT w/o Contrast
  - Indication: Dizziness
  - Impressions: Nonspecific white matter changes consistent with small-vessel ischemic disease. Mild cerebral volume loss.

LABS:

- HEMOGLOBIN AIC

Collection DT	Spec	HbAIc	Ref Range
00/00/2000 09:38	BLOOD	9.2 H	<5.7%

- GLUCOSE (non-fasting)

Collection DT	Spec	Glucose	Ref Range
00/00/2000 09:38	BLOOD	55 L	72-108 mg/dl

==

CT = computerized tomography; w/o = without; AIC = hemoglobin AIC (glycohemoglobin)

Fig. 4.16 Neuroimaging and labs

==

PRIOR COGNITIVE EVALUATIONS:

- Cognitive screen (Inpatient Neurology 00/00/1995)
  - MoCA: 24/30, points missed on attention and recall tasks
- No prior comprehensive neuropsychological evaluations

==

MoCA = Montreal Cognitive Assessment

Fig. 4.17 Prior cognitive evaluations

Neuroimaging and Labs

Neuropsychologists, particularly those working in medical settings, must have a working knowledge of diagnostic studies such as PET, MRI, and basic laboratory panels. Documentation of findings should be transcribed verbatim from the conclusions of the radiology report (Fig. 4.16). If direct review is not available, the results should be clearly caveated as being based on third-party report. Neuroimaging should be displayed chronologically with the date, type of scan, purposes, and whether contrast was used (if applicable). Lab values—if recent enough to be relevant to the referral question—should be documented with the corresponding reference range, and organized by both date and function (e.g., viral, hepatic, endocrine). Other types of clinical tests, such as electrograms, are often included in this section as well.

Prior Cognitive Evaluations

Any prior cognitive evaluations should be documented with results of such clearly stated and/or summarized (Fig. 4.17). Evaluation can include conclusions of prior cognitive screenings, bedside examinations, or large comprehensive assessments conducted in inpatient or outpatient settings. When direct quoting of prior studies is infeasible (e.g., due to lengthy reports, non-standard interpretations), the neuropsychologist is encouraged act within their expertise in interpreting findings.

Developmental and Social History

An inpatient's developmental and social history need not be extensively documented, as is common in outpatient settings. However, some brief information about their background and current circumstance can help inform treatment planning and

=======================================

DEVELOPMENTAL AND SOCIAL HISTORY:

- Born: Born in Boston, MA
- Development: No reported issues
  - Language of Origin: English
- Education: 14 Years, Associate's degree
- Occupation: Schoolteacher (retired)
- Social:
  - Relationship Status: Widowed for the past 8 years
  - Children: One adult son living in California, good relationship
  - Religious Affiliation: non-practicing Christian
  - Hobbies: Playing cards with her neighbor
- Housing/Economic: Lives alone in a home she owns in the suburbs of Boston with her cats, fixed retirement income.
- Legal: No legal issues pending

=======================================

MA = Massachusetts

Fig. 4.18 Developmental and social history

recommendations (Fig. 4.18). For example, a patient of limited economic means may not be eligible for costly in-home care services following discharge. Should this patient also have limited social support, investigation of community pro bono services would be warranted during discharge planning. Similarly, understanding a patient's unique background can help providers maintain cultural sensitivity when recommending interventions, such as blood transfusions or other procedures contraindicated in certain cultural or religious traditions.

Procedures and Data

Risk Assessment

Documentation of suicidal and homicidal risk assessment is a required component of all written reports. Each institution or setting may have a standardized template that is preferred. In general, risk assessment includes endorsement or denial of suicidal/homicidal ideation, plan, and any prior attempt (Fig. 4.19). In inpatient settings, we recommend caveating risk pathology when endorsed or displayed in the context of delirium or confusion, as this may not reflect true underlying intent. In such cases, family or team member distress is not uncommon—especially in patients with no prior psychiatric history, and recommendations made in the report can be used to provide education on the nature of acute confusion. Interventions in place to reduce risk of self-harm should also be included in this section, such as whether a patient is on a 1:1 observation or the use of any physical restraints. If the neuropsychologist identifies an at-risk patient, the medical treatment team should be immediately notified to implement such interventions.

Lastly, in inpatient settings, providers are more likely to encounter acute or severe illnesses which may have an attenuated or terminal prognosis. As such, it is critical for the inpatient neuropsychologist to differentiate between patients with true suicidal ideation and those that are facing the reality of a limited lifespan. Regards, we always encourage documentation of risk/protective factors identified via clinical interview or in patient history.

SUICIDE RISK ASSESSMENT:

Gender:	Female
Race & Age:	White, non-Hispanic, 76y
Marital Status:	Widowed
Social Support:	Adequate
Disabling illness:	Yes, suspected cognitive decline
Mood disorder:	No
Psychotic disorder:	No
Active substance abuse:	No
Recent stressors:	Hospitalization
Religious prohibition:	Unknown
Future-Oriented:	Yes
Suicidal Ideation:	Denied
Suicidal Plan:	Denied
History of Suicide Attempt:	Denied
Estimate of Risk (Current):	Low

HOMICIDE RISK ASSESSMENT:

Ideation:	Denied
Plan:	Denied
History:	Denied

Fig. 4.19 Risk assessment

BEHAVIORAL OBSERVATIONS:

Ms. G was seen alone in her hospital room. Collateral was not present. She was found asleep, laying supine in bed though was easily aroused with auditory and light stimulation. She maintained adequate arousal and eye-contact throughout the visit. She was appropriately groomed in a hospital gown. Hearing and vision (corrected with reading glasses) were adequate for testing purposes. Speech was somewhat hypophonic and mild word-finding difficulties were observed. Comprehension appeared intact for simple language, though she had difficulty with complex language, requiring simplification of instructions or questions. Mood was euthymic with full range of affect. No evidence of psychosis. Insight was fair. Cognitive and constructional perseverations were observed. She was a poor historian, often confusing details of her recent circumstance or medical history. She was also easily distracted by hallway noise, though able to be redirected back to task.

Ms. G was cooperative with the evaluation and appeared to put forth good effort.

Fig. 4.20 Behavioral observations

Behavioral Observations

Stylistic documentation of behavioral observations often depends on the abundance and complexity of qualitative data. Bullet points are an efficient means of recording select behaviors in an otherwise normal patient encounter. However, narrative descriptions, which more easily convey longitudinal or contingent information, are recommended in situations where such observations are heavily used to augment objective data, fortify hypothesized conclusions, or describe observations across multiple visits (Fig. 4.20). Narrative descriptions of modifications to test administration can be included in this section or, if more appropriate, when reporting test results (see section below).

PROCEDURES: Chart review, Clinical interview with patient and collateral (with consent), Montreal Cognitive Assessment (MoCA), Tampa Neurobehavioral Test (TNT), Beck Depression Inventory -II (BDI-2)

PREMORBID: Based on her educational and occupational history, Ms. G's estimated premorbid baseline likely falls in the upper end of the average range.

RESULTS:

Beck Depression Inventory-II: 2/30, no significant symptoms of depression

Neurobehavioral Exam (TNT)

- Orientation: See Below
- Inattention/Neglect: Negative
- R/L Orientation: Intact
- Diadochokinesias: Intact Bilaterally
- 3-Step Sequencing: Impaired bilaterally with perseverative errors
- Motor Inhibition: Impaired with right (dom) hand
- Mathematics: Intact
- Repetitions: Intact
- Sustained Attn: See Below
- Ideational Praxis: Intact
- Ideomotor Praxis: Intact
- Commands: Intact for simple commands, impaired for complex
- Abstract Reasoning: Impaired
- Dysexecutive Behaviors: Mild disinhibition

MoCA = 15/30, Impaired

- Orientation: 5/6 (off by one date)
- Attention: 4/6 (points missed on digits backward and serial 7s)
- Language: 2/3 (fluency = 4 words in 60 sec)
- Naming: 1/3 ("bear" [tiger], "hippo" [rhino])
- Abstraction: 1/2 ("you can watch a ruler")
- Delayed Recall: 1/5 (additional two one with multiple choice cues)
- Executive/Visuospatial: 1/5
    - Trails: unable to complete one switch, appeared confused about instructions
    - Cube: constructional perseverations, internal lines incorrectly drawn
    - Clock: intact contour, misnumbered, hands incorrectly placed at "10" and "11"

TNT = Tampa Neurobehavioral Test; MoCA = Montreal Cognitive Assessment; BDI-2 = Beck Depression Inventory-II; dom = dominant

Fig. 4.21 Testing results

Testing Procedures and Results

Test results for formal psychiatric and neuropsychological measures should be clearly displayed along with normative scores and/or qualitative descriptors, when available (Fig. 4.21). Even if formal measures of premorbid cognitive abilities are not administered, we recommend including a short statement estimating the patient's baseline to use as comparison when interpreting their performance. If not documented as behavioral observations, neurobehavioral and qualitative data can be included in this section alongside corresponding tests.

Conclusions

One of the most important parts of the inpatient neuropsychology report, the conclusions section offers an opportunity to synthesize data across *all* sources and yield a refined diagnostic differential (Fig. 4.22). We recommend keeping conclusions brief and tailored to the original referral question, with the ultimate goal of being *helpful* to the treatment team. While the formatting of conclusions and diagnoses can vary, we recommend clear, concise language that is formatted in a way that is easy for treatment teams to understand and use.

SUMMARY: Ms. G demonstrated impairments in executive dysfunction (reasoning, set-switching, fluency), complex attention, comprehension of complex language, confrontation naming, and memory recall. Behaviorally, she was observed to be perseverative and mildly impulsive with limited insight into the severity of her cognitive and functional disability.

CONCLUSIONS: Given medical comorbidities and course, Ms. G's presentation best represents acute-on-chronic cognitive decline. While some improvements in cognition may be demonstrated with medical stabilization and increased recovery time post-SDH, there is concern for an underlying neurodegenerative process. Prognosis for full recovery of cognitive abilities is poor. She is likely to require ongoing assistance to ensure safe completion of IADLs following discharge.

DIAGNOSES:

// Major Neurocognitive Disorder, mild severity, due to multiple etiologies
- Left frontal subdural hematoma (resolving) s/p fall versus syncope 2/2 hypoglycemia
- Acute TME (resolving) in the setting of DMII noncompliance
- 2+ year history of impaired IADLs raises concern for underlying dementia process

// Depression, by history
- No significant symptoms during the current evaluation

// Alcohol Abuse Disorder (in sustained remission)

DMII = diabetes mellitus II; TME = toxic-metabolic encephalopathy; SDH = subdural hematoma; IADLs = instrumental activities of daily living; s/p = status post; 2/2 = secondary to

Fig. 4.22 Conclusions

RECOMMENDATIONS:

1. Underlying neurocognitive disease increases susceptibility to TME, especially in the context of noncompliance with DMII treatment. Recommend discharge home with support (e.g., home health) or consideration of ALF. If patient is amenable, discussion of discharge options with family (son) is encouraged.
2. Outpatient neuropsychological evaluation in 3-6 months is recommended for diagnostic clarity following discharge and resolution of TME, as well as recovery from acute SDH.
3. Driving is contraindicated until cleared via formal driving evaluation or by her treating medical provider.
4. Consideration of ACHE inhibitor, if not medically contraindicated. Consider consult to outpatient geriatric psychiatry following discharge.
5. Although patient denied depression during this visit, we encourage regular monitoring of mood given history. Should ALF be considered, we encourage placement in pet-friendly facility where Ms. G can bring her pet cats.
6. Recommend Social Work follow-up with Ms. G to establish Advanced Directive, if not already completed.
7. Consult dietician given reduced appetite and recent weight loss.
8. Ongoing delirium and fall precautions during admission.

Thank you for the opportunity to participate in this patient's care.
Please reach out to me directly at 555-010-6789 with any questions.

ALF = assisted living facility; DMII = diabetes mellitus type II; TME = toxic-metabolic encephalopathy; ACHE = acetylcholinesterase; SDH = subdural hematoma

Fig. 4.23 Recommendations

Recommendations

The recommendations section is another crucial part of the inpatient report. Recommendations should address a patient's prognosis and offer suggestions to compensate for (or overcome) barriers to care. Recommendations should be individualized and feasible to implement given each patients' unique circumstance (Fig. 4.23). The inpatient neuropsychologist can help advance care by identifying what other referrals or data may be useful in clarifying the clinical picture (imaging, certain labs, EEG, etc.) and/or assist the treatment team in initiating behavioral interventions prior to discharge. Lastly, we always encourage revisiting the patient's own goals and values when formulating recommendations as this helps ensure patient-centric practice.

Report Writing in Special Circumstances

Serial Monitoring

Serial monitoring, or the repeated assessment of a patient's cognitive abilities during admission is commonly used when patients are experiencing any acute change in mental status. Repeated assessment over time serves a dual purpose, assisting in both differential diagnosis and prognosis. First, serial monitoring can detect fluctuation of, or decline in, a patient's mental status, thus assisting in identification of occult pathology (e.g., undetected metabolic derangements). Similarly, temporal correlation of contextual factors can further clarify mediating variables that can refine the diagnostic differential. For example, observation of altered mental status exclusive to evening hours may elevate concern for underlying dementia (e.g., sundowning in AD).

Second, serial monitoring is also used to ascertain when patients recover (i.e., clear, emerge) from a confusional state. As fluctuations are inherent to a delirium, serial monitoring at regular intervals can reinforce the clinical utility of an improved score on a cognitive screening measure. In certain fields of literature, such as TBI, specific protocols have been developed to standardize how emergence from confusion is operationally defined (Dahdah et al. 2017; Dismuke-Greer et al. 2023; Nakase-Richardson et al. 2011; Sherer et al. 2020), as research shows differential recovery trajectories depending on the duration of posttraumatic confusion. For instance, emerging from a minimally conscious state within 3 months post-TBI is associated with better recovery (Sazbon and Groswasser 1991; Luaute et al. 2010), while individuals with protracted durations of post-traumatic confusion show worse long-term outcomes (Nakase-Richardson et al. 2007; Wilde et al. 2006). When working with these populations, accurately gaging the duration of posttraumatic confusion via serial monitoring will allow the inpatient neuropsychologist to provide more specific prognostic information and give patients and families education and recommendations about expected functional outcomes.

As with any serial evaluation, the impact of test novelty and the possibility of practice effects should be considered, especially once patients emerge from confusional states. Some bedside screening instruments, such as the MoCA, offer multiple forms. Others limit screening items to circumstantial information, such as orientation questions, which are not likely to be corrupted by repeated administration. Regardless of the specific measures used, serial monitoring should be clearly documented in a manner that allows for quick assessment of data over time. See Fig. 4.24 for an example of how to document serial monitoring data.

SERIAL MONITORING TRACKING

PRIOR EVALUATIONS:

DATE	TIME	O-LOG	COG-LOG	ABS
1/1/2020	0900	21/30	14/30	16
1/3/2020	0830	19/30	12/30	15
1/5/2020	1300	20/30	13/30	16
1/9/2020	1100	10/30	----	14
1/11/2020	1400	7/30	----	16
1/12/2020	1000	8/30	----	17
1/15/2020	0930	15/30	13/30	16
1/17/2020	1300	18/30	----	15
1/20/2020	0900	23/30	----	14
1/24/2020	1045	26/30	----	15
1/26/2020	1300	25/30	----	14
1/27/2020	0900	27/30	18/30	14
1/31/2020	1030	26/30	17/30	14
2/3/2020	1400	28/30	----	14

CURRENT EVALUATION:

DATE	TIME	O-LOG	COG-LOG	ABS
2/6/2020	1330	27/30	19/30	14

O-log = Orientation Log; Cog-Log = Cognitive Log; ABS = Agitated Behavior Scale

Fig. 4.24 Documenting serial monitoring data

Epilepsy Monitoring and Wada's Testing

Epilepsy treatment centers are another area of specialized practice for inpatient neuropsychologists. As an essential part of epilepsy interdisciplinary systems of care, neuropsychologists in these settings routinely conduct bedside cognitive evaluations, assist with intraoperative brain mapping, or perform memory and language testing as part of Wada procedures (Kirlin and Locke 2014). Cognitive data are compiled alongside other clinical work-up to determine candidacy for surgery in patients with refractory epilepsy. Neuropsychologists are also often asked to provide education and recommendations to patients and families regarding etiology and course of cognitive symptoms and prognosis. Please note that conclusions may heavily depend on the specifics of a patient's seizures and the proposed surgery (i.e., extent of resection or ablation, stimulator placement).

Bedside cognitive testing in these contexts may be administered in conjunction with long-term or video EEG monitoring to link clinical symptoms to epileptiform discharges, characterize psychogenic symptoms not associated with EEG activity (e.g. psychogenic non-epileptic attacks or PNEA), or identify transient clinical pathology that may otherwise go undetected without EEG correlation (Lee 2010). In contrast with the brevity recommended previously in this chapter when evaluating medical inpatients, neuropsychological evaluation of hospitalized epilepsy patients is typically comprehensive and comparable to traditional outpatient exams. This is done for several reasons. First, comprehensive baseline performance across many domains must be obtained to reliably predict postoperative cognitive outcomes from any planned surgery, particularly those involving tissue resection or ablation. Second, as neuropsychological evaluations are used to help determine lateralization and seizure focus, the planned battery should be sensitive and broad enough to allow the best chance for detection of possible pathology. Lastly, robust data spanning a range of cognitive complexity yield the best ecological forecasting of functional abilities, such as return to work or school.

Intraoperative brain mapping and Wada procedures include brief assessment of language and memory functioning in real time during medically induced, neurologic stimulation. The former is conducted during an awake craniotomy where the brain is systematically stimulated to precisely identify eloquent areas, particularly those with preserved functioning, with the purpose of guiding the surgical plan by differentiating epileptogenic areas from healthy tissue. Patients may be asked to perform memory, naming or fluency tasks, follow commands, or mimic hand movements as brain tissue is electrically stimulated by a surgeon (electrocortical stimulation [ECS]). Changes in cognitive or behavioral responses are noted corresponding to the stimulated brain region, thus mapping the individualized anatomic locations of these abilities (Leuthardt et al. 2013).

During a Wada procedure, an intravenous fast-acting barbiturate (such as methohexital or amobarbital) is injected into the internal carotid artery via femoral catheter to temporarily anesthetize brain tissue perfused by the anterior and middle cerebral arteries of each cerebral hemisphere, in succession. Brief cognitive evaluation is conducted for each cerebral hemisphere to determine lateralization of memory and language dominance. Patients are asked to recall information presented during anesthetization once they return to cognitive baseline (~8 to 10 min after the initial injection) and comparison of performance across injection trials yields valuable prognostic information about anticipated postsurgical cognitive outcome (Lee 2010). See Fig. 4.25 below for an example of a brief Wada testing report.

For more extensive information regarding the role of an inpatient neuropsychologist in epilepsy treatment centers, we recommend referencing the text *Neuropsychology of Epilepsy and Epilepsy Surgery* by Gregory Lee (2010) which includes detailed information on epilepsy syndromes, antiepileptic drugs (AEDs), seizure nosology, and perioperative evaluation, as well additional example reports.

NEUROPSYCHOLOGY NOTE
Intracarotid Methohexital (Wada's) Test

Patient Name:
Date of Procedure:
Date of Report:
Referring Physician:

HISTORY: Ms. W is a 30-year-old right-handed female with history of medically intractable epilepsy. EEG showed left sided onset.

PROCEDURES:
Left Injection: 5 mg of methohexital (Brevital) were injected in the left common carotid artery approximately 30 seconds apart, 3 mg then 2 mg. There was complete right hemiparesis. EEG was read as demonstrating adequate left hemisphere slowing.
Language: Complete aphasia
Memory: Spontaneous Recall was 0/8. Recognition was 2/8

Right Injection: 5 mg of methohexital (Brevital) were injected in the right common carotid artery approximately 30 seconds apart, 3 mg then 2 mg. There was complete left hemiparesis. EEG was read as demonstrating adequate right hemisphere slowing.
Language: No speech disruption. Some mild dysarthria
Memory: Spontaneous Recall was 6/8. Recognition was 7/8.

IMPRESSION:
Wada test was valid.
- Patient appears left hemisphere dominant for language.
- Asymmetry of memory would support left mesial temporal dysfunction.
- The right hemisphere function would suggest adequate post-surgical memory support.

Thank you for the opportunity to assist in the care of this patient.

EEG = electroencephalogram; mg = milligram

Fig. 4.25 Wada's testing report

SUMMARY

Inpatient neuropsychology plays a critical role in the assessment and management of patients within diverse medical settings, including both acute and post-acute care. Both bedside neuropsychological evaluation and documentation in the inpatient setting require efficiency, flexibility, and the ability to synthesize medical, psychological, and functional complexities. As such, neuropsychologists in this setting should focus on targeted evaluations and clear communication of recommendations to enhance patient-centered care, assist in treatment decisions, and engage as effective members of the interdisciplinary team.

Appendices

Appendix 1: Common Bedside Cognitive Assessments

Test	Items	Population Considerations	Domains Assessed	Duration of Administration
Galveston Orientation and Amnesia Test (GOAT)	Verbal responses on 10 items	Delirium, early TBI/stroke recovery	Orientation, Posttraumatic amnesia	5-10 min
Written GOAT	*Written responses on items 1-3 and 6-10*	*Patients with intact comprehension and speech difficulties (e.g., dysarthria, hypophonia)*	*Orientation*	
Modified GOAT	*Multiple-choice responses on items 1-3 and 6-10*	*Patients with aphasia/speech and motor difficulties*	*Orientation*	
Orientation Log (O-LOG)	10 items with options for spontaneous response and cueing	Delirium, early TBI/stroke recovery	Orientation, Posttraumatic amnesia	~5 min
Cognitive Log (COG-LOG)	10 items across 5 domains	Rehabilitation settings, TBI/stroke recovery	Orientation, simple attention, working memory, executive control, abstract reasoning, memory recall	~8-10min
**Often used in serial assessments as a companion measure with the O-LOG.*				
Confusion Assessment Method (CAM)	Observational data and verbal/non-verbal responses across 4 domains	Delirium, disorders of consciousness	Fluctuations in cognitive status, attention, disorganized thinking, abstract reasoning, command following, alterations of consciousness	<5 min
**Available in 11 languages*				

CAM-Intensive Care Unit (ICU)	*Observational data and non-verbal responses only*	*Patients with difficulty communicating due to mechanical ventilation*	*Above*	
CAM for Emergency Departments	*Observational data and verbal/non-verbal*	*Emergency settings*	*Above with additional attention tasks and questions pertaining to patient history*	
Agitated Behavior Scale (ABS)	Observational data on 14 items	Disorders of consciousness, delirium, or posttraumatic amnesia	Behavioral agitation (observational)	<5 min
Mini-Mental Status Examination (MMSE)	Written and verbal responses on 11 items	Cognitive decline, dementia	Orientation, basic attention, working memory, naming, visuoconstruction, basic comprehension, command following, repetition	~10 min
Montreal Cognitive Assessment (MoCA)	Written and verbal responses across 8 domains	Cognitive decline, dementia, commonly used in medical settings	Orientation, attention, working memory, naming, visuoconstruction, memory recall, basic comprehension, abstract reasoning, verbal fluency, executive functioning	~8-10 min
	**Multiple forms, available in over 100 languages*			
MoCA-Blind	*Omission of visually mediated items, verbal responses only*	*Patients with visual impairments*	*Orientation, attention, working memory, memory recall, abstract reasoning, verbal fluency*	
Saint Louis University Mental Status Exam (SLUMS)	Written and verbal responses on 11 items	Cognitive decline, dementia, commonly used in VA settings	Orientation, attention, working memory, calculations, abstract reasoning, visuospatial abilities	~5 min

**Available in over 15 languages*				
Mini-Cog	Verbal and written responses on 3 items	Cognitive decline, dementia	Word registration, memory recall, executive functioning (clock drawing test)	~3 min
Mississippi Aphasia Screening Test (MAST)	Verbal and written responses across 9 subtests	Aphasia	Naming, automatic speech, repetition, object recognition, command following, reading, writing, verbal fluency, yes/no accuracy	5-15 min
Frontal Assessment Battery (FAB)	Verbal and motor responses across 6 domains	Cognitive decline, post TBI/stroke, dementia	Similarities, verbal fluency, motor sequencing, command following, response inhibition	5-10 min
**Available in 8 languages*				
Tampa Neurobehavioral Test (TNT)	Verbal, written, and motor responses across 19 items	Cognitive decline, post TBI/stroke, dementia, delirium	Orientation, visual fields, visual tracking, visual neglect, gross motor strength, right/left orientation, motor coordination, motor sequencing, response inhibition, calculation, repetition, attention, working memory, ideational/ideomotor praxis, abstract reasoning, command following, executive functioning (clock drawing test, ramparts)	~15 min

Appendix 2: Tampa Neurobehavioral Test

TAMPA NEUROBEHAVIORAL TEST

Handedness: R / L / A

1) **Orientation** (/4)
(Person: √ / X Place: √ / X Date/Day: √ / X Situation: √ / X)

Date: Day of Week: Month: Year: State: Town/City:
County: Place: Situation: Time of Day:

2) **Visual Fields** (CN II)

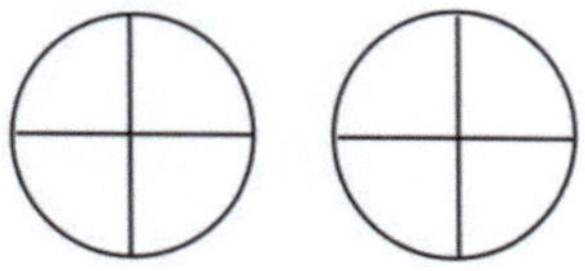

3) **Visual Tracking** (CN's III, IV, VI)
Right EOM Intact / Impaired
Left EOM Intact / Impaired
Right Nystagmus Yes / No
Left Nystagmus Yes / No

4) **Extinctions**
Visual R / L Tactile R / L
Auditory R / L

5) **Right / Left Orientation** (/3)
Intact / Mild / Severe

6) **Gross Strength** (/6)
Right Intact / Mild / Severe
Left Intact / Mild / Severe

7) **Finger-to-Nose** (/6)
Right Intact / Mild / Severe
Left Intact / Mild / Severe

8) **Rapid Alternating Movements** (/6)
Right Intact / Mild / Severe
Left Intact / Mild / Severe

9) **3-Step Motor Sequencing** (/6)
Right Intact / Mild / Severe
Left Intact / Mild / Severe

10) **Motor Inhibition Tasks** (/8)

a. Automatic Response Inhibition 1-1-2-1-2-2-2-1-2-2-1-1
0 Errors 1-2 Errors 3-4 Errors >4 Errors
Total Errors:
Set Loss Errors:
Perseverative Errors:

b. Go-No Go 1-1-1-2-1-2-2-2-2-1-1-2
0 Errors 1-2 Errors 3-4 Errors >4 Errors
Total Errors:
Set Loss Errors:
Perseverative Errors

11) **Mathematics** (/3)
28/7= 6x8= 37-19= 13 -5= 16+7= 5+7=
Intact Mild (dyscalculia) Severe (acalculia)

12) **Ideational Praxis** (/3) (Make PB&J)
Intact / Mild / Severe
Praxic Errors:
Sequencing Errors:

13) **Repetitions** (/4)
Seven: Massachusetts: Methodist Episcopal:
The cat was eaten by the mouse:

14) **Ideomotor Praxis** (/4)
Dominant – (Screwdriver, scissors)
Intact / Mild / Severe

Non-Dominant - (Hammer, comb)
Intact / Mild / Severe

15) **Sustained Attention** (/3)
Tap on 'E' A-B-E-D-E-E-J-O-Y-T-E-R-B-E-I-E
Intact (0 errors) Mild (1-2 errors) Severe (>2 errors)

16) **Conceptual Similarities** (/6)
Arm/leg: Laugh/Cry: Eat/Sleep:

17) **Commands** (/8)

a. Multi-Step: (3) Touch ear, point to wall, open mouth: (2) Floor then door: (1) Ceiling:

b. Complex: (5) Do what I do, touch your ear (touch nose):
(4) Before touching your ear, open your mouth:

18) Ramparts (/ 9)

a. Square/Triangle (/ 3)
Intact / Mild / Severe

b. M's and N's (/ 3)
Intact / Mild / Severe

c. Loops (/ 3)
Intact / Mild / Severe

19) Clock (/ 12)

a. Free Draw (/ 4)
Gestalt Intact / Impaired
Numbers Intact / Mild / Severe
Hands Intact / Mild / Severe

b. Copy (/ 8)
Gestalt Intact / Impaired
Numbers Intact / Mild / Severe
Hands Intact / Mild / Severe

REFERENCES

Alderson AL, Novack TA. Reliable serial measurement of cognitive processes in rehabilitation: the cognitive log. Arch Phys Med Rehabil. 2003;84(5):668–72.

American Geriatrics Society Beers Criteria® Update Expert Panel. American Geriatrics Society 2023 updated AGS beers criteria® for potentially inappropriate medication use in older adults. J Am Geriatr Soc. 2023;71(7):2052–81.

Bijnsdorp FM, Schouten B, Reyners AKL, Wagner C, Francke AL, van Schoten SM. Measurement and documentation of quality indicators for the end-of-life care of hospital patients a nationwide retrospective record review study. BMC Palliat Care. 2023;22(1):174.

Borson S, Scanlan J, Brush M, Vitaliano P, Dokmak A. The mini-cog: a cognitive 'vital signs' measure for dementia screening in multi-lingual elderly. Int J Geriatr Psychiatry. 2000;15(11):1021–7.

Boudreaux ED, Clark S, Camargo CA Jr. Mood disorder screening among adult emergency department patients: a multicenter study of prevalence, associations, and interest in treatment. Gen Hosp Psychiatry. 2008;30(1):4–13.

Corrigan JD. Development of a scale for assessment of agitation following traumatic brain injury. J Clin Exp Neuropsychol. 1989;11(2):261–77.

Centers for Disease Control and Prevention. CDC's core infection prevention and control practices for safe healthcare delivery in all settings. 2024.

Dahdah MN, Hofmann M, Pretz C, An V, Barnes SA, Bennett M, et al. Factors associated with rehabilitation outcomes after traumatic brain injury: comparing functional outcomes between TBIMS centers using hierarchical linear modeling. J Head Trauma Rehabil. 2017;32(4):E1–E10.

Daray FM, Hunter F, Anastasia A, Fornaro M. Psychometric properties of the patient health questionnaire nine-item version (PHQ-9) for use among hospitalized non-psychiatric medical patients. Gen Hosp Psychiatry. 2019;61:69–75.

Delaney LJ, Van Haren F, Lopez V. Sleeping on a problem: the impact of sleep disturbance on intensive care patients-a clinical review. Ann Intensive Care. 2015;5:1–10.

Dismuke-Greer CE, Almeida EJ, Silva MA, Dams-O'Connor K, Rocek G, Phillips LM, et al. Effect of post-traumatic amnesia duration on traumatic brain injury (TBI) first year hospital costs: a veteran's affairs traumatic brain injury model systems study. Arch Phys Med Rehabil. 2023;104(7):1007–15.

Doğan O, Ertekin Ş, Doğan S. Sleep quality in hospitalized patients. J Clin Nurs. 2005;14(1):107–13.

Doraciak Y. Aspects of sleep. Nurs Times. 1990;86:38–40.

Dubois B, Slachevsky A, Litvan I, Pillon BFAB. The FAB: a frontal assessment battery at bedside. Neurology. 2000;55(11):1621–6.

Duggan MC, Wang L, Wilson JE, Dittus RS, Ely EW, Jackson JC. The relationship between executive dysfunction, depression, and mental health-related quality of life in survivors of critical illness: results from the BRAIN-ICU investigation. J Crit Care. 2017;37:72–9.

Folstein M, Folsten S, McHugh P. Mini-mental state: a practical method for grading the cognitive state of patients for the clinician. J Psychiatry Res. 1975;12:189–98.

Garrigues E, Janvier P, Kherabi Y, Le Bot A, Hamon A, Gouze H, et al. Post-discharge persistent symptoms and health-related quality of life after hospitalization for COVID-19. J Infect. 2020;81(6):e4–6.

Gillen R, Tennen H, McKee TE, Gernert-Dott P, Affleck G. Depressive symptoms and history of depression predict rehabilitation efficiency in stroke patients. Arch Phys Med Rehabil. 2001;82(12):1645–9.

Goodglass H, Kaplan E. The assessment of aphasia and related disorders. Philadelphia: Lea and Febiger; 1983.

Gusmao-Flores D, Salluh JIF, Chalhub RÁ, Quarantini LC. The confusion assessment method for the intensive care unit (CAM-ICU) and intensive care delirium screening checklist (ICDSC) for the diagnosis of delirium: a systematic review and meta-analysis of clinical studies. Crit Care. 2012;16:1–10.

Harmon EY, Gillen RW. Comparison of the Brief Interview for Mental Status (BIMS) and Montreal Cognitive Assessment (MoCA) for identifying cognitive impairments and predicting rehabilitation outcomes in an inpatient rehabilitation facility. PM R. 2023;15(9):1083–91.

Hawkins MA, Gathright EC, Gunstad J, Dolansky MA, Redle JD, Josephson R, et al. The MoCA and MMSE as screeners for cognitive impairment in a heart failure population: a study with comprehensive neuropsychological testing. Heart Lung. 2014;43(5):462–8.

Inouye S, Van Dyck C, Alessi C, Balkin S, Siegal A, Horwitz R. Clarifying confusion: the confusion assessment method. Ann Intern Med. 1990;113(12):941–8.

IsHak WW, Wen RY, Naghdechi L, Vanle B, Dang J, Knosp M, et al. Pain and depression: a systematic review. Harv Rev Psychiatry. 2018;26(6):352–63.

Jackson WT, Novack TA, Dowler RN. Effective serial measurement of cognitive orientation in rehabilitation: the Orientation Log. Arch Phys Med Rehabil. 1998;79(6):718–21.

Jain N, Layton BS, Murray PK. Are aphasic patients who fail the GOAT in PTA? A modified Galveston Orientation and Amnesia Test for persons with aphasia. Clin Neuropsychol. 2000;14(1):13–7.

Jiang YN, Zhou W, Zhao XH, Hong X, Wei J. Post-traumatic stress disorder in convalescent patients of severe acute respiratory syndrome:(1) H-MRS study. Zhonghua Yi Xue Za Zhi. 2013;93(5):366–9.

Kessler HR. The bedside neuropsychological examination. In: Snyder PJ, Nussbaum PD, Robins DL, editors. Clinical neuropsychology: a pocket handbook for assessment. 2nd ed. Washington, DC: American Psychological Association; 2006. p. 75–101.

Kim AM, Rossi KC, Jetté N, Yoo JY, Hung K, Dhamoon MS. Increased risk of hospital admission for mood disorders following admission for epilepsy. Neurology. 2018;91(9):e800–10.

Kirlin KA, Locke DE. The role of neuropsychology on an epilepsy monitoring unit: a peek behind the "do not disturb" sign. Neurodiag J. 2014;54(3):289–98.

Kroenke K, Spitzer RL, Williams JB. The PHQ-9: validity of a brief depression severity measure. J Gen Intern Med. 2001;16(9):606–13.

Lee GP. Neuropsychology of epilepsy and epilepsy surgery. Oxford: Oxford University Press; 2010.

Lenze EJ, Munin MC, Dew MA, Rogers JC, Seligman K, Mulsant BH, Reynolds CF III. Adverse effects of depression and cognitive impairment on rehabilitation participation and recovery from hip fracture. Int J Geriatr Psychiatry. 2004;19(5):472–8.

Leuthardt EC, Roland J, Breshears J, Bandt SK, Shimony JS. Listening to the brain: new techniques in intraoperative brain mapping. Neurosurgery. 2013;60:64–9.

Levin HS, O'Donnell VM, Grossman RG. The Galveston Orientation and Amnesia Test: a practical scale to assess cognition after head injury. J Nerv Ment Dis. 1979;167(11):675–84.
Losier B, Jongsma K, King JP, McNeely HE. Cognitive assessment in medical settings. In: McMaster textbook of internal medicine. Kraków: Medycyna Praktyczna; 2020. https://empendium.com/mcmtextbook/chapter/B31.II.21.29.
Luaute JMDP, Maucort-Boulch DMDP, Tell L, Quelard F, Sarraf T, Iwaz J, et al. Long-term outcomes of chronic minimally conscious and vegetative states. Neurology. 2010;75(3):246–52.
Manea L, Gilbody S, McMillan D. A diagnostic meta-analysis of the Patient Health Questionnaire-9 (PHQ-9) algorithm scoring method as a screen for depression. Gen Hosp Psychiatry. 2015;37(1):67–75.
Morley J, Tumosa N. Saint Louis University mental status examination (SLUMS). Aging Success. 2002;12(1):4.
Nakase-Richardson R, Yablon SA, Sherer M. Prospective comparison of acute confusion severity with duration of post-traumatic amnesia in predicting employment outcome after traumatic brain injury. J Neurol Neurosurg Psychiatry. 2007;78(8):872–6.
Nakase-Richardson R, Sherer M, Seel RT, Hart T, Hanks R, Arango-Lasprilla JC, et al. Utility of post-traumatic amnesia in predicting 1-year productivity following traumatic brain injury: comparison of the Russell and Mississippi PTA classification intervals. J Neurol Neurosurg Psychiatry. 2011;82(5):494–9.
Nakase-Thompson R, Manning E, Sherer M, Yablon SA, Gontkovsky SLT, Vickery C. Brief assessment of severe language impairments: initial validation of the Mississippi aphasia screening test. Brain Inj. 2005;19(9):685–91.
Nasreddine ZS, Phillips NA, Bédirian V, Charbonneau S, Whitehead V, Collin I, et al. The Montreal Cognitive Assessment, MoCA: a brief screening tool for mild cognitive impairment. J Am Geriatr Soc. 2005;53(4):695–9.
Pavol MA. Inpatient neuropsychological assessment in older adults. In: Handbook on the neuropsychology of aging and dementia. Cham: Springer; 2019. p. 89–103.
Raymond I, Nielsen TA, Lavigne G, Manzini C, Choinière M. Quality of sleep and its daily relationship to pain intensity in hospitalized adult burn patients. Pain. 2001;92(3):381–8.
Reid E. Factors affecting how patients sleep in the hospital environment. Br J Nurs. 2001;10(14):912–5.
Rentsch D, Dumont P, Borgacci S, Carballeira Y, Detonnac N, Archinard M, Andreoli A. Prevalence and treatment of depression in a hospital department of internal medicine. Gen Hosp Psychiatry. 2007;29(1):25–31.
Sazbon L, Groswasser Z. Time-related sequelae of TBI in patients with prolonged post- comatose unawareness (PC-U) state. Brain Inj. 1991;5(1):3–8.
Schnider A. Neuropsychological testing: bedside approaches. Handb Clin Neurol. 2008;88:137–54.
Sherer M, Katz DI, Bodien YG, Arciniegas DB, Block C, Blum S, et al. Post-traumatic confusional state: a case definition and diagnostic criteria. Arch Phys Med Rehabil. 2020;101(11):2041–50.
Siegel JD, Rhinehart E, Jackson M, Chiarello L, Healthcare Infection Control Practices Advisory Committee. 2007 Guideline for isolation precautions: preventing transmission of infectious agents in health care settings. Atlanta: CDC; 2019, May.
Tate JA, Happ MB. The confusion assessment method for the ICU (CAM-ICU). Aust Crit Care. 2018;24(2):126–32.
Teng E, Chui H. The modified mini-mental state examination (3MS). Can J Psychiatr. 1987;41(2):114–21.
Toglia J, Fitzgerald KA, O'Dell MW, Mastrogiovanni AR, Lin CD. The Mini-Mental State Examination and Montreal Cognitive Assessment in persons with mild subacute stroke: relationship to functional outcome. Arch Phys Med Rehabil. 2011;92(5):792–8.
Waszynski CM. The confusion assessment method (CAM). Ann Intern Med. 2003;113(12):941–8.
Wilde EA, Bigler ED, Pedroza C, Ryser DK. Post-traumatic amnesia predicts long-term cerebral atrophy in traumatic brain injury. Brain Inj. 2006;20(7):695–9.
Worsham CM, Banzett RB, Schwartzstein RM. Dyspnea, acute respiratory failure, psychological trauma, and post-ICU mental health: a caution and a call for research. Chest. 2021;159(2):749–56.
Zigmond AS, Snaith RP. The hospital anxiety and depression scale. Acta Psychiatr Scand. 1983;67(6):361–70.

5 Diversity Considerations for Inpatient Neuropsychologists

Martina Azar, Sheryl Berardinelli, Karina Loyo,
and Claire McGrath

ABBREVIATIONS

ABI	Acquired Brain Injury
AHRQ	Agency for Healthcare Research and Quality
APA	American Psychological Association
CALPS	Cognitive and Academic Language Proficiency Skills
CDC	Centers for Disease Control and Prevention
ESL	English as a Second Language
ICF	International Classification of Functioning, Disability, and Health
LEP	Limited English Proficiency
LOS	Length of Stay
SCI	Spinal Cord Injury
SDOH	Social Determinants of Health
TBI	Traumatic Brain Injury
U.S.	United States
WHO	World Health Organization

INTRODUCTION

Healthcare disparities in diagnosis, care, and outcomes in inpatient and rehabilitation settings are well documented (Chavez et al. 2024; Doshi et al. 2017; Flores et al. 2020; Odonkor et al. 2021; Wheeler et al. 2004). These disparities present additional barriers for patients with cognitive deficits, as their symptoms can make it more challenging for them to understand their care and advocate for their needs in a manner that aligns with their cultural values and needs (Boltz et al. 2021;

Authors contributed equally to the chapter and are listed in alphabetical order.

M. Azar (✉)
Brown University Health, Providence, RI, USA
e-mail: mazar2@brownhealth.org

S. Berardinelli
Jefferson Moss-Magee Rehabilitation, Elkins Park, PA, USA
e-mail: sheryl.berardinelli@jefferson.edu

K. Loyo
Department of Rehabilitation Psychology/Neuropsychology, TIRR Memorial Hermann Health System, Houston, TX, USA
e-mail: Karina.loyo@memorialhermann.org

C. McGrath
Bancroft, Mt. Laurel, NJ, USA
e-mail: Claire.Mcgrath@bancroft.org

E. K. Bailey et al. (eds.), *Inpatient Neuropsychology*, https://doi.org/10.1007/978-3-032-05880-5_5

Campbell et al. 2014; Chen et al. 2021; Pereira et al. 2022). As psychologists committed to patient well-being, justice, and the respect of individual rights and dignity, we have a responsibility to address these disparities to avoid perpetuating disadvantage (American Psychological Association [APA] 2017). This task is especially challenging in the fast-paced environment of hospital and rehabilitation settings. Neuropsychologists must possess a foundational understanding of diverse cultures, clinical skills that enable culturally respectful interactions, and the self-awareness to recognize their own cultural limitations and biases. These three factors form the basis for culturally competent care (Fig. 5.1).

This chapter is intended to provide a pathway toward culturally competent neuropsychological care in acute hospital and inpatient rehabilitation settings. We aim to help readers appreciate the complex interplay between contextual-level factors (cultural influences that affect groups of people as a whole) and individual-level factors (cultural influences filtered through each person's unique experiences) that affect the healthcare services a person receives, the way these services are provided, and how the person engages in treatment. When neuropsychologists approach their care from a culturally informed lens, they can build rapport with patients from diverse backgrounds, convey assessment findings in a meaningful and understandable manner, guide team members to consider cultural factors, and assist patients in strategizing ways to integrate recommendations into their lives post discharge. In other words, neuropsychologists can espouse care that is respectful, responsive, and tailored to patients' cultural backgrounds and personal needs.

This chapter covers a range of topics, including illness perception, the influence of individualism/collectivism, health literacy, disability, aging, the impact of language (person/identity-first language, proficiency, use of translators), and social determinants of health. The broad goal of this chapter is to provide introductory information surrounding these multi-layered topics to help foster consideration and discussion of culturally informed care for all individuals (Fig. 5.2). The chapter con-

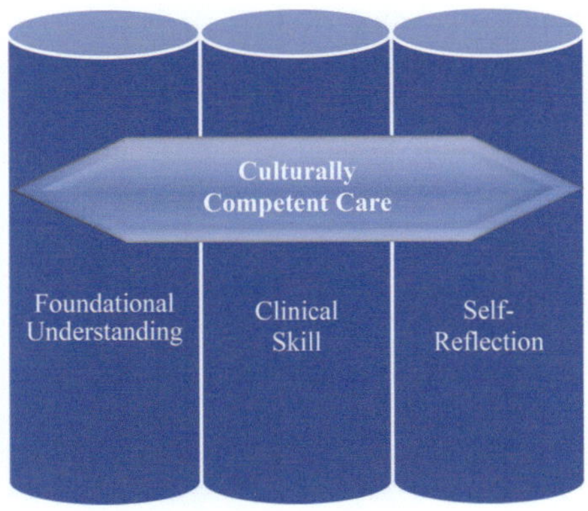

Fig. 5.1 Three pillars of culturally competent care

Fig. 5.2 Chapter goals

cludes with pragmatic recommendations to assist the inpatient neuropsychologist in providing equitable and effective treatment.

We, the authors of this chapter, recognize the power of language and the potential harm caused by using terms that reinforce marginalization, discrimination, and stereotyped beliefs. When available, we use the terms provided in published papers to describe samples or refer to the nation of origin. When not available, we adhere to the APA Inclusive Language Guidelines (APA 2023) when referring to ethnic, racial, and cultural groups. To be inclusive, we use gender neutral pronouns (they/them/theirs), we select identity-first or person-first language based on group preferences within the disability community, and we avoid terms that promote ageism and stereotypes of older individuals (APA 2014). Paralleling the ever-changing diversity of the United States, the research outlined below is expected to evolve in the years to come. As part of cultural humility and to deepen understanding of these complex topics, readers are encouraged to augment this chapter with contemporary resources as appropriate.

Scope of the Problem

Research has identified cultural disparities in medical conditions and healthcare outcomes across various neurologic injuries and diagnoses (Williams and Rucker 2000). In acute care and rehabilitation settings, large-scale studies find a higher incidence of stroke among Mexican Americans, American Indians, and Blacks compared to Whites (Tsao et al. 2022). Black patients, in particular, often experience greater functional impairment upon admission to acute stroke rehabilitation programs, and these discrepancies persist even after rehabilitation (Go et al. 2013). Additionally, Black patients report poorer functional outcome one-year post stroke (Ellis et al. 2014) and are at higher risk for dementia 5 years after stroke compared to Whites (Clark et al. 2018). Higher post-stroke disability is also found among Hispanic patients (Cruz-Flores et al. 2011; Simmonds et al. 2021). In the traumatic brain injury (TBI) literature, racial and ethnic differences are documented with TBI-related hospitalizations and deaths highest among American Indian/Alaskan Native people, followed by non-Hispanic Whites (Center for Disease Control and Prevention [CDC] 2023). Hispanic, non-Hispanic Black, and Asian patients with TBI are discharged from inpatient brain injury rehabilitation facilities with lower functional scores (Warren and García 2022); non-Hispanic Black and Hispanic patients are less likely to receive follow-up care and rehabilitation after a TBI (CDC 2023; Dismuke et al. 2015; Meagher et al. 2015); and racial and ethnic minoritized patients often experience poorer psychological, functional, and employment-related outcomes after TBI (CDC 2023; Arango-Lasprilla et al. 2009; Gary et al. 2009). On the other hand, studies on spinal cord injury (SCI) raise concerns about poorer recovery of self-care and mobility skills in Black patients compared to Hispanic and White patients, although these differences may not persist long-term (Fyffe et al. 2014).

Ethnic and racial differences in healthcare access are also evident (Ramsoondar et al. 2023) among stroke (Levine et al. 2020), TBI (Gao et al. 2018), and joint replacement (Freburger et al. 2011) populations. Even after accounting for patient demographics, diagnosis, severity of the condition, and hospital factors, significant disparities in hospital length of stay (LOS) based on race/ethnicity and socioeconomic status remain (Ghosh et al. 2021). Black patients consistently face longer hospital LOS compared to White patients, although White patients with Medicaid tend to stay in the hospital longer than Black, Hispanic, and other patients from underrepresented groups with Medicaid when the discharge destination is to home. Frebruger and colleagues (2011) found patients from historically marginalized communities, those with Medicaid insurance, individuals living in rural areas, and people from lower socioeconomic backgrounds often receive less intensive care and have reduced access to institutional care (e.g., hospitals), inpatient rehabilitation facilities, skilled nursing facilities, and home health services following joint replacement surgery.

Racial and ethnic disparities in healthcare are not entirely explained by differences in access to care, or financial resources and healthcare insurance (Nelson 2002). Cultural factors including limited health literacy and English proficiency without access to interpreters, ableism, and stigma interfere with effective communication and pose barriers to patient engagement and follow-up after discharge (Schillinger 2020). As discussed later in this chapter, discrepancies between the health beliefs of a patient and those of the healthcare provider or system can result in misunderstandings, leading to mistrust, dissatisfaction, and dismissal of treatment recommendations (Hwang et al. 2008; Mpofu et al. 2012). Neuropsychologists in acute care and rehabilitation settings who are aware of these cultural challenges can collaborate with patients and other team members to tailor interventions and assessments, contributing to inclusive and effective patient care.

CONTEXTUAL FACTORS

Cultural context encompasses the social, historical, and cultural influences that shape an individual's beliefs, and their values and behaviors. Cultural norms, traditions, customs, language, and socioeconomic status play a significant role in shaping an individual's identity and worldview, and how they interact with others. In a healthcare setting, understanding these cultural factors gives providers valuable insights into a patient's expectations for treatment, their general expectation for healthcare services, and how they may respond to health-related experiences. For example, individuals from cultures where doctors are highly esteemed may expect the admitting doctor to be involved in most details of their care. A collaborative and multidisciplinary approach may be confusing, and lead to frustration or anxiety. In this section, cultural beliefs and norms that are particularly relevant to health and healthcare services are explored, including individualism and collectivism, cultural perceptions of illness, health literacy, disability, and aging.

Individualism and Collectivism

Individualism and collectivism are cultural paradigms by which a person views themselves in comparison to others. These normative processes are often anchored in the larger cultural context and have implications for how people engage with a healthcare system. Suggestions for how to practice sensitivity with respect to these concepts are outlined in Fig. 5.3.

Individualism

The concept of individualism highlights the importance of independence and autonomy, viewing the self as distinct from others (Fatehi et al. 2020). Prevalent in countries like the United States and Western European nations, this perspective deferentially prioritizes an individual's rights and needs (Hofstede 2011). In these cultures, healthcare often adopts a shared decision-making model, where providers and patients collaborate by exchanging information to reach a mutual agreement (Elwyn et al. 2012; Makoul and Clayman 2006; Mead et al. 2013). The focus is on addressing the patient's medical needs and symptoms through direct, concise, fact-based discussions with the patient (Crawford and Brown 2011). Family or community involvement is not assumed, and patients are encouraged to make timely decisions about their care based on the information offered by the providers. Those who disagree with recommendations or fail to follow through on discharge instructions may be considered "non-compliant", which can lead to further estrangement (Weiner 2021).

General Recommendations

- Engage with patients and their families in a respectful and open manner.
- Avoid pathologizing terms (e.g., "enmeshment" versus the values of familism, allocentrism, and interdependence).
- Encourage family participation that aligns with the patient's needs and preferences.

Individualism

- Communicate with patients directly (unless another personal preference is indicated).
- Be open to obtaining less input from - or having separate conversations with - caregivers or family members.
- Expect an individual-focus of symptom reporting and coping strategies.
- If follow-up occurs, direct communication to the patient directly.

Collectivism

- Ask collateral/family members or friends for input on symptoms and during follow-up treatment (with patient consent).
- Ask about the patient's role within the family and community and consider how these roles influence the patient's values and goals for treatment.
- Consider how indirect communication strategies may be affecting information relayed by the patient.
- Be aware of possible underreporting of symptoms and somatic presentations of emotional symptoms.

Fig. 5.3 Individualism and collectivism in patient care

Collectivism

A collectivistic perspective emphasizes the interconnectedness of individuals, particularly with families and communities (Fatehi et al. 2020). Sometimes referred to as familism, collectivism views the self as an integral part of a larger group, highlighting strong bonds of "attachment, responsibility, and reciprocity" among family members (Sagbakken et al. 2018). This perspective is prevalent in Asian (Mahoney et al. 2005), Black (Berwald et al. 2016; Mahoney et al. 2005), Hispanic/Latino (CDC 2012; Hanssen and Tran 2019; Mahoney et al. 2005), Pacific Islander (Braun et al. 2015), Goan communities from India (Hanssen and Tran 2019; Patel and Prince 2001), and some Middle Eastern communities (Fatehi et al. 2020; Al-Amrani 2021).

In these cultures, communication and decision-making often rely on indirect methods, such as use of non-verbal cues, avoiding specific topics or details, and deferring to others for harmony (Fujii 2018). Patients and their families may minimize symptoms or compensate for difficulties to maintain a sense of well-being and cohesion (Berwald et al. 2016; Crist et al. 2009). These cultural nuances can influence interactions between patients and healthcare providers in acute care and rehabilitation settings. Healthcare providers should recognize that patients may not verbalize their concerns directly and are encouraged to engage in thoughtful communication to determine what information is most pertinent to an individual's decisions about their medical care (Hawley and Morris 2017; Zisman-Ilani et al. 2023).

Symptom presentation and reporting also differ in collectivistic cultures. Fujii (2018) reports that Japanese patients, influenced by Eastern philosophies, religions, and values that emphasize appropriate conduct, deference to elders and authority figures, and stoic endurance of hardships, may avoid questioning or disagreeing with providers out of politeness and respect. Emotional symptoms might be minimized due to cultural stigma, with somatic symptoms sometimes masking underlying emotional distress (Takemura et al. 2011). Moreover, patients from collectivistic cultures might be mindful particularly of how their symptoms affect their family or community (Derrington et al. 2018). Therefore, it is crucial to consider a patient's role within their family when providing healthcare services in the acute care and inpatient rehabilitation settings. For example, if a patient is a family matriarch responsible for meal preparation, recommending that she avoid using the stove for safety reasons might inadvertently threaten her valued family role. Other family members may feel they have no authority to restrict her activities in the kitchen. In such cases, involving family members in the assessment process and collaborating on care plans can lead to recommendations that respect cultural roles while promoting safety, such as identifying alternative cooking methods or offering additional training for family members who assist with cooking.

It is important to be aware of the ways in which patients from collectivistic cultures may feel alienated or disrespected within the US healthcare system. Research has found that communities of color report lower involvement in healthcare decision-making because of their experiences of discrimination, language barriers, and socio-demographic risk factors (Hall and Heath 2021; Hawley and Morris 2017). Some Hispanic communities, for example, may view medical care as potentially conflicting with traditional family-centered care (Hanssen and Tran 2019; Mahoney et al. 2005) and may have concerns about institutional or long-term care placements. Cognitive assessments can be particularly daunting if there are fears of institutionalization. Neuropsychologists may be able to alleviate these fears by building rapport with patients and their family members before beginning a neuropsychological assessment and ensuring that they understand the purpose of personal questions and how the information will be used. When neuropsychologists position themselves as advocates and trusted members of the treatment team, patients and families are more likely to share their concerns and be receptive to recommendations. By integrating cultural knowledge into neuropsychological care, providers can develop inclusive and effective treatment plans that align with and respect the patient's cultural beliefs and values.

Illness Perception

Illness perception refers to personal understanding and beliefs about various aspects of an illness, including its cause, timeline, controllability, potential cures, seriousness, and consequences (Broadbent et al. 2015; Leventhal et al. 2016; Rosenstock 1974).

It is a multifaceted concept shaped by cultural beliefs, values, and personal experiences (Okoro 2023). Research indicates that illness perception is closely related to health behaviors, which can vary across different cultures and nationalities (Abubakari et al. 2013; Alyami et al. 2020; Patel et al. 2016) and is, therefore, crucial when addressing health behavior changes in patient care. While a detailed review of illness perception models is beyond the scope of this chapter, Table 5.1 presents several of the widely known models.

Studies have found ethnic differences in illness perception and management of chronic health conditions such as diabetes (Abubakari et al. 2013; Shiyanbola et al. 2022), hypertension (Levine et al. 2020), kidney disease (Kim et al. 2012), and neurodegenerative conditions (Lee et al. 2022) as well as engagement in preventive healthcare practices (Fenton et al. 2009). For example, among adults with Type 2 Diabetes in the United Kingdom, Abubakari et al. (2013) found that Black-Caribbean and Black-African individuals were more likely to perceive their condition as being short term and benign compared to White-British adults. Black-Africans also identified fewer negative consequences from the illness than other groups. In the United States, focus groups with African American adults with Type 2 Diabetes found that beliefs about poverty, provider intentions, and stigma were associated with avoidance and denial about the condition. There was also concern that diabetes management would interfere with important family mealtime events and cultural practices (Shiyanbola et al. 2022; Shiyanbola et al. 2018).

Cultural differences in illness perception also extend to neurologic conditions where individuals and their family members may offer subjective interpretations or moral attributions for cognitive, behavioral, or psychiatric symptoms (Gray et al. 2009; Patel and Prince 2001; Withers et al. 2019). Some communities perceive dementia as a "white person's illness," which can lead to disparities in medical interventions and care (Berwald et al. 2016). The normalization of cognitive decline in older adults is more common among Black, Asian, and Hispanic communities, potentially contributing to decreased engagement in medical interventions and genetic testing among these groups (Alzheimer's Association 2024; Chin et al. 2011; Gray et al. 2009; Withers et al. 2019).

Identification of Illness Perception

Standardized assessments such as the *Brief Illness Perception Questionnaire* (Broadbent et al. 2006) and the *Revised Illness Perception Questionnaire* (Moss-Morris et al. 2002) have been developed to evaluate illness perception, although cross-cultural validity and reliability of these measures is unclear. Research indicates that illness perceptions can be modified through open conversation, education, and culturally sensitive approaches (Hu et al. 2014; Sawyer et al. 2019). In the fast-paced acute care and inpatient rehabilitation settings, neuropsychologists may have the most opportunity to establish rapport and gather insights into healthcare beliefs as they engage with patients and their family members to

Table 5.1 Models of illness perception

Model	Description
Commonsense Model of Illness Perception	Individuals form mental representations of their illness by drawing on readily available information, which helps develop common sense beliefs about their experiences.[a, b]
Self-Regulation Model	Individuals seek to manage perceived threats related to illness by targeting both objective features (e.g., specific symptoms) and the resulting emotional impact (e.g., stress associated with these symptoms). By integrating external information about a disease with their personal experiences, individuals derive meaning from their illness across five dimensions: illness identity (the label given to the disease), timeline (the course of symptoms), consequences (the values and roles at risk due to the disease), causes, and cure/controllability.[c]
Health Belief Model	An individual's beliefs and perceptions about a health issue influence their decision to engage in health-promoting behaviors. Components include perceived susceptibility, severity, benefits, barriers, cues to action, and self-efficacy.[d]

[a] *Hagger et al., 2017;* [b] *Okoro et al., 2023;* [c] *Leventhal et al., 2016;* [d] *Rosenstock, 1974*

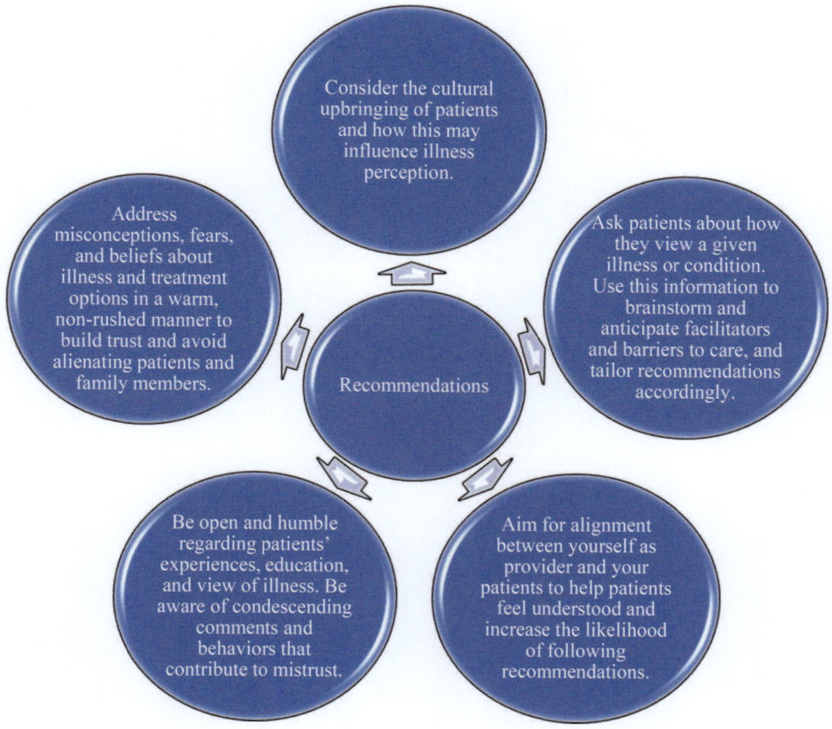

Fig. 5.4 Addressing illness perception

discuss interventions, assessments, and results. By identifying misconceptions, fears, and beliefs about illness and treatment options, offering education and resources in an open manner, and collaborating with patients and family members on strategies to promote health behaviors, neuropsychologists can develop recommendations that address the patient's genuine concerns and be implemented effectively after discharge (Fig. 5.4).

Health Literacy

Personal health literacy is defined as "the ability to find, understand, and use information and services to inform decisions regarding health care for oneself and others" (CDC 2020). It is influenced by an individual's skills, abilities, and access to health information; the complexity of health information; and the response of healthcare organizations to the needs of the individual. To this end, organizational health literacy refers to "the degree to which organizations equitably enable individuals to engage in personal health literacy" (CDC 2020) and research suggests that the ability to participate in care depends on how easy it is for patients to understand and act on information provided by the hospital (Glick et al. 2019).

Limited health literacy affects 77 million adults in the United States, or 36% of the population; moreover, limited health literacy contributes to health disparities and is related to income, race, ethnicity, and age (CDC 2009; Chesser et al. 2016; Glick et al. 2019; Yin et al. 2009). A nationwide survey of patients at federally qualified health centers and large/urban medical centers found that most responders (one-third of whom had limited English proficiency) identified health literacy challenges due to difficulty understanding written healthcare information and lack of confidence completing medical forms by themselves (Wynia and Osborn 2010). Other studies indicate that health literacy is affected by cultural factors such as communication barriers, mistrust in healthcare, and provider misunderstandings of patient expectations and needs (Baumeister et al. 2021). Older adults are disproportionately affected by poor health literacy (Chesser et al. 2016).

Fortunately, there is evidence that interventions targeting poor health literacy are beneficial (Fig. 5.5). Mahajan et al. (2022) created a virtual shared medical appointment program for older adults and their family members aimed at improving provider-patient communication and increasing patient knowledge about management of cerebrovascular disease. Feedback from participants was overwhelmingly positive, with most indicating they would recommend the program to others. Other recommendations for improvement include assessing patient health literacy and identifying needs, simplifying information, offering information in formats that are most accessible to patients, investing in organizational health

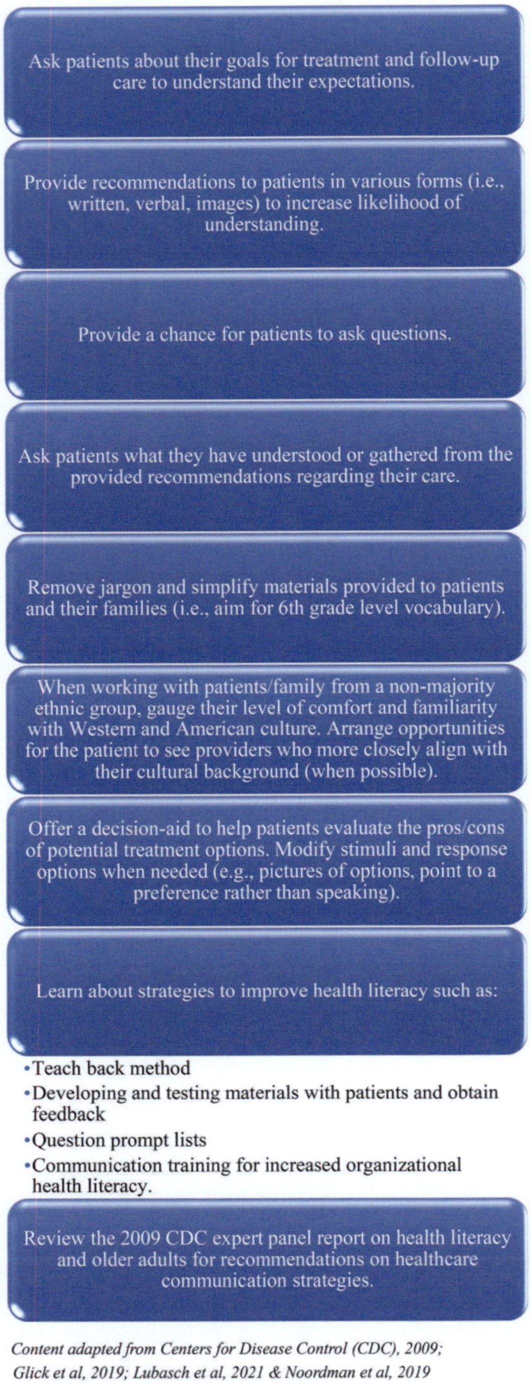

Fig. 5.5 Targeting and improving health literacy

literacy, and promoting ethnic concordance between providers and patients (Baumeister et al. 2021; Diamond et al. 2019). Given that the quality of communication between providers and patients is important, enhancing patient-centered communication in healthcare organizations could provide an opportunity to improve patient outcomes.

In an acute rehabilitation setting, various health literacy issues may arise. For example, consider an East Asian patient who is having difficulty following a prescribed dysphagia diet and is not eating well in the hospital. The family brings in congee or jook—a rice porridge dish customarily consumed during illness—and the patient eats this familiar and comforting dish. The rehabilitation team is alarmed to learn that the patient consumed food that did not meet the prescribed diet. In this situation, there is a gap between the patient and family's understanding of the risks of aspiration and the cultural practice of offering congee to those who are unwell. If handled in a culturally insensitive manner by the treatment team (i.e., interpreting this as an example of non-adherence to medical recommendations), the event could lead to conflict, alienation, and mistrust. Instead, team members could appreciate the significance of offering congee as a gesture of care and respect and provide education about dysphagia and diet recommendations in a respectful manner that acknowledges this important cultural factor. In this example, the neuropsychologist could facilitate discussion among the patient, family, and speech therapist to explore ways to adjust the consistency and temperature of the congee, so it is safe for the patient to consume. The neuropsychologist could discuss other cultural values and health knowledge to identify additional opportunities for education and modification of recommendations that enhance the patient's care.

Neuropsychologists can play a vital role in identifying patients with limited health literacy by asking patients and family members about their language proficiency, determining their understanding of the patient's medical condition and treatment, and facilitating access to educational materials and other team members who can support a treatment plan that considers psychosocial and cultural information. Information about the patient's cognitive status from a neuropsychological assessment will inform ways in which educational material should be discussed/shared with the patient, and whether family members will need to be responsible for remembering and implementing the recommendations after discharge. Noordman et al. (2019) offer strategies to facilitate communication with patients who have low health literacy, such as using jargon-free language, employing the teach-back method, involving patients in the development and testing of materials, providing patient decision aids, and using question prompt lists. Glick et al. (2019) provide detailed recommendations on health literacy-friendly strategies for patients and their families, and Lubasch et al. (2021) offer suggestions for communication training to build organizational health literacy. The Agency for Healthcare Research and Quality (AHRQ), among other organizations, offers toolkits to guide providers on the use of health literacy tools with patients (AHRQ 2020).

Disability

Disability refers to physical, emotional, and/or cognitive impairments which hinder daily activity and engagement in one's environment (CDC 2024). While this definition considers physical and functional aspects of disability, it does not incorporate individual strengths and personal growth that can follow a medical or neurological event. Therefore, different models have been proposed to explain how the experience of disability is affected by individual factors (e.g., moral and medical model), environmental factors (e.g., social model), and the interaction between an individual and their environment (e.g., WHO ICF model). While disability is a unique aspect of identity that is often conceptualized as an individual-level factor, we propose disability also as a contextual-level factor because of the system-level approaches and biases that may shape the individual's experience. While not exhaustive, Table 5.2 presents a selection of prominent models of disability that are widely discussed in the field.

The experience of disability is shaped by cultural, social, and systemic influences. Negative attitudes toward disability are common in many cultures and such attitudes mask unhelpful beliefs about the causes of illness/injury, as well as recovery and future well-being (e.g., reflection of punishment for wrongdoing, personal weakness). These societal attitudes toward disability may be internalized by patients and/or their family members, leading to feelings of worthlessness and low self-esteem, which can interfere with the patient's ability to participate in their recovery and access resources following discharge. For example, perceived stigma during the early phase following acquired neurological injury independently predicts long-term community reintegration (Ownsworth et al. 2024). Given its relationship with psychological well-being and social participation, perceived stigma should be assessed and considered in treatment planning (Poritz et al. 2019). For example, a patient from a culture that views illness and disability as signs of divine displeasure may experience overwhelming guilt and shame after a stroke, perceiving it as a fair punishment for wrongdoing. The patient may believe they have brought suffering upon themselves, or their family, and that they deserve their fate. Moreover, the family may avoid seeking assistance and resources due to the shame associated with the patient's illness. In such cases, the neuropsychologist may face challenges engaging the patient in assessments and motivating both the patient and family to follow recommendations, as attempts to improve the patient's condition might be viewed as futile or even contrary to divine will. Conversely, cultural beliefs that see illness and hardships as opportunities for perseverance and growth can foster a sense of purpose, determination, and engagement in treatment among patients and their families.

Disability-related bias is found among healthcare providers as well as patients and family members (Meidert et al. 2023), although providers tend to deny such prejudicial beliefs. VanPuymbrouck et al. (2020) found implicit preferences for non-disabled individuals to be most notable among older providers, male providers, and East Asian providers. Negative explicit

Table 5.2 Models of disability

Model	Description
Moral Model	Views disability as a moral failing, punishment for wrongdoing, or karma. Places blame and responsibility on the individual with a disability. Leads to stigmatization, discrimination, and exclusion. [a]
Medical Model	Perceives disability as a defect of a body function. Focuses on the impairment or condition itself. Emphasizes the need for medical interventions, treatments, and rehabilitation to fix or cure the disability.[a]
Social Model	Sees disability as a result of societal barriers and discrimination. Recognizes that disability is influenced by physical, social, and attitudinal barriers. Views disability as a social construct created by society's organization and beliefs. Emphasizes the need to remove barriers and create an inclusive and accessible society. Requires collective action and societal change. Highlights the importance of addressing systemic issues rather than solely focusing on individual impairments.[b]
Diversity Model	Embraces and honors the wide range of experiences and identities within the disability community. Encourages a sense of pride. Celebrates connection and identity within the disability culture. Proponents often prefer to identify as "disabled" individuals as opposed to person-first language. Promotes the idea that disability is a social identity that intersects with other aspects of a person's identity.[c, d, e]
World Health Organization's (WHO) International Classification of Functioning Disability and Health (ICF) Model	Framework for understanding health and disability from a biopsychosocial perspective. Recognizes that health and disability are influenced by a complex interaction of biological, psychological, and social factors. Consists of two main components: functioning and disability (further divided into three domains: body functions and structures, activities and participation, environmental factors), and contextual factors (divided into personal factors and environmental factors). Emphasizes the importance of considering the broader social and environmental factors that can either enable or restrict an individual's participation and inclusion in society. Widely used in healthcare, research, and policy development.[f]

Olkin, 2017[a]; Oliver, 2013[b]; Altman, 2001[c]; Olkin, 2002[d]; Andrews, 2020[e]; WHO, 2002[f]

biases were more prevalent among Black, East Asian, South Asian, and Native Hawaiian/Pacific Islanders and less so among Indigenous and Latinx providers, compared to White providers. The relationship between implicit bias and healthcare treatment is unclear, with some studies finding no differences and others suggesting that treatment decisions may be more closely related to group affiliation (Meidert et al. 2023). Regardless, these findings reinforce the importance of neuropsychologists understanding their own explicit and implicit disability biases to ensure equitable care. Without this awareness and self-reflection, neuropsychologists may underestimate the cognitive and functional potential of individuals with disabilities seen in acute care and rehabilitation settings. They may fail to account for the unique sensory, motor, and communication challenges faced by individuals with disabilities, leading to biased interpretations of standardized neuropsychological tests and inaccurate conclusions about a patient's cognitive status. Additionally, there is potential for bias in assuming that disability in one domain extends to disability in other areas (Dembo et al. 1975). For example, a provider may incorrectly assume that a patient with language impairment also has other cognitive deficits, or conversely, that a patient with intact language abilities has relatively preserved memory, reasoning, and problem-solving skills. Providers may also presuppose that individuals with disabilities are unhappy, depressed, or less satisfied with their lives simply because of their disability, overlooking the person's actual emotional state or personal experience.

Every individual, including someone with a disability, has the right to make choices about their own life, even if those choices involve some level of risk. Taking measured risks is a fundamental part of living a full and meaningful life, yet the "dignity of risk" is often dismissed by ableist approaches to healthcare that overlook the importance of autonomy and capacity for self-determination (Banner 2012; Stanley et al. 2022). Social perceptions that paint individuals with disabilities as inherently vulnerable or incapable of managing their own care can impact a provider's judgment regarding discharge recommendations, such as the level of ongoing care and support necessary for a patient's safety and well-being. Neuropsychologists should be aware of these biases and consider ways in which recommendations can reinforce strengths and support autonomy. For example, rather than assuming that 24-h supervision is the only appropriate discharge recommendation for a patient with cognitive impairment, a neuropsychologist can engage in discussions with the patient and family members about realistic and practical ways to enhance safety while fostering independence for the patient and minimizing undue stress and strain on family relationships. Options might include intermittent supervision with telephone checks, a medical alert button, and indoor cameras in common areas (e.g., kitchen, hallway, stairs, living room); scheduling activities that increase the risk for falls (e.g., personal care, toileting) when a caregiver is present; and preparing meals for the patient so they do not need to cook without supervision. Locating living quarters on the ground floor can reduce the risk of falls and homecare therapies may provide additional support immediately after discharge. Community resources such as neighbors or volunteer groups from places of worship may also be able to visit the patient during the day. This collaborative approach results in discharge recommendations that reflect the realities of patients' lives, prioritize cultural and personal preferences, and empower patients to thrive in their communities.

In summary, neuropsychologists who are aware of cultural biases regarding disability can anticipate ableist beliefs internalized by patients and family members, creating opportunities to offer education, support, and guidance in a warm and open manner. This approach helps to minimize misconceptions and assists patients with developing skills to navigate stigma that negatively affects self-esteem, sense of belonging, and social interactions. Sensitivity to cultural norms surrounding disability is crucial for providing interventions and recommendations that are accepted and respected by patients. Neuropsychologists are encouraged to learn about their own explicit and implicit biases with respect to disability and take action to minimize the effects on patient care. Ways to self-reflect on disability bias are included in Fig. 5.6 and outlined in the APA Guidelines for Assessment and Intervention with Persons with Disabilities (APA 2022).

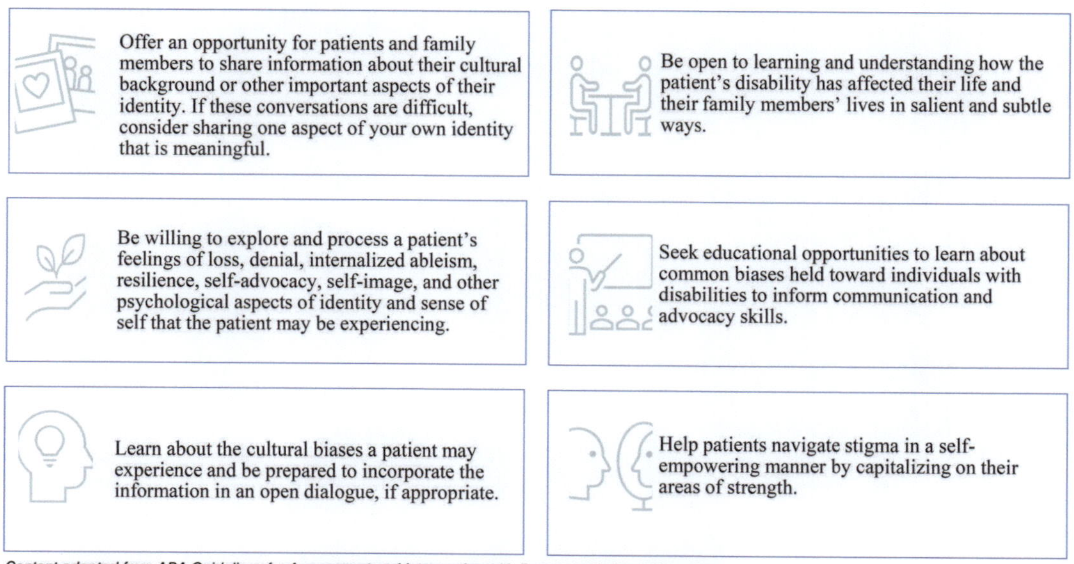

Content adapted from APA Guidelines for Assessment and Intervention with Persons with Disabilities, 2022

Fig. 5.6 Disability awareness

Aging

According to the U.S. Census Bureau (2023), the adult population is expected to double from 46 million to 98 million by 2060, with adults aged 65 and older living longer and experiencing more chronic health conditions. Longer life expectancy has resulted in higher demand for hospitalization and rehabilitative care for older adults, requiring providers to adapt to the unique needs and complexities of this population. Neuropsychologists working in adult acute care and rehabilitation settings should be competent in assessing and treating older adults (APA 2024), given the nuances of how aging affects healthcare engagement (Fig. 5.7). Similar to the discussion of disability, the authors acknowledge that aging is a unique and individual process. We also recognize the influence of system-level and societal views on aging and how this affects patients and healthcare providers in the inpatient setting. Thus, we consider aging as a contextual-level factor of diversity.

Ageism, or discrimination and prejudice against older individuals, has negative effects on patient well-being, healthcare utilization, and patient outcomes (APA 2020). Aging also brings sensory changes and functional limitations that interact with external barriers (e.g., inaccessible physical environments) and discriminatory attitudes and practices to limit older adults' access to healthcare and support services, and contributes to social exclusion (Gwinnutt et al. 2023). Unfortunately, ageism is a culturally accepted prejudice in the United States (Weir 2023) and the biased mindset is evident even in treatment facilities specifically geared toward older adults (Heyman et al. 2020; Gaynor et al. 2014). Research finds that the failure to appreciate and recognize older adults' unique strengths and skills in these settings, along with the application of biased and stereotypical beliefs about aging, contribute to discriminatory practices. For example, the stereotypical belief that older adults are frail and dependent lowers expectations for rehabilitation potential, which in turn leads to limiting access to treatments and interventions.

The "paradox of aging" refers to the fact that many older adults lead fulfilling and active lives (Li et al. 2022). Psychological and emotional resilience develops over time, and older individuals accumulate experiences, knowledge, and coping strategies (Bowling 2007; Merchant et al. 2022) that support well-being when an individual is adjusting to chronic illness, disability, decline in physical or cognitive abilities, and social isolation. Interventions that support positive emotional growth in later life include making sense of events (Janoff-Bulman and Yopyk 2004), finding benefit in experiences (Pakenham and Cox 2009), cognitive reappraisal, positive reframing, seeking support, and posttraumatic growth (Calhoun 2001). Individuals

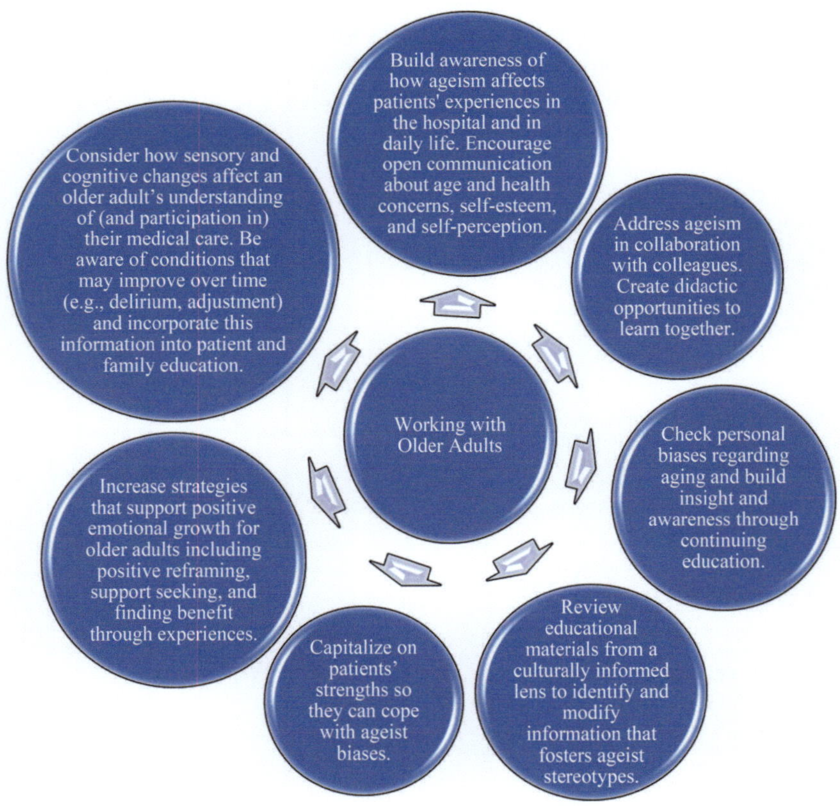

Fig. 5.7 Working with older adults

who can change priorities in response to age-related stressors, such as focusing on meaningful relationships and personal interests rather than external achievements or material possessions, are better equipped to face the challenges that come with older age (Waldinger and Schulz 2023). We recommend that inpatient neuropsychologists become familiar with these interventions and use a strength-based approach to assessment that considers how patients have historically coped with stressors. Spending time with the patient (and family) after the assessment to review recommendations and brainstorm ways in which the patient's personal and social strengths can offset limitations to support higher quality of life is also encouraged.

Specialized areas of neuropsychological care for older adults in acute care and rehabilitation settings include capacity evaluations and assistance with end-of-life decisions and planning. While a comprehensive discussion of culturally informed practice in these areas is beyond the scope of this chapter, it is important to realize that cultural beliefs and values inform preferences regarding disclosure of diagnosis, decisions regarding medical care, the role of family in palliative and end-of-life care, and the process of a "good death" (Evans and Ume 2012; Saeed et al. 2023). Neuropsychologists should be aware of these cultural norms to provide compassionate, culturally competent capacity assessments and patient-centered end-of-life care.

INDIVIDUAL FACTORS

In the previous section, we reviewed contextual factors that characterize the general beliefs, lifestyles, and experiences of various cultural groups. While broad factors provide a framework for understanding health-related behavior, it is important to recognize that generalizing about individuals based on group affiliation can be reductionistic and may lead to stereotyping. Cultural competency involves appreciating the complex interplay between cultural context and individual factors, including each person's unique abilities, skills, challenges, and preferences. In this next section, we offer an overview of individual considerations that are critical to providing culturally component acute care and rehabilitation neuropsychological practice.

Identity Language

Discrimination against transgender individuals is a common issue in healthcare settings, leading many to delay seeking care (Jaffee et al. 2016). To provide effective care, neuropsychologists must be culturally informed and respectful of transgender, gender-expansive, and intersex individuals' use of gender pronouns, names, and courtesy titles. Unfortunately, electronic health records may not reflect a patient's identified name, gender, or pronoun, leading to a practice known as "deadnaming" where a person is called by the name used prior to transitioning, usually their birth or given name (Lau et al. 2020). This can be distressing or alienating for individuals and negatively affect the therapeutic relationship and trust between patients and providers. Using an individual's identified name and asking for the pronouns they use is recommended, rather than making assumptions based on stereotypes or assigned sex. Alternatives like "Mx." (pronounced *em eks*) or "M." instead of "Mr." or "Ms." can be appropriate if providers are uncomfortable using first names. The use of gender-neutral pronouns like "they/them/theirs" is generally accepted, and some providers display their own pronouns to support open discussions about gender identity. We also recommend avoiding phrases that may misgender others. For example, rather than saying "Good Morning Ma'am" or "ladies and gentlemen," a neuropsychologist might say "Good day" or "Hello everyone" to avoid gender-specific assumptions and possible alienation.

Identity language and disability is complicated and varies among individuals, disability communities, and advocacy groups. The general recommendation is to use person-first language (e.g., "patients with diabetes" or "people with mental illness") rather than terms that refer to the medical condition (e.g., "diabetic" or "mentally ill") to avoid medicalizing and dehumanizing individuals with these conditions. However, there are questions about whether person-first language reinforces negative connotations about medical conditions and disability by assuming that individuals feel ashamed or want to distance themselves from the condition (Gernsbacher 2017). Some disability communities advocate for Identity-first language such as "autistic person" or "deaf person," because this language places the condition at the center of identity as a source of pride and community, rather than a condition that requires treatment (Whyte et al. 2013). Interestingly, the identity language individuals use may not be consistent with healthcare provider preferences. For example, studies find that autistic adults generally prefer Identity-first language, while providers generally prefer Person-first language (Taboas et al. 2023), and this difference contributes to a sense of alienation for autistic people. We encourage neuropsychologists to review recommendations provided by disability communities and talk directly with patients about their preferred identity language. These conversations can foster more collaborative patient-provider relationships and demonstrate respect and understanding for the complexity of identity language (Fig. 5.8).

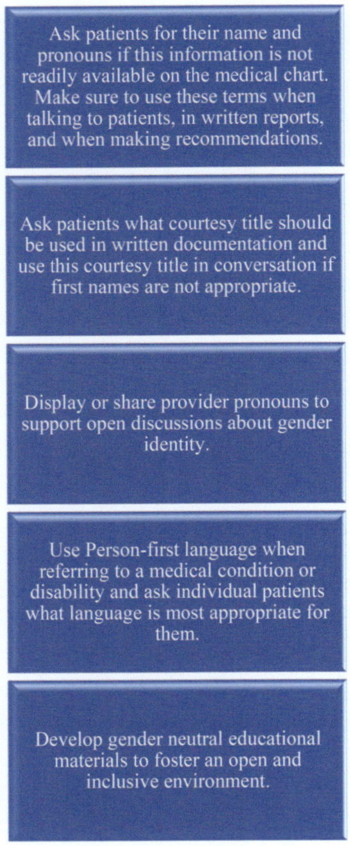

Fig. 5.8 Incorporating identity language

Limited Language Proficiency (LEP)

Limited English proficiency (LEP) is common in the United States, with approximately 68 million people speaking a language other than English at home and 25 million people reporting they do not speak English well (U.S. Census Bureau 2024). LEP is associated with poorer understanding of medical conditions such as stroke, negative rehabilitation experiences, and lower quality of life after discharge from the hospital (Clark et al. 2022). LEP is also related to higher emergency department and hospital admissions (Njeru et al. 2015; Squires et al. 2022). Indeed, language barriers during the provision of both routine nursing care (e.g., medication administration, pain assessments) and urgent medical intervention (Van Rosse et al. 2016) can increase the risk of negative outcomes as communication about important health information can be affected. Fortunately, language-concordant services are associated with improved quality of care, satisfaction with care, understanding of medical information, and communication with healthcare providers (Diamond et al. 2019). The use of professional interpreters can improve a patient's understanding of their diagnosis and prognosis, as well as optimize the management of symptoms such as pain and anxiety (Silva et al. 2016).

Interpreter Services and Improving Communication

Early detection and rapid accommodations to mitigate language barriers (e.g., interpreter services, simplified language, use of gestures) by the medical team have been found to be helpful for patients and family members with LEP (Maletsky et al. 2023). By identifying when interpreter services are needed, inpatient neuropsychologists can advocate for the use of these services to the medical team. Simply asking patients or family members if they need interpreter services is often not sufficient, as some individuals fluent in conversational English may lack the nuance needed to discuss complex healthcare information, reliably complete cognitive assessments, and make fully informed healthcare decisions. In addition, patients and family members with LEP may not feel comfortable requesting interpreter services. Neuropsychologists can practice culturally competent care for patients with LEP by conducting assessments using validated tools and appropriate norms, and by requesting translation of educational materials and neuropsychological reports into the patient's first language (Fig. 5.9).

In-person interpreters are recommended over remote or telephone-based services, when possible (Heath et al. 2023), especially for patients with sensory or cognitive-linguistic deficits such as dysarthria, dysphonia, apraxia, aphasia, hearing

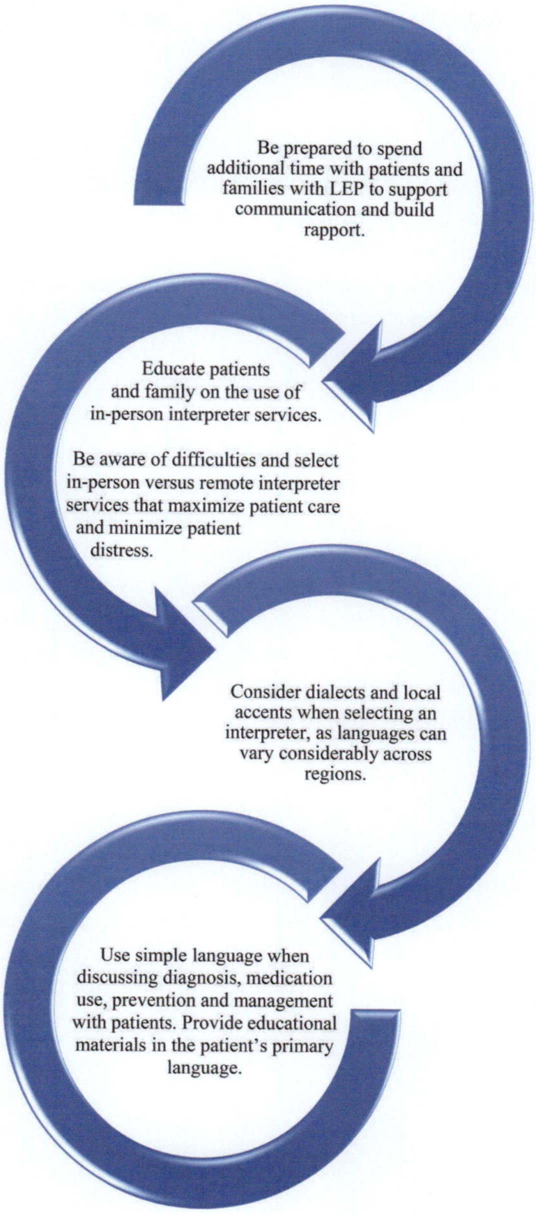

Fig. 5.9 Considerations for LEP and interpreter use

loss, low vision, and impaired attention/distractibility. However, in-person interpreters may not be available, or the interpreter may not be familiar with the patient's dialect or accent as languages vary considerably across regions. For example, a Spanish-speaking interpreter from Puerto Rico may not be able to communicate with a patient from a rural area of Mexico. Cultural factors also need to be considered with the use of in-person interpreter services. For example, a patient may not be willing to discuss symptoms with an interpreter from a different caste, social class, or sex; conflict in the country of origin may be relevant if the interpreter and patient are from different ethnic groups; and if the patient lives in a small cultural enclave, they may be hesitant to share information with an interpreter whom they could encounter in the community. In these types of situations, working with remote or telephone-based interpreter services may be less disruptive to care.

Neuropsychologists can promote the effective use of interpreter services in the acute care and inpatient rehabilitation setting by meeting with the interpreter before the session to explain the purpose of the encounter and their role (e.g., reinforce the importance of accurate interpretation rather than "correcting" a patient's mistakes) and review what questions may be asked and specific terminology that may arise. General information about the patient's cultural background can be discussed to identify potential conflicts or difficulties. The neuropsychologist can create an atmosphere that encourages open dialogue with the patient about interpreter services by asking the patient if they feel understood, have questions, or need clarification. When possible, it is also helpful to supplement verbal communication with visual aids, diagrams, or written materials in the patient's primary language to clarify complex information.

Cognitive Evaluation in the Context of LEP

Neuropsychological testing of patients with LEP requires validated and reliable tests, appropriate norms, and an understanding of whether the assessment process itself is culturally appropriate. Some measures are not valid or reliable for individuals with limited functional literacy such as dementia screening tools (Maher and Calia 2021) or LEP (Fujii 2017, 2018) and neuropsychologists must be aware of this information before conducting an evaluation. Readers are referred to Julayanont and Ruthirago (2018) for recommendations regarding assessment and communication strategies.

Factors such as the patient's country of origin, age of exposure to English, and use of English in different settings can help determine an individual's level of proficiency. Information about the patient's country of origin can clarify whether English is used in formal instruction/educational settings, and whether this differs by social class or socioeconomic status. The use of English may vary by region within a country (e.g., Belize, Jamaica, Philippines) and some countries may teach English reading and writing but not the spoken language (e.g., Japan). Age of exposure to English also affects proficiency. Children learning English as a Second Language (ESL) can build basic communication skills within a year, but they may require several more years to develop Cognitive and Academic Language Proficiency Skills (CALPS). Adults learning a second language may develop uneven skills such that they are able to communicate more effectively in social situations or in settings where they frequently use English (e.g., work, grocery store), and conversational abilities may not reflect a person's skills with reading and writing in English. When interpreter services are warranted, it is beneficial to educate interpreters on their role in test administration prior to beginning of an assessment. When available, working with an interpreter who has certified experience in neuropsychological assessment and rehabilitation can further ensure confidence in results and conclusions.

Social Determinants of Health

Understanding social determinants of health (SDOH) is important in acute care and rehabilitation because these factors significantly influence patients' recovery, access to care, outcomes, and overall well-being. SDOH refer to the "nonmedical factors that influence health outcomes" and "are the conditions in which people are born, grow, work, live, worship, and age" (Hacker et al. 2022; see Fig. 5.10). Health equity, health literacy, and SDOH are the three priority

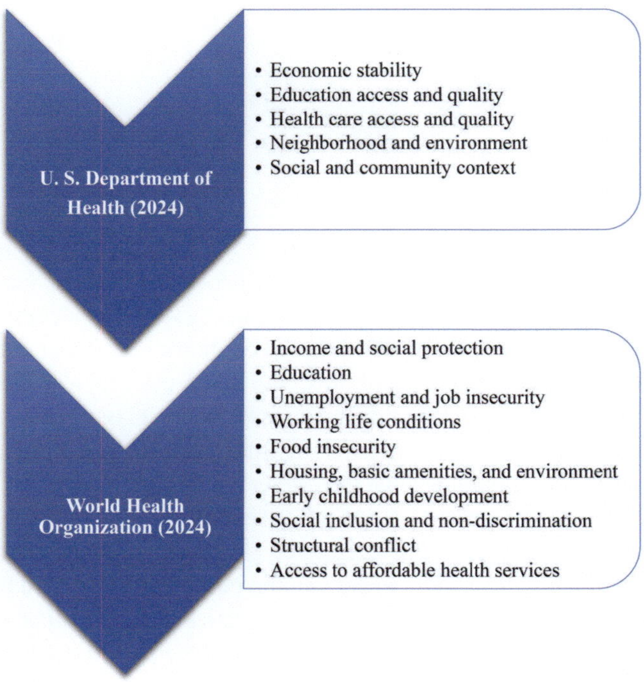

Fig. 5.10 Examples of social determinants of health

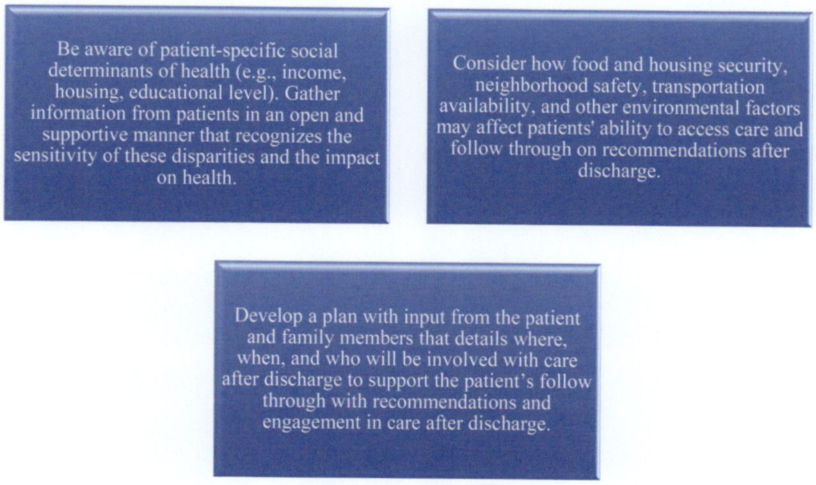

Fig. 5.11 SDOH in practice

areas for the Office of Disease Prevention and Health Prevention's Healthy People 2030 initiative. Socio-environmental disadvantages, such as financial instability, lower educational attainment, and housing insecurity, are generally associated with increased risk of morbidity and mortality (Compton and Shim 2015). Environmental factors also impact the risk of developing and maintaining disability. A study by Jaja et al. (2013) demonstrated a significant and inverse relationship between personal income and mortality risk among inpatient stroke patients in the United States. Higher socioeconomic status has also been associated with better long-term recovery, particularly with motor functioning and self-reported health status after inpatient rehabilitation (Wolf et al. 2022). Factors such as minoritized status, insufficient health insurance, lack of care support, and premorbid unemployment are independently associated with poorer functional recovery patterns at a 12-month follow-up after acute inpatient cardiac rehabilitation (Agarwal et al. 2022). On an individual level, unfavorable social conditions can limit available resources or options and increase the risk of poor health behaviors, such as unhealthy diets, substance use, and risky behaviors (Compton and Shim 2015). Common issues affecting attendance and sustained participation in cardiac and cancer rehabilitation, such as employment and income, social support, transportation, housing, and food security, can similarly affect neurological populations (Frier et al. 2017). For instance, Lundine et al. (2022) found that resources such as transportation, health insurance, childcare availability, and time off from work were instrumental in ensuring follow-up attendance for families of pediatric brain injury patients.

Neuropsychologists offer more equitable, effective, and holistic care when social determinants of health (U.S. Department of Health and Human Services, n.d.; Wickstrom et al. 2024) are integrated into acute rehabilitation (Fig. 5.11). This includes tailoring treatment recommendations to incorporate support services such as medical transportation assistance, resources for legal aid services, and access to affordable assistive devices or sliding-scale treatment centers. Addressing health inequities, enhancing social support, and allocating resources to patients can maximize their rehabilitation during inpatient admission and after discharge. This approach not only improves individual patient outcomes but also contributes to broader public health goals by addressing the root causes of health disparities.

A transdisciplinary approach involving various healthcare providers including case managers, social work services, neuropsychologists, physical/occupational/speech therapists, nurses, physicians, recreational therapists, and hospital chaplains can further assist with identifying and addressing obstacles to care as well as support patients in navigating financial, logistical, social, and emotional challenges commonly encountered in inpatient settings.

Family and Caregivers

Family and caregiver/care partner involvement is routine in acute care and inpatient rehabilitation settings. Culture plays a significant role in influencing a caregiver's willingness to provide care, approaches to caregiving roles, communication within the family about caregiving needs, the use of support services by patients and caregivers, and the process of regaining balance within the family system following an injury or illness (Sander et al. 2007). Understanding cultural aspects of family and caregiving is crucial not only for supporting assessment and treatment during the hospital stay but also for offering culturally informed discharge recommendations (Fig. 5.12). For example, some cultures (e.g., Asian and Hispanic) may have a

Fig. 5.12 Integrating families and caregivers

stronger inclination to assist a family or community member due to strong familial responsibility (Scharlach et al. 2008), familism, allocentrism, and interdependence (Zea et al. 1994).

However, other social factors, such as income and acculturation, may also contribute to family interdependence. For instance, Hooker et al. (2023) found that family income mediated the relationship between ethnicity and familial interdependence, but only for certain ethnic groups (i.e., European Americans and US-born Hispanic Americans), highlighting the complex relationship between ethnicity, social class, and familial interdependence. This research underscores the importance of examining cultural expectations and personal factors, which can shape the degree of available informal support resources and guide inpatient clinicians in recommending the most helpful and patient-centric interventions (e.g., family training). Taken together, sensitivity to the environmental context enhances healthcare providers' ability to be attuned to the diverse social circumstances of patients, to best meet their unique needs and promote functional outcomes.

CONCLUSIONS AND RECOMMENDATIONS

It is important to acknowledge that while progress has been made in recognizing and embracing diversity within inpatient acute and rehabilitation settings, there is still much work to be done to fully integrate cultural factors into a strengths-based approach to care. Although not exhaustive, this chapter seeks to initiate discussions on these topics, describing the disparities that exist between majority and minoritized groups and offering practical recommendations for culturally competent practice in acute care or rehabilitation settings. We emphasize the need to consider the unique and complex interplay between individual and contextual factors, such as cultural and community influences, which may intersect and affect patient care and outcomes during hospitalization, outpatient care, and beyond. We encourage providers to utilize the available resources to educate themselves and their fellow neuropsychology trainees and colleagues, fostering a sense of cultural humility and empathy towards diverse perspectives and preferences. By doing so, we believe we can strengthen the practice of neuropsychology and therapeutic alliance by delivering care that respects and considers diverse cultural backgrounds. In Table 5.3, we summarize recommendations for how neuropsychologists in inpatient practice can provide culturally informed and tailored care with the intent for all individuals to receive equitable and effective treatment.

Table 5.3 Recommendations for culturally sensitive practice

Topic	Clinical Recommendations	Related Resources
Pronoun Use	• Use pronouns and names as directed by the patient, including gender-neutral pronouns when appropriate (they/them/their, Mx. or M as alternative to Mr. and Ms.). Providers may disclose their own pronouns to foster an open dialogue. Select Person-first or Identity-first language based on group preferences within the disability community to encourage inclusivity.	• APA Inclusive Language Guide - https://www.apa.org/about/apa/equity-diversity-inclusion/language-guidelines • APA Guidelines for Psychological Practice with Transgender and Gender Nonconforming People - https://www.apa.org/practice/guidelines/transgender.pdf • APA Guidelines for Psychological Practice with Sexual Minority Persons - https://www.apa.org/about/policy/psychological-sexual-minority-persons.pdf
Illness Perception	• Understand how the patient perceives their illness (e.g., medical, moral, spiritual). • Be aware of potential discrepancies in health beliefs and address them openly and in a culturally informed manner.	• APA Health Disparities Fact sheet - https://www.apa.org/topics/racism-bias-discrimination/health-disparities-defined.pdf • APA Report of the 2021 American Psychological Association Presidential Task Force on Psychology and Health Equity - https://www.apa.org/pi/health-equity/report.pdf • National Institute on Minority Health and Health Disparities - https://www.nimhd.nih.gov/
Individualism and Collectivism	• Consider the patient's individualistic or collectivistic tendencies as part of their self-identified group, as these approaches may influence support system, help seeking, report of symptoms, and medication and/or treatment preferences.	• APA Guidelines on Race and Ethnicity in Psychology https://www.apa.org/about/policy/guidelines-race-ethnicity.pdf • APA Multicultural Guidelines: An Ecological Approach to Context, Identity, and Intersectionality - https://www.apa.org/about/policy/multicultural-guidelines.pdf
Caregiver/ social support	• Consider the role of caregiving and common coping strategies within the patient's cultural group and how these may be influenced by stressors and resources in family dynamics, specifically adaptability, cohesion, and communication. • Gather collateral information to understand the patient's baseline functional status and background. Be attentive to collateral reports of social and role changes. Be sensitive to questions regarding the patient's mental health history.	• APA Resources for caregivers of diverse populations and specific age groups - https://www.apa.org/pi/about/publications/caregivers/resources/populations • APA Resources for caregivers: Specific Health issues - https://www.apa.org/pi/about/publications/caregivers/resources/health-issues • APA caregiver resources - https://www.apa.org/pi/about/publications/caregivers/resources/psychologists?tab=1

Table 5.3 (continued)

	• Learn about the cultural aspects of family and caregiving (e.g., filial piety, interdependence, familism) to support assessment, treatment, and discharge recommendations. Use a tailored approach for caregiver education that incorporates patient preference regarding caregiver involvement. • Consider cultural disparities and family roles in advance directives, medical interventions, and resource allocation.	• Final Report of the APA 2010 Presidential Task Force on Caregivers - https://www.apa.org/pi/about/publications/caregivers/final-report-2010
English Proficiency	• Consider the patient's English proficiency and cultural factors when using an interpreter. • Gather information about the patient's country of origin and preferred mode of information provision (written, oral). • Discuss interpreter preferences with the patient, including use of in-person interpreters. Consider whether sensory and/or cognitive-linguistic deficits may limit the effectiveness of video or telephone interpreters. • Avoid relying on family members to interpret, as it may raise ethical concerns and assumes an advanced level of English fluency. • Informal assessments such as asking open-ended questions or asking patients to describe a picture in English and in their native language (for comparison of fluency) may help estimate English proficiency. Avoid using published tests if they are culturally biased.	• APA Guidelines for Psychological Assessment and Evaluation - https://www.apa.org/about/policy/guidelines-psychological-assessment-evaluation.pdf • US Department of Health and Human Services Limited English Proficiency website with resources - https://www.hhs.gov/civil-rights/for-individuals/special-topics/limited-english-proficiency/index.html • APA Guidelines for Providers of Psychological Services to Ethnic, Linguistic, and Culturally Diverse Populations - https://www.apa.org/pi/oema/resources/policy/provider-guidelines
Interpreter Use	• Consider the gender and age of the interpreter in relation to the patient's culture. • Whenever possible, share the name of the interpreter with the patient or family prior to the first meeting.	• Council of National Psychological Associations for the Advancement of Ethnic Minority Interests (CNPAAEMI) Testing and Assessment with Persons & Communities of Color - https://www.apa.org/pi/oema/resources/testing-assessment-monograph.pdf
	• Educate the interpreter about the purpose of the assessment or treatment and emphasize the importance of accurate interpretation.	• APA Ethical principles of Psychologists and Code of Conduct - https://www.apa.org/ethics/code

Table 5.3 (continued)

Socioeconomic Status	• Consider socioeconomic status when making treatment recommendations.	• APA Guidelines for Psychological Practice for People with Low-Income and Economic Marginalization https://www.apa.org/about/policy/guidelines-low-income.pdf • Report of the APA Task Force on Socioeconomic Status - https://www.apa.org/pi/ses/resources/publications/task-force-2006.pdf • APA Policy Statement on Socioeconomic Status: Resolution on Poverty and Socioeconomic Status - https://www.apa.org/about/policy/poverty-resolution • APA Office on Socioeconomic Status Resources and Publications - https://www.apa.org/pi/ses/resources
Health Literacy	• Use technology to improve health literacy and communication with older adults. • Consider health literacy when providing care. For example, use jargon-free language, employ the teach-back method, involve patients in the development and testing of materials, provide patient decision aids, and use question prompt lists (Lusbach et al., 2021).	• APA Guidelines for Psychological Practice for People with Low-Income and Economic Marginalization - https://www.apa.org/about/policy/guidelines-low-income.pdf • CDC Older Adults Health Literacy - https://www.cdc.gov/healthliteracy/developmaterials/audiences/olderadults/index.html
Ageism	• Avoid ageist terms and stereotypes when referring to older adults. • Recognize diversity within the older adult population and take a strengths-based approach rather than focusing on deficits.	• APA Inclusive Language Guide - https://www.apa.org/about/apa/equity-diversity-inclusion/language-guidelines • APA Guidelines for the Evaluation of Dementia and Age-Related Cognitive Change - https://www.apa.org/practice/guidelines/guidelines-dementia-age-related-cognitive-change.pdf • APA Guidelines for Psychological Practice with Older Adults - https://www.apa.org/practice/guidelines/older-adults • ASPE Office of Behavioral Health, Disability, and Aging Policy https://aspe.hhs.gov/about/offices/bhdap • Administration for Community Living - https://acl.gov/

Table 5.3 (continued)

Disability	• Understand the culture of disability and adjust your approach accordingly. • Familiarize yourself with different types of disabilities and their effects on patients' lives. • Take a collaborative transdisciplinary team approach to support the decision-making capacity assessments of people with disability. • Explore testing accommodations, such as use of communication aids, picture-based tools, functional everyday tasks, and behavioral observations. • Support a patient's disability identity development by emphasizing their strengths when summarizing neuropsychological findings in reports. • Recognize that disability identity development is a personal process and adapt therapeutic techniques accordingly.	• APA Inclusive Language Guide - https://www.apa.org/about/apa/equity-diversity-inclusion/language-guidelines • APA Guidelines for Assessment and Intervention with Persons with Disabilities - https://www.apa.org/about/policy/guidelines-assessment-intervention-disabilities.pdf • World Health Organization International Classification of Functioning, Disability and Health (ICF) - https://www.who.int/standards/classifications/international-classification-of-functioning-disability-and-health • WHO ICF Practice Manual - https://cdn.who.int/media/docs/default-source/classification/icf/drafticfpracticalmanual2.pdf?sfvrsn=8a214b01_4&download=true • ASPE Office of Behavioral Health, Disability, and Aging Policy - https://aspe.hhs.gov/about/offices/bhdap • Administration for Community Living - https://acl.gov/
General	• Be aware of, reflect on, and understand your own cultural characteristics and biases. • Be aware of stigmatizing views and biases that may affect the quality of care. • Be aware of the juxtaposition of the medical model of rehabilitation in the US and the conventional rehabilitation practices that promote values such as personal choice and self-reliance.	• APA Multicultural Guidelines: An Ecological Approach to Context, Identity, and Intersectionality - https://www.apa.org/about/policy/multicultural-guidelines.pdf • APA Health Disparities Fact sheet - https://www.apa.org/topics/racism-bias-discrimination/health-disparities-defined.pdf • APA Report of the 2021 American Psychological Association Presidential

Table 5.3 (continued)

	• Consider how cultural lenses influence health information interpretation and treatment preferences. • Address cultural challenges related to disparities in healthcare access and services (e.g., admission rates and access to institutional care, inpatient rehabilitation, skilled nursing facilities, home health services, and outpatient therapies). • Collaborate with other disciplines such as physical, occupational, recreational therapists, speech and language pathologists, nurses, social workers, chaplains, and physicians to address deficits and set patient-centered goals.	Task Force on Psychology and Health Equity - https://www.apa.org/pi/health-equity/report.pdf • National Institute on Minority Health and Health Disparities - https://www.nimhd.nih.gov/

REFERENCES

Abubakari AR, Jones MC, Lauder W, Kirk A, Anderson J, Devendra D, Naderali EK. Ethnic differences and socio-demographic predictors of illness perceptions, self-management, and metabolic control of type 2 diabetes. Int J Gen Med. 2013;6:617–28.

Agarwal S, Shen EQ, Tincher IM, Abukhadra SL, Roh D, Park S, et al. Acute inpatient rehabilitation and social determinants of health are associated with functional recovery patterns at 12 months relative to hospital discharge from cardiac arrest. Circulation. 2022;146(Suppl_1):A227.

AHRQ Health Literacy Universal Precautions Toolkit. Agency for healthcare research and quality. 2020. Available from: https://www.ahrq.gov/health-literacy/improve/precautions/index.html

Al-Amrani, S. N.. The impact of Arab collectivist culture on willingness to communicate in a second language: a critical review. 2021. Available at SSRN 3926968.

Altman BM. Disability definitions, models, classification schemes, and applications. In: Handbook of disability studies. London: Sage; 2001. p. 97–122.

Alyami M, Serlachius A, Mokhtar I, Broadbent E. The association of illness perceptions and God locus of health control with self-care behaviours in patients with type 2 diabetes in Saudi Arabia. Health Psychol Behav Med. 2020;8(1):329–48.

Alzheimer's Association. Race, ethnicity and Alzheimer's in America [Internet]. 2024. Available from: https://alz.org/media/Documents/alzheimers-facts-and-figures-special-report-2021.pdf

American Psychological Association. Guidelines for psychological practice with older adults. Am Psychol. 2014;69(1):34–65.

American Psychological Association. Ethical principles of psychologists and code of conduct. 2017. Retrieved from https://www.apa.org/ethics/code

American Psychological Association. APA resolution on aging. 2020. Retrieved from https://www.apa.org/about/policy/resolution-ageism.pdf

American Psychological Association. APA Task Force on guidelines for assessment and intervention with persons with disabilities. 2022. Retrieved from https://www.apa.org/about/policy/guidelines-assessment-intervention-disabilities.pdf

American Psychological Association. Inclusive language guide. 2nd ed. 2023. Retrieved from https://www.apa.org/about/apa/equity-diversity-inclusion/language-guidelines.pdf

American Psychological Association. Office of aging. 2024. Retrieved from https://www.apa.org/pi/aging

Andrews EE. Disability as diversity: developing cultural competence. Oxford: Oxford University Press; 2020.

Arango-Lasprilla JC, Ketchum JM, Gary K, Hart T, Corrigan J, Forster L, Mascialino G. Race/ethnicity differences in satisfaction with life among persons with traumatic brain injury. NeuroRehabilitation. 2009;24(1):5–14.

Banner NF. Unreasonable reasons: normative judgements in the assessment of mental capacity. J Eval Clin Pract. 2012;18(5):1038–44.

Baumeister A, Chakraverty D, Aldin A, Seven ÜS, Skoetz N, Kalbe E, Woopen C. "The system has to be health literate, too"-perspectives among healthcare professionals on health literacy in transcultural treatment settings. BMC Health Serv Res. 2021;21:1–16.

Berwald S, Roche M, Adelman S, Mukadam N, Livingston G. Black African and Caribbean British Communities' perceptions of memory problems: "We don't do dementia". PLoS One. 2016;11(4):e0151878.

Boltz M, BeLue R, Resnick B, Kuzmik A, Galik E, Jones JR, et al. Disparities in physical and psychological symptoms in hospitalized African American and white persons with dementia. J Aging Health. 2021;33(5–6):340–9.

Bowling A. Aspirations for older age in the 21st century: what is successful aging? Int J Aging Hum Dev. 2007;64(3):263–97.

Braun KL, Thomas WL Jr, Domingo JLB, Allison AL, Ponce A, Haunani Kamakana P, et al. Reducing cancer screening disparities in medicare beneficiaries through cancer patient navigation. J Am Geriatr Soc. 2015;63(2):365–70.

Broadbent E, Petrie KJ, Main J, Weinman J. The brief illness perception questionnaire. J Psychosom Res. 2006;60(6):631–7.

Broadbent E, Wilkes C, Koschwanez H, Weinman J, Norton S, Petrie KJ. A systematic review and meta-analysis of the brief illness perception questionnaire. Psychol Health. 2015;30(11):1361–85.

Calhoun LG. Posttraumatic growth: the positive lessons of loss. In: Meaning reconstruction and the experience of loss. Washington, DC: American Psychological Association; 2001.

Campbell NL, Cantor BB, Hui SL, Perkins A, Khan BA, Farber MO, et al. Race and documentation of cognitive impairment in hospitalized older adults. J Am Geriatr Soc. 2014;62(3):506–11.

Centers for Disease Control and Prevention. What is health literacy. 2020. Retrieved from https://www.cdc.gov/healthliteracy/learn/index

Centers for Disease Control and Prevention. Understanding health disparities in traumatic brain injury (TBI) and concussion. 2023. Retrieved from https://www.cdc.gov/traumaticbraininjury/pdf/CDC_TBI_Health_Disparities-508.pdf

Centers for Disease Control and Prevention. Improving health literacy for older adults: expert panel report 2009. 2009. Retrieved from https://www.cdc.gov/healthliteracy/pdf/olderadults-508.pdf

Centers for Disease Control and Prevention. Cultural insights; communicating with Hispanics/Latinos. CDC Office for the Associate Director of Communication, Division of Communication Services; 2012. Retrieved from https://stacks.cdc.gov/view/cdc/13183

Centers for Disease Control and Prevention. Disability and health overview. 2024. Retrieved from https://www.cdc.gov/ncbddd/disabilityand-health/disability.html

Chavez AA, Simmonds KP, Venkatachalam AM, Ifejika NL. Health care disparities in stroke rehabilitation. Phys Med Rehabil Clin. 2024;35(2):293–303.

Chen J, Benjenk I, Barath D, Anderson AC, Reynolds CF III. Disparities in preventable hospitalization among patients with Alzheimer diseases. Am J Prev Med. 2021;60(5):595–604.

Chesser AK, Keene Woods N, Smothers K, Rogers N. Health literacy and older adults: a systematic review. Gerontol Geriatr Med. 2016;2:2333721416630492.

Chin AL, Negash S, Hamilton R. Diversity and disparity in dementia: the impact of ethnoracial differences in Alzheimer disease. Alzheimer Dis Assoc Disord. 2011;25(3):187–95.

Clark DG, Boan AD, Sims-Robinson C, Adams RJ, Amella EJ, Benitez A, et al. Differential impact of index stroke on dementia risk in African-Americans compared to whites. J Stroke Cerebrovasc Dis. 2018;27(10):2725–30.

Clark JR, Shlobin NA, Batra A, Liotta EM. The relationship between limited English proficiency and outcomes in stroke prevention, management, and rehabilitation: a systematic review. Front Neurol. 2022;13:790553.

Compton MT, Shim RS. The social determinants of mental health. Focus. 2015;13(4):419–25.

Crawford P, Brown B. Fast healthcare: brief communication, traps and opportunities. Patient Educ Couns. 2011;82(1):3–10.

Crist JD, McEwen MM, Herrera AP, Kim SS, Pasvogel A, Hepworth JT. Caregiving burden, acculturation, familism, and Mexican American elders' use of home care services. Res Theory Nurs Pract. 2009;23(3):165.

Cruz-Flores S, Rabinstein A, Biller J, Elkind MS, Griffith P, Gorelick PB, et al. Racial-ethnic disparities in stroke care: the American experience: a statement for healthcare professionals from the American Heart Association/American Stroke Association. Stroke. 2011;42(7):2091–116.

Dembo T, Leviton GL, Wright BA. Adjustment to misfortune: a problem of social-psychological rehabilitation. Rehabil Psychol. 1975;22(1):1.

Derrington SF, Paquette E, Johnson KA. Cross-cultural interactions and shared decision-making. Pediatrics. 2018;142(Supplement_3):S187–92.

Diamond L, Izquierdo K, Canfield D, Matsoukas K, Gany F. A systematic review of the impact of patient–physician non-English language concordance on quality of care and outcomes. J Gen Intern Med. 2019;34:1591–606.

Dismuke CE, Gebregziabher M, Egede LE. Racial/ethnic disparities in VA services utilization as a partial pathway to mortality differentials among veterans diagnosed with TBI. Global J Health Sci. 2015;8(2):260.

Doshi RP, Aseltine RH, Sabina AB, Graham GN. Racial and ethnic disparities in preventable hospitalizations for chronic disease: prevalence and risk factors. J Racial Ethn Health Disparities. 2017;4:1100–6.

Ellis C, Hyacinth HI, Beckett J, Feng W, Chimowitz M, Ovbiagele B, et al. Racial/ethnic differences in poststroke rehabilitation outcomes. Stroke Res Treat. 2014;2014(1):950746.

Elwyn G, Frosch D, Thomson R, Joseph-Williams N, Lloyd A, Kinnersley P, et al. Shared decision making: a model for clinical practice. J Gen Intern Med. 2012;27:1361–7.

Evans BC, Ume E. Psychosocial, cultural, and spiritual health disparities in end-of-life and palliative care: where we are and where we need to go. Nurs Outlook. 2012;60(6):370–5.

Fatehi K, Priestley JL, Taasoobshirazi G. The expanded view of individualism and collectivism: one, two, or four dimensions? Int J Cross-cult Manag. 2020;20(1):7–24.

Fenton JJ, Tancredi DJ, Green P, Franks P, Baldwin LM. Persistent racial and ethnic disparities in up-to-date colorectal cancer testing in Medicare enrollees. J Am Geriatr Soc. 2009;57(3):412–8.

Flores LE, Verduzco-Gutierrez M, Molinares D, Silver JK. Disparities in health care for Hispanic patients in physical medicine and rehabilitation in the United States: a narrative review. Am J Phys Med Rehabil. 2020;99(4):338–47.

Freburger JK, Holmes GM, Ku LJE, Cutchin MP, Heatwole-Shank K, Edwards LJ. Disparities in post–acute rehabilitation care for joint replacement. Arthritis Care Res. 2011;63(7):1020–30.

Frier A, Barnett F, Devine S. The relationship between social determinants of health, and rehabilitation of neurological conditions: a systematic literature review. Disabil Rehabil. 2017;39(10):941–8.

Fujii D. Conducting a culturally informed neuropsychological evaluation. American Psychological Association; 2017.

Fujii DE. Developing a cultural context for conducting a neuropsychological evaluation with a culturally diverse client: the ECLECTIC framework. Clin Neuropsychol. 2018;32(8):1356–92.

Fyffe DC, Deutsch A, Botticello AL, Kirshblum S, Ottenbacher KJ. Racial and ethnic disparities in functioning at discharge and follow-up among patients with motor complete spinal cord injury. Arch Phys Med Rehabil. 2014;95(11):2140–51.

Gao S, Kumar RG, Wisniewski SR, Fabio A. Disparities in health care utilization of adults with traumatic brain injuries are related to insurance, race, and ethnicity: a systematic review. J Head Trauma Rehabil. 2018;33(3):E40–50.

Gary KW, Arango-Lasprilla JC, Stevens LF. Do racial/ethnic differences exist in post-injury outcomes after TBI? A comprehensive review of the literature. Brain Inj. 2009;23(10):775–89.

Gaynor EJ, Geoghegan SE, O'Neill D. Ageism in stroke rehabilitation studies. Age Ageing. 2014;43(3):429–31.
Gernsbacher MA. Editorial perspective: the use of person-first language in scholarly writing may accentuate stigma. J Child Psychol Psychiatry. 2017;58(7):859–61.
Ghosh AK, Geisler BP, Ibrahim S. Racial/ethnic and socioeconomic variations in hospital length of stay: a state-based analysis. Medicine. 2021;100(20):e25976.
Glick AF, Brach C, Yin HS, Dreyer BP. Health literacy in the inpatient setting: implications for patient care and patient safety. Pediatr Clin. 2019;66(4):805–26.
Go AS, Mozaffarian D, Roger VL, Benjamin EJ, Berry JD, Borden WB, et al. Heart disease and stroke statistics—2013 update: a report from the American Heart Association. Circulation. 2013;127(1):e6–245.
Gray HL, Jimenez DE, Cucciare MA, Tong HQ, Gallagher-Thompson D. Ethnic differences in beliefs regarding Alzheimer disease among dementia family caregivers. Am J Geriatr Psychiatry. 2009;17(11):925–33.
Gwinnutt JM, Norton S, Hyrich KL, Lunt M, Combe B, Rincheval N, et al. Influence of social support, financial status, and lifestyle on the disparity between inflammation and disability in rheumatoid arthritis. Arthritis Care Res. 2023;75(5):1026–35.
Hacker K, Auerbach J, Ikeda R, Philip C, Houry D. Social determinants of health—an approach taken at CDC. J Public Health Manag Pract. 2022;28(6):589–94.
Hagger MS, Koch S, Chatzisarantis NL, Orbell S. The common sense model of self-regulation: meta-analysis and test of a process model. Psychol Bull. 2017;143(11):1117.
Hall GL, Heath M. Poor medication adherence in African Americans is a matter of trust. J Racial Ethn Health Disparities. 2021;8(4):927–42.
Hanssen I, Tran PTM. The influence of individualistic and collectivistic morality on dementia care choices. Nurs Ethics. 2019;26(7–8):2047–57.
Hawley ST, Morris AM. Cultural challenges to engaging patients in shared decision making. Patient Educ Couns. 2017;100(1):18–24.
Heath M, Hvass AMF, Wejse CM. Interpreter services and effect on healthcare-a systematic review of the impact of different types of interpreters on patient outcome. J Migr Health. 2023;7:100162.
Heyman N, Osman I, Ben Natan M. Ageist attitudes among healthcare professionals and older patients in a geriatric rehabilitation facility and their association with patients' satisfaction with care. Int J Older People Nursing. 2020;15(2):e12307.
Hofstede G. Dimensionalizing cultures: the Hofstede model in context. Online Read Psychol Cult. 2011;2(1):8.
Hooker ED, Corona K, Guardino CM, Schetter CD, Campos B. What predicts interdependence with family? The relative contributions of ethnicity/race and social class. Cult Divers Ethnic Minor Psychol. 2023;31(1):12–22.
Hu D, Juarez DT, Yeboah M, Castillo TP. Interventions to increase medication adherence in African-American and Latino populations: a literature review. Hawai'i J Med Public Health. 2014;73(1):11.
Hwang WC, Myers HF, Abe-Kim J, Ting JY. A conceptual paradigm for understanding culture's impact on mental health: the cultural influences on mental health (CIMH) model. Clin Psychol Rev. 2008;28(2):211–27.
Jaffee KD, Shires DA, Stroumsa D. Discrimination and delayed health care among transgender women and men: implications for improving medical education and health care delivery. Med Care. 2016;54(11):1010–6.
Jaja BN, Saposnik G, Nisenbaum R, Schweizer TA, Reddy D, Thorpe KE, Macdonald RL. Effect of socioeconomic status on inpatient mortality and use of postacute care after subarachnoid hemorrhage. Stroke. 2013;44(10):2842–7.
Janoff-Bulman R, Yopyk DJ. Random outcomes and valued commitments. In: Handbook of experimental existential psychology. New York: Guilford Press; 2004. p. 122–38.
Julayanont P, Ruthirago D. The illiterate brain and the neuropsychological assessment: from the past knowledge to the future new instruments. Appl Neuropsychol Adult. 2018;25(2):174–87.
Kim Y, Pavlish C, Evangelista LS, Kopple JD, Phillips LR. Racial/ethnic differences in illness perceptions in minority patients undergoing maintenance hemodialysis. Nephrol Nurs J. 2012;39(1):39.
Lau F, Antonio M, Davison K, Queen R, Devor A. A rapid review of gender, sex, and sexual orientation documentation in electronic health records. J Am Med Inform Assoc. 2020;27(11):1774–83.
Lee S, Kim D, Lee H. Examine race/ethnicity disparities in perception, intention, and screening of dementia in a community setting: scoping review. Int J Environ Res Public Health. 2022;19(14):8865.
Leventhal H, Phillips LA, Burns E. The Common-Sense Model of Self-Regulation (CSM): a dynamic framework for understanding illness self-management. J Behav Med. 2016;39:935–46.
Levine DA, Duncan PW, Nguyen-Huynh MN, Ogedegbe OG. Interventions targeting racial/ethnic disparities in stroke prevention and treatment. Stroke. 2020;51:3425.
Li J, Lai DW, Chappell NL. The older, the happier?: a cross-sectional study on the "paradox of aging" in China. Res Aging. 2022;44(1):34–43.
Lubasch JS, Voigt-Barbarowicz M, Ernstmann N, Kowalski C, Brütt AL, Ansmann L. Organizational health literacy in a hospital—insights on the patients' perspective. Int J Environ Res Public Health. 2021;18(23):12646.
Lundine JP, Ciccia AH, Koterba C, Guernon A. Factors that influence follow-up care for families of children with acquired brain injury: a scoping review. Brain Inj. 2022;36(4):469–78.
Mahajan UV, Sharma N, Maynard M, Kang L, Labak CM, Raghavan A, et al. Inpatient virtual shared medical appointments to improve health literacy, increase patient self-efficacy, and reduce provider burnout in acute cerebrovascular pathology patients and their caregivers: a pilot study. Neurosurg Focus. 2022;52(6):E12.
Maher C, Calia C. The effect of illiteracy on performance in screening tools for dementia: a meta-analysis. J Clin Exp Neuropsychol. 2021;43(10):945–66.
Mahoney DF, Cloutterbuck J, Neary S, Zhan L. African American, Chinese, and Latino family caregivers' impressions of the onset and diagnosis of dementia: cross-cultural similarities and differences. Gerontologist. 2005;45(6):783–92.
Makoul G, Clayman ML. An integrative model of shared decision making in medical encounters. Patient Educ Couns. 2006;60(3):301–12.
Maletsky KD, Worsley D, Tran Lopez K, Del Valle Mojica C, Ortiz P, Bonafide CP, Tenney-Soeiro R. Communication experiences of caregivers using a language other than English on inpatient services. Hosp Pediatr. 2023;13(6):471–9.
Mead EL, Doorenbos AZ, Javid SH, Haozous EA, Alvord LA, Flum DR, Morris AM. Shared decision-making for cancer care among racial and ethnic minorities: a systematic review. Am J Public Health. 2013;103(12):e15–29.

Meagher AD, Beadles CA, Doorey J, Charles AG. Racial and ethnic disparities in discharge to rehabilitation following traumatic brain injury. J Neurosurg. 2015;122(3):595–601.

Meidert U, Dönnges G, Bucher T, Wieber F, Gerber-Grote A. Unconscious bias among health professionals: a scoping review. Int J Environ Res Public Health. 2023;20(16):6569.

Merchant RA, Aprahamian I, Woo J, Vellas B, Morley JE. Resilience and successful aging. J Nutr Health, Aging. 2022;26(7):652–6.

Moss-Morris R, Weinman J, Petrie K, Horne R, Cameron L, Buick D. The revised illness perception questionnaire (IPQ-R). Psychol Health. 2002;17(1):1–16.

Mpofu E, Chronister J, Johnson ET, Denham G. Aspects of culture influencing rehabilitation and persons with disabilities. In: The Oxford handbook of rehabilitation psychology. Oxford: Oxford University Press; 2012.

Nelson A. Unequal treatment: confronting racial and ethnic disparities in health care. J Natl Med Assoc. 2002;94(8):666.

Njeru JW, St. Sauver JL, Jacobson DJ, Ebbert JO, Takahashi PY, Fan C, Wieland ML. Emergency department and inpatient health care utilization among patients who require interpreter services. BMC Health Serv Res. 2015;15:1–9.

Noordman J, van Vliet L, Kaunang M, van den Muijsenbergh M, Boland G, van Dulmen S. Towards appropriate information provision for and decision-making with patients with limited health literacy in hospital-based palliative care in Western countries: a scoping review into available communication strategies and tools for healthcare providers. BMC Palliat Care. 2019;18:1–10.

Odonkor CA, Esparza R, Flores LE, Verduzco-Gutierrez M, Escalon MX, Solinsky R, Silver JK. Disparities in health care for black patients in physical medicine and rehabilitation in the United States: a narrative review. PM R. 2021;13(2):180–203.

Okoro ON. Developing a culturally responsive sexual and reproductive health curriculum for HIV prevention among African immigrant women. Med Res Arch [Online]. 2023;11(3). https://doi.org/10.18103/mra.v11i3.3646.

Oliver M. The social model of disability: thirty years on. Disabil Soc. 2013;28(7):1024–6.

Olkin R. Could you hold the door for me? Including disability in diversity. Cult Divers Ethn Minor Psychol. 2002;8(2):130.

Olkin R. Disability-affirmative therapy. In: The professional counselor's desk reference. Cham: Springer; 2009. p. 355–69.

Ownsworth T, Bates A, Watter K, Morgan C, Bell R, Griffin J, et al. Reclaiming agency in care decisions and barriers from the perspectives of individuals with acquired brain injury and their family members. Health Expect. 2024;27(3):e14109.

Pakenham KI, Cox S. The dimensional structure of benefit finding in multiple sclerosis and relations with positive and negative adjustment: a longitudinal study. Psychol Health. 2009;24(4):373–93.

Patel V, Prince M. Ageing and mental health in a developing country: who cares? Qualitative studies from Goa, India. Psychol Med. 2001;31(1):29–38.

Patel I, Erickson SR, Caldwell CH, Woolford SJ, Bagozzi RP, Chang J, Balkrishnan R. Predictors of medication adherence and persistence in Medicaid enrollees with developmental disabilities and type 2 diabetes. Res Soc Adm Pharm. 2016;12(4):592–603.

Pereira C, LaRoche A, Arredondo B, Pugh E, Disbrow E, Reekes TH, et al. Evaluating racial disparities in healthcare system utilization and caregiver burden among older adults with dementia. Clin Neuropsychol. 2022;36(2):353–66.

Poritz JM, Harik LM, Vos L, Ngan E, Leon-Novelo L, Sherer M. Perceived stigma and its association with participation following traumatic brain injury. Stigma Health. 2019;4(1):107.

Ramsoondar N, Anawati A, Cameron E. Racism as a determinant of health and health care: rapid evidence narrative from the SAFE for health institutions project. Can Fam Physician. 2023;69(9):594.

Rosenstock IM. The health belief model and preventive health behavior. Health Educ Monogr. 1974;2(4):354–86.

Saeed F, Eneanya A, Allen RJ, Tamura MK, Fiscella KA. Addressing racial injustice, developing cultural humility, and fostering rapport-building communication skills to improve disparities in end-of-life planning. J Pain Symptom Manag. 2023;66(3):e451–3.

Sagbakken M, Spilker RS, Ingebretsen R. Dementia and migration: family care patterns merging with public care services. Qual Health Res. 2018;28(1):16–29.

Sander AM, Davis LC, Struchen MA, Atchison T, Sherer M, Malec JF, Nakase-Richardson R. Relationship of race/ethnicity to caregivers' coping, appraisals, and distress after traumatic brain injury. NeuroRehabilitation. 2007;22(1):9–17.

Sawyer AT, Harris SL, Koenig HG. Illness perception and high readmission health outcomes. Health Psychol Open. 2019;6(1):2055102919844504.

Scharlach AE, Giunta N, Chow JCC, Lehning A. Racial and ethnic variations in caregiver service use. J Aging Health. 2008;20(3):326–46.

Schillinger D. The intersections between social determinants of health, health literacy, and health disparities. In: Health literacy in clinical practice and public health. Amsterdam: IOS Press; 2020. p. 22–41.

Shiyanbola OO, Ward E, Brown C. Sociocultural influences on African Americans' representations of type 2 diabetes: a qualitative study. Ethn Dis. 2018;28(1):25.

Shiyanbola OO, Rao D, Kuehl S, Bolt D, Ward E, Brown C. Psychometric evaluation of a culturally adapted illness perception questionnaire for African Americans with type 2 diabetes. BMC Public Health. 2022;22(1):741.

Silva MD, Genoff M, Zaballa A, Jewell S, Stabler S, Gany FM, Diamond LC. Interpreting at the end of life: a systematic review of the impact of interpreters on the delivery of palliative care services to cancer patients with limited English proficiency. J Pain Symptom Manag. 2016;51(3):569–80.

Simmonds KP, Luo Z, Reeves M. Race/ethnic and stroke subtype differences in poststroke functional recovery after acute rehabilitation. Arch Phys Med Rehabil. 2021;102(8):1473–81.

Squires A, Ma C, Miner S, Feldman P, Jacobs EA, Jones SA. Assessing the influence of patient language preference on 30-day hospital readmission risk from home health care: a retrospective analysis. Int J Nurs Stud. 2022;125:104093.

Stanley M, Van Kessel G, Murray CM, Forsythe D, Mackintosh S. Occupational therapists and physiotherapists weighing up the dignity of risk for people living with a brain injury: grounded theory. Disabil Rehabil. 2022;44(23):7145–51.

Taboas A, Doepke K, Zimmerman C. Preferences for identity-first versus person-first language in a US sample of autism stakeholders. Autism. 2023;27(2):565–70.

Takemura S, Yoshimasu K, Fukumoto J, Yamamoto H, Miyashita K. Subjective symptoms related to depression and suicidal risk in a Japanese community: a cross-sectional study. Open J Psychiatry. 2011;1(02):40.

Tsao CW, Aday AW, Almarzooq ZI, Alonso A, Beaton AZ, Bittencourt MS, et al. Heart disease and stroke statistics—2022 update: a report from the American Heart Association. Circulation. 2022;145(8):e153–639.

U.S. Census. America is getting older [Internet]. 2023. Available from: https://www.census.gov/newsroom/press-releases/2023/population-estimates-characteristics.html#:~:text=New%20Population%20Estimates%20Highlight%20Increase,by%20the%20U.S.%20Census%20Bureau
U.S. Census Bureau. Language spoken at home [Internet]. Census.gov. [Cited 2024 Dec 10]. Available from: https://www.census.gov/programs-surveys/acs/
U.S. Department of Health and Human Services. Social Determinants of Health – Healthy People 2030. n.d. | odphp.health.gov
U.S. Department of Health and Human Services. Social determinants of health – healthy people 2030. odphp.health.gov [Internet]. [Cited 2024 Dec 10]. Available from: https://odphp.health.gov/healthypeople/priority-areas/social-determinants-health
Van Rosse F, de Bruijne M, Suurmond J, Essink-Bot ML, Wagner C. Language barriers and patient safety risks in hospital care. A mixed methods study. Int J Nurs Stud. 2016;54:45–53.
VanPuymbrouck L, Friedman C, Feldner H. Explicit and implicit disability attitudes of healthcare providers. Rehabil Psychol. 2020;65(2):101.
Waldinger R, Schulz M. The good life: lessons from the world's longest scientific study of happiness. New York: Simon and Schuster; 2023.
Warren KL, García JJ. Centering race/ethnicity: differences in traumatic brain injury inpatient rehabilitation outcomes. PM R. 2022;14(12):1430–8.
Weiner SJ. Advancing health equity by avoiding judgmentalism and contextualizing care. AMA J Ethics. 2021;23(2):91–6.
Weir K. Ageism is one of the last socially acceptable prejudices. Psychologists are working to change that. Monit Psychol. 2023;54(2):36.
Wheeler EC, Klemm P, Hardie T, Plowfield L, Birney M, Polek C, Lynch KG. Racial disparities in hospitalized elderly patients with chronic heart failure. J Transcult Nurs. 2004;15(4):291–7.
Whyte AK, Aubrecht AL, McCullough CA, Lewis JW, Thompson-Ochoa D. Understanding deaf people in counseling contexts. Couns Today. 2013;56(4):38–45.
Wickstrom J, Leone EJ, Sasson N, Morris HC, Livinski AA, Camarillo N D, et al. Impact of social determinants of health on rehabilitation service use and outcomes in adults in the USA: a scoping review protocol. BMJ Open. 2024;14(11):e087254.
Williams DR, Rucker TD. Understanding and addressing racial disparities in health care. Health Care Financ Rev. 2000;21(4):75–90.
Withers M, Sayegh P, Rodriguez-Agudelo Y, Ernstrom K, Raman R, Montoya L, et al. A mixed-methods study of cultural beliefs about dementia and genetic testing among Mexicans and Mexican-Americans at-risk for autosomal dominant Alzheimer's disease. J Genet Couns. 2019;28(5):921–32.
Wolf S, Holm SE, Ingwersen T, Bartling C, Bender G, Birke G, et al. Pre-stroke socioeconomic status predicts upper limb motor recovery after inpatient neurorehabilitation. Ann Med. 2022;54(1):1265–76.
World Health Organization. Social determinants of health. [Cited 2024 Dec 10]. Available from: https://www.who.int/health-topics/social-determinants-of-health
World Health Organization ICF. ICF beginner's guide: towards a common language for functioning, disability, and health. 2002. Available from: https://www.who.int/publications/m/item/icf-beginner-s-guide-towards-a-common-language-for-functioning-disability-and-health
Wynia MK, Osborn CY. Health literacy and communication quality in health care organizations. J Health Commun. 2010;15(S2):102–15.
Yin HS, Johnson M, Mendelsohn AL, Abrams MA, Sanders LM, Dreyer BP. The health literacy of parents in the United States: a nationally representative study. Pediatrics. 2009;124(Supplement_3):S289–98.
Zea MC, Quezada T, Belgrave FZ. Latino cultural values: their role in adjustment to disability. J Soc Behav Pers. 1994;9(5):185–200.
Zisman-Ilani Y, Khaikin S, Savoy ML, Paranjape A, Rubin DJ, Jacob R, et al. Disparities in shared decision-making research and practice: the case for black American patients. Ann Fam Med. 2023;21(2):112–8.

Part II

Clinical Syndromes and Applied Assessment

Delirium I: Pathophysiology, Treatment, and Outcomes

6

Joel E. Kamper, Ryan Van Patten, Emily Hallowell, and Michelle Fox

ABBREVIATIONS

3MS	Modified Mini-Mental State Examination
ABI	Acquired Brain Injury
APACHE	Acute Physiology and Chronic Health Evaluation
BBB	Blood-Brain Barrier
C&L	Consultation & Liaison
CAM	Confusion Assessment Method
CNS	Central Nervous System
DOS	Delirium Observation Scale
DRS	Delirium Rating Scale
DSM	Diagnostic and Statistical Manual
ED	Emergency Department
ESRD	End-Stage Renal Disease
GABA	Gamma-Aminobutyric Acid
GI	Gastrointestinal
H&P	History and Physical
HPA	Hypothalamic-Pituitary-Adrenal
ICU	Intensive Care Unit
IQCODE	Informant Questionnaire on Cognitive Decline in the Elderly
LTP	Long-Term Potentiation
MMSE	Mini-Mental State Examination
MoCA	Montreal Cognitive Assessment

J. E. Kamper (✉)
James A. Haley Veterans' Hospital, Mental Health and Behavioral Science Service (MS/BSS), Tampa, FL, USA
e-mail: Joel.Kamper@va.gov

R. Van Patten
Center for Neurorestoration & Neurotechnology, VA Providence Healthcare System, Providence, RI, USA

Mental Health & Behavioral Sciences Service, VA Providence Healthcare System, Providence, RI, USA

Department of Psychiatry and Human Behavior, Warren Alpert Medical School of Brown University, Providence, RI, USA
e-mail: Ryan_van_Patten@brown.edu

E. Hallowell
Mental Health & Behavioral Sciences Service, VA Providence Healthcare System, Providence, RI, USA

Department of Psychiatry and Human Behavior, Warren Alpert Medical School of Brown University, Providence, RI, USA
e-mail: Emily_Hallowell@brown.edu

M. Fox
Mental Health Service, Lt. Col. Luke Weathers Jr. VA Medical Center, Memphis, TN, USA
e-mail: Michelle.Fox@va.gov

E. K. Bailey et al. (eds.), *Inpatient Neuropsychology*, https://doi.org/10.1007/978-3-032-05880-5_6

PICS	Post-Intensive Care Syndrome
PICS-F	Post-Intensive Care Syndrome in Families
Rx	Drugs
TBI	Traumatic Brain Injury

INTRODUCTION

Delirium is defined as an acute change in mental status due to an underlying medical condition. Although clinical presentation is varied, disruptions in attention form the core hallmark of delirium. Delirium is common, with the annual incidence outpacing that of other common disorders such as Alzheimer's disease in patients older than 65 (Maldonado 2013). It is particularly frequent in medical inpatients, with rates as high as 85% in some inpatient populations (Maldonado 2008; Partridge et al. 2014). Despite this high prevalence, delirium is often under-identified, with one study suggesting that only 11% of cases are properly recognized (Titlestad et al. 2024). While delirium has a myriad of potential etiologies, the clinical presentation, measurement, treatment, and outcomes are broadly similar, and will be discussed in this chapter. Specific etiologies and associated features are covered elsewhere and are outside the scope of this chapter.

Before embarking on this topic, it is important to understand the assorted (and at times redundant) nomenclature used in the healthcare lexicon. Unlike the widely agreed upon term "dementia," the clinical syndrome of delirium has been characterized by a multitude of names that can interfere with one's grasp of the topic. For example, terms such as *acute encephalopathy*, *acute confusional state*, *acute brain dysfunction*, and *altered mental status* (AMS) may be encountered, despite being poorly defined and lacking diagnostic utility (Slooter et al. 2020). To aid the reader, we recommend use of the updated nomenclature proposed by Slooter et al. (2020), which can be found in Fig. 6.1.

Clinical Presentation and Diagnosis

Most simply, delirium is a transient disturbance in attention. As the foundation for other cognitive functions, disturbances in attention typically lead to impairments in higher-order aspects of cognition (see the section on assessment below), but at its core delirium is best understood as a disorder of attention. Here, it is important to note that attention is not monolithic but is reflective of a duality. Selective attention, as frequently impaired in subcortical dementias and other neurocognitive syndromes, and as measured by most neuropsychological instruments of attention, is a top-down process dependent on higher-order cortical areas. This type of attention encapsulates intentional focus or directed attention. In contrast, basic arousal is a more fundamental type of attention, a bottom-up process dependent on the reticular formation and the ascending reticular activating system. This attentional pathway is responsible for maintaining alertness and consciousness, and is impaired in coma and pathologies affecting the brainstem. This dual model of attention is exemplified in the two hallmark phenotypic subtypes: *hypoactive* and *hyperactive* delirium.

Hypoactive Delirium

Impairments in bottom-up attention may manifest as somnolence or frank obtundation. Such cases of hypoactive delirium may be mediated by certain neurotransmitter derangements and can mimic disordered consciousness at the most severe stage, with patients that are hard to rouse or to keep awake during even a brief exam. There is some evidence that hypoactive

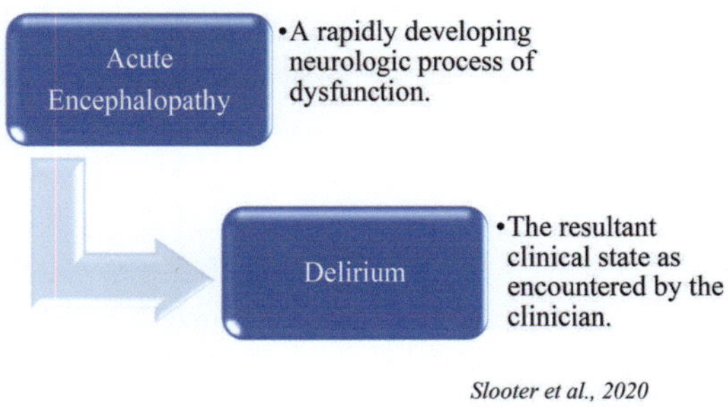

Fig. 6.1 Acute encephalopathy and delirium

delirium has a worse prognosis, perhaps due to delayed identification and treatment, as hyperactive delirium is often much more noticeable to families and other laypersons (Kiely et al. 2007; Yang et al. 2009). While outside the scope of this chapter, we find that reviewing a patient's medication list is a good place to start with hypoactive delirium, as oversedation can occur even from well-meaning medical teams.

Hyperactive Delirium

Conversely, hyperactive delirium involves poor attentional control, which often implicates increased sympathetic nervous system activity. At the mild end of the spectrum, this can manifest as irritability and anxiety, and patients may appear inattentive and be difficult to redirect. However, given the systemic nature of most encephalopathies, these attentional deficits may be subtle and not be grossly observable in brief clinical encounters, though can be easily quantified on cognitive testing. With increased density of delirium, patients may exhibit akathisia, restlessness, agitation/aggression, and be difficult to engage in a meaningful exam (Maldonado 2017). Use of soft restraints and other behavioral management techniques may be employed in such cases. Despite agitation, avoiding sedating medications is strongly encouraged due to the high risk of iatrogenic conversion to a hypoactive delirium.

Expanded Phenotypic Classification

While the simplicity of the dichotomized nosology described above is clinically useful, extreme manifestations of symptoms may not be fully captured. As such, Maldonado (2017) has offered a nuanced phenotypic classification model (Table 6.1).

Table 6.1 An expanded model of delirium

Phenotype	Description	Implications	Able to Participate in Bedside Exam?
Subsyndromal	Underrecognized; patients incompletely meet diagnostic criteria	Have similar risk factors and outcomes as those that meet DSM criteria	Yes
Hypoactive	Somnolence, lethargy, obtundation; impairments in "bottom-up" attention	Less frequently recognized so may have worse outcomes	Yes
Hyperactive	Agitated, irritable, labile, hallucinations, delusions; impairments in "top-down" attention	Sedating medications should be avoided as they can prolong delirium	Yes
Mixed	Mixed hyperactive and hypoactive symptoms	Commonly features waxing/waning 'fluctuations'	Yes
Protracted	Delayed recovery or permanent delirium-like state	May be synonymous with PICS (see below)	Yes, may present as a nonspecific dementia
Catatonic	Extreme end of hypoactive delirium	Longer recovery and higher risk of permanent changes	No (too somnolent)
Excited	Extreme end of hyperactive delirium	Longer recovery and higher risk of permanent changes	No (too agitated)

DSM = Diagnostic and Statistical Manual; PICS = Post-Intensive Care Syndrome; Adapted from Maldonado, 2017

These core features hint at the other major hallmark sign of delirium: symptom fluctuations due to the dynamism and instability of the underlying neurological dysfunction (American Psychiatric Association 2022; Keenan and Jain 2022). Fluctuations are more common in patients with a mixed subtype, as those at extreme ends of the hypoactive or hyperactive continuum are less likely to demonstrate the classic waxing/waning course. The cause of fluctuations may include poor neurologic coping to noxious stimuli, reflecting a failure in brain homeostasis that becomes more common with aging (Maldonado 2017, 2018).

Delirium Diagnosis

While various other frameworks for diagnosis exist, we find the description given in the *Diagnostic and Statistical Manual-5th Edition-Text Revision* (DSM-V-TR) to be the most helpful, as well as the most widely used (Table 6.2; APA 2022).

Epidemiology and Risk Factors

Delirium is the most common neuropsychiatric syndrome encountered in inpatient medical settings (Maldonado 2017). Due to the diverse pathophysiology and clinical presentation, international studies across cultures and healthcare systems indicate a wide prevalence, with rates as low as 10% of general medical inpatients to as high as 85% in those with advanced cancer (Maldonado 2008). Thus, while delirium is quite common, prevalence varies significantly depending on a patient's underlying health, the nature/severity of the deliriogenic factor, and the underlying organ system or disease that is leading to the neurobehavioral symptoms. While specific etiologies and types of delirium will be covered in detail elsewhere, Maldonado (2017) provides a helpful mnemonic for identifying causes of delirium: END ACUTE BRAIN FAILURE (Table 6.3).

Table 6.2 DSM diagnostic criteria for delirium

	DSM-V-TR Diagnostic Criteria For Delirium
A	Disturbance in attention with reduced awareness of the environment
B	Onset of inattention is acute, reflects a change from baseline, and typically fluctuates
C	Attentional fluctuations are accompanied by other cognitive decrements
D	These disturbances are not better explained by other neurocognitive disorders, and do not occur in the context of disorder of consciousness (e.g., coma)
E	History, physical exam, laboratory findings, or other datapoints can establish that this attentional change is resultant from another medical condition(s)
Specify if:	• Acute: Hours to days • Persistent: Weeks to months
Specify if:	• Hyperactive: Increased psychomotor activity, mood lability, agitation, irritability • Hypoactive: Lethargy, obtundation, somnolence • Mixed activity: Normal arousal and motor activity, or fluctuating between the above

Adapted from American Psychiatric Association, 2022

Table 6.3 A list of predisposing and precipitating causes of delirium

END	ACUTE	BRAIN	FAILURE
E Electrolyte imbalances and dehydration	A Age and Sex	B Behavioral/Psych (e.g. emotional distress, substance abuse, mood symptoms, psychosis, etc.)	F Failure (organ)
N Neurologic disorders and injuries	C Cognition (baseline impairment)	R Rx (drugs) and toxins (e.g. prescribed medications, carbon monoxide, solvents, heavy metals, poisons, etc.)	A APACHE score (severity of illness)
D Deficiencies (nutritional)	U U-Tox (intoxication and withdrawal)	A Decreased oxygenation (e.g. anemia, anoxia, hypoxia, low perfusional state, etc.)	I Isolation and immobility
	T Trauma (e.g. physical, heat stroke, burns, surgery, hypothermia, etc.)	I Infections	L Light, sleep, and circadian dysfunctions
	E Endocrine disturbances	N Noxious stimuli (pain)	U Uremia and other metabolic disorders (e.g. acidosis, hyperammonemia, glucose, etc.)
			R Restraints (for limbs, ventilators, catheters, casts, traction devices)
			E Emergence (from medication-induced sedation, coma, or paralysis)

APACHE = Acute Physiology and Chronic Health Evaluation; Adapted from Maldonado, 2018

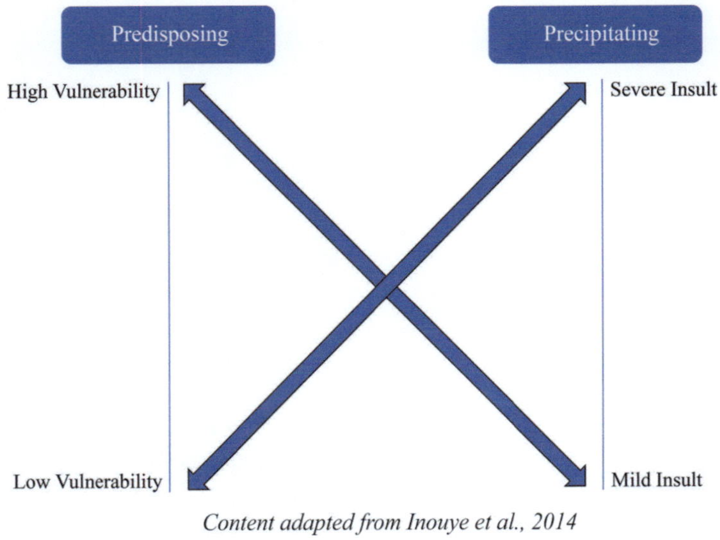

Fig. 6.2 Chiastic model of delirium susceptibility

Predisposing and Precipitating Factors

Causes of delirium can be classified as either *predisposing* (background comorbidities, distal causes) or *precipitating* (acute/subacute effects, proximal causes) (Inouye et al. 2014a; Wilson et al. 2020). Common predisposing risk factors include older age, cerebral atrophy, baseline cognitive impairment/dementia, frailty, chronic medical comorbidities, mental health disorders, vision/hearing loss, and others (Laurila et al. 2008; Magny et al. 2018; Mattison 2020). Precipitating factors are wide-ranging and include medications with cognitive side effects, drug intoxication/withdrawal, various acute medical conditions (e.g., sepsis, anemia, dehydration, infection), metabolic disturbances, physical trauma, pain, psychosocial stress, and others, as covered in Table 6.3 (Baller et al. 2020; Maldonado 2017; Traube et al. 2017; Zaal et al. 2015). For those interested in further discussion, we recommend consulting a review article by Ormseth et al. (2023) for an exhaustive list of predisposing and precipitating factors beyond those described above.

While the distinction between predisposing and precipitating factors appears to present a bifurcated and relatively linear pathophysiology, it is rarely this straightforward. Many of the abovementioned causes can serve as either a predisposing factor or precipitant, depending on the circumstance. Consider, for example, a patient with poorly controlled diabetes and anemia who presents to the ED in acute delirium after suffering a fall. In this case, diabetes and anemia would serve as predisposing factors, while the fall would be the precipitant. However, in a patient with a gastrointestinal bleed or poor diet, anemia or fluctuations in blood glucose could instead precipitate delirium. Predisposing and precipitating factors can also be classified as *modifiable* or *unmodifiable* (Fong et al. 2009). For patients presenting to an acute care or rehabilitation setting with delirium, identification of modifiable factors can inform potential treatment targets, while recognizing unmodifiable factors can help qualify risk.

Further complicating this distinction is the chiastic relationship between predisposing and precipitating factors, with more severe deliriants needed to cause delirium in those with lower predisposition (Fig. 6.2; Inouye et al. 2014b). For example, perhaps the most salient nonmodifiable factor is age, as those with advanced age require less severe precipitating insult to trigger delirium (Fong et al. 2009; Inouye et al. 2014b; Maldonado 2018).

PATHOPHYSIOLOGY OF DELIRIUM

Although delirium can be caused by direct insult to the central nervous system (CNS) (e.g., stroke), it differs from other neurocognitive disorders in that most predisposing and precipitating etiologies are outside the CNS (Shaw et al. 2019). This necessitates the inpatient neuropsychologist to have a broader knowledge base and a more expansive view of the potential systemic contributors to cognitive impairment. While certain precipitating medical disorders can be easily treated (e.g., fluids for dehydration, deprescribing anticholinergic mediations, correcting a vitamin deficiency), many cases of delirium are not so easily diagnosed and treated, and even straightforward precipitants often involve multiple, complex biological systems (Fig. 6.3; Rengel et al. 2018; Wilson et al. 2020). Multiple hypotheses have been proposed to explain the underlying patho-

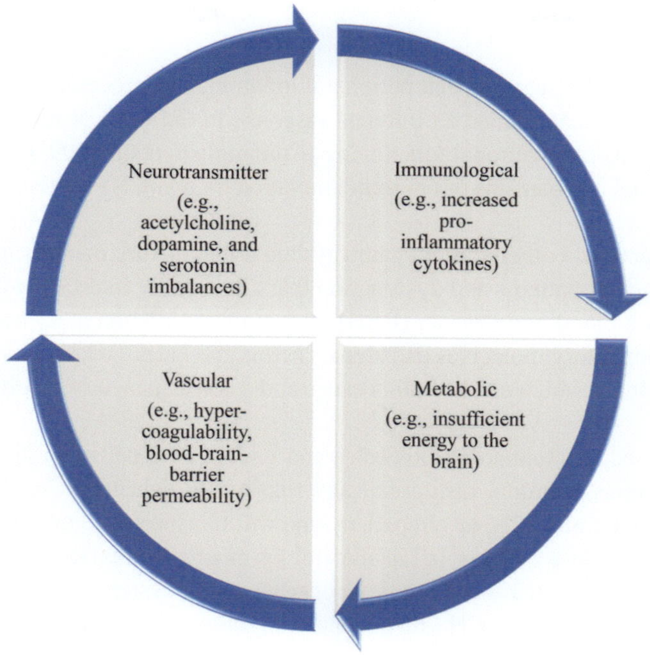

Fig. 6.3 Biological systems contributing to deliriogenesis

physiological mechanisms of delirium, and are discussed in greater detail below. It is noteworthy that the hypothesized mechanisms interact with one another in complex ways rather than being mutually exclusive, and contributions to delirium are multidetermined in many patients. Regardless of the specifics, the final common pathway for deliriogenesis is neural network dysfunction, impairments in functional connectivity in the brain, variability in integrating and processing sensory information, and variable motor responses to internal and external cues, which lead to the observed clinical syndrome (Maldonado 2018; Van Montfort et al. 2019).

Neuronal Aging Hypothesis

The Neuronal Aging Hypothesis suggests that physiologic aging confers a unique and independent risk for development of delirium due to loss of physiologic reserve (Maldonado 2013; Troncale 1996). While the idea of physiologic reserve is seemingly conceptual, age-related physiological changes such as decreased cerebrovascular perfusion, alterations in stress-regulating neurotransmitters, decreased vascular density, neuronal loss, and other mechanisms that accompany aging are well documented and negatively impact the efficiency of allostatic adjustments in response to stressors. Just as increasing biological age is associated with greater risk for dementia, so neuronal aging is associated with greater propensity to develop delirium (Maldonado 2013; Pandharipande et al. 2006).

Aging is also a salient risk factor for developing dementia, and the presence of baseline cognitive impairment catalyzes the relationship between neuronal aging and the likelihood of developing delirium (Maldonado 2013). In addition to increasing the risk for developing delirium, the presence of baseline cognitive impairment in aged populations also increases the likelihood of incomplete recovery from delirium, and other negative outcomes (Maldonado 2018). Morphological data support this relationship, as research shows that patients with reduced brain volumes (often associated with dementia) have extended durations of delirium (Gunther et al. 2012; Juraska et al. 2012). However, even subtle declines in executive functioning and other cognitive functions—below what would even constitute a diagnosable cognitive disorder, are predictive of postoperative delirium in surgical cohorts (Smith et al. 2009), suggesting that one need not have an overt dementia for baseline cognitive symptoms to increase the likelihood of delirium. Furthermore, this relationship is bidirectional, as developing delirium significantly increases the risk of future cognitive decline (Rockwood et al. 1999; Pandharipande et al. 2013). In fact, Pandharipande et al. (2013) found that older patients with delirium admitted to intensive care units (ICUs) had a 63% chance of developing dementia up to 4 years later, while those patients who did not develop delirium only saw an 8% incidence of dementia at that time point.

Neuroinflammatory Hypothesis

The Neuroinflammatory Hypothesis suggests that peripheral inflammatory processes (e.g., infections, surgery) or major trauma (e.g., heart bypass, hypoxia) can induce an immune response in the CNS, leading to the release of inflammatory signals such as cytokines, which disrupt neuronal functioning (Cunningham et al. 2009; Cunningham 2011; de Rooij et al. 2007; Maldonado 2018). Proposed mechanisms of neuroinflammation by non-CNS factors include:

- Activation of the vagus nerve and other afferent pathways via inflammatory markers in the body.
- Pathogen-associated molecular patterns and dysfunction that can directly interface with the choroid plexus and other parts of the CNS that lack a blood-brain barrier (BBB), and cause proinflammatory behavior.
- Active movement of inflammatory molecules and signals across the BBB.
- BBB breakdown or disruption, which can be both a cause and a consequence of inflammation.

Once the BBB is compromised, inflammatory molecules and white blood cells can directly infiltrate the brain, which often leads to further inflammation, neuronal disruption, and subsequent delirium (Cunningham 2011; Elie et al. 1998; Maldonado 2018). Consider the case of a 72-year-old patient who was transported to the Emergency Department (ED) after sustaining partial thickness (second degree) burns to his bilateral lower extremities. During work-up and early in his admission, the patient was noted to have altered mental status consistent with systemic inflammatory-induced delirium. This delirium occurred despite the lack of primary CNS involvement, highlighting the impact of distal causes on the functional integrity of the brain.

Oxidative Stress Hypothesis

The Oxidative Stress Hypothesis proposes that delirium is related to a defect in cerebral metabolism caused by poor perfusion, decreased oxygen availability, and/or increased brain metabolism (Engel and Romano 1959; Maldonado 2013). Reductions in oxidative metabolism can, in turn, lead to neurotransmitter abnormalities (including an overabundance of glutamate and dopamine) which disrupt action potentials (Iijima et al. 1998; Maldonado 2018; Moghaddam et al. 1987). Oxygen-dependent synthesis of acetylcholine, norepinephrine, and other neurotransmitters is reduced, further compounding the imbalance (Beresin 1988; Maldonado 2018). This deliriogenic pathway can be seen in cases of hypoxia or hypoxemia, among others.

Neuroendocrine Dysregulation Hypothesis

The Neuroendocrine Dysregulation Hypothesis suggests that delirium reflects a physiological reaction to acute or chronic physical or psychological stress (Maldonado 2018). Through a complex neuroendocrine cascade, the hypothalamic-pituitary-adrenal (HPA) axis promotes the release of cortisol and other glucocorticoids from the adrenal gland, which under normal circumstances is beneficial for short-term management of stress. However, with prolonged limbic-HPA activation, levels of glucocorticoids remain elevated, with chronic activation of receptors in the brain. This can lead to CNS dysfunction, most notably in the hippocampus, which contains the highest concentration of these receptors (Sapolsky 1996; Seckl and Olsson 1995). This chronic excess of glucocorticoids has widespread effect, including impairments in long-term potentiation (LTP), excess glutamate release, and impaired neurogenesis (Maldonado 2013)—thus reducing the resiliency of neurons, leading to increased vulnerability when the brain encounters an insult via one of the other pathways (Maldonado 2018).

Circadian Dysregulation Hypothesis

The Circadian Dysregulation Hypothesis proposes that disruptions in sleep architecture and circadian rhythm contribute to the development of delirium (BaHammam 2006; Maldonado 2008, 2013). It is well known that poor sleep is a major stressor on the body, with a host of negative endocrine, metabolic, and immunological effects. Neuropsychologists working in traditional outpatient settings will also recognize the impact of poor sleep on attention and other cognitive processes

(Bernstein et al. 2018). This hypothesis goes further and implicates poor sleep as a predisposing risk factor for frank delirium, with both physiological and clinical evidence of this relationship documented in the literature, including reduced melatonin in delirious patients (Helton et al. 1980; Johns et al. 1974; Pandharipande and Ely 2006).

Neurotransmitter Derangements

Neurotransmitter derangements are both a primary and a mediating cause of delirium, serving as one common pathway by which disparate etiologies can synergistically raise deliriogenic risk (Maldonado 2013, 2018). Many of the abovementioned theories feature neurotransmitter abnormalities as an indirect result of disparate physiological changes (Maldonado 2018), while other distal causes of delirium (e.g., metabolic derangements, hypoxic brain injury, medical illness) can directly affect neurotransmitter production and function.

The most common neurotransmitter changes include reduced acetylcholine; increased dopamine, norepinephrine, and glutamate; and alterations in serotonin, GABA, and others (Maldonado 2018; van der Mast 1998). These neurotransmitter changes are directly associated with the clinical presentation of delirium. For example, withdrawal from alcohol or other CNS depressants causes increased acetylcholine, dopamine, and glutamate with reduced GABA, which helps explain the hyperactive presentation typically seen in these patients. Electrolyte imbalances (e.g., as seen in hyponatremia or dehydration) have a similar neurotransmitter profile but with increased GABA, which may explain the mixed pattern of clinical symptoms broadly seen in this etiology. Due to the central role of neurotransmitter dysregulation in several hypothesized theories of delirium, dopamine antagonists are often included in pharmacological treatment, as described in more detail below. For more specific information, the reader is directed to two papers by Maldonado (2013, 2018) which elucidates this topic in greater detail.

Network Dysconnectivity Theory

As discussed above, delirium is primarily a disorder of attention. More so than many other cognitive abilities, attention is a diffuse function that relies on complimentary networks of top-down and bottom-up activation. Thus, it is reasonable to suggest that disruption of these networks can lead to dysregulation and the neurobehavioral phenotypes of delirium. The network dysconnectivity theory suggests that failure in the brain's ability to integrate and process information in synchrony is both a primary and a mediating cause of delirium (Maldonado 2013), which is tightly bound to the neurotransmitter derangements described above. However, this theory further distinguishes between baseline dysconnectivity (influenced by aging and other non-modifiable risk factors for delirium), and the loss of "inhibitory tone" from neurotransmitter derangements (Maldonado 2013; Sanders 2011). In this way, the inpatient neuropsychologist can think of network dysconnectivity as reflective of the interaction between the predisposing and precipitating factors described above.

Systems Integration Failure Hypothesis

The hypotheses and theories of delirium outlined above offer specific, yet somewhat insular mechanisms of cause. However, nearly all patients will have some involvement of multiple pathologic pathways. For this reason, conceptualizing these proposed mechanisms comprehensively, as contributing and integrating within a broad system, is recommended. This top-down view has been termed the Systems Integration Failure Hypothesis, which synthesizes each of these individual pathways and conceptualizes delirium as a complex, neuropathophysiological crisis or "acute brain failure" that impedes neuronal homeostasis and the brain's ability to cope with noxious stimuli (Maldonado 2017, 2018). Thus, delirium is ultimately classified as a neurobehavioral syndrome caused by dysregulation of neuronal activity due to systemic disturbances (Maldonado 2013).

As delirium is often incorrectly viewed as a monolithic entity, use of the Systems Integration Failure Hypothesis when working with patients can help preserve a more global perspective of the possible pathophysiological mechanisms of, and etiological contributors to, their specific symptoms. This in turn allows for increased accuracy of diagnosis, more nuanced prognosis, and overall higher quality service to both patients and medical teams. In the following section, we provide guidance to build fluency and diagnostic comfort with delirium and offer recommendations for treatment interventions.

ASSESSMENT AND INTERVENTION

Successful management of delirium involves identifying the syndrome, establishing predisposing and precipitating factors, and addressing those factors in collaboration with the medical team. When approaching the case of a patient with potential delirium, it is important to remember that, while attention serves as the bedrock of other cognitive functions, the clinical presentation of delirium can be quite heterogenous and deficits in other areas (e.g., language, memory, executive functions) are extremely common. We therefore encourage a multifaceted evaluation of all salient symptoms, not just basic attention and arousal. Maldonado (2018) proposed the following as clinically pertinent domains of dysfunction in delirium:

- Attention: Disturbances in consciousness and difficulty directing, shifting, or sustaining attention.
- General cognitive impairment: Involvement of other cognitive domains, including perception.
- Psychomotor: Hypoactive or hyperactive, often not goal directed.
- Emotional: Perplexity, fear, anxiety, irritability, and/or anger.
- Sleep: Circadian disruptions leading to fragmented sleep-wake cycle.

Given the often dynamic nature of the inpatient setting, as well as the acuity of patients' illnesses, brief bedside exams are recommended when evaluating delirium. We advise against extensive evaluations or feedback/intervention sessions such as those resembling traditional outpatient neuropsychological batteries, as this practice risks obtaining floor effects of little clinical value as well as frustrating both patient and provider. Additional clinical recommendations are provided in the Role of the Inpatient Neuropsychologist section below. While comprehensive discussion of bedside exams is discussed elsewhere, here we briefly cover assessment and intervention as they pertain to delirium.

Assessment

When evaluating a patient for potential delirium, differential diagnoses may include a progressive neurodegenerative disorder or a primary psychotic disorder, among others. Data gathered about a patient's presentation prior to the time of the current evaluation is highly valuable to determine the onset and course of symptoms. This may come from a clinical interview or informant-report measure of perceived cognitive changes completed by the patient's loved one, ideally with whom they live. Examples of the latter include the Informant Questionnaire on Cognitive Decline in the Elderly (IQCODE), which screens for premorbid cognitive disturbances (Jorm and Jacomb 1989). Providers may also review notes in a patient's chart in the interest of identifying fluctuation in orientation, disturbances in sleep-wake cycles, or signs of psychomotor agitation during their hospitalization. If outpatient provider notes are available, recent documentation may also help determine whether symptom onset was acute or gradual.

Unless patients are extremely hyperactive or hypoactive, bedside evaluations consist of a standardized instrument to characterize the presence and severity of symptoms. Most cognitive screening measures take 15 minutes or less to administer and commonly evaluate symptoms across relevant domains including orientation, memory, psychomotor activity, attention, executive functions, perceptual disturbances, disturbance in level of consciousness, and alterations in sleep/wake cycle (Grover and Kate 2012). Jones et al. (2019) identified six high-quality measures in their systematic review: the Confusion Assessment Method (CAM)-Severity Score, Confusional State Examination, Delirium-O-Meter, Delirium Observation Scale (DOS), Delirium Rating Scale (DRS), and Memorial Delirium Assessment Scale (Breitbart et al. 1997; de Jonghe et al. 2005; Inouye et al. 2014b; Robertsson et al. 1997; Schuurmans et al. 2003; Trzepacz et al. 2001). Using a combination of two clinical delirium assessment tools or two family/collateral measures can improve diagnostic accuracy (Moss et al. 2022). More general screening tools can also be of use, with something like the Mini-Mental Status Examination (MMSE), its expanded derivative the 3MS, or the Montreal Cognitive Assessment (MoCA) being particularly useful (Danquah et al. 2024; Folstein et al. 1975; Nasreddine et al. 2005; Teng and Chui 1987).

Intervention Strategies

Foundational treatment of delirium involves correcting modifiable predisposing and precipitating factors. While this is likely to be implemented by the medical team, identification and communication of treatment targets is integral to the services provided by the inpatient neuropsychologist. When communicating with medical teams, it is important to note that in most cases of delirium, clinical improvements usually lag physiologic recovery by days to (up to) weeks or months following correction or treatment of the underlying cause. For example, McCusker et al. (2003) identified an average ~11-day length of delirium, much of which occurring after correction of underlying medical contributors. In our experience, most referring inpatient providers expect delirium to clear rapidly with medical intervention and appreciate ongoing feedback about the often-latent nature of improvements in mentation. Failure to appreciate this lag can lead to unintended consequences, such as further expense and time chasing additional deliriogenic factors that do not exist. However, while full clearance takes time, patients who show no response at all to treatment should be evaluated for causes or alternative diagnoses that may have been overlooked.

While correction of physical factors causing delirium is typically managed by the medical team, additional environmental and behavioral interventions to reduce the prevalence, duration, and severity of delirium should be discussed and/or implemented by the inpatient neuropsychologist. These interventions are intended to reduce iatrogenic and other precipitating risk factors such as sensory deprivation, circadian disruption, poor sleep, psychological stress, loneliness, disorientation, and others. Empirical support and appreciation for multidomain, nonpharmacological strategies to manage delirium has been documented in the literature (Grassi et al. 2015; Hshieh et al. 2015; Kotfis et al. 2022; Ludolph et al. 2020). Some modern intensive care units are starting to incorporate environmental modifications through research-driven architectural design that allows access to sunlight (large windows), reduces unwanted sounds (physical barriers to noisy medical devices), and promotes interpersonal interaction (screens for videoconferencing, space for family and friends; Kotfis et al. 2022). Given the multifactorial pathophysiology of most delirium cases, targeting and preventing specific precipitants as well as managing expected behavioral changes, such as agitation, will have maximum efficacy. Neuropsychologists should be familiar with these prevention and intervention strategies and, when possible, facilitate the use of these techniques through patient/team education and advocacy. See Fig. 6.4 for examples of behavioral intervention strategies used in inpatient settings.

Pharmacologic treatment of delirium is also critical. Most inpatient settings will have a Consultation and Liaison (C&L) Psychiatry service, whose practitioners are a valuable resource to evaluate for medication indication. Treatment often revolves around correction of underlying neurotransmitter derangements and can thus be somewhat more individualized depending on the predisposing and precipitating factors (Maldonado 2018). For example, dopamine antagonists, such as Haldol or similar agents, may be administered given the often-elevated levels of dopamine in hyperactive delirium subtypes.

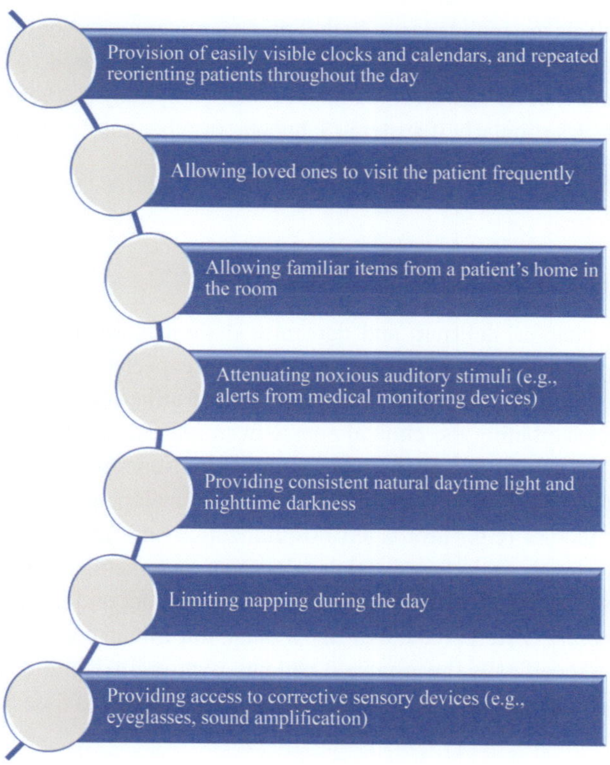

Fig. 6.4 Example behavioral intervention strategies

OUTCOMES AND PROGNOSIS

Delirium is typically thought of as an acute, transient disorder, which resolves to clinical baseline within days-to-weeks of correction of the underlying precipitant. While recent research has increasingly identified poor chronic outcomes associated with discrete episodes of delirium, most patients do recover. For example, in a prospective sample of 193 older general medical inpatients, McCusker et al. (2003) identified an average post-admission length of delirium as 7 days (+/−9), with an average of ~11 days from initial symptom onset to improvement. At hospital discharge (mean of 17.5 days), 89% of those without preexisting cognitive impairment had either fully or nearly cleared delirium.

However, as will be discussed in the following sections, delirium does not always resolve and can itself be considered a poor health outcome due to its reflection of health abnormalities and increased risks for repeated delirium or more permanent cognitive change. Not only is delirium associated with distress in patients and their loved ones, it is a sign of acute brain failure that inherently assumes the presence of at least some predisposing and precipitating factor(s) (Breitbart et al. 2002; Grover et al. 2015; Maldonado 2018).

Delayed Recovery

Despite delirium being traditionally conceptualized as an acute or subacute condition, with the disease course lasting days to weeks, a substantial subset of patients (about one in five) show a prolonged course, with symptoms continuing for weeks to months before improving (Cole et al. 2008). Such presentations tend to violate expectations for an acute process associated with secondary factors, and may be misdiagnosed as a senescent neurodegenerative condition (Fong et al. 2015). Moreover, patients exhibiting a protracted course of delirium have a greater number or severity of deliriogenic factors, which raises the risk for incomplete recovery of the syndrome and generally poorer outcomes from other medical comorbidities (Inouye 2014a). While failure to fully return to baseline is possible in these cases, active and ongoing evaluation and management of causative factors, as well as patience, is recommended.

In addition to those with delayed or incomplete recover, there is a minority of patients who experience *static encephalopathy,* or chronic alterations in mental status due to a persistent medical condition. For example, a patient with kidney failure may experience persistent delirium due to end-stage renal disease (ESRD). While modest, temporary improvements in cognition may occur following regular dialysis treatments, this patient may never fully or consistently clear delirium and return to their "cognitive baseline" as dialysis is a maintenance (rather than curative) treatment for ESRD. In the case of this patient, kidney transplant may serve to improve cognition, but the duration of delirium would reduce the likelihood of a full return to historic baseline. In our experience, other causes of static encephalopathy may include severe untreated sleep apnea, persistent bacteria or fungal infections (particularly if systemic), hepatic or other solid organ dysfunction, or chronic substance use.

Risk for Dementia

As mentioned throughout this chapter, delirium and dementia share a bidirectional relationship, with each condition increasing risk for the other (Fong et al. 2015; Goldberg et al. 2020). As expected, co-occurrence is frequent (Morandi et al. 2017), increasing clinical complexity and the risk of diagnostic error. Furthermore, delirium has been shown to alter the expected trajectory of dementia, exacerbating the clinical course even after acute alterations in mental status are resolved. For example, Partridge et al. (2014) found atypically high rates of cognitive and functional decline following hospital admissions among patients with pre-existing cognitive impairment who had developed delirium while hospitalized. Further, the duration of delirium and related factors (e.g., prolonged hospitalization) are independently associated with progressive declines in cognition following discharge, including distal development of dementia (Binder and Robins 1990; Hartley et al. 2017; McCusker et al. 2003; Wilson et al. 2012). Given the detrimental impact on long-term prognosis, rapid identification and treatment of delirium is paramount. Differential diagnosis of dementia among hospitalized patients is usually determined by the course of symptoms. More specifically, hospitalized patients with pre-existing dementia will usually exhibit stable cognitive symptoms without gross fluctuations or recovery, as well as true functional decline that precedes the deliriogenic condition (see Table 6.5).

Post-ICU Syndrome

While the relationship between delirium and dementia has been well described, a characterization or precise description of the type of dementia encountered has been less well elucidated. Typically, persistent or new onset cognitive decline after a hospital discharge is simply termed "dementia." However, a diagnosis of dementia is nonspecific regarding the underlying etiology of impairment, and further characterization of any features specific to the phenomenon of permanent (or worsening) cognitive change after delirium is beneficial for treatment planning and prognosis.

One attempt to better describe these post-discharge changes is Post-Intensive Care Syndrome (PICS), which was first described in 2012 as progressive cognitive, psychiatric, and/or physical symptoms following hospitalization (Needham et al. 2012). While PICS was originally termed to describe new-onset or worsening symptoms following a critical illness, it can also be considered to encompass residual, stable, and often permanent cognitive dysfunction due to prolonged delirium (Needham et al. 2012).

Causes and Risk Factors for PICS

The primary common risk factor for PICS is prolonged delirium (Pandharipande et al. 2013; Marra et al. 2018), particularly in the context of hypoxia, metabolic or endocrine disruption, immobilization during hospital stay, sleep/wake disruptions, use of sedatives, and respiratory failure, among other conditions (Table 6.4; Needham et al. 2012). More specifically, the added insult of an "acute brain failure" episode with prolonged delirium in the context of these comorbidities appears to moderate the development of PICS. This suggests a synergistic relationship between these health conditions and delirium in causing permanent cognitive change. Outlined in Table 6.4, the reader will recognize many of the conditions associated with PICS as those also associated with delirium development more generally. However, given the incomplete overlap between

Table 6.4 Identified risk factors for PICS

Type of Risk Factor	Risk Factors
Predisposing Personal Factors	• Age • Fewer years of education • Preexisting dementia • Lower cognitive reserve
Predisposing Health Factors	• Glucose dysregulation • Congestive Heart Failure • Obstructive Sleep Apnea • Alcohol or substance abuse
Precipitating Factors	• Hypotension • Duration of delirium • Need for mechanical ventilation • Duration of coma • Sepsis • Surgery (particularly cardiac) • Number of hospital stays (particularly ICU) • Hypoxia/hypoxemia • Respiratory failure • Duration of mechanical ventilation • Mean daily dose of sedative or opioid • Sepsis duration • Cardiac arrest • Length of hospital (particularly ICU)

ICU = Intensive Care Unit;
Needham et al., 2012; Marra et al., 2018; Pandharipande et al., 2013; Sakusic, et al., 2018

these causes and the wider array of deliriants, it is likely that not all causes of delirium are equally potent contributors to post-discharge cognitive change.

Cognitive Impairment in PICS

Most outcome research in post-delirium cognitive change has followed cohorts up to 12 months post-hospitalization. The rates of distal impairment vary across studies but appear to be higher in those with respiratory involvement during hospitalization. For example, in a large sample of 1040 patients with either shock or respiratory failure, approximately 24% demonstrated cognitive impairment at 12 months post-discharge (Girard et al. 2010). In a smaller study of 821 mechanically ventilated patients, persistent cognitive impairment was found in approximately one-third of participants at the same follow-up interval (Pandharipande et al. 2013), suggesting a relatively greater percentage of persistent impairment in those with respiratory dysfunction. While lower rates of persistent cognitive impairment have been shown in heterogenous inpatient cohorts (Sakusic et al. 2018), this may be due to the wide and heterogeneous spectrum of critical illness represented (as not all are associated with PICS), as well as differences in methodology (i.e., diagnosis made via chart review versus formal cognitive testing). Further, the fact that much of the research on PICS comes from critical care medicine and other sources that are less neuropsychologically nuanced may also explain the variability in long-term cognitive outcomes.

Studies using comprehensive neuropsychological batteries to examine long-term cognitive change are less numerous, but tend to describe primary impairments in attention, executive functioning, and aspects of memory (Jackson et al. 2003; Pandharipande et al. 2013; Sukantarat et al. 2005). These findings are broadly reflective of the nonspecific subcortical pattern seen with many neuropsychological disorders, as would be expected given the diffuse nature of delirium. Other research supports this subcortical phenotype of cognitive impairments in PICS. For example, a recent study of 156 post-ICU patients showed persistent cognitive decrements after 3 months in up to ~47% of the sample (Fernández-Gonzalo et al. 2020). Further analysis in this study identified three symptom clusters, two of which were characterized by subcortical patterns of deficits in learning, memory retrieval, processing speed, and executive functions; the third cluster was mostly normal.

ROLE OF THE INPATIENT NEUROPSYCHOLOGIST

The clinical care of patients with delirium is inherently interdisciplinary, with involvement from multiple stakeholders. This has traditionally included internal medicine physicians, critical care specialists, surgeons, anesthesiologists, nurses, oncologists, physiatrists, physical/occupational therapists, C&L psychiatrists, and others (Mudge et al. 2013; Neufeld et al. 2011). However, Tieges et al. (2018) advocate for increased involvement of neuropsychologists in the evaluation of patients with alterations in mental status. Neuropsychologists hold advanced competencies in brain-behavior relationships that yield precision in differential diagnosis of cognitive impairment. Further, the holistic and comprehensive scope of neuropsychological practice lends itself well to diagnosing and tracking delirium subtypes, as well as differentially diagnosing delirium when there are questions of comorbid cognitive impairment. When negotiating for referrals on a busy medical team, the neuropsychologist should reference literature showing that neuropsychological findings have high association with eventual placement and other outcomes (Bishop et al. 2003), although in our experience most medical teams are quite happy for the help and collaboration.

While focused on acute delirium, this chapter has touched on several related topics, including the prolonged courses of delirium, static encephalopathy, PICS, and preexisting cognitive impairments. Differentiating these syndromes is often difficult but doing so has important implications for prognosis (e.g., "*Is this patient expected to cognitively recover or decline over time?*"). Table 6.5 explains how one might differentiate between these interrelated syndromes. Further, for neuropsychologists less familiar with diagnosing delirium, the following diagnostic procedural pathway is recommended:

1. Clarify the referral question as needed to understand what is being asked.
2. Perform a thorough chart review, particularly for records proximal to the hospital admission. This is useful in establishing the course of cognitive symptoms, a patient's predisposing risk factors for delirium, and perhaps also precipitating factors. An admission or History and Physical (H&P) note is often particularly helpful.
3. Evaluate the patient. Collect robust behavioral observations, and note any somnolence, agitation, fluctuations in attention, disorientation, or other signs of delirium. Formal assessment measures are described above but remember that reliance on non-standardized tools is often needed (e.g., neurobehavioral testing).
4. Speak with nursing staff, the referring provider, or other floor staff to get a sense of a patient's in-hospital baseline. For example: Is the patient typically this confused? Are they typically more confused but have been improving?

Table 6.5 Differential diagnosis of delirium

	Typical Delirium	Prolonged Delirium	Static Encephalopathy	PICS	Underlying Dementia
Onset	Rapid	Rapid	Typically rapid	Insidious	Insidious
Course	Improvement and resolution within days of medical stability.	Initial improvements may be seen, but taper over time (weeks-months).	Some initial improvements, but chronic symptoms likely.	Insidious decline or stable impairment after discharge.	Insidious decline before hospitalization.
Cause	Acute precipitating medical change.	Acute precipitating medical change with greater burden of predisposing factors.	Incomplete resolution of underlying precipitating factors.	Delirium combined with preexisting vulnerabilities.	Independent syndrome the precedes a delirium.
Description	Primarily a disorder of attention; hypoactive, hyperactive, or mixed presentation.	Primarily a disorder of attention; hypoactive, hyperactive, or mixed presentation.	Primarily a disorder of attention like delirium, milder in severity.	Typically has a subcortical pattern; frank impairments in alertness or basic attention should be absent.	Varies depending on the cause of the dementia.
Outcome	Good, likely full return to premorbid baseline.	Fair, improvements will be slow and may be incomplete.	Contingent on the underlying medical precipitant, but typically good.	Poor, continued impairment or further decline expected.	Poor, further decline expected.
Pre-hospital cognitive impairments?	No	Typically no, may have very subtle dysfunction.	Typically no	May have subtle dysfunction.	Yes
Residual medical derangements?	No	Typically no, may be slow to resolve.	Yes	No	No

5. When scoring objective measures, pay attention to patterns that don't make sense, or other subtler signs of attentional fluctuations.
6. Discussion with family or other collateral is highly recommended. Ask questions about the onset and course of cognitive dysfunction, as well as any new-onset behaviors that accompanied the cognitive change. For example: Did the patient become both confused and hypersomnolent around the same time?
7. Review recent labs, neuroimaging, and other studies. Consult Table 6.3 to identify possible deliriogenic differentials.
8. If needed, consider making a list of deliriogenic predisposing and precipitating factors. Pay attention to whether these factors are well controlled (or not), as this will affect their salience to cause delirium (Table 6.3).
9. Create your differential diagnosis (Table 6.5).
10. If the presence of delirium is unclear or equivocal, consider serial evaluations over several days to check for fluctuations in attention.
11. When diagnosing delirium, be as specific as possible as to the likely predisposing and precipitating factors. For example: "Delirium due to dehydration and hyponatremia in the context of poorly controlled diabetes and advanced age."
12. If possible, give a prognosis or suspected course. For example: "Previously healthy patient with dense delirium on admission 2 weeks ago but now appears to be improving, as is expected given his aggressive ongoing treatment. Prognosis for ultimate return to cognitive baseline is good, although given the length of delirium some residual symptoms may persist."

13. Provide additional recommendations as needed. Common considerations include:
    - Discharge
        - e.g., Is the patient cognitively safe to discharge or do they need ongoing inpatient care to treat the underlying cause of their delirium?
    - Behavioral interventions that can be employed by the treatment team.
        - e.g. Techniques to manage agitation and confusion
    - Recommended consults or requests for additional workup if needed.
    - Noting any deliriogenic medications that should be evaluated.
    - Whether (or not) an outpatient neuropsychology evaluation following discharge would be helpful.

In addition to the above diagnostic procedure, consider the following pearls when working with this patient population:

- While the diagnosis of delirium is important, many referring providers are primarily interested in how cognition may impact hospital course, medical treatment, and discharge planning. Providing helpful recommendations on these topics, including prognosis of the delirium, will be appreciated.
- Referring providers will likely lack a nuanced appreciation for brain-behavior relationships. As experts in this area, neuropsychologists are encouraged to clarify the referral and to provide direct feedback to referring providers, which can build trust and rapport.
- Don't forget about the "question behind the question," and to address any clinical questions that may not have been posed but need answering. For example, while a referral may be for identification of delirium, elucidating the presence of other neuropsychological disorders (e.g., underlying dementia) will be helpful.
- If unfamiliar with lab values, neuroimaging, and other medical data, consider referencing additional materials in this book or consulting with a C&L psychiatrist or other inpatient physician.
- Remember that the presence of deliriogenic predisposing and precipitating factors does not automatically mean a patient is delirious. How well something is managed or controlled makes a big difference. For example, a patient with well-controlled diabetes and hypertension would be less likely to become delirious from those factors unless there is a sufficiently noxious precipitant.
- As dementia and other baseline cognitive impairment is a potent predisposing factor for delirium, remember that there very well may be an "acute-on-chronic" cognitive disorder; that is, delirium overlying a dementia. Such patients will often experience a worsening of the dementia post-discharge, particularly if the delirium is prolonged.
- Conversely, delirium is also a risk factor for developing dementia (PICS). For patients with prolonged confusion and disorientation despite stable medical status, consider whether what you are seeing is a dementia.
- When delineating what is causing the delirium, remember that there is no magic formula (e.g., "hyponatremia is ALWAYS hypoactive"). It may be helpful to be inclusive as to possible precipitants to ensure adequate treatment is provided.
- Diagnosing someone with delirium who is medically stable may inadvertently alarm the medical team. Be careful with language to prevent this. Alternatively, if a delirium truly is suspected, be prepared to give recommendations for further diagnostic workup. These topics are described in detail elsewhere.
- It is absolutely within the neuropsychologist's purview to comment on medications that may be causing or prolonging the delirium. While dosages and medical indications should be left to physicians and other colleagues, identifying medications that negatively impact cognition can be very helpful to the medical team. Of particular concern in the elderly are anticholinergics, opioids, and other sedating drugs.

SUMMARY

Delirium is an acute change in mental status that is common in inpatient medical settings. Subtypes include hypoactive, hyperactive, and mixed phenotypes. While caused by a wide spectrum of predisposing and precipitating factors, delirium can be thought of as "acute brain failure" (Maldonado 2017), with multiple plausible and intersecting pathophysiological pathways that have been consolidated as the Systems Failure Integration Hypothesis (Maldonado 2018). Treatment for delirium includes correcting the underlying factors, reorienting patients, regulating circadian rhythm, and other nonpharmacologic interventions. While long assumed to resolve back to premorbid baseline, delirium has variable outcomes, with the length of time spent in delirium being a primary risk factor for incomplete recovery. Those with residual cognitive impairment may meet criteria for dementia, which has been conceptualized as Post-ICU Syndrome (Needham et al. 2012). Neuropsychologists play a critical role in recognizing and diagnosing delirium, providing treatment and prognostic recommendations, and differentially diagnosing dementia and other comorbid neurocognitive disorders that are commonly associated with delirium.

REFERENCES

American Psychiatric Association. Diagnostic and statistical manual of mental disorders. 5th ed., text rev.). Washington, DC: APA; 2022.

BaHammam A. Sleep in acute care units. Sleep Breath = Schlaf Atmung. 2006;10(1):6–15. https://doi.org/10.1007/s11325-005-0044-8.

Baller EB, Hogan CS, Fusunyan MA, Ivkovic A, Luccarelli JW, Madva E, Nisavic M, Praschan N, Quijije NV, Beach SR, Smith FA. Neurocovid: pharmacological recommendations for delirium associated with COVID-19. Psychosomatics. 2020;61(6):585–96. https://doi.org/10.1016/j.psym.2020.05.013.

Beresin EV. Delirium in the elderly. J Geriatr Psychiatry Neurol. 1988;1(3):127–43. https://doi.org/10.1177/089198878800100302.

Bernstein JPK, Calamia M, Keller JN. Multiple self-reported sleep measures are differentially associated with cognitive performance in community-dwelling nondemented elderly. Neuropsychology. 2018;32(2):220–9. https://doi.org/10.1037/neu0000407.

Binder EF, Robins LN. Cognitive impairment and length of hospital stay in older persons. J Am Geriatr Soc. 1990;38(7):759–66. https://doi.org/10.1111/j.1532-5415.1990.tb01466.x.

Bishop CL, Temple RO, Tremont G, Westervelt HJ, Stern RA. Utility of the neuropsychological evaluation in an acute medical hospital. Clin Neuropsychol. 2003;17(4):468–73. https://doi.org/10.1076/clin.17.4.468.27944.

Breitbart W, Rosenfeld B, Roth A, Smith MJ, Cohen K, Passik S. The memorial delirium assessment scale. J Pain Symptom Manag. 1997;13(3):128–37. https://doi.org/10.1016/s0885-3924(96)00316-8.

Breitbart W, Gibson C, Tremblay A. The delirium experience: delirium recall and delirium-related distress in hospitalized patients with cancer, their spouses/caregivers, and their nurses. Psychosomatics. 2002;43(3):183–94. https://doi.org/10.1176/appi.psy.43.3.183.

Cole MG, Ciampi A, Belzile E, Zhong L. Persistent delirium in older hospital patients: a systematic review of frequency and prognosis. Age Ageing. 2008;38(1):19–26. https://doi.org/10.1093/ageing/afn253.

Cunningham C. Systemic inflammation and delirium: important co-factors in the progression of dementia. Biochem Soc Trans. 2011;39(4):945–53. https://doi.org/10.1042/BST0390945.

Cunningham C, Campion S, Lunnon K, Murray CL, Woods JF, Deacon RM, Rawlins JN, Perry VH. Systemic inflammation induces acute behavioral and cognitive changes and accelerates neurodegenerative disease. Biol Psychiatry. 2009;65(4):304–12. https://doi.org/10.1016/j.biopsych.2008.07.024.

Danquah MO, Yan E, Lee JW, Philip K, Saripella A, Alhamdah Y, He D, Englesakis M, Chung F. The utility of the Montreal cognitive assessment (MoCA) in detecting cognitive impairment in surgical populations – a systematic review and meta-analysis. J Clin Anesth. 2024;97:111551. https://doi.org/10.1016/j.jclinane.2024.111551.

de Jonghe JF, Kalisvaart KJ, Timmers JF, Kat MG, Jackson JC. Delirium-O-Meter: a nurses' rating scale for monitoring delirium severity in geriatric patients. Int J Geriatr Psychiatry. 2005;20(12):1158–66. https://doi.org/10.1002/gps.1410.

de Rooij SE, van Munster BC, Korevaar JC, Levi M. Cytokines and acute phase response in delirium. J Psychosom Res. 2007;62(5):521–5. https://doi.org/10.1016/j.jpsychores.2006.11.013.

Elie M, Cole MG, Primeau FJ, Bellavance F. Delirium risk factors in elderly hospitalized patients. J Gen Intern Med. 1998;13(3):204–12. https://doi.org/10.1046/j.1525-1497.1998.00047.x.

Engel GL, Romano J. Delirium, a syndrome of cerebral insufficiency. J Chronic Dis. 1959;9(3):260–77. https://doi.org/10.1016/0021-9681(59)90165-1.

Fernández-Gonzalo S, Navarra-Ventura G, Bacardit N, Gomà Fernández G, de Haro C, Subirà C, López-Aguilar J, Magrans R, Sarlabous L, Aquino Esperanza J, Jodar M, Rué M, Ochagavía A, Palao DJ, Fernández R, Blanch L. Cognitive phenotypes 1 month after ICU discharge in mechanically ventilated patients: a prospective observational cohort study. Crit Care. 2020;24(1):618. https://doi.org/10.1186/s13054-020-03334-2.

Folstein MF, Folstein SE, McHugh PR. "Mini-mental state". A practical method for grading the cognitive state of patients for the clinician. J Psychiatr Res. 1975;12(3):189–98. https://doi.org/10.1016/0022-3956(75)90026-6.

Fong TG, Tulebaev SR, Inouye SK. Delirium in elderly adults: diagnosis, prevention and treatment. Nat Rev Neurol. 2009;5(4):210–20. https://doi.org/10.1038/nrneurol.2009.24.

Fong TG, Davis D, Growdon ME, Albuquerque A, Inouye SK. The interface between delirium and dementia in elderly adults. Lancet Neurol. 2015;14(8):823–32. https://doi.org/10.1016/S1474-4422(15)00101-5.

Girard TD, Jackson JC, Pandharipande PP, Pun BT, Thompson JL, Shintani AK, Gordon SM, Canonico AE, Dittus RS, Bernard GR, Ely EW. Delirium as a predictor of long-term cognitive impairment in survivors of critical illness. Crit Care Med. 2010;38(7):1513–20. https://doi.org/10.1097/CCM.0b013e3181e47be1.

Goldberg TE, Chen C, Wang Y, Jung E, Swanson A, Ing C, Garcia PS, Whittington RA, Moitra V. Association of delirium with long-term cognitive decline: a meta-analysis. JAMA Neurol. 2020;77(11):1373. https://doi.org/10.1001/jamaneurol.2020.2273.

Grassi L, Caraceni A, Mitchell AJ, Nanni MG, Berardi MA, Caruso R, Riba M. Management of delirium in palliative care: a review. Curr Psychiatry Rep. 2015;17(3):13. https://doi.org/10.1007/s11920-015-0550-8.

Grover S, Kate N. Assessment scales for delirium: a review. World J Psychiatry. 2012;2(4):58–70. https://doi.org/10.5498/wjp.v2.i4.58.

Grover S, Ghosh A, Ghormode D. Experience in delirium: is it distressing? J Neuropsychiatry Clin Neurosci. 2015;27(2):139–46. https://doi.org/10.1176/appi.neuropsych.13110329.

Gunther ML, Morandi A, Krauskopf E, Pandharipande P, Girard TD, Jackson JC, Thompson J, Shintani AK, Geevarghese S, Miller RR 3rd, Canonico A, Merkle K, Cannistraci CJ, Rogers BP, Gatenby JC, Heckers S, Gore JC, Hopkins RO, Ely EW, VISIONS Investigation, VISualizing Icu SurvivOrs Neuroradiological Sequelae. The association between brain volumes, delirium duration, and cognitive outcomes in intensive care unit survivors: the VISIONS cohort magnetic resonance imaging study. Crit Care Med. 2012;40(7):2022–32. https://doi.org/10.1097/CCM.0b013e318250acc0.

Hartley P, Gibbins N, Saunders A, Alexander K, Conroy E, Dixon R, Lang J, Luckett J, Luddington T, Romero-Ortuno R. The association between cognitive impairment and functional outcome in hospitalised older patients: a systematic review and meta-analysis. Age Ageing. 2017;46(4):559–67. https://doi.org/10.1093/ageing/afx007.

Helton MC, Gordon SH, Nunnery SL. The correlation between sleep deprivation and the intensive care unit syndrome. Heart Lung. 1980;9(3):464–8.

Hshieh TT, Yue J, Oh E, Puelle M, Dowal S, Travison T, Inouye SK. Effectiveness of multicomponent nonpharmacological delirium interventions: a meta-analysis. JAMA Intern Med. 2015;175(4):512. https://doi.org/10.1001/jamainternmed.2014.7779.

Iijima T, Shimase C, Iwao Y, Sankawa H. Relationships between glutamate release, blood flow and spreading depression: real-time monitoring using an electroenzymatic dialysis electrode. Neurosci Res. 1998;32(3):201–7. https://doi.org/10.1016/s0168-0102(98)00090-x.

Inouye SK, Westendorp RG, Saczynski JS. Delirium in elderly people. Lancet (Lond, Engl). 2014a;383(9920):911–22. https://doi.org/10.1016/S0140-6736(13)60688-1.

Inouye SK, Kosar CM, Tommet D, Schmitt EM, Puelle MR, Saczynski JS, Marcantonio ER, Jones RN. The CAM-S: development and validation of a new scoring system for delirium severity in 2 cohorts. Ann Intern Med. 2014b;160(8):526–33. https://doi.org/10.7326/M13-1927.

Jackson JC, Hart RP, Gordon SM, Shintani A, Truman B, May L, Ely EW. Six-month neuropsychological outcome of medical intensive care unit patients. Crit Care Med. 2003;31(4):1226–34.

Johns MW, Large AA, Masterton JP, Dudley HA. Sleep and delirium after open heart surgery. Br J Surg. 1974;61(5):377–81. https://doi.org/10.1002/bjs.1800610513.

Jones RN, Cizginer S, Pavlech L, Albuquerque A, Daiello LA, Dharmarajan K, Gleason LJ, Helfand B, Massimo L, Oh E, Okereke OI, Tabloski P, Rabin LA, Yue J, Marcantonio ER, Fong TG, Hshieh TT, Metzger ED, Erickson K, Schmitt EM, et al. Assessment of instruments for measurement of delirium severity: a systematic review. JAMA Intern Med. 2019;179(2):231–9. https://doi.org/10.1001/jamainternmed.2018.6975.

Jorm AF, Jacomb PA. The informant questionnaire on cognitive decline in the elderly (IQCODE): socio-demographic correlates, reliability, validity and some norms. Psychol Med. 1989;19(4):1015–22. https://doi.org/10.1017/s0033291700005742.

Juraska JM, Lowry NC. Neuroanatomical changes associated with cognitive aging. Curr Top Behav Neurosci. 2012;10:137–62. https://doi.org/10.1007/7854_2011_137.

Keenan CR, Jain S. Delirium. Med Clin North Am. 2022;106(3):459–69. https://doi.org/10.1016/j.mcna.2021.12.003.

Kiely DK, Jones RN, Bergmann MA, Marcantonio ER. Association between psychomotor activity delirium subtypes and mortality among newly admitted postacute facility patients. J Gerontol Ser A Biol Med Sci. 2007;62(2):174–9. https://doi.org/10.1093/gerona/62.2.174.

Kotfis K, Van Diem-Zaal I, Williams Roberson S, Sietnicki M, Van Den Boogaard M, Shehabi Y, Ely EW. The future of intensive care: delirium should no longer be an issue. Crit Care. 2022;26(1):200. https://doi.org/10.1186/s13054-022-04077-y.

Laurila JV, Laakkonen M-L, Laurila JV, Timo SE, Reijo TS. Predisposing and precipitating factors for delirium in a frail geriatric population. J Psychosom Res. 2008;65(3):249–54. https://doi.org/10.1016/j.jpsychores.2008.05.026.

Ludolph P, Stoffers-Winterling J, Kunzler AM, Rösch R, Geschke K, Vahl CF, Lieb K. Non-pharmacologic multicomponent interventions preventing delirium in hospitalized people. J Am Geriatr Soc. 2020;68(8):1864–71. https://doi.org/10.1111/jgs.16565.

Magny E, Le Petitcorps H, Pociumban M, Bouksani-Kacher Z, Pautas É, Belmin J, Bastuji-Garin S, Lafuente-Lafuente C. Predisposing and precipitating factors for delirium in community-dwelling older adults admitted to hospital with this condition: a prospective case series. PLoS One. 2018;13(2):e0193034. https://doi.org/10.1371/journal.pone.0193034.

Maldonado JR. Delirium in the acute care setting: characteristics, diagnosis and treatment. Crit Care Clin. 2008;24(4):657–722, vii. https://doi.org/10.1016/j.ccc.2008.05.008.

Maldonado JR. Neuropathogenesis of delirium: review of current etiologic theories and common pathways. Am J Geriatr Psychiatry. 2013;21(12):1190–222. https://doi.org/10.1016/j.jagp.2013.09.005.

Maldonado JR. Acute brain failure: pathophysiology, diagnosis, management, and sequelae of delirium. Crit Care Clin. 2017;33(3):461–519. https://doi.org/10.1016/j.ccc.2017.03.013.

Maldonado JR. Delirium pathophysiology: an updated hypothesis of the etiology of acute brain failure. Int J Geriatr Psychiatry. 2018;33(11):1428–57. https://doi.org/10.1002/gps.4823.

Marra A, Pandharipande PP, Girard TD, Patel MB, Hughes CG, Jackson JC, Thompson JL, Chandrasekhar R, Ely EW, Brummel NE. Co-occurrence of post-intensive care syndrome problems among 406 survivors of critical illness. Crit Care Med. 2018;46(9):1393–401. https://doi.org/10.1097/CCM.0000000000003218.

Mattison MLP. Delirium. Ann Intern Med. 2020;173(7):ITC49–64. https://doi.org/10.7326/AITC202010060.

McCusker J, Cole M, Dendukuri N, Han L, Belzile E. The course of delirium in older medical inpatients: a prospective study. J Gen Intern Med. 2003;18(9):696–704. https://doi.org/10.1046/j.1525-1497.2003.20602.x.

Moghaddam B, Schenk JO, Stewart WB, Hansen AJ. Temporal relationship between neurotransmitter release and ion flux during spreading depression and anoxia. Can J Physiol Pharmacol. 1987;65(5):1105–10. https://doi.org/10.1139/y87-173.

Morandi A, Davis D, Bellelli G, Arora RC, Caplan GA, Kamholz B, Kolanowski A, Fick DM, Kreisel S, MacLullich A, Meagher D, Neufeld K, Pandharipande PP, Richardson S, Slooter AJC, Taylor JP, Thomas C, Tieges Z, Teodorczuk A, et al. The diagnosis of delirium superimposed on dementia: an emerging challenge. J Am Med Dir Assoc. 2017;18(1):12–8. https://doi.org/10.1016/j.jamda.2016.07.014.

Moss SJ, Hee Lee C, Doig CJ, Whalen-Browne L, Stelfox HT, Fiest KM. Delirium diagnosis without a gold standard: evaluating diagnostic accuracy of combined delirium assessment tools. PLoS One. 2022;17(4):e0267110. https://doi.org/10.1371/journal.pone.0267110.

Mudge A, Maussen C, Duncan J, Denaro C. Improving quality of delirium care in a general medical service with established interdisciplinary care: a controlled trial. Intern Med J. 2013;43(3):270–7.

Nasreddine ZS, Phillips NA, Bédirian V, Charbonneau S, Whitehead V, Collin I, Cummings JL, Chertkow H. The Montreal Cognitive Assessment, MoCA: a brief screening tool for mild cognitive impairment. J Am Geriatr Soc. 2005;53(4):695–9. https://doi.org/10.1111/j.1532-5415.2005.53221.x.

Needham DM, Davidson J, Cohen H, Hopkins RO, Weinert C, Wunsch H, Zawistowski C, Bemis-Dougherty A, Berney SC, Bienvenu OJ, Brady SL, Brodsky MB, Denehy L, Elliott D, Flatley C, Harabin AL, Jones C, Louis D, Meltzer W, Muldoon SR, et al. Improving long-term outcomes after discharge from intensive care unit: report from a stakeholders' conference. Crit Care Med. 2012;40(2):502–9. https://doi.org/10.1097/CCM.0b013e318232da75.

Neufeld KJ, Joseph Bienvenu O, Rosenberg PB, Mears SC, Lee HB, Kamdar BB, Sieber FE, Krumm SK, Walston JD, Hager DN, Touradji P, Needham DM. The Johns Hopkins delirium consortium: a model for collaborating across disciplines and departments for delirium prevention and treatment. J Am Geriatr Soc. 2011;59(s2). https://doi.org/10.1111/j.1532-5415.2011.03672.x.

Ormseth CH, LaHue SC, Oldham MA, Josephson SA, Whitaker E, Douglas VC. Predisposing and precipitating factors associated with delirium: a systematic review. JAMA Netw Open. 2023;6(1):e2249950. https://doi.org/10.1001/jamanetworkopen.2022.49950.

Pandharipande P, Ely EW. Sedative and analgesic medications: risk factors for delirium and sleep disturbances in the critically ill. Crit Care Clin. 2006;22(2):313–27, vii. https://doi.org/10.1016/j.ccc.2006.02.010.

Pandharipande P, Shintani A, Peterson J, Pun BT, Wilkinson GR, Dittus RS, Bernard GR, Ely EW. Lorazepam is an independent risk factor for transitioning to delirium in intensive care unit patients. Anesthesiology. 2006;104(1):21–6. https://doi.org/10.1097/00000542-200601000-00005.

Pandharipande PP, Girard TD, Jackson JC, Morandi A, Thompson JL, Pun BT, Brummel NE, Hughes CG, Vasilevskis EE, Shintani AK, Moons KG, Geevarghese SK, Canonico A, Hopkins RO, Bernard GR, Dittus RS, Ely EW, BRAIN-ICU Study Investigators. Long-term cognitive impairment after critical illness. N Engl J Med. 2013;369(14):1306–16. https://doi.org/10.1056/NEJMoa1301372.

Partridge JS, Dhesi JK, Cross JD, Lo JW, Taylor PR, Bell R, Martin FC, Harari D. The prevalence and impact of undiagnosed cognitive impairment in older vascular surgical patients. J Vasc Surg. 2014;60(4):1002–11.e3. https://doi.org/10.1016/j.jvs.2014.04.041.

Rengel KF, Pandharipande PP, Hughes CG. Postoperative delirium. Presse Med. 2018;47(4):e53–64. https://doi.org/10.1016/j.lpm.2018.03.012.

Robertsson B, Karlsson I, Styrud E, Gottfries CG. Confusional state evaluation (CSE): an instrument for measuring severity of delirium in the elderly. Br J Psychiatry J Ment Sci. 1997;170:565–70. https://doi.org/10.1192/bjp.170.6.565.

Rockwood K, Cosway S, Carver D, Jarrett P, Stadnyk K, Fisk J. The risk of dementia and death after delirium. Age Ageing. 1999;28(6):551–6. https://doi.org/10.1093/ageing/28.6.551.

Sakusic A, Gajic O, Singh TD, O'Horo JC, Jenkins G, Wilson GA, Petersen R, Fryer JD, Kashyap R, Rabinstein AA. Risk factors for persistent cognitive impairment after critical illness, nested case-control study. Crit Care Med. 2018;46(12):1977–84. https://doi.org/10.1097/CCM.0000000000003395.

Sanders RD. Hypothesis for the pathophysiology of delirium: role of baseline brain network connectivity and changes in inhibitory tone. Med Hypotheses. 2011;77(1):140–3. https://doi.org/10.1016/j.mehy.2011.03.048.

Sapolsky RM. Why stress is bad for your brain. Science (New York, N.Y.). 1996;273(5276):749–50. https://doi.org/10.1126/science.273.5276.749.

Schuurmans MJ, Shortridge-Baggett LM, Duursma SA. The delirium observation screening scale: a screening instrument for delirium. Res Theory Nurs Pract. 2003;17(1):31–50. https://doi.org/10.1891/rtnp.17.1.31.53169.

Seckl JR, Olsson T. Glucocorticoid hypersecretion and the age-impaired hippocampus: cause or effect? J Endocrinol. 1995;145(2):201–11. https://doi.org/10.1677/joe.0.1450201.

Shaw RC, Walker G, Elliott E, Quinn TJ. Occurrence rate of delirium in acute stroke settings: systematic review and meta-analysis. Stroke. 2019;50(11):3028–36. https://doi.org/10.1161/STROKEAHA.119.025015.

Slooter AJC, Otte WM, Devlin JW, Arora RC, Bleck TP, Claassen J, Duprey MS, Ely EW, Kaplan PW, Latronico N, Morandi A, Neufeld KJ, Sharshar T, MacLullich AMJ, Stevens RD. Updated nomenclature of delirium and acute encephalopathy: statement of ten societies. Intensive Care Med. 2020;46(5):1020–2. https://doi.org/10.1007/s00134-019-05907-4.

Smith PJ, Attix DK, Weldon BC, Greene NH, Monk TG. Executive function and depression as independent risk factors for postoperative delirium. Anesthesiology. 2009;110(4):781–7. https://doi.org/10.1097/aln.0b013e31819b5bc2.

Sukantarat KT, Burgess PW, Williamson RCN, Brett SJ. Prolonged cognitive dysfunction in survivors of critical illness. Anaesthesia. 2005;60(9):847–53.

Teng EL, Chui HC. The modified mini-mental state (3MS) examination. J Clin Psychiatry. 1987;48(8):314–8.

Tieges Z, Evans JJ, Neufeld KJ, MacLullich AMJ. The neuropsychology of delirium: advancing the science of delirium assessment. Int J Geriatr Psychiatry. 2018;33(11):1501–11. https://doi.org/10.1002/gps.4711.

Titlestad I, Haugarvoll K, Solvang SH, Norekvål TM, Skogseth RE, Andreassen OA, Årsland D, Neerland BE, Nordrehaug JE, Tell GS, Giil LM. Delirium is frequently underdiagnosed among older hospitalised patients despite available information in hospital medical records. Age Ageing. 2024;53(2):afae006. https://doi.org/10.1093/ageing/afae006.

Traube C, Silver G, Gerber LM, Kaur S, Mauer EA, Kerson A, Joyce C, Greenwald BM. Delirium and mortality in critically ill children: epidemiology and outcomes of pediatric delirium. Crit Care Med. 2017;45(5):891–8. https://doi.org/10.1097/CCM.0000000000002324.

Troncale JA. The aging process. Physiologic changes and pharmacologic implications. Postgrad Med. 1996;99(5):111–22.

Trzepacz PT, Mittal D, Torres R, Kanary K, Norton J, Jimerson N. Validation of the delirium rating scale-revised-98: comparison with the delirium rating scale and the cognitive test for delirium. J Neuropsychiatry Clin Neurosci. 2001;13(2):229–42. https://doi.org/10.1176/jnp.13.2.229.

van der Mast RC. Pathophysiology of delirium. J Geriatr Psychiatry Neurol. 1998;11(3):138–58. https://doi.org/10.1177/089198879801100304.

Van Montfort SJT, Van Dellen E, Stam CJ, Ahmad AH, Mentink LJ, Kraan CW, Zalesky A, Slooter AJC. Brain network disintegration as a final common pathway for delirium: a systematic review and qualitative meta-analysis. NeuroImage Clin. 2019;23:101809. https://doi.org/10.1016/j.nicl.2019.101809.

Wilson RS, Hebert LE, Scherr PA, Dong X, Leurgens SE, Evans DA. Cognitive decline after hospitalization in a community population of older persons. Neurology. 2012;78(13):950–6. https://doi.org/10.1212/WNL.0b013e31824d5894.

Wilson JE, Mart MF, Cunningham C, Shehabi Y, Girard TD, MacLullich AMJ, Slooter AJC, Ely EW. Delirium. Nat Rev Dis Primers. 2020;6(1):90. https://doi.org/10.1038/s41572-020-00223-4.

Yang FM, Marcantonio ER, Inouye SK, Kiely DK, Rudolph JL, Fearing MA, Jones RN. Phenomenological subtypes of delirium in older persons: patterns, prevalence, and prognosis. Psychosomatics. 2009;50(3):248–54. https://doi.org/10.1176/appi.psy.50.3.248.

Zaal IJ, Devlin JW, Peelen LM, Slooter AJC. A systematic review of risk factors for delirium in the ICU. Crit Care Med. 2015;43(1):40–7. https://doi.org/10.1097/CCM.0000000000000625.

Delirium II—Common Etiologies

7

Ryan Van Patten, Emily Hallowell, Joel E. Kamper, Kyle J. Jennette, and Robert Kanser

ABBREVIATIONS

A1C	Hemoglobin A1C (Glycohemoglobin)
ACBS	Anticholinergic Cognitive Burden Scale
ACEP	American College of Emergency Physicians
ACTH	Adrenocorticotropic Hormone
ADH	Antidiuretic Hormone
ADLs	Activities of Daily Living
ADS	Anticholinergic Drug Scale
AGS	American Geriatrics Society
AI	Adrenal Insufficiency
AKI	Acute Kidney Injury
AKD	Acute Kidney Disease
AMA	American Medical Association
AMS	Altered Mental Status
APA	American Psychological Association

R. Van Patten (✉)
Center for Neurorestoration & Neurotechnology, VA Providence Healthcare System, Providence, RI, USA

Mental Health & Behavioral Sciences Service, VA Providence Healthcare System, Providence, RI, USA

Department of Psychiatry and Human Behavior, Warren Alpert Medical School of Brown University, Providence, RI, USA
e-mail: Ryan_van_patten@brown.edu

E. Hallowell
Mental Health & Behavioral Sciences Service, VA Providence Healthcare System, Providence, RI, USA

Department of Psychiatry and Human Behavior, Warren Alpert Medical School of Brown University, Providence, RI, USA
e-mail: Emily_hallowell@brown.edu

J. E. Kamper
James A. Haley Veterans' Hospital, Mental Health and Behavioral Science Service (MS/BSS), Tampa, FL, USA
e-mail: Joel.Kamper@va.gov

K. J. Jennette
Department of Psychiatry, University of Illinois at Chicago – College of Medicine, Chicago, IL, USA

Department of Neurology and Rehabilitation, University of Illinois at Chicago – College of Medicine, Chicago, IL, USA
e-mail: Kjennett@uic.edu

R. Kanser
Department of Physical Medicine and Rehabilitation, University of North Carolina School of Medicine, Chapel Hill, NC, USA
e-mail: Robert_kanser@med.unc.edu

E. K. Bailey et al. (eds.), *Inpatient Neuropsychology*, https://doi.org/10.1007/978-3-032-05880-5_7

ARLD	Alcohol-Related Liver Disease
ARS	Anticholinergic Risk Scale
ATP	Adenosine Triphosphate
AVP	Arginine Vasopressin
BBB	Blood-Brain Barrier
BCAA	Branched-Chain Amino Acids
BMP	Basic Metabolic Panel
CABG	Coronary Artery Bypass Grafting
CB1	Cannabinoid Type 1
CBD	Cannabidiol
CHF	Congestive Heart Failure
CIWA	Clinical Institute Withdrawal Assessment for Alcohol
CKD	Chronic Kidney Disease
CMP	Comprehensive Metabolic Panel
CNS	Central Nervous System
CRH	Corticotropin-Releasing Hormone
CSF	Cerebrospinal Fluid
CT	Computerized Tomography
DEA	Drug Enforcement Agency
DKA	Diabetic Ketoacidosis
DT	Delirium Tremens
DWI	Diffusion-Weighted Imaging
ECMO	Extracorporeal Membrane Oxygenation
ED	Emergency Department
EEG	Electroencephalogram
eGFR	Estimated Glomerular Filtration Rate
ESRD	End-Stage Renal Disease
ExDS	Excited Delirium Syndrome
GABA	Gamma-Aminobutyric Acid
GCS	Glasgow Coma Scale
GI	Gastrointestinal
HE	Hepatic Encephalopathy
HHS/HHNS	Hyperosmolar (Non-Ketotic) Hyperglycemia Syndrome
HIV	Human Immunodeficiency Virus
HPA	Hypothalamic-Pituitary-Adrenal
HPT	Hypothalamic-Pituitary-Thyroid
ICU	Intensive Care Unit
IV	Intravenous
KS	Korsakoff's Syndrome
LOC	Loss of Consciousness
LOLA	L-ornithine L-aspartate
MAI	Medication Appropriateness Index
MAOI	Monoamine Oxidase Inhibitor
MCI	Mild Cognitive Impairment
MHE	Minimal Hepatic Encephalopathy
MINDS	Minnesota Detoxification Scale
MoCA	Montreal Cognitive Assessment
MRI	Magnetic Resonance Imaging
mTBI	Mild Traumatic Brain Injury
MVA	Motor Vehicle Accident
NAFLD	Nonalcoholic Fatty Liver Disease

NCSE	Nonconvulsive Status Epilepticus
NMDA	N-methyl-D-aspartate
NMS	Neuroleptic Malignant Syndrome
ODS	Osmotic Demyelination Syndrome
OHE	Overt Hepatic Encephalopathy
ORIF	Open Reduction and Internal Fixation
PAI	Primary Adrenal Insufficiency
PIP	Potentially Inappropriate Prescribing
POCD	Postoperative Cognitive Decline
POD	Postoperative Delirium
PRN	As Needed (Latin = pro re nata)
PSE	Portosystemic Encephalopathy
RRT	Renal Replacement Therapy
RVD	Regulatory Volume Decrease
SAE	Sepsis-Associated Encephalopathy
SAI	Secondary Adrenal Insufficiency
SIADH	Syndrome of Inappropriate ADH
SOFA	Sequential Organ Failure Assessment
SS	Serotonin Syndrome
SSRI	Selective Serotonin Reuptake Inhibitor
START	Screening Tool to Alert Doctors to Right Treatment
STOPP	Screening Tool or Older Persons' Prescriptions
TAI	Tertiary Adrenal Insufficiency
TBI	Traumatic Brain Injury
TCA	Tricyclic Antidepressant
THC	Delta-9-Tetrahydrocannibinol
TIVA	Total Intravenous Anesthesia
TME	Toxic/Metabolic Encephalopathy
TRH	Thyrotropin-Releasing Hormone
TSH	Thyroid Stimulating Hormone
US	United States
USDeN	U.S. Deprescribing Research Network
WE	Wernicke's Encephalopathy
WHO	World Health Organization

INTRODUCTION

Normal brain functioning relies on a delicate physiological balance to maintain homeostasis. While the brain has some allostatic ability to regulate disruptions, rapid or significant changes in this balance can lead to acute brain failure or delirium (Maldonado 2017). Although delirium is theoretically reversible, predisposing factors such as aging and underlying disease can exacerbate the severity and chronicity of clinical symptoms (Inouye et al. 2014; Maldonado 2008; Maldonado 2018). Delirium has many subtypes and causes, the literatures for which run the gamut of organ systems, physiological processes, and medical specialties. Although an exhaustive discussion of all possible deliriogenic precipitants is beyond the purview of this chapter, we will review common causes of delirium that may be encountered by the clinician.

There is often disagreement regarding the terminology used to classify and identify delirium, which can cause confusion for researchers and clinicians. Thus, throughout this chapter we will adhere to the nomenclature proposed by Slooter et al. (2020) defining *acute encephalopathy* as a rapidly developing neurologic process of dysfunction, and *delirium* as the resultant clinical syndrome. We will also refer to *predisposing factors*, which refer to chronic or preexisting medical, cognitive, or functional conditions that lower the threshold for delirium, as well as *precipitating factors*, which are the proximal causes for deliriogenesis.

Toxic-Metabolic Encephalopathy

One major category of delirium precipitants is a type of brain dysfunction called toxic-metabolic encephalopathy (TME). Broadly, TME refers to a disruption in the balance of metabolic and physiological processes in the brain, which causes direct effects on neurons and neuronal functioning as well as on neurotransmitter production and usage. These effects are more pronounced when there is increased oxidative stress or neuroendocrine (e.g., glucocorticoid) dysfunction—both of which can affect the brain's ability to utilize electrolytes, glucose, and other metabolic necessities (Maldonado 2018). Thus, TME is both a cause of delirium as well as a pathophysiologic pathway by which other etiologies can cause brain dysfunction.

Clinically, TME can be characterized by hyperactive and/or hypoactive delirium, depending on the specific cause. Besides typical symptoms of delirium, most types of TME can also present with myoclonus, which may include *asterixis*, a "wing flapping" tremor often incorrectly described as being pathognomonic for hepatic dysfunction (Sanders et al. 2025). Seizures may be seen in severe cases of TME and a consult to neurology can be helpful in differentiating TME-related myoclonic abnormalities from true ictal activity. This is particularly useful in ruling out non-convulsive status epilepticus (NCSE), or ictal activity without convulsions, which can mimic delirium.

TME encompasses many deliriogenic precipitants, including nutrient and other metabolic deficiencies, iatrogenic medication effects, toxic substances such as drugs of abuse, infections, and some types of solid organ dysfunction. These etiologies are covered in detail below. Other common causes of delirium discussed in the following sections include neuroendocrine dysfunction and post-surgical delirium; while these are sometimes grouped with TME, they have distinct pathophysiological processes. Types of delirium *not* covered below include those due to trauma (e.g., posttraumatic confusion), hypoxic-ischemic injuries (e.g., hypoxic respiratory failure, myocardial infarction), hypertension, primary autoimmunity, or rare diseases.

IMBALANCES OF ESSENTIAL NUTRIENTS

Nutrients are chemical compounds needed to sustain life. These include electrolytes, carbohydrates, proteins, fats, vitamins, and minerals. While deficiencies in any of these can result in delirium, electrolyte and vitamin deficiencies, as well as glucose abnormalities, are the most common precipitants. There are myriad causes of deficiency, ranging from simple dietary insufficiencies to the secondary or tertiary effects of a distal disease process. Table 7.1 provides an overview of nutrients important for cognition, their normative reference ranges, critical values associated with clinical symptoms, and common causes of abnormalities.

Electrolyte Derangements

Electrolytes are essential minerals that break down into ions when dissolved in fluid, such as water or blood. In the body, these charged particles are crucial for adequate metabolic processes, enabling the electrical currents required for maintaining intra- and extracellular homeostasis and regulating nerve and muscle functions. Electrolytes are typically measured extracellularly through blood analysis, usually via a basic or comprehensive metabolic panel (BMP and CMP, respectively). Derangements in electrolytes do not carry equal risk of inducing delirium. For example, abnormalities in calcium, magnesium, and potassium are common, though far less likely to cause encephalopathy (Angel and Young 2011). In contrast, sodium is the most abundant electrolyte (in blood and other extracellular fluid) and the most likely to cause delirium when imbalanced.

Sodium is required for normal functioning of neuronal action potentials, among other critical brain operations. Appropriate levels of sodium are maintained by the kidneys, where sodium is either retained or expelled in the urine (Moritz and Ayus 2003). When even small (i.e., 1–2%) increases in plasma osmolality are detected, special neurons in the anterior hypothalamus stimulate thirst and synthesize the arginine vasopressin (AVP) or antidiuretic (ADH) hormones, which act on the kidneys to balance sodium levels by retaining water from urine. Similarly, should sodium levels decrease below normal limits, plasma osmolality is rebalanced by restricting ADH production and decreasing thirst. Failure of this system can result in hyponatremia or hypernatremia, abnormally low or high concentrations of sodium, respectively. Although both disorders can cause delirium, hyponatremia is more common (Dineen et al. 2017).

Table 7.1 Common nutritional precipitants of delirium

Nutrient	Normal Range	Pathological State	Abnormality Threshold	Common Causes
Sodium	135-145 mEQ/L	Hyponatremia	<125 mEq/L (may be higher)	Iatrogenic, poor diet, SIADH, ACTH deficiency, hypothyroidism, hyperglycemia, cirrhosis, CHF, other organ failure.
		Hypernatremia	>155-160 mEq/L	Water loss, diabetes insipidus, excess salt intake.
Calcium	4.3-5.2 mEq/L (2.1-2.6 mmol/L)	Hypocalcemia	<1 mEq/L (<0.5 mmol/L)	Hypoparathyroidism, low vitamin D, pancreatitis, rhabdomyolosis, osteoblastic syndromes.
		Hypercalcemia	>6 mEq/L (>3 mmol/L)	Multiple myeloma and other bone lysis syndromes, metastatic cancers, hyperparathyroidism, adrenal insufficiency.
Magnesium	0.8-1.05 mmol/L	Hypomagnesemia	<0.5 mmol/L	Protein-calorie malnutrition, malabsorption, diabetic ketoacidosis, sepsis, alcohol abuse.
		Hypermagnesemia	>2-3.5 mmol/L	Renal failure.
Phosphorus	>2.5 mg/dL (0.83 mmol/L)	Hypophosphatemia	<1.5 mg/dL (<0.5 mmol/L)	Refeeding syndrome, respiratory alkalosis, mechanical ventilation, increased urinary excretion, decreased intestinal absorption.
Glucose	70-100 mg/dL	Hypoglycemia	<55 mg/dL	Poorly managed diabetes, renal failure, sepsis, substance abuse, malnutrition, heart failure, excess insulin.
		Hyperglycemia	Varies, at least 300 mg/dL	Uncontrolled diabetes, diabetic ketoacidosis, hyperosmolar hyperglycemia.
Vitamin D	30-100 ng/mL	Deficiency	<20 ng/mL	Lack of sunlight exposure, kidney disease, poor absorption, cystic fibrosis
		Excess	Varies	Alternative diets, overuse of supplements
Vitamin B12	>300 pg/mL	Deficiency	<200 pg/mL	Poor diet, anemia, malabsorption
Vitamin B1 (Thiamine)	70-180 nmol/L	Deficiency	~70 or less nmol/L	Poor diet, alcohol abuse, malabsorption,

ACTH = Adrenocorticotropic Hormone; CHF = Congestive Heart Failure;
SIADH = Syndrome of Inappropriate Diuretic Hormone;
Angel & Young, (2011); Dineen et al., (2017); Morgan & Weinsier, (1998)

Hyponatremia

Hyponatremia is defined as plasma sodium levels below 135 mEQ/L and is the most common electrolyte disorder, as well as a common cause of delirium (Adrogué et al. 2022). Hyponatremia causes delirium through hypoosmolar stress (driving water out of cells), which affects neuronal functioning and neurotransmitter balances. The brain responds via Regulatory Volume Decrease (RVD), using electrolytes and organic osmolytes (some of which are neurotransmitters) to rebalance homeostasis and move water back into cells. However, if sodium decreases too rapidly or too severely, these adaptive mechanisms are overwhelmed. This can further affect neuronal functioning and neurotransmitter balances but can also lead to frank cerebral edema (Angel and Young 2011; Maldonado 2018).

Hyponatremia can be seen in up to 30% of hospitalized patients and can be due to a myriad of distal etiologies (Dineen et al. 2017; Moritz and Ayus 2003; Upadhyay et al. 2006). When hyponatremia is identified, medical teams will typically comment on the water volume, urine osmolality, and overall urine output—the combination of which can lead to classification of the underlying metabolic derangement. For example, cases of hyponatremia can be subdivided by the comorbid water volume in the body (low water volume = *hypovolemia*, normal water volume = *euvolemia*, excess water volume = *hypervolemia*). Hypovolemia is the result of both water and sodium loss or depletion, with relatively greater loss in the latter, and may be caused by vomiting/diarrhea, reduced dietary intake, prolonged exercise without replenishing electrolytes, diuretic overuse, and some illicit drugs. Euvolemia is due to sodium depletion without loss of water, and can be due to hypothyroidism, pituitary dysfunction, and excess anti-diuretic hormone (e.g., syndrome of inappropriate anti-diuretic hormone or SIADH), as well as beer potomania (e.g., overconsumption of beer in the context of poor dietary intake, which results in sodium depletion and retained water volume). SIADH is caused by the over release of antidiuretic hormone from the pituitary gland, which leads to too much water reabsorption by the kidneys, leading to hyponatremia. Finally, hypervolemia is the result of increased water, which dilutes sodium levels, and may be caused by liver cirrhosis, heart failure, kidney dysfunction, or excess water consumption.

Despite the high prevalence of hyponatremia, delirium resulting from this condition is less common, with associated clinical symptoms occurring in only ~3% of general medical populations (or ~ 10% of those with hyponatremia; Angel and Young 2011). Delirium associated with hyponatremia usually occurs when serum sodium drops below 125 mEQ/L in otherwise healthy patients (Dineen et al. 2017). This cutoff is flexible, however, and the presence of predisposing deliriogenic factors (including premorbid neurocognitive dysfunction) may increase the threshold for pathological symptoms to between 125–135 mEQ/L. While those with mild chronic hyponatremia may be asymptomatic, hyponatremic delirium tends to present as hypoactive in phenotype, with frank obtundation and so-called "catatonic delirium" possible in values of ~115 mEq/L and lower (Angel and Young 2011). Further declines in sodium levels can eventually lead to seizures, coma, brain herniation, and death (Ellis 1995; Maldonado 2017; Sterns 2015).

Hypernatremia

Hypernatremia, defined as plasma sodium levels above 145 mEq/L, is less commonly encountered due to the automatic release of ADH and prominent thirst response generated by high sodium serum levels that promote water intake (Seay et al. 2020). Individuals with significant baseline cognitive impairments, however, may be unable to perceive or act on this thirst response, and are at increased risk of hypernatremia. When serum sodium is too high, water moves out of the brain parenchyma into intercellular space, leading to reduction in brain volume that is then corrected by movement of water from CSF into the parenchyma (Pullen et al. 1987). While this compensatory mechanism serves to preserve intracellular water balances, severe or acute increases in sodium are likely to overwhelm the brain's ability to compensate, leading to delirium. Distal causes of hypernatremia include water loss (e.g., excess sweating, urinary output), GI disruptions, and sodium surplus. Although less common, AVP disorders can cause hypernatremia due to deficiencies in AVP production (Central Diabetes Insipidus or AVP-D) or renal AVP resistance (Nephrogenic Diabetes Insipidus or AVP-R). Delirium due to hypernatremia includes muscle fasciculations, irritability, and lethargy. Seizures and coma are possible with very high sodium levels (Moder and Hurley 1990).

Addressing Delirium Due to Electrolyte Derangements

Electrolyte abnormalities are often easily corrected. In patients with deficiencies, simply repleting the electrolyte in question (either via oral or IV supplements) is sufficient. However, supplementation may need to continue after discharge in many cases due to the risk of recurrence. In patients with excess electrolytes, expanding intravascular volume with free water or saline may be used to promote dilution, as well as administration of hormones or other substances to promote rebalance. As with all types of delirium, improvements in cognition typically lag correction of the underlying precipitating factor, sometimes by days to weeks.

While treatment of most electrolyte imbalances is relatively straightforward, treatment of sodium abnormalities requires special consideration, particularly with hyponatremia, due to the risk of osmotic demyelination syndrome (ODS; previously called central pontine myelinolysis). ODS involves typically irreversible demyelination in the pons and elsewhere caused by rapid correction of sodium levels, and is characterized by rapid confusion, quadriplegia, locked-in syndrome, bulbar palsies, and other symptoms (Dineen et al. 2017; Moritz and Ayus 2003). For this reason, hyponatremia is corrected slowly over a period of several days. While perhaps less risky than raising low sodium levels, gradual titration is still advised when treating hypernatremia, as rapid correction can also cause demyelinating lesions (Chang et al. 2005).

Glucose

Glucose is a carbohydrate necessary for proper absorption of electrolytes. Although only accounting for ~2% of body weight, the brain consumes ~20% of the body's glucose-derived energy (Mergenthaler et al. 2013). Glucose is critical for neurotransmitter production and maintenance of both neurons and non-neuronal cells but is not directly synthesized or stored in the brain, thus causing heightened vulnerability of brain homeostasis with glucose abnormalities. Individuals with diabetes, particularly those with type 1 diabetes, are more at risk for developing abnormal or labile glucose levels (Frier 2004).

Hypoglycemia

Delirium associated with hypoglycemia is well described in the literature and frequently associated with negative outcomes (Angel and Young 2011). In hypoglycemia, a counterregulatory response (release of glucagon and epinephrine) attempts to rebalance glucose levels. However, like the electrolyte imbalances described above, particularly rapid or severe declines in glucose overwhelm the brain's ability to compensate, leading to neuroglycopenia and alterations in mental status (Rickels et al. 2007). This pathophysiology is further catalyzed by the comorbid presence of oxidative stress and excess glucocorticoids, both of which can impair the transport and use of glucose in the brain (Maldonado 2018). Hypoglycemic risk factors include impaired counterregulatory hormone functions, reduced C-peptide levels, and a history of labile glucose (Auer et al. 1984).

Clinical symptoms of hypoglycemia include tremor (usually an initial symptom), diaphoresis, agitation, drowsiness, and other signs of epinephrine release, which can progress to delirium if uncorrected (Frier 2004). Delirium associated with hypoglycemia is more likely to be hypoactive. Seizures, coma, and death can occur with extremely low glucose levels. While rare, there have been cases of permanent neurologic and neurocognitive impairment from a single episode of severe hypoglycemia. Rodent studies suggest that the hippocampal formation, basal ganglia, and cerebral cortex are at relatively greater risk in such cases, with EEG showing increased delta and theta activity and DWI abnormalities common on MRI (Angel and Young 2011).

Hyperglycemia

Delirium associated with hyperglycemia is incompletely described in the literature, but appears to be confined to those with diabetes, particularly when the condition is poorly managed (Angel and Young 2011; See Figs. 7.1 and 7.2 for a relevant case example). Common complications of hyperglycemic delirium include diabetic ketoacidosis (DKA), or impaired glucose metabolism causing ketosis and changes in blood pH, as well as nonketotic hyperosmolar hyperglycemia (HHS)—with the latter being more common. HHS occurs when the kidneys attempt to regulate elevated levels of glucose by increasing urine production (osmotic diuresis), resulting in a loss of water and, often, concomitant hyponatremia.

Studies examining clinical manifestations of DKA and HHS describe neurotransmitter derangements and seizures, although delirium due to hyperglycemia without these complications is likely more common (Angel and Young 2011; Maldonado 2018). While the presence of DKA or HHS may yield hypoactive or hyperactive delirium, in our experience delirium in uncomplicated hyperglycemia tends to present with hyperactive symptoms, particularly when patients exhibit labile glucose over several days' time or when other metabolic abnormalities are present.

CASE EXAMPLE #1:
DELIRIUM DUE TO HYPERGLYCEMIA, INTRODUCTION

Demographics: 74-year-old married white man with 20 years of education who is a retired university professor.

Admission: Presented to emergency department independently. Patient is homeless and was living at a hotel; he primarily eats fast food. Vital signs were stable, but he was noted to be agitated and confused. Patient was a vague historian but reported a recent fall that precipitated his hospitalization.

Medical History: Prostate cancer, obstructive sleep apnea, type II diabetes, peripheral vascular disease, hypothyroidism, and a concussion 4 years ago following a car accident.

Psychiatric History: Depression by history

Substance Use History: Amphetamine usage, most recently prior to his fall

Medications: Not taking any medications for some time. Started on sertraline, metformin, and insulin in the hospital.

Bedside Exam: Neuropsychology noted mild fluctuations in attention, consistent with delirium (subtype unspecified)

Question 1. What are at least two predisposing factors and at least two precipitating factors for delirium in this patient?

Question 2. What medical workup would be helpful in refining the differential?

Fig. 7.1 Delirium due to hyperglycemia, introduction

CASE EXAMPLE #1:
DELIRIUM DUE TO HYPERGLYCEMIA, CONCLUSION

Question 1.

Predisposing factors: Age, likely cerebrovascular disease, depression, poor diet, sleep apnea, type II diabetes, hypothyroidism.

Precipitating factors: Physical trauma (fall), subdural hemorrhage from fall, amphetamine usage, hyperglycemia (diabetes).

Question 2.

Imaging: MRI Brain 3 months prior showed mild white matter disease and a 5mm focus of hemosiderin in the left frontal lobe. CT Head on admission was negative for acute findings.

***Labs*:** Glucose was >700 with an A1C>14.0. UDS was positive for amphetamines. Metabolic labs indicated that patient was in DKA. Hyperammonia was noted d/t liver injury from the DKA.

Conclusions: Patient diagnosed with mild, resolving delirium since the neuropsychological exam was not markedly abnormal. Predisposing factors included poor diabetes management, sleep apnea, and age. Precipitating factors included severe hyperglycemia with DKA (leading to other metabolic derangements), a recent subdural hemorrhage from fall, amphetamine usage, and hyperglycemia. Prognosis for recovery to cognitive baseline was good given his estimated high cognitive reserve. Low suspicion for dementia or other chronic cognitive disorder. Hemosiderin deposition was chronic and incidental.

CT = computerized tomography; MRI = magnetic resonance imaging; UDS = urine drug screen; A1C = hemoglobin A1C (glycohemoglobin); DKA = diabetic ketoacidosis

Fig. 7.2 Delirium due to hyperglycemia, conclusion

Vitamin Deficiencies

Vitamin deficiencies are a common but often overlooked cause of delirium. Among the most clinically relevant for neuropsychological practice are vitamins D, B12, and B1 (thiamine), which are essential for cellular metabolism, neurotransmitter function, and regulation of oxidative stress. Deficiencies can result in delirium that is more likely to include neuropsychiatric features such as apathy, depression, and psychosis. Delirium due to vitamin deficiencies is often misdiagnosed as dementia, perhaps due to the nonspecific nature of the neuropsychiatric symptoms and the higher prevalence of both vitamin deficiencies and dementia in acutely ill and older adults (Sechi and Serra 2007).

Vitamin D

Vitamin D is critical for balancing calcium and phosphate, as well as for promoting calcium absorption in the intestinal tract. Vitamin D also supports bone mineralization and immune system functions. When dietary vitamin D is metabolized into 1,25-dihydroxyvitamin D (calcitriol), it binds to vitamin D receptors to regulate calcium metabolism, reduce inflammation, and promote cellular growth (Holick 2007). Vitamin D receptors are also widely distributed in the CNS, most densely in the hippocampus, cerebellum, cingulate, substantia nigra, and prefrontal cortex (Eyles et al. 2013). In the brain, vitamin D helps support neuron and glial cell functions, synaptic plasticity, and neurotransmitter synthesis –specifically dopamine, serotonin, and GABA (Bivona et al. 2019). Vitamin D is also neuroprotective, buffering against neuronal injury and degeneration by modulating antioxidant enzyme and cytokine expression to reduce systemic and neuroinflammation (Garcion et al. 2002).

Deficiency of vitamin D is common, with rates as high as 50% in certain populations, particularly those with limited sun exposure and chronic illness (Annweiler et al. 2010). Vitamin D is fat-soluble and can be stored in adipose tissue and the liver for several months, thus deficiency generally develops gradually and asymptomatically as body stores decrease faster than they are repleted. In general, serum concentrations under 20 ng/mL are associated with an onset of clinical symptoms (Holick 2007). Vitamin D deficiencies can cause dysregulation of calcium, neuroinflammation, disruptions to the blood-brain barrier (BBB), oxidative stress, and increased beta-amyloid deposition, which secondarily increases risk of neurologic disorders, such as Alzheimer's disease, Parkinson's disease, multiple sclerosis, or cerebrovascular disease, as well as other metabolic diseases (Annweiler et al. 2010; Di Somma et al. 2017; Zhao et al. 2021).

Nonspecific, early symptoms of vitamin D deficiency include fatigue, muscle and bone pain, weakness, and mood changes. In older adults, even mild decreases in muscle strength can increase risk for falls and fractures, which compounds the risk for delirium and other negative outcomes. Factors that accelerate onset or exacerbate severity of clinical symptoms associated with vitamin D deficiency include advanced age, darker skin pigmentation, and conditions that disrupt synthesis, storage, or activation of vitamin D, such as kidney or liver disease, malabsorption syndromes, and obesity (Holick 2007; Eyles et al. 2013). If vitamin D deficiency remains untreated or worsens, secondary complications may emerge in response to insufficient calcium. These complications include hyperparathyroidism (which can lead to muscle and bone pain), imbalance, and osteomalacia (softening of the bone due to impairments in bone metabolism from inadequate mineralization). With more severe and prolonged deficiency, individuals can develop myopathy and gait instability, further increasing the risk of falls and fractures (Bischoff-Ferrari et al. 2004).

Repletion of vitamin D is typically achieved with cholecalciferol (vitamin D3) or calcifediol. Calcifediol is specifically indicated for (and most effective in) patients with hepatic or renal dysfunction (Quesada-Gomez et al. 2022). Since correcting vitamin D deficiencies can improve cognitive functioning, routine screening is recommended in high-risk populations, including older adults or patients with new-onset cognitive complaints, chronic liver disease, mood symptoms, frailty, or fatigue. Neuropsychologists should be aware of the role vitamin D in contributing to a patient's current mental status, as well as risk of future neurodegenerative disease and advocate for early identification and supplementation when evaluating high-risk patients.

Vitamin B12

Vitamin B12 has two primary purposes: 1) the conversion of homocysteine into methionine, which is used to facilitate DNA synthesis and to maintain neuronal myelination; and 2) the conversion of methylmalonyl-CoA into succinyl-CoA, which facilitates energy metabolism of mitochondria. Deficiencies in B12 are relatively uncommon in high-income countries, though may occur in vulnerable populations such as older adults, infants, and pregnant women, as well as in individuals with malabsorption, vegetarian diets, or chronic alcohol use, which can hinder intracellular metabolism and transport (Green et al. 2017; O'Leary and Samman 2010; Stabler 2013). Adults with B12 deficiency due to malabsorption or inadequate diet who are otherwise healthy may experience an indolent course, as the liver can store water-soluble B12 thus delaying the onset of clinical symptoms for months to years (O'Leary and Samman 2010).

If B12 deficiency continues, significant hematological and irreversible neurological symptoms may occur. Severe, prolonged deficiency may be initially heralded by sensory neuropathy and anemia. Following these symptoms, demyelination in the spinal cord is typically the first CNS sign, which can lead to spasticity, impaired proprioception, and ataxia (Healton et al. 1991). Additional CNS involvement may then occur; neuroimaging studies in patients with B12 deficiency show frontal, medial temporal, and posterior cingulate atrophy (Douaud et al. 2013). Most cases of B12 deficiency will present as a mild delirium. However, if chronic deficiencies continue to go untreated, elevated homocysteine can cause more prominent white matter lesions and gray matter loss (greater than would otherwise be expected for a patient's age), as well as increase future risk for dementia due to Alzheimer's disease (Smith et al. 2010; Hooshmand et al. 2010).

Early intervention with high-dose B12 replacement can reverse neurobehavioral symptoms and many neurological symptoms. Full recovery to premorbid baseline is less common when treatment is initiated more than six months after symptom onset, as the demyelination and other neurologic signs are typically permanent (Stabler 2013). Oral supplementation is often effective in most cases, however parenteral treatment (i.e., intramuscular, intravenous, or subcutaneous treatments that bypass the gastrointestinal tract) is indicated when malabsorption is the suspected etiology rather than deficits in dietary intake (Carmel 2000). Neuropsychologists should work with medical teams to screen for B12 status in patients with unexplained cognitive or psychiatric symptoms, particularly in older adults, vegetarians, or those with gastrointestinal disorders. Collaborative care should emphasize the neurocognitive risks of untreated deficiency, the importance of thorough follow-up, and use of sensitive diagnostic tests. Given the nonspecific symptom profile in early stages and the considerable overlap with other neuropsychiatric conditions, B12 deficiency should always be considered in the differential diagnosis of cognitive decline, especially in older adults with elevated homocysteine or low serum B12.

Vitamin B1 (Thiamine)

Vitamin B1, or thiamine, is critical for multiple enzymatic pathways that facilitate glucose metabolism and mitochondrial function. This includes the production of ATP via the tricarboxylic acid cycle, synthesis of acetylcholine, and modulation of neurotransmitters such as glutamate and serotonin (Ott and Werneke 2020). Critically, thiamine cannot be synthesized by the human body (it must be obtained from dietary sources) and can only be stored in organs for ~2–3 weeks without repletion. Thiamine deficiency has long been associated with alcohol use disorder, though can also be seen in other patient populations at risk for malnourishment such as older adults, individuals with eating disorders or inflammatory bowel disease, those with hyperemesis gravidarum (severe morning sickness) or who are undergoing cancer treatments, as well as among patients following bariatric surgery (Isenberg-Grzeda et al. 2012; Oudman et al. 2018; Oudman et al. 2019; Oudman et al. 2021; Yu et al. 2023).

The onset and clinical course of thiamine deficiency varies considerably as a function of the underlying cause. Insidious progression may occur in cases of chronic malnutrition or alcohol use. In contrast, acute medical events, such as prolonged vomiting, infection, and critical illness—particularly in the setting of metabolic distress—can lead to a more rapid course where symptoms emerge over hours to days. Early clinical signs are nonspecific and include fatigue, irritability, abdominal discomfort, and reduced appetite—symptoms that are easily misattributed to other causes in medically complex patients and older adults (Butterworth 2009). Further, age-related reductions in thiamine absorption and increased vulnerability to oxidative stress and excitotoxicity may accelerate neurologic deterioration in older adults and put these individuals at increased risk for irreversible symptoms if not emergently identified and treated (Butterworth 2009). This, coupled with higher rates of general brain atrophy and cerebrovascular disease in older adults, makes sensitive and thoughtful differential diagnosis essential to minimize adverse outcomes. Further compounding this risk, hospitalized or institutionalized older adults experience higher rates of suboptimal nutrition, polypharmacy, or prolonged intravenous feeding that does not allow for adequate thiamine repletion, further shortening the window for differential identification.

There is no quick or sensitive biological assay for thiamine deficiency, and delaying treatment for confirmatory laboratory tests raises the risk of permanent disability. While a brain MRI may support diagnosis, findings are usually nonspecific (Isenberg-Grzeda et al. 2012). Instead, clinical suspicion, particularly for at-risk individuals presenting with altered mental status, is typically adequate to prompt prophylactic thiamine supplementation. In neurologic and psychiatric contexts, thiamine deficiency should be considered in cases of acute confusion, memory impairment, or unexplained motor signs. Medical guidelines recommend acute administration of parenteral thiamine, with monitored oral maintenance thereafter (Sechi and Serra 2007). Even in the absence of acute symptoms, routine thiamine supplementation is recommended in high-risk populations, such as individuals with alcohol use disorder, malnutrition, hyperemesis, critical illness, and those undergoing bariatric surgery. Neuropsychologists can be helpful in recognizing those at risk for thiamine deficiency and advocating for supplementation, when appropriate, to attain the highest probability of meaningful recovery.

Addressing Vitamin-Related Delirium

Timely identification of vitamin-related encephalopathy and resultant delirium is critical, as many of these conditions are reversible with early treatment but can progress to permanent damage if intervention is delayed. However, the onset of clinical symptoms associated with vitamin-related encephalopathy is near-universally subtle, and easily confounded by general medical and psychiatric syndromes. While deficiencies may be confirmed with a blood test, the risk of adverse iatrogenic effects is low with initial supplementation in suspected cases of higher risk patients such as older adults, patients who are post-bariatric surgery, individuals with multiple underlying chronic medical illnesses, or those with a history of alcohol use, eating disorders, or restrictive diets (Sinha et al. 2019). Early administration of thiamine (sometimes augmented with magnesium) is particularly encouraged for patients with a history of heavy alcohol use, as chronic thiamine deficiencies can lead to Korsakoff's Syndrome (KS; discussed later in this chapter), an irreversible dementia. Effective dose quantity, frequency, duration, and route of thiamine administration have not been formally established and may depend on the root cause of the deficiency (Smith et al. 2021).

Long-term intervention to reduce future risk of vitamin-related encephalopathy includes prophylactic vitamin supplementation and education on the causes and consequences of deficiency. Neuropsychologists should always consider vitamin deficiencies in their differential diagnosis when evaluating patients with atypical or rapidly progressive cognitive symptoms, especially in high-risk patients. Engaging with treatment teams and encouraging laboratory screening or supplementation in higher-risk patients can be critical for establishing accurate diagnosis and initiating treatment to reverse symptoms and prevent progression.

MEDICATION-INDUCED DELIRIUM

Considering all precipitating risk factors for delirium, medications may be the most common (Marcantonio 2017), attributed to approximately 12–39% of delirium cases in one study (Alagiakrishnan and Wiens 2004). There are a wide variety of medications that affect the central nervous system (CNS) and that may increase the risk for delirium in vulnerable brains. For this reason, one general method of investigating medication-induced delirium risk is by identifying and tracking polypharmacy.

Polypharmacy has various definitions in the literature, including the use of more than one drug, taking more drugs than are clinically recommended, or using ≥5 medications total (Benetos et al. 2015; Midão et al. 2018). Each of these definitions has limitations, including subjectivity (e.g., "clinically recommended") and arbitrariness (e.g., a cutoff value of 5), among others. A more precise conceptualization is that of *inappropriate polypharmacy*, which specifically implies iatrogenesis (i.e., the phenomenon whereby medical procedures or interventions cause negative outcomes in patients). When prescribed drugs are found to be excessive/unnecessary, especially when multiple psychoactive medications are taken contemporaneously, one frequent negative outcome (particularly in older adults) is delirium (Kurisu et al. 2020; McCoy et al. 2021; Mueller et al. 2020).

Instances of polypharmacy have risen over the past few decades, which may explain some portion of the high prevalence rates of delirium in inpatient settings. For example, in a study of over 13,000 older adults in the United States who were followed from 1988 to 2010, the authors reported a doubling of the median number of medications taken per patient, from 2 to 4, and a threefold increase in the proportion of the sample taking ≥5 total medications (12.8% to 39.0%; Charlesworth et al. 2015). Consistent with these findings, international data suggest increases in polypharmacy in a variety of countries worldwide (Wastesson et al. 2018).

As reviewed above, older age is an independent risk factor for delirium due to a reduction in functional reserve (Bugiani 2021). Simultaneously, older adults process drugs less efficiently than younger adults due to physiological changes during aging; namely, alterations in *pharmacokinetics*: the manner in which the body acts upon the drug (e.g., absorption, distribution, metabolism, excretion), and *pharmacodynamics*: the ways in which the drug acts upon the body and brain (e.g., binding to receptors, causing action potentials; Stahl 2021). Consequently, older adults are both more susceptible to delirium in general and more vulnerable to adverse impacts from medications, some of which can have cognitive and behavioral side effects that precipitate delirium. This likely explains why polypharmacy is typically (though not always) a significant risk factor for delirium, particularly for older adults (Heinrich et al. 2021; Kurisu et al. 2020; McCoy et al. 2021; Mueller et al. 2020).

Although crude metrics such as the overall number of medications prescribed are easy to calculate and have some clinical utility, these criteria mask important delirium-relevant variance in risk—both within and across specific medication classes. It is well established that certain types of drugs confer disproportionate risk for cognitive side effects (American Geriatrics Society [AGS] 2019), including delirium (Clegg and Young 2011). In the following sections, we provide a non-exhaustive review highlighting five

CASE EXAMPLE #2:
MEDICATION INDUCED DELIRIUM, INTRODUCTION

Demographics: 67-year-old married white woman with 12 years of education who is a retired administrative assistant.

Admission: Presented to a Level 1 Trauma Center two weeks earlier as an unaccompanied driver involved in a left driver side motor vehicle accident at high speed (she was "T-boned" by a drunk driver who ran a red light). She had a mild TBI (LOC for several minutes, GCS = 13 in the ED) and orthopedic injuries (two fractured ribs, a broken left fibula, and facial contusions).

Hospital Course and Referral: The patient underwent a head CT scan for the mTBI (results were negative) and an open reduction and internal fixation (ORIF) surgery for the broken fibula. She recovered well from surgery, per orthopedic and nursing notes. However, several days following surgery, she developed confusion/agitation and began yelling at staff members and pulling on her IV lines. These fluctuating symptoms continued for the next ~7 days. Neuropsychology was consulted to determine the extent/severity of her acute confusion, to help identify predisposing and precipitating factors, and for treatment/discharge recommendations.

Medical History: Hyperlipidemia, hypertension, type 2 diabetes, hypothyroidism, osteoporosis, arthritis, fall risk

Psychiatric History: Generalized anxiety disorder, depression

Substance Use History: Unremarkable

Medications (Longstanding): Atorvastatin, lisinopril, levothyroxine, amitriptyline, metformin, omeprazole, mirabegron, lorazepam PRN

Medications (During Hospitalization): Codeine (pain), olanzapine (agitation)

Instrumental ADLs: Living independently with her husband in a two-story home; socially active

---

Question 1. What are at least three predisposing factors and at least three precipitating factors for delirium in this patient?

Question 2. Assuming the neuropsychologist determines that the patient is in a state of delirium, what are 2-3 recommendations to consider?

mTBI = mild traumatic brain injury; LOC = loss of consciousness; GCS = Glasgow Coma Scale; ED = Emergency Department
CT = computerized tomography; IV = intravenous; ORIF = open reduction and internal fixation;
PRN = pro re nata (Latin = as needed); ADLs = activities of daily living

Fig. 7.3 Medication-induced delirium, introduction

categories with relevance to inpatient neuropsychologists: benzodiazepines, anticholinergics, opioids, antipsychotics, and antidepressants. The case example in Figs. 7.3 and 7.7 provides additional context and clinical questions for consideration.

Medication Classes

Benzodiazepines

Benzodiazepines have multiple potential therapeutic effects relevant to the inpatient setting, including sedative/hypnotic, anxiolytic, anticonvulsant, muscular relaxant, and amnesic, with the first two being the most common. Benzodiazepines' mechanism of action is to increase the effectiveness of the neurotransmitter gamma-aminobutyric acid (GABA), which exerts an inhibitory effect on the brain via the opening of chloride channels, allowing this negatively charged ion into the neuron, thereby causing a membrane hyperpolarization and making an action potential less likely (Stahl 2021). When both benzodiazepines and GABA are attached to the GABA-A receptor, they work in concert to increase the frequency with which the chloride channels open to a greater extent than GABA alone (meaning that benzodiazepines are a GABA agonist).

Of the many brain networks affected by GABA, one example is the extended amygdala, and GABA's inhibitory effect on the amygdala can reduce fear and anxiety, explaining a portion of these medications' therapeutic effects. However, widespread inhibition across the CNS can also lead to mental slowness, inattention, and general cognitive inefficiencies (Ferreira et al. 2022; Hanlon et al. 1998; Tannenbaum et al. 2012), which likely reduces functional reserve and precipitates delirium in people who are already at risk (Blair et al. 2019; Clegg and Young 2011; Kok et al. 2018; Zaal et al. 2015). Critically, these negative effects also apply to the so-called "non benzodiazepines" or "z drugs," which function as agonists at the GABA-A receptor and are typically used for sedative hypnotic (sleep promoting) indications: zolpidem (Ambien), eszopiclone (Lunesta), and zaleplon (Sonata).

Anticholinergics

The term *anticholinergic* does not refer to a medication class per se, but instead to a specific mechanism of action (acetylcholine antagonism). These anticholinergic effects can be either the primary effect of the drug or a secondary (side) effect. Anticholinergic medications cut across traditional drug classes, and include some tricyclic antidepressants, non-dopaminergic medications used to treat motor symptoms of Parkinson's disease, atypical (second generation) antipsychotics, and even several over-the-counter cold remedies, among others. Because of the pharmacologic diversity of these chemicals, they can have a wide variety of therapeutic indications, and an important concept in this space is that of *anticholinergic burden*, or the cumulative effect of taking multiple medications that block the action of acetylcholine (O'Donnell et al. 2017).

Regarding mechanism of action, there are two primary subtypes or classes of acetylcholine receptor: nicotinic, with an affinity for nicotine, and muscarinic, which are classically stimulated by muscarine (a mushroom alkaloid). The latter (muscarinic receptors) are the target of anticholinergics. Because of the importance of cholinergic projections from basal forebrain networks throughout the cerebral cortex in learning/memory and overall cognitive functioning, the anticholinergic effect can lead to cognitive dysfunction such as mild cognitive impairment/dementia (Campbell et al. 2018; Grossi et al. 2019; Ruxton et al. 2015) and delirium (Campbell et al. 2009; Egberts et al. 2017; Mueller et al. 2020), particularly when the overall anticholinergic burden is high. This is relevant for inpatient neuropsychologists in that many hospitalized patients with cognitive dysfunction may be on medications with anticholinergic effects, and recognition of this as a contributing factor could lead to interventions to improve cognition (e.g., altering a medication regimen). Additionally, inpatient neuropsychologists should be cognizant of the potential cumulative effect of anticholinergics prescribed prior to hospitalization (e.g., oxybutynin, quetiapine, dicyclomine, benztropine), as well as those that may have been administered acutely for management of allergy (e.g., diphenhydramine), nausea/vomiting (e.g., scopolamine), during surgical procedures (e.g., glycopyrrolate), or agitation (e.g., olanzapine).

Opioids

Opioids are a group of drugs that can be naturally derived from the opium poppy plant (e.g., codeine, morphine) or either fully (e.g., fentanyl, methadone) or semi-synthetically (e.g., hydrocodone and oxycodone) produced to mimic a similar chemical compound. It is now widely appreciated that opioids have a high risk for abuse, dependence, and overdose. The abuse potential of opioids is due to their predilection to hijack the brain's mesocorticolimbic dopamine circuitry and the resultant impact on motivated behavior and reward sensations (Koob and Le Moal 2008; Stahl 2021). In response to the well-publicized opioid epidemic in the United States, strict guidelines were created to outline clinical indication for appropriate use of these medications (e.g., short-term management of moderate/severe pain) (Dowell et al. 2022; Lee et al. 2021; Manchikanti 2012).

The clinical use of opioid medications can be a double-edged sword. The primary indication—analgesia—is typically effective in controlling pain. As moderate-severe pain can precipitate delirium, the judicious use of opioids, together with other pain management strategies, may actually improve cognition in certain individuals (Morrison et al. 2003). However, opioids in and of themselves are risk factors for delirium due to the psychoactive CNS effects (Clegg and Young 2011; Swart et al. 2017). This paradox can lead to complex management decisions (e.g., palliative cancer care) in many inpatient scenarios, and neuropsychologists should be aware of cost-benefit considerations in these situations (Paice and Ferrell 2011; Wood et al. 2018).

Antipsychotics

Antipsychotic medications, also known as neuroleptics, were developed to manage symptoms associated with schizophrenia, bipolar disorder, and agitation. First-generation antipsychotics were developed in the 1950s and led to critical improvements in the care of patients with severe mental illness through deinstitutionalization and provision of community-based treatment (Pow et al. 2015). However, due to problematic side effects (e.g., the motor syndrome, *tardive dyskinesia*), these medications were largely replaced by second-generation (or *atypical*) antipsychotics. While both generations act as dopamine antagonists via blockade of D2 receptors in the CNS, first-generation antipsychotics do so via consistent occupancy of D2 receptors, whereas atypical antipsychotics do so transiently, thus allowing for normal dopamine neurotransmission, fewer side effects, and increased safety (Ameer and Saadabadi 2023). However, long-term use of atypical antipsychotics has been associated with problematic side effects as well, particularly related to metabolic derangement and risk for diabetes

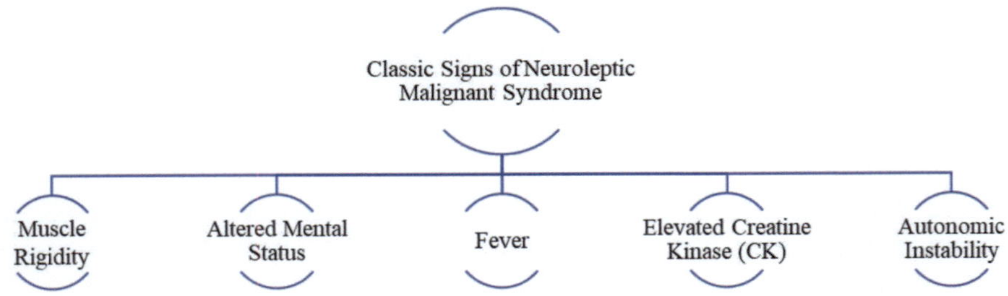

Fig. 7.4 Neuroleptic malignant syndrome

(Hirsch et al. 2017). In inpatient settings, antipsychotics may be acutely administered for management of aggressive behavior, positive symptoms of psychosis, and to treat symptoms of hyperactive delirium. To date, evidence is equivocal as to their true efficacy (Neufeld et al. 2016), with some research advising against use when possible (Meagher et al. 2018).

In addition to frequent side effects, some patients are at risk for the uncommon but clinically significant adverse reaction to antipsychotics known as *neuroleptic malignant syndrome* (NMS, Fig. 7.4), a potentially life-threatening condition characterized by delirium, muscle rigidity (and associated skeletal muscle damage), hyperthermia, and autonomic disruption (Gurrera et al. 2011). The pathophysiology of NMS is not entirely clear, although it has been conceptualized as a reaction to the blockade of D2 dopamine receptors in the hypothalamus and brainstem (Ware et al. 2018). Risk factors include prescription of multiple antipsychotic medications, rapid dose titration, or high daily dose, as well as systemic factors including malnutrition, dehydration, trait-level CNS dopaminergic hypoactivity, iron deficiency, and a prior episode of NMS (Belvederi Murri et al. 2015; Strawn et al. 2007; Ware et al. 2018).

Diagnosis of NMS is one of exclusion and is challenging due to the absence of specific hallmark laboratory abnormalities. While neuroimaging is typically normal, EEG may demonstrate generalized slowing and labs may reveal leukocytosis, metabolic acidosis, and elevations in serum creatinine phosphokinase (Ware et al. 2018). Although there is no standard of treatment for NMS, common intervention involves discontinuation of the offending neuroleptic medication and supportive medical care for management of associated symptoms, including managing dehydration or administrating benzodiazepines or dopaminergic agents (Strawn et al. 2007).

Antidepressants

Antidepressant medications such as monoamine oxidase inhibitors (MAOIs), tricyclic antidepressants (TCAs), and selective-serotonin re-uptake inhibitors (SSRIs) function by modifying the levels of available neurotransmitters, including norepinephrine, serotonin, and dopamine. In most cases, antidepressants act as agonists by inhibiting the breakdown or reuptake of serotonin. Given that these medications are frequently used to treat mood disorders, together with the fact that mental health symptoms represent a common risk factor for delirium, it is not unusual for patients who are vulnerable to delirium to be taking antidepressants. Fortunately, most antidepressants (apart from the anticholinergic effects of TCAs and select SSRIs like paroxetine) do not directly increase the risk for cognitive impairment/delirium and tend to have a wide therapeutic index—indicating low risk for overdose. However, excessive central and peripheral serotonin can, in rare cases, result in CNS toxicity. This excess occurs as a result of concurrent use of several serotonergic agents, overdose of a serotonergic medication, and/or accumulation due to lack of typical pharmacokinetic activity (e.g., renal dysfunction; Scotton et al. 2019).

The cluster of signs and symptoms associated with serotoninergic toxicity is known as *serotonin syndrome* (SS), which is a unique subtype of delirium (Fig. 7.5). SS is marked by mental status change (e.g., confusion, anxiety, agitation, coma), neuromuscular dysfunction (e.g., myoclonus, hyperreflexia, tremor, incoordination), and autonomic hyperactivity (e.g., sinus tachycardia, flushing, hyperthermia, diarrhea; Spadaro et al. 2022; Talton 2020). Diagnosis of SS is made based on a thorough clinical history that establishes ingestion of a serotonin-elevating, substance and clinical presentation using established diagnostic criteria (i.e., the Hunter, Sternbach, or Radomski criteria; Dunkley et al. 2003; Radomski et al. 2000; Sternbach 1991). Treatment depends on the severity of toxicity and symptom presentation but should prioritize cessation of the medication associated with the reaction and supportive care to stabilize the patient. In more severe cases it may be appropriate to use a benzodiazepine for sedation, IV fluids for dehydration or hypotension, mechanical ventilation in cases of extreme muscle rigidity, or the use of a serotonin antagonist (Scotton et al. 2019; Talton 2020). In most situations, however, symptoms resolve within 24 hours of stopping the offending drug, with good prognosis (Scotton et al. 2019).

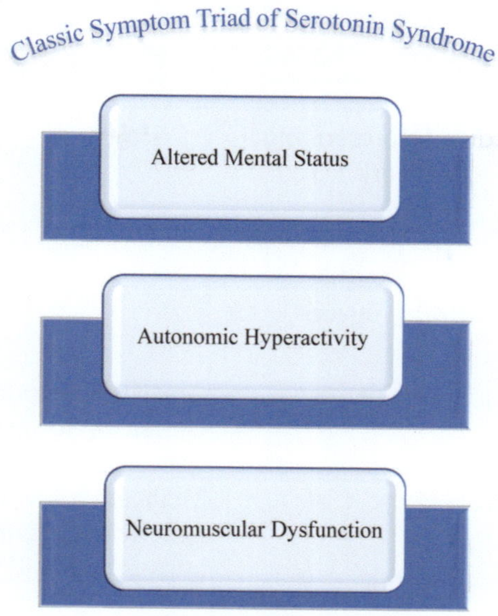

Fig. 7.5 Serotonin syndrome

Deprescribing

Issues surrounding increases in polypharmacy, overprescribing, toxic syndromes, and the negative side effects of various medications have garnered a great deal of attention in the medical community. Indeed, an entire area of inquiry, labeled *deprescribing,* has the stated goal of combating inappropriate polypharmacy, often including the psychoactive medications described above (Scott et al. 2015; Steinman 2016; Steinman and Landefeld 2018). The process of deprescribing involves an examination of the cost-benefit ratio of each medication so that only those drugs that are doing more harm than good are considered for removal. In this manner, deprescribing can be conceptualized as an intervention in its own right; just as the addition of the right medication for the right person can be therapeutic, the removal of the wrong medication can also have clinical benefit. This is highly relevant to delirium, where medications are the most common modifiable risk factor (Marcantonio 2017; Mattison 2020).

Much of the effort in deprescribing has focused on geriatrics, given the disproportionate effects of polypharmacy on older adults. For example, the US Deprescribing Research Network (USDeN) aims to generate and disseminate knowledge about effective deprescribing strategies in older adults via grant funding, investigator development, stakeholder engagement, and other efforts (Thompson et al. 2022). Clinically, several tools, guidelines, and initiatives have been created to help frontline providers make informed decisions about medications and avoid potentially inappropriate prescribing (PIP). For example, the American Geriatrics Society (AGS) Beers List criteria (AGS 2019), the Screening Tool or Older Persons' Prescriptions/Screening Tool to Alert Doctors to Right Treatment (STOPP/START) criteria (O'Mahony et al. 2014), and the Medication Appropriateness Index (MAI; Hanlon and Schmader 2013) identify medications that may pose a greater risk of harm (versus benefit) in populations who are at increased risk for delirium. Table 7.2 provides helpful resources for identifying specific medications that may be contributing to delirium.

The method for approaching deprescribing varies by setting and provider, but often includes consulting the AGS Beers criteria and/or similar publications. Some of these references provide tables identifying deliriogenic drugs and suggest alternative treatment considerations (Marcantonio 2017), while others outline decision algorithms for identifying and minimizing the deliriogenic effect of high-risk medications (Campbell et al. 2009). As evidence for the benefit of deprescribing, van Velthuijsen et al. (2018) examined the impact of a medication reconciliation (or medication review) on clinical outcomes in 218 older adults with delirium. The study team found that medication review (which led to a reduction in medications in most cases) was associated with a shorter length of delirium relative to a control condition.

Table 7.2 Resources to address medication-induced delirium

Purpose	Name/Title	Reference
Education, patient resources, and research funding mechanisms related to deprescribing	Canadian Deprescribing Network	https://www.deprescribingnetwork.ca/
	U.S. Deprescribing Research Network	https://deprescribingresearch.org/
Identification of potentially deliriogenic medications	AGS Beers' List	American Geriatrics Society. (2019). American Geriatrics Society 2019 Updated AGS Beers Criteria® for Potentially Inappropriate Medication Use in Older Adults. Journal of the American Geriatrics Society, 67(4), 674–694. https://doi.org/10.1111/jgs.15767
	STOPP/START criteria	O'Mahony, D., O'Sullivan, D., Byrne, S., O'Connor, M. N., Ryan, C., & Gallagher, P. (2014). STOPP/START criteria for potentially inappropriate prescribing in older people: Version 2. Age and Ageing, 44(2), 213–218. https://doi.org/10.1093/ageing/afu145
	Medication Appropriateness Index	Hanlon, J. T., & Schmader, K. E. (2013). The Medication Appropriateness Index at 20: Where It Started, Where It Has Been, and Where It May Be Going. Drugs & Aging, 30(11), 893–900. https://doi.org/10.1007/s40266-013-0118-4
Calculation of anticholinergic burden	Anticholinergic Cognitive Burden Scale (ACBS)	Boustani, M., Campbell, N., Munger, S., Maidment, I., & Fox, C. (2008). Impact of anticholinergics on the aging brain: A review and practical application. Aging Health, 4(3), 311–320. https://doi.org/10.2217/1745509X.4.3.311
	Anticholinergic Risk Scale (ARS)	Rudolph, J. L., Salow, M. J., Angelini, M. C., & McGlinchey, R. E. (2008). The anticholinergic risk scale and anticholinergic adverse effects in older persons. Archives of Internal Medicine, 168(5), 508–513. https://doi.org/10.1001/archinternmed.2007.106
	Anticholinergic Drug Scale (ADS)	Carnahan, R. M., Lund, B. C., Perry, P. J., Pollock, B. G., & Culp, K. R. (2006). The Anticholinergic Drug Scale as a measure of drug-related anticholinergic burden: Associations with serum anticholinergic activity. Journal of Clinical Pharmacology, 46(12), 1481–1486. https://doi.org/10.1177/0091270006292126

Medication Review: *"The patient is currently prescribed [insert medication name], which can precipitate delirium in vulnerable individuals. It is recommended that the patient, her family, and her prescriber meet to discuss the cost-benefit ratio of this medication, and to determine if an alternative treatment with lower risk of cognitive side effects is indicated."*

Medication Review: *"A review of the patient's medications and supplements may be helpful to determine the extent to which polypharmacy and/or medication interactions may be contributing to his cognitive symptoms (i.e., delirium). I recognize that the analysis of medications involves cost-benefit considerations, and I defer to the expertise of his medical team in this matter."*

Fig. 7.6 Example recommendations for medication review

The specific role of a neuropsychologist in the deprescribing process is nuanced. Most neuropsychologists are not trained as prescribers and should defer to the prescribing provider in collaboration with the patient for final decisions regarding adding/removing medications. However, this does not necessitate a passive or uninvolved approach, and inpatient neuropsychologists may consider requesting a medication reconciliation or review conducted by a pharmacist as an intervention for delirium when polypharmacy is a concern. Further, interdisciplinary and patient-centered care includes close communication between various specialists on the healthcare team and it is well within the scope of neuropsychologists' expertise to raise concerns about potential contributors to cognitive deficits, including medications. This can occur in the neuropsychological report and/or during direct consultation with medical colleagues. With direct communication that is appropriately professional, the neuropsychologist can remain within the bounds of their competence while also advocating for best practices in patient care. Figure 7.6 provides two examples of recommendations for medication reconciliation that an inpatient neuropsychologist might include in their report.

Addressing Medication Effects

Neuropsychologists are well versed in recognizing, prioritizing, and ameliorating risk factors for cognitive impairment, which can include addressing inappropriate polypharmacy (Fig. 7.7). This necessitates remaining informed and up to date on the potential cognitive side effects of common medications used for inpatients with delirium. If there is concern for possible iatrogenic effect of a medication, we recommend:

- Discussing questions/concerns directly with the prescribing provider (e.g., via a phone call, an electronic message, or at daily clinical rounds).
- Providing patients and families with education about cognition and prescribed medication.
- Providing patients and families with strategies with which to self-advocate with their prescribers (e.g., make a list of current medications, write down questions for providers in advance, follow up with providers if there are inconsistencies in the medication regimen).
- Addressing the cognitive impact of medications in documentation and including tailored recommendations for an interdisciplinary audience.
- Placing a referral for a medication review to the Pharmacy Service. This will initiate a review of the patient's chart by a pharmacist who will provide clinically informed recommendations to the patient's treatment team without actively changing the medication list.

CASE EXAMPLE #2:
MEDICATION INDUCED DELIRIUM, CONCLUSION

Question 1.

Predisposing factors: Age, likely cerebrovascular disease, frailty, long term ingestion of a benzodiazepine (lorazepam), mental health symptoms.

Precipitating factors: Physical trauma (MVA), surgery with anesthesia, acute pain, opioid medication (codeine), possible situational factors in the ICU (e.g., lack of circadian cues, poor sleep, psychosocial stress, isolation, and loneliness).

Question 2.

Recommendations may include:

A) Medical

- Optimize nonpharmacological management of pain and minimize use of opioids where possible.
- Consider a medication review to determine if polypharmacy can be reduced; note that several medications she was taking prior to the MVA do not have associated medical conditions in her chart.
- Check lab results for other reversible causes of delirium.

B) Hospital-Based Behavioral/Environmental Interventions

- Ensure exposure to natural light and provision of large clocks and calendars to facilitate orientation.
- Minimize interruptions to the patient's sleep at night.
- Limit the use of physical restraints and limit antipsychotic medications for hyperactive delirium.
- Consider a physical therapy consultation for promotion of daily activity, as medically indicated.
- Provide access to sensory correction (e.g., eyeglasses, hearing aids) if needed.
- Allow frequent visits from friends/family (video call if necessary); encourage the placement of pictures of loved ones in the patient's room based on her wishes.

C) Disposition

- Many people recover fully from delirium and return to their baseline level of cognitive functioning. With treatment and resolution of her delirium, it is likely that the patient will be able to return to independent living with her husband. However, delirium is also an independent risk factor for cognitive decline and dementia. Due to her pre-injury cognitive risk factors (particularly cerebrovascular) and the medical complexity/severity of her MVA and hospital stay, she is at risk for cognitive decline. Therefore, her cognition should be monitored following resolution of delirium and discharge from the hospital.

MVA = motor vehicle accident; ICU = intensive care unit

Fig. 7.7 Medication-induced delirium, conclusion

DRUGS OF ABUSE

Use of CNS-altering substances of abuse such as alcohol and illicit drugs represents a unique and important risk factor for delirium. Chronic heavy substance use has been identified as a predisposing factor for delirium, while acute health complications associated with misuse (e.g., withdrawal, poor nutrition, metabolic dysfunction) serve as precipitating factors (Wilson et al. 2020). Notably however, given the illicit nature of many substances of abuse and corresponding constraints in research (e.g., due to U.S. Drug Enforcement Agency [DEA] scheduling), evidence pertaining to the link between use, abuse, intoxication, and withdrawal is more limited than for prescribed medications. In the following sections, we provide a review of substances thought to be most relevant to inpatient neuropsychologists: alcohol, cannabis, psychostimulants, and narcotics. The case example in Fig. 7.8 provides additional context and clinical questions while reading through the sections below.

Alcohol

Alcohol consumption is widespread, with global estimates of approximately 2.3 billion current drinkers, driven largely by high rates of consumption (> 50% of the population) in the Americas, Europe, and Western Pacific (World Health Organization [WHO] 2018). In the United States, the prevalence of alcohol consumption, episodes of binge drinking, and related problems (e.g., emergency department visits, deaths) have risen, particularly among middle-aged and older adults (Grucza et al. 2018). Recently, the COVID-19 pandemic further exacerbated this trend (Barbosa et al. 2021).

CASE EXAMPLE #3:
SUBSTANCE INDUCED DELIRIUM, INTRODUCTION

Demographics: 57-year-old, single, white man with 10 years of education who is currently unemployed.

Admission: Brought to a Level 1 Trauma Center by the police after he was found wandering in a nearby street. At the time of admission, he struggled to engage in intake interview. He was profoundly disoriented, inattentive, and forgetful. Gait was slow and wide-based when he was found but upon arrival to the hospital, he was unable to ambulate independently.

Hospital Course and Referral: Initial assessment revealed gait ataxia and oculomotor dysfunction, including nystagmus and pupillary sluggishness. Laboratory assay revealed very low levels of circulating thiamine, but complete blood count and comprehensive metabolic panel were within normal limits. Brain MRI showed decreased T1 signal in the medial thalamus, superior and inferior colliculi (i.e., tectal plate), and mammillary bodies. Intravenous thiamine was administered, and oculomotor dysfunction resolved. Gait ataxia and confusion gradually improved over the next 5 days; however, the patient continued to demonstrate impaired memory. He regularly expressed a belief that he was at a hotel in another city. Neuropsychology was consulted to assess for predisposing and precipitating factors, and to comment on prognosis and long-term recommendations.

Question 1. During the patient's hospitalization a family member arrived. What information would be important to gather from the patient's family member about predisposing or precipitating factors of delirium?

Question 2. What are prognostic considerations and long-term recommendations that you may have?

MRI = magnetic resonance imaging

Fig. 7.8 Substance-induced delirium, introduction

Acutely, alcohol serves as a CNS depressant through increased potentiation of GABA receptors/GABAergic neurotransmission and suppression of glutamatergic activity (Grover and Ghosh 2018). Chronic use of alcohol results in adaptive changes in the CNS, including down-regulation of inhibitory GABA receptors and up-regulation of excitatory N-methyl-D-aspartate (NMDA) glutamate receptors (Calvo et al. 2018). Sudden cessation of alcohol following chronic or heavy use exposes this neurotransmitter imbalance, resulting in a state of acute CNS hyperstimulation and excitotoxicity, known more commonly as alcohol withdrawal (Long et al. 2017). Typically, symptoms of withdrawal, such as nausea/vomiting, tremulousness, headache, sweating, palpitations, tachycardia, and hypertension, arise within hours of last use and subside within 1–7 days (Manasco et al. 2012).

The timely assessment and treatment of patients admitted with signs of alcohol withdrawal syndrome is vital to reducing the risk of more severe complications such as seizures or delirium. Symptom management and intervention is informed by the course and severity of symptoms. This can be measured using instruments such as the Clinical Institute Withdrawal Assessment for Alcohol (CIWA-Ar; Sullivan et al. 1989) or the Minnesota Detoxification Scale (MINDS; DeCarolis et al. 2007). These brief measures use a combination of interview and observation to assess for the presence and severity of the most common withdrawal symptoms. While considered the clinical gold standards, these measures can be limited in intensive care settings where patients may be unable to respond to prompts due to intubation, sedation, or delirium (Steel et al. 2021). When administered, ratings inform the provider's choice of intervention, including the need for supportive interventions for mild cases (e.g., fluid resuscitation, thiamine supplementation, nutritional/electrolyte intake), optimal administration of benzodiazepines (fixed-dose versus symptom-triggered), or the benefit of alternative agents that may target specific symptoms or reduce the risk of seizure or other neurologic injury (e.g., valproic acid, gabapentin; Farrokh et al. 2021; Pribék et al. 2021).

Delirium Tremens

Repeated assessment of mental status early in an admission for severe alcohol withdrawal is critical, as individuals may develop *delirium tremens* (DT; Fig. 7.9). DT is a rapid-onset delirium accompanied by autonomic instability (e.g., tachycardia, sweating, hyperthermia), tremor, and occasionally, psychosis, hallucinations, seizures, coma, and death (Long et al. 2017). Although physical symptoms can onset quickly, alterations in mental status typically develop within 24–72 hours and worsen in the evening (Mainerova et al. 2015).

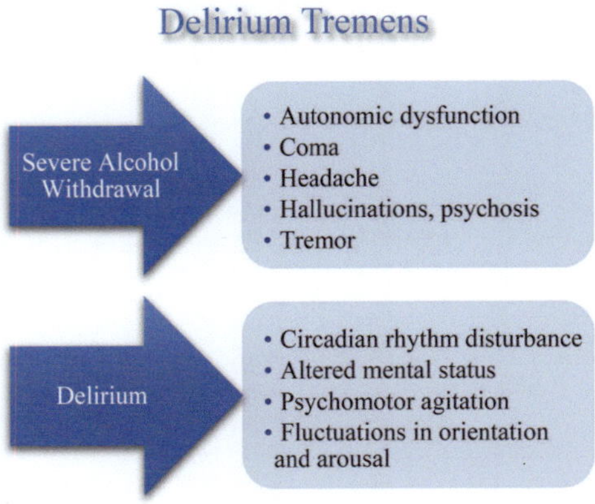

Fig. 7.9 Delirium tremens

Treatment of DT requires interdisciplinary cooperation, as multiple body systems are affected. The inpatient team should rule out, or identify and manage, other contributors to delirium (e.g., metabolic dysfunction, head injury, infection, mineral/electrolyte imbalance), as well as withdrawal symptoms. Inpatient neuropsychology can address mental health and behavioral symptoms, assess cognitive status, and monitor for complications (Grover and Ghosh 2018). With appropriate intervention, acute symptoms of DT typically resolve within 1–8 days, though other symptoms of withdrawal may remain present for several days thereafter.

Wernicke's Encephalopathy

Chronic and heavy alcohol consumption is also a risk factor for Wernicke's encephalopathy (WE), a life-threatening condition that stems from a thiamine (vitamin B1) deficiency. Thiamine deficiency has been linked to brain damage, including but not limited to lesions in the bilateral thalami, mammillary bodies, periaqueductal gray, and dorsal midbrain, with mammillary body atrophy being relatively specific for WE (Manzo et al. 2014). As neural and glial functioning requires immense amounts of energy, these anatomical areas are particularly sensitive to the effects of limited ATP (Magistretti and Allaman 2015). In addition, disruption in metabolic processes that require adequate levels of thiamine can lead to lactic acidosis, neural cell excitotoxity and damage, as well as inflammation that degrades the integrity of the BBB (Kohnke and Meek 2021; Ott and Werneke 2020). The three classic features of a clinical diagnosis of WE include delirium (e.g., confusion, disorientation, drowsiness, inability to concentrate), ataxia, and oculomotor dysfunction (e.g., nystagmus), though all three components are not required to make a diagnosis (Fig. 7.10). In fact, the full triad is seen in less one one-third of patients on initial examination (Ota et al. 2020).

In individuals with alcohol use disorder, WE can occur because of inadequate nutritional intake, decreased absorption from the gastrointestinal tract, and problems with cellular uptake and utilization. However, while alcohol abuse is the most common cause in the United States (Sinha et al. 2019), WE can develop from any condition that impacts the intake or absorption on thiamine. The clinical presentation of non-alcohol-related WE differs from alcohol-related WE, in that the latter has a greater likelihood of oculomotor abnormality and less frequent cerebellar signs (Chamorro et al. 2017), which may delay diagnosis.

Korsakoff Syndrome

When WE goes unrecognized and thiamine is not administered, patients can develop subsequent Korsakoff syndrome (KS, Fig. 7.10), an irreversible cognitive disorder characterized by dense anterograde and retrograde amnesia with confabulation and anosognosia (Kohnke and Meek 2021). Despite this classically defined triad, all features do not invariably occur, with confabulations less likely to be present (Eckardt and Martin 1986). Although initial symptoms of WE may be missed clinically, patients with KS will still demonstrate WE-associated mammillary body lesions on MRI or autopsy (Harper 1983). In addition, volume loss in the medial thalami, corpus callosum, anterior thalamus, and basal forebrain are highly suspicious for KS, though absence of atrophy does not disqualify a KS diagnosis in the context of supportive clinical signs (Pitel et al. 2012; Salmon and Butters 1987).

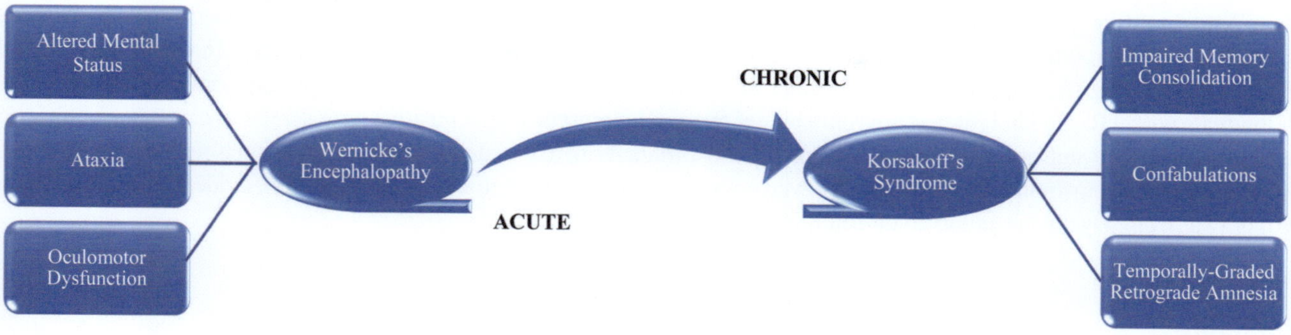

Fig. 7.10 Wernicke's encephalopathy and Korsakoff's syndrome

Misdiagnosis of KS is not uncommon. In our experience, medical teams are quick to consider WE for any patient with alcohol use history, though they are less adept at identifying KS. Consequently, these patients may be misdiagnosed having a primary psychiatric disorder (as confabulations may be misconstrued as delusions) or delirium given broad disturbances in attention. This may lead to eventual neuropsychological consultation due to a working hypothesis of cryptogenic delirium or other neuropsychiatric change, which then reveals the presence of KS. To avoid these pitfalls, we recommend a two-pronged approach: a) thorough clinical interview with a trusted family member or other collateral and review of all available records to identify discrepant information that may signify a confabulation and, b) education to the medical team on the difference between confabulations, delirium, and psychosis—the latter of which is unlikely to idiopathically emerge later in life. Finally, the dense amnesia characteristic of KS may be misattributed to dementia due to Alzheimer's disease. In differentiating between these two syndromes, language testing can elucidate deficits in semantic networks, which are common in AD but not KS, while the presence of confabulations is more unique to KS. For these reasons, we recommend considering KS in patients with unexplained impairment who have a history of known risk factors (e.g., heavy alcohol use, homelessness or transience).

Psychostimulants

Psychostimulant medications are widely used in both clinical and recreational settings worldwide. Clinically, psychostimulants (i.e., amphetamine, methylphenidate, modafinil) are prescribed for their ability to increase alertness, improve attention, reduce psychomotor agitation, and enhance cognition, such as in individuals with ADHD or narcolepsy. Recreationally, psychostimulants (e.g., cocaine, methamphetamine) are sought for increased energy and feelings of euphoria (McCreary et al. 2015). Within the brain, psychostimulants appear to act on the prefrontal cortex and the broader frontal-striatal circuitry largely via increased dopaminergic and noradrenergic activity with downstream effects throughout the cortex (Berridge and Devilbiss 2011).

Chronic or excessive use of psychostimulants can have neurotoxic effects via alteration to dopamine signaling, oxidative stress, neuroinflammation, and glutamatergic overactivity with subsequent excitotoxicity (Gonçalves et al. 2014). In the case of cocaine toxicity, this can result in acute symptoms (e.g., hypertension, tachycardia, headache, nausea, agitation, restlessness, confusion) that can progress to full delirium and other central nervous system symptoms (e.g., seizures, areflexia, coma, loss of vital functions; Richards and Le 2023). Notably, some have linked excessive use of stimulants, such as cocaine or methamphetamine, to acute hyperactive or excited delirium syndrome (ExDS). This clinical syndrome has been characterized as AMS paired with psychomotor agitation, violence, aggression, paranoia, unexpected physical strength, hyperthermia, and even sudden death (Gonin et al. 2018; Springer 2023; Takeuchi et al. 2011). Unfortunately, the nature of ExDS and best course for management remains somewhat unclear as there is no consensus-based diagnosis. ExDS is recognized by the American College of Emergency Physicians (ACEP 2009) but not by major healthcare societies including the American Psychological Association (APA), American Medical Association (AMA), or World Health Organization (WHO; Saadi et al. 2022; Walsh et al. 2023).

Emerging literature suggests that even at non-toxic levels, use of stimulants may contribute to an increased risk of developing delirium when paired with other brain injuries. For example, Safdar et al. (2024) found that among patients who presented to the hospital following a traumatic brain injury (TBI), those who screened positive for psychostimulants (i.e., amphetamine, methamphetamine, cocaine, and ecstasy) had higher odds of developing a delirium during their hospitalization. While not thought to reflect causation, the authors point to the complex interaction that may occur between TBI and stimulants at the structural, biochemical, and cellular levels as a possible explanation for this finding. Inpatient staff, including neuropsychologists, should therefore consider this connection and consider advocating for early drug screening among individuals with TBI, and heightened surveillance for those who present with a positive drug screen to carefully monitor for the development of delirium. Withdrawal from psychostimulants is associated with symptoms such as agitation, fatigue, anxiety, dysphoria, insomnia/hypersomnia, slowed cognition, and psychomotor slowing. However, evidence directly linking withdrawal from psychostimulants to delirium is limited at present.

Cannabis

Cannabis is one of the most widely consumed psychoactive substances, having been used by an estimated 4% of the global population (approximately 209 million people) between the ages 15 and 64 in 2020 (United Nations Office on Drugs and Crime 2022) and an estimated 18.7% of individuals over the age of 12 in United States in 2021(Substance Abuse and Mental Health Services Administration 2021). The *Cannabis sativa* plant contains more than 100 phytocannabinoids, including cannabidiol (CBD) and delta-9-tetrahydrocannibinol (THC). When consumed, THC exhibits an agonist effect on the endogenous cannabinoid receptor system via cannabinoid type 1 (CB1) receptors which results in acute euphoria but long-term downregulation of CNS CB1 receptors (Lu and Mackie 2021; Spindle et al. 2021). Despite this known impact on the endocannabinoid system, cannabis-induced delirium is poorly understood and may be impacted by misdiagnosis of cannabis-induced psychosis with agitation (Gorelick 2023). Available case reports suggest that a hyperactive delirium characterized by hyperactivity, agitation, autonomic instability, disorientation, and hallucinations is more common than hypoactive delirium (Leikin and Amusina 2017). Other predisposing factors at the time of THC ingestion likely increase risk for cannabis-induced delirium, as was suspected in a recent case study that described admission and treatment of a 94-year-old woman who experienced altered mental status and hallucinations after ingesting THC in the context of dehydration and poor oral intake (Moussa et al. 2023).

Addressing Substance-Induced Delirium

Inpatient neuropsychologists are well-positioned to assess for substance-induced delirium within the interdisciplinary team (Fig. 7.11). If a neuropsychologist observes that substance use may be contributing to a patient's delirium, there are several specific ways in which they can contribute to the relevant clinical decisions:

- Conduct a thorough history of recently or regularly ingested substances with collateral reporters, when possible. Gather information about types, quantities, recent changes, and sources from which the patient acquired their substances of abuse.
- Discuss concerns about a patient's substance use history with their medical providers to ensure appropriate laboratory workup has been conducted. It may be important to advocate for review of prescribed medications or for the completion of a tox screen in individuals presenting with delirium or at risk of delirium due to other brain injury (e.g., TBI), as these substances may exert an additive or exacerbating effect resulting in onset of or prolonging of delirium.
- Communicate with all relevant stakeholders about short- and long-term recommendations related to substance use in the neuropsychological report. This should include discussion of both the acute presentation of delirium and the potential long-term consequences. Discussion should be specific to the underlying mechanism of delirium. For example, if a patient presented with a Wernicke's encephalopathy, it would be important to inform care partners of their elevated risk of persistent cognitive deficits (i.e., Korsakoff syndrome). Neuropsychologists should present information on the typical course of recovery, and common symptoms of dementia (e.g., rapid forgetting, the potential for confabulation) for which family should monitor.
- Provide outpatient resources and referrals for substance use treatment.
- Clearly delineate "next steps" for family, including the need for repeat evaluation in the future (ideally following a period of abstinence from substances), compensatory strategies that may be specifically beneficial to the patient/family, and behavioral management techniques.

CASE EXAMPLE #3:
SUBSTANCE INDUCED DELIRIUM, CONCLUSION

Question 1.

Predisposing factors: Age, history of substance use (duration, frequency, substance of use), dietary history (i.e., evaluate for poor nutritional status), past episodes of altered mental status, surgical history.

Precipitating factors: Changes to substance intake, modifications to diet, recent weight loss, acute stressors, trauma (e.g., involvement in fights).

Question 2.

Given ongoing memory complaints after thiamine administration, there is concern for the development of Korsakoff Syndrome. This may include ongoing memory difficulties, confabulations, or (in rare cases) hallucinations.

Recommendations may include:
- Continued abstinence from alcohol.
- Oral thiamine (vitamin B1) supplementation.
- Eating a well-rounded diet.
- Assistance for instrumental activities of daily living (as appropriate given the patient's status at discharge).
- Outpatient neuropsychological evaluation following discharge to assess memory.

Fig. 7.11 Substance-induced delirium, conclusion

INFECTION

Infection refers to the invasion and multiplication of microorganisms (usually bacterial or viral) in biological tissue. Humans and other mammals possess an immune system that is designed to identify and combat infections, comprised of both innate and adaptive components. The innate immune system enacts a quick, but nonspecific defense to infection, followed by the adaptive immune system, which provides targeted responses to specific infections via specialized cells and processes. While the human immune response is critical for fighting infections, some aspects, including inflammation, can also lead to adverse effects.

Inflammation is part of the innate immune response, with the goal of combating the offending microorganism. Inflammation is characterized by the dilation of blood vessels, which aids in flooding infected tissue with blood plasma containing immune cells that identify and destroy pathogens, as well as releasing signaling molecules (called cytokines) and other messengers. Infections and other causes of inflammation, particularly if prolonged, can affect the brain in multiple ways, including compromise of the blood-brain barrier (BBB) and infiltration of immune cells into the CNS causing neural damage (Cunningham 2011; Elie et al. 1998; Maldonado 2018).

Thus, while infections are typically localized to one area or region, the immune response can impact organ systems throughout the body, including the brain. For example, urinary tract infections (UTIs), are common localized infections that frequently cause delirium in older adults and account for up to 25% of hospital admissions (Dutta et al. 2022). While UTIs may be confined to the urinary tract, the inflammatory cascade described above can result in delirium. However, while localized infections are much more frequent, infections can also be widespread. Systemic infection, known as sepsis, is a less common, yet potent cause of delirium and is a syndrome likely less familiar to the inpatient neuropsychologist.

Sepsis

Sepsis is a life-threatening disorder caused by a robust and dysregulated autoimmune response to an infection, leading to organ and tissue injury. It is a leading cause of intensive care unit (ICU) admissions and a costly health concern (Rezende et al. 2008; Salluh et al. 2010). In the US alone, the annual financial burden of sepsis and associated complications is estimated to exceed $24 billion (Martin et al. 2003; Torio and Moore 2016). Globally, sepsis and its derivatives (see below) are estimated to contribute to ~20% of all deaths and remain a leading cause of death in ICUs, with an associated mortality rate of ~20–40% (Levy et al. 2012; Rudd et al. 2020; Sprung et al. 1990).

Although sepsis can arise from a localized infection, it is more common in instances of systemic infection from pathogens circulating in the blood (called *septicemia*). Fungal and viral infections can cause sepsis; however, bacteria are the most common cause (termed bacteremia), usually due to *Staphylococcus aureus, Enterococcus faecium, Acinetobacter* spp., *Pseudomonas aeruginosa,* or *Stenotrophomonas maltophilia* (Martin et al. 2003; Zhang et al. 2012). This systemic infection triggers an inflammatory cascade (i.e., a cytokine storm), which can directly injure organs, vasculature, and other tissue via apoptosis and microvascular lesions. Moreover, disseminated intravascular coagulation can occur, whereby small blood clots cause microthrombi, further damaging organs (Iba et al. 2020). Alterations in the BBB can lead to disruptions in CNS cell signaling and neurotransmission, as well as parasympathetic nervous system derangements and increased cortisol (Iacobone et al. 2009; Maldonado 2017; Maldonado 2018).

Diagnostic criteria for sepsis have rapidly evolved over time. Current consensus guidelines define sepsis as life-threatening organ dysfunction due to a dysregulated host response to infection (Gül et al. 2017). Organ dysfunction is operationalized as an increase of two more points on the Sequential Organ Failure Assessment (SOFA) tool, which includes decreased blood oxygen, decreased blood platelets, increased bilirubin, worsening hypotension, lower GCS score, and increased creatinine as input metrics. When lactate levels remain elevated and blood pressure is unable to be maintained despite administration of fluids, patients are considered to have *septic shock,* which confers a greater risk of mortality (>40%) than sepsis alone (>10%) (Singer et al. 2016; Vincent et al. 1998).

Altered mental status is a core symptom, and often one of the earliest signs of sepsis. Sepsis-associated encephalopathy (SAE) is considered the leading cause of delirium in the ICU, with approximately 30–70% of patients with sepsis experiencing altered mental status (Ely et al. 2004; Lu et al. 2023; Siddiqi et al. 2006; Sonneville et al. 2017). SAE occurs on a spectrum from mild delirium to coma, with evidence suggesting hypoactive delirium as being more common than hyperactive presentations (90% versus 10%, respectively; Ely et al. 2004; Sonneville et al. 2017). In addition to AMS, a neurologic work-up of SAE often reveals paratonic rigidity (i.e., velocity-dependent limb resistance to passive movement) and neuromyopathy, the latter affecting approximately 70% of patients with SAE in the subsequent days to weeks (Eidelman et al. 1996). Evidence of CNS dysfunction is often observed through EEG (e.g., diffuse slowing) and evoked potentials (Hosokawa et al. 2014). SAE is important from a prognostic standpoint, as its presence and severity are associated with increased morbidity and mortality (Sprung et al. 1990; Eidelman et al. 1996; Zhang et al. 2012).

Addressing Sepsis-Associated Encephalopathy

Treatment for SAE consists of identifying the causative infectious agent, followed by addressing the underlying infection and immunological response (e.g., IV antibiotics, fluid resuscitation, vasopressors). Rapid assessment is critical in determining the presence of an infectious vs non-infectious etiology and determining the clinical indication for antibiotics (Evans et al. 2021). International guidelines for the treatment/management of sepsis and septic shock were updated in 2021, although intravenous antibiotic therapy remains the mainstay of treatment for most causes of sepsis (Evans et al. 2021). In those with suspected sepsis due to bacteremia, broad-spectrum antibiotics should be administered within the first hour of recognition as time-to-administration is key to reducing mortality (Ferrer et al. 2014; Gaieski et al. 2010). General behavioral and environmental strategies to manage delirium are also recommended. Neuropsychologists can be helpful in identifying the mental status alterations central to sepsis, as well as tracking patients' mentation over time to evaluate for improvement and clearance from delirium.

SOLID ORGAN DYSFUNCTION

Liver Disease

Liver disease refers to typically progressive declines in liver functioning, most often due to longstanding heavy alcohol use, nonalcoholic fatty liver disease, or infections such as hepatitis (Sharma and Nagalli 2023). While in rare instances it can occur acutely, most cases of liver disease are chronic in nature, which is the focus of this chapter. End-stage liver disease is known as cirrhosis, the most of severe type of which is termed "decompensated cirrhosis" and is characterized by symptoms including ascites, variceal bleeding, and jaundice.

Hepatic encephalopathy (HE), also referred to as portosystemic encephalopathy (PSE), is delirium due to liver disease, which is commonly seen with cirrhosis (Amodio 2018; D'Amico et al. 2006). Episodes of HE are typically triggered by precipitants related to hepatic dysfunction and treatment, including infection, gastrointestinal bleeding, dehydration, electrolyte imbalances such as hyponatremia and hypokalemia, and the use of sedative medications such as benzodiazepines (Weissenborn 2019). While outside the scope of this chapter, HE is medically graded on a 1–4 scale called the West Haven Criteria, with higher scores indicative of more severe disease and less favorable prognosis (Vilstrup et al. 2014).

HE is difficult to objectively measure and track longitudinally, and precise prevalence and incidence rates vary in the literature. Some studies estimate that up to 30–45% of patients with cirrhosis will experience delirium and other significant signs of HE, while more subtle neurobehavioral changes may be present in up to 80% of patients with cirrhosis over the course of the disease (Rudler et al. 2020). HE prevalence also varies by liver disease etiology and is both more common and severe in alcohol-related liver disease (ARLD) relative to viral and nonalcoholic fatty liver disease (NAFLD; Tapper 2019; Sanyal et al. 2021).

Like prevalence, the precise pathophysiology of HE is also unclear, although factors such as neurotoxins, systemic inflammation, and derangements in brain metabolism are commonly implicated (Ferenci 2017). These factors can result in oxidative stress, increased glucocorticoid production, and both direct and indirect neurotransmitter changes (Maldonado 2018). At a minimum, the pathophysiology of liver dysfunction in HE includes hepatocellular failure and/or vascular changes that allow blood to bypass the liver, preventing normal detoxification and filtering (portosystemic shunting).

Elevated blood ammonia (i.e., hyperammonemia) is the prevailing mechanistic hypothesis for HE (Romero-Gómez et al. 2015). Ammonia is predominantly produced in the gastrointestinal tract as a byproduct of nitrogen metabolism, which is typically metabolized by the liver via the urea cycle. Ammonia can penetrate the BBB and is implicated in at least two neurotoxic processes—the disruption of neurotransmission and the swelling of astrocytes—which can lead to edema (Häussinger et al. 2000; Blei and Larsen 1999; Raabe 1993). Multiple neurotransmitter systems can be affected, including GABA, dopamine, serotonin, and glutamate (Ferenci 2017). However, despite the known neurotoxic effects of ammonia, lab tests that analyze the quantity of circulating ammonia are equivocal in diagnosing HE, suggesting a nonlinear or more complicated pathophysiological relationship (Rudler et al. 2020).

Clinically, HE is dichotomized by phenotype. With covert or minimal HE (MHE), patients may exhibit subtle clinical symptoms that are only detectable through formal neuropsychological testing or based on observation from collateral informants familiar with the individual's baseline behavior (Amodio 2018). Personality changes are not uncommon in MHE and may include apathy, irritability, and mood lability. Patients can vacillate between irritable, depressed, and anxious states, with some exhibiting social disinhibition and poor judgment. Sleep-wake cycles are also often disturbed, with many patients developing insomnia at night and hypersomnolence during the day. Sleep-wake inversion has been identified as an early warning sign of HE even before overt confusion presents. As severity progresses into acute or overt HE (OHE), cognitive and behavioral deficits become more pronounced with symptoms ranging from disorientation and lethargy to seizures and coma (Tapper 2019). Deficits in complex executive functions and anosognosia are common, and patients can either present with severe hypo- or hyperactive delirium phenotypes.

In addition to cognitive and psychiatric changes, motor abnormalities are common in HE due to the associated neurotoxicity. The classic motor sign, asterixis, often appears in moderate stages of HE and in the context of known cirrhosis and is suggestive of HE. However, it can also be seen in other types of TME and is ultimately nonspecific. Other potential motor symptoms include hyperreflexia, muscle rigidity or spasticity, and a positive Babinski sign due to impaired cortical inhibition. In higher grade HE, muscle tone and deep tendon reflexes can decline, leading to hypotonia and coma. Other possible motor symptoms in HE can resemble extrapyramidal disorders (i.e., Parkinson's disease) and may include shuffling gait, bradykinesia, monotonic speech, and masked facies that are attributed to basal ganglia dysfunction exacerbated by manganese deposition (Rudler et al. 2020). Patients can also develop other types of tremors and in rare cases choreiform movements or ataxia. Focal and lateralizing neurologic deficits and seizures are uncommon in HE and their presence should raise concern for other etiologies, such as stroke or alcohol withdrawal.

The use of EEG and MRI may support diagnosis, particularly in cases where neurocognitive deficits are subtle or not clearly attributable to HE based solely on clinical observation (Córdoba 2011). EEG may evidence triphasic waves, a finding that is relatively specific for metabolic encephalopathy, including HE (Emmady and Murr 2023). However, nonspecific EEG slowing, as well as other general patterns of encephalopathy, may also be present. As such, EEG findings should not be used to definitively diagnose HE. Structural neuroimaging may also be nonspecific or even normal; however, MRI can show T1 signal hyperintensity in the basal ganglia, which may be secondary to manganese accumulation and is relatively specific for HE (Rudler et al. 2020).

Addressing Hepatic Encephalopathy

The effective diagnosis and management of HE requires collaboration across various specialties, including hepatology, neurology, psychiatry, neuropsychology, and nursing. Goals of HE management encompass reducing the frequency and severity of episodes, minimizing the functional impact, and preventing recurrence. A detailed assessment of the etiology of liver dysfunction, as well as the context, severity, and course can help inform appropriate intervention (Vilstrup et al. 2014).

Pharmacological treatment to control ammonia, such as non-absorbable disaccharides (particularly lactulose) are first-line agents used to acidify the colon, convert ammonia to ammonium, and promote excretion by increasing bowel movements. The antibiotic Rifaximin is frequently added as a secondary preventive agent and has been shown to significantly reduce recurrence rates of overt HE and hospitalizations (Bass et al. 2010). Other agents, such as L-ornithine L-aspartate (LOLA), branched-chain amino acids (BCAAs), and probiotics, have varying degrees of efficacy (Plauth et al. 2006). Malnutrition and sarcopenia are key modifiable risk factors for HE. Nutritional counseling should focus on maintaining protein intake, avoiding protein restriction, fiber intake, and encouraging small, frequent meals with a late-evening carbohydrate-rich snack to reduce muscle catabolism (Plauth et al. 2006). Supervised physical activity can help maintain muscle mass, enhance ammonia clearance, and improve functional status.

Unfortunately, even with optimal treatment some patients will experience recurrent HE, usually defined as two or more episodes within six months. Regardless of subtype, recurrent HE is associated with higher 1-year mortality compared with single-episode OHE (Weissenborn 2019). Recurrent HE often signifies that the patient's liver disease or precipitating factors are not fully controlled, necessitating long-term treatment with lactulose. However, this medication is subject to poor compliance due to increased bowel movements which may impact quality of life, and recurrent HE is frequently refractory to treatment, the combination of which yield a high risk of recurrence (upwards of 40% within one year) (Vilstrup et al. 2014). While lactulose can help manage HE, without treatment of the underlying liver dysfunction HE recurrence typically continues over time in conjunction with other complications of liver failure, contributing to poor long-term outcomes (Ferenci 2017). The only definitive treatment for HE, particularly for severe or persistent HE unresponsive to medical treatment, is liver transplantation (Vilstrup et al. 2014; García-Martínez et al. 2022).

While medical interventions are critical to addressing the underlying liver dysfunction, nonpharmacological methods are also helpful in preventing future recurrence or complications. Structured education for patients and caregivers on disease mechanisms, symptom monitoring, medication adherence (especially for lactulose and rifaximin), and safety precautions (e.g., fall and driving risk) have been shown to reduce HE recurrence and hospitalization rates (European Association for the Study of the Liver 2022). Family and other caregivers are often the first to detect behavioral and cognitive changes such as apathy, irritability, or altered sleep–wake cycles, which are valuable predictors of overt HE episodes. Active engagement with caregivers during routine evaluations and discharge planning can thus improve early detection and prevention.

For neuropsychologists working in inpatient settings, it is important to remember that delirium caused by HE is nonspecific. The recommended practice approach, covered in greater detail elsewhere, includes first identifying the presence of a delirium through a typical bedside examinattion, with the addition of tests for asterixis and other neurobehavioral signs. As referenced above, asterixis is nonspecific but supportive of an HE etiology to the delirium, although HE may very well still be present if this sign is absent. After delirium is identified, a detailed evaluation of a patient's labs and other clinical datapoints can help rule out other causes of delirium, and rule in hepatic encephalopathy.

Finally, it is important to remember that OHE is identified as carrying particular risk for mortality, hospitalization, and quality of life, even among patients with MHE who have experienced a prior episode of now-resolved OHE (D'Amico et al. 2006; Elsaid et al. 2020; Tapper 2019). Thus, even in the absence of overt symptoms during admission, there may be clinical indication for serial evaluation of possible MHE in what may otherwise be considered stable cases of liver disease (Patidar et al. 2014). Several psychometric measures have been developed and validated specifically for the assessment of MHE or adapted from common neuropsychological measures (Amodio et al. 2008; Randolph et al. 2009). While discussion of outpatient or more comprehensive neuropsychological workup of MHE is outside the scope of this chapter, readers are directed to Randolph et al. (2009) for additional details.

Kidney Disease

The kidneys regulate and maintain multiple critical functions, including the concentration of fluids and electrolyte balance, excretion of metabolic waste products, and blood pressure homeostasis. Kidney injury or renal dysfunction can be caused by a spectrum of disease and have systemic consequences, including delirium. Renal disease is classified as either acute or chronic, with specific diagnostic and treatment considerations, respectively (Fig. 7.12).

CASE EXAMPLE #4:
DELIRIUM DUE TO KIDNEY DISEASE, INTRODUCTION

Demographics: 66-year-old single black male with 12 years of education

Admission: Patient presented to hospital from his independent living facility with acute-on-chronic body aches and worsening cough. He has had multiple recent hospitalizations for ESRD and unstable housing. Collateral (niece) accompanied him and noted that, while he lives independently, he has had progressive difficulty managing his medications, bills, and other instrumental tasks over the past ~2-3 years.

Hospital Course and Referral: Admission workup revealed leukocytosis, high urea nitrogen, very high creatinine (8.1), high procalcitonin, and high c-reactive protein. Due to concern for pneumonia, he was started on antibiotics. He was also found to be anemic and to have severe malnutrition. Blood glucose was high (175) but A1C was only mildly elevated (7.4). He receives dialysis 3 times per week but missed his last appointment. Neuropsychology was consulted due to noted confusion, with request for differential diagnosis.

Medical History: Kidney cancer 15 years ago with nephrectomy, but later found to have metastatic disease to both lungs. He also has a history of type 2 diabetes, heart failure, and hypertension.

Psychiatric History: Unremarkable.

Substance Use History: Smokes 1 pack per day. History of cocaine abuse in sustained remission.

Neuroimaging: MRI Brain revealed diffuse volume loss and confluent white matter disease.

Bedside Exam: Neuropsychology initially saw him Monday before dialysis and he was noted to be somnolent and confused. Reevaluation was done later that day following treatment. MoCA was +18/30, with most points lost for visuospatial/executive, speeded tasks (e.g., verbal fluency), working memory, and memory recall (recalled +1/5 words, but got +4/5 with category cues). Mild fluctuations in attention also noted.

Question 1. What are at least three predisposing factors and at least three precipitating factors for delirium in this patient?

Question 2. What is the most likely diagnostic formulation? Consider whether he may have a delirium, dementia, or both?

ESRD = end-stage renal disease; A1C = hemoglobin A1C (glycohemoglobin); MRI = magnetic resonance imaging; MoCA – Montreal Cognitive Assessment

Fig. 7.12 Delirium due to kidney disease, Introduction

Acute Kidney Injury

Acute kidney injury (AKI) is the result of sudden ischemic, toxic, or obstructive insult to the kidneys causing retention of nitrogenous waste, fluid overload, and electrolyte imbalance. AKI is defined by a rapid decline in kidney function within hours to days. Depending on the cause, AKI may be fully reversible, however it may lead to persistent kidney injury, also referred to as acute kidney disease (AKD), and increase the probability of chronic kidney disease (CKD). AKI is typically categorized according to the mechanism of injury: prerenal due to hypovolemia or cardiac failure; intrinsic due to acute tubular necrosis or glomerulonephritis; or postrenal due to obstruction by stones or prostatic hypertrophy. AKI occurs in up to 20% of hospitalized patients, and approximately half of patients admitted to ICUs, making it a leading cause of mortality in hospitals (Hoste et al. 2015; Kung and Chou 2023; Susantitaphong et al. 2013). This syndrome disproportionately affects older adults, those who have had cardiac surgery, and those with sepsis. AKI is a significant predictor of increased risk of CKD and mortality, especially with recurrent or prolonged episodes (Coca et al. 2012). Symptoms of AKI, especially in the hospital setting, include decreased urine output, fluid overload, electrolyte disturbances (e.g., hyperkalemia), and azotemia (increased urea and creatinine in the blood). In severe cases or when AKI develops in the context of comorbid critical illness, patients can experience AMS, hypotension, and multi-organ dysfunction (Hoste et al. 2015; Maldonado 2017).

Chronic Kidney Disease

In contrast, chronic kidney disease (CKD) is characterized by a progressive and irreversible decline in kidney function, characterized by damage to the filtering units of the kidneys (nephrons), resulting in a sustained reduction in the ability of the kidneys to filter and remove waste from the blood (a syndrome called uremia). Uremic toxins, specifically indoxyl sulfate and p-cresyl sulfate, have neurotoxic effects and lead to oxidative stress and endothelial dysfunction (Viggiano et al. 2020). These toxins cross the blood-brain barrier and accumulate in the central nervous system, resulting in disrupted neurotransmission and synaptic plasticity. Neuroinflammation (and resultant cerebrovascular compromise) further incites glial and neuronal dysfunction. Primary and secondary small-vessel ischemia and arteriosclerosis are also very common due to both primary kidney dysfunction and frequent comorbidities (e.g., hypertension, diabetes), and increase the risk for white matter lesions and lacunar infarcts in the brain. Mental status changes associated with CKD are believed to be due to disruptions in fluid and electrolyte regulation, acid–base homeostasis, and red-blood cell production—the latter of which can contribute to anemia and metabolic acidosis. These metabolic factors can further disrupt oxygen delivery and neuronal excitability, exacerbating the risk for delirium and other neurocognitive dysfunction (Tanaka and Okusa 2020).

CKD is subdivided into five stages: Stages 1–4 describe progressive deterioration of kidney function, while stage 5 (also called end-stage renal disease or ESRD) indicates complete organ failure, necessitating dialysis or transplantation for survival. CKD severity is staged by estimated glomerular filtration rate (eGFR), which refers to the volume of blood plasma that is filtered through the glomeruli of nephrons per unit of time, commonly expressed as mL/min/1.73 m^2 (Duffield 2014). CKD is characteristically asymptomatic until eGFR declines below 60 (i.e., Stage 3a). however, the onset of subtle, measurable cognitive dysfunction can occur as early as CKD Stages 1–2 (Torres et al. 2017). While CKD progression is typically gradual, factors that can accelerate decline include proteinuria, prolonged or recurrent AKI, and uncontrolled or poorly managed chronic health conditions (Levey et al. 2005).

The global prevalence of CKD is estimated at approximately 10–15% of the adult population, with significant variation by region, age group, and diagnostic criteria (Hill et al. 2016). In the United States, >785,000 individuals were living with ESRD as of 2021, with higher rates among Black men compared with other demographic groups (Johansen et al. 2022). The most common causes of CKD are diabetes type II, which induces glomerular damage due to chronic hyperglycemia, and hypertension, which leads to nephrosclerosis through arteriolosclerosis and ischemia (Vaidya and Aeddula 2024). Other causes include glomerulonephritis, polycystic kidney disease, and obstructive uropathy, which cause structural and inflammatory damage (Webster et al. 2017).

Addressing Uremic Encephalopathy

The goal of AKI management is early identification and focused treatment of exacerbating factors. Priorities include discontinuing nephrotoxic agents, correcting electrolyte imbalances, and serial renal replacement therapy (RRT) in severe or unresponsive cases as needed. Management of AKI is often collaborative, requiring close communication with other providers to reduce the risk of exacerbating damage, as even mild AKI can increase CKD and cardiovascular disease risk (Coca et al. 2012).

While early CKD may cause subtle changes in cognitive abilities, it is unlikely to independently cause delirium (Vanderlinden et al. 2019). However, even earlier stage CKD should not be overlooked as a potential contributor to delirium when other deliriants are present. For example, a patient who develops pneumonia may be more likely to become delirious if they have CKD as a comorbidity. In such cases, CKD serves as a predisposing factor; identifying and treating the acute precipitant may be insufficient to promptly clear the delirium, and a holistic approach is often required to manage CKD and other predisposing factors. AKI or later-stage CKD, however, may be potent enough to trigger delirium on their own. Termed *uremic encephalopathy*, such cases present with a nonspecific phenotype, but in our experience may be more likely to exhibit hypoactive symptoms.

Treatment and management of CKD is targeted toward slowing progression, preventing complications, and preparing for transition to renal replacement therapy. Key interventions aimed at slowing progression include managing blood pressure, diabetes, and other comorbidities (Vaidya and Aeddula 2024). Other interventions include dietary modifications such as sodium restriction and phosphate control, and the use of erythropoietin-stimulating agents which target anemia. For those with diabetes, strict glycemic control may reduce the rate of neurocognitive deterioration. These interventions not only work to reduce inflammation, but also help mitigate cognitive dysfunction.

Patients with ESRD undergoing dialysis can exhibit cognitive deficits that are dependent on dialysis-specific factors (e.g., duration, frequency) and typically fluctuate within the treatment cycle due to changes in the accumulation of uremic toxins. These symptoms are often worse in patients with longer interdialytic intervals, such as the two-day weekend gap (Murray et al. 2006). Following dialysis, many patients report a period of “dialysis fog” or mental fatigue, that partially recovers in the hours following treatment, though typically fails to reach pre-CKD baseline. Thus, patients with ESRD on dialysis are likely at their cognitive best several hours post-treatment, with mentation slowly declining until the next dialysis session; such cases may be termed to have a static encephalopathy (Fig. 7.13).

CASE EXAMPLE #4:
DELIRIUM DUE TO KIDNEY DISEASE, CONCLUSION

Question 1.

Predisposing factors: Severe cerebrovascular disease, kidney cancer with lung metastases, frailty, anemia, heart failure. Diabetes could serve as a predisposing factor but is relatively well controlled. Hypertension could be a predisposing or precipitating factor if poorly controlled, otherwise is noncontributory. Concern for dementia, which would significantly lower the bar for developing delirium.

Precipitating factors: Active infection, malnutrition, antibiotics, exacerbations in kidney disease. His ESRD could also be the precipitating factor, and would likely show a static course, with weekly fluctuations in tandem with dialysis.

Question 2.

Conclusions: Patient was diagnosed with acute-on-chronic neurocognitive dysfunction. MoCA scores were broadly reflective of subcortical dysfunction and there was enough evidence to diagnose a mild vascular dementia due to small-vessel disease. He was also diagnosed with overlying static delirium due to ESRD, active infection, and malnutrition in the context of multiple significant health concerns.

ESRD = end-stage renal disease; MoCA – Montreal Cognitive Assessment

Fig. 7.13 Delirium due to kidney disease, conclusion

ENDOCRINE-INDUCED DELIRIUM

Besides the various types of TME described above, delirium can also result from endocrine dysfunction. As compared with TME, delirium due to endocrine dysfunction involves dysregulation in hormones and their associated organs, which can both directly and indirectly affect brain functioning. While some of these indirect causes (e.g., inflammation, secondary metabolic derangements) will be familiar to readers of this chapter, the pathophysiology of endocrine-induced delirium is unique, and is often considered independent of the broad constellation of disorders and dysfunction associated with TME. While exhaustive discussion of endocrine dysfunction is outside the scope of this chapter, we will cover two major systems and their associated deliriums.

Adrenal Dysfunction

The adrenal glands are essential endocrine organs responsible for producing three main types of steroid hormones: glucocorticoids (primarily cortisol), mineralocorticoids (primarily aldosterone), and adrenal androgens. These hormones play critical roles in regulating energy metabolism, fluid and electrolyte balance, immune responses, and physiological manifestations in response to stress (Barthel et al. 2019; Husebye et al. 2018). The release of cortisol is regulated by the hypothalamic–pituitary–adrenal (HPA) axis via a feedback loop involving the release of corticotropin-releasing hormone (CRH) by the hypothalamus, stimulation of adrenocorticotropic hormone (ACTH) from the anterior pituitary gland, and downstream activation of the adrenal glands. The activity of the HPA axis is linked to circadian rhythms, with peak cortisol levels typically occurring in the early morning and lower levels in the evening (Martin-Grace et al. 2024). Disruption at any stage can lead to adrenal insufficiency (AI) during times of physiological stress (Barthel et al. 2019; Husebye et al. 2018). Neuroendocrine derangements from AI may be partial or complete and reversible or permanent, depending on their etiology, nature of exposure, and duration.

Primary adrenal insufficiency (PAI), also known as Addison's disease, involves direct damage to the adrenal glands, most commonly due to autoimmune inflammation, leading to decreased production of glucocorticoids, mineralocorticoids, and adrenal androgens (Betterle et al. 2019). Contemporary prevalence rates of PAI range from 100 to 140 cases per million, and are highly skewed toward women, especially when due to autoimmunity. While most causes of PAI are due to autoimmunity, other causes include infections (e.g., tuberculosis, HIV), bilateral adrenalectomy, metastatic disease, adrenal hemorrhage, and rare genetic syndromes such as congenital adrenal hyperplasia and adrenoleukodystrophy (Barthel et al. 2019). Over time the susceptibility to adrenal crisis increases, particularly during stress, infection, or injury (Bleicken et al. 2010).

Secondary adrenal insufficiency (SAI) results from failure of the pituitary gland to secrete adrenocorticotropic hormone (ACTH), typically due to tumors, surgery, or radiotherapy (Husebye et al. 2018). SAI is estimated to be approximately twice as common as PAI, with prevalence estimated at 150–280 per million (Husebye et al. 2018). SAI is more common in older adults and individuals with cancer or other chronic illnesses (Martin-Grace et al. 2024).

Tertiary adrenal insufficiency (TAI) is the most common form of AI, although precise estimations of prevalence are limited due to diagnostic challenges and under recognition (Martin-Grace et al. 2024). TAI is most often the consequence of prolonged exogenous glucocorticoid exposure that suppresses corticotropin-releasing hormone and ACTH via negative feedback, resulting in adrenal atrophy and low cortisol output (Martin-Grace et al. 2024). Unlike systemic steroids, non-oral glucocorticoids—including inhaled, topical, and injection directly into joints—are also associated with significant risk of TAI, which is most concerning in older adults with multiple comorbidities and higher rates of polypharmacy (Barthel et al. 2019).

Symptom onset in AI is typically gradual but can be fulminant during episodes of physiological stress that increase cortisol demand, such as infection or surgery. Regardless of the course, common symptoms across subtypes include fatigue, weakness, weight loss, nausea, muscle and joint pain, depression, and labile mood. Hyponatremia and reduced stress tolerance can also be seen early in the clinical course. Many patients remain asymptomatic or present with prolonged non-specific symptoms, which does not preclude AI serving as a predisposing contributor to delirium.

Regardless of subtype, severe symptoms are usually associated with *adrenal crisis*, a life-threatening emergency characterized by severe weakness, hypotension, nausea and vomiting, abdominal and back pain, hypoglycemia, dehydration, electrolyte disturbances, and altered mental status (Allolio 2015; Bornstein et al. 2016). If not treated emergently, patients can go into shock and coma (Martin-Grace et al. 2024). Approximately one-half of patients with PAI and one-third of patients with SAI or TAI will experience at least one episode of adrenal crisis during their lifetime.

Neurobehavioral changes in AI are attributed to several factors. First, cortisol is modulated through two receptor systems: mineralocorticoid receptors, which are densely found in the hippocampus; and glucocorticoid receptors, which are distributed more broadly and modulate stress response and memory consolidation. Both high and low cortisol can be problematic and are associated with impaired neurocognition, especially memory and executive functioning (Lupien et al. 2007). In patients with AI, this relationship is exacerbated by disruption of natural circadian patterns of cortisol release that can increase neuronal stress and result in delirium or more subtle cognitive dysfunction (Harbeck et al. 2009). Like many causes of delirium, AI is more likely to serve as a predisposing factor but can also serve as the primary precipitant when alterations in cortisol and other adrenal hormones are severe or rapidly develop. In addition to typical, nonspecific symptoms, when AI is associated with ACTH-secreting pituitary tumors delirium can present with significant apathy with blunted affect (Pertichetti et al. 2020).

Thyroid Dysfunction

The thyroid gland is a butterfly-shaped endocrine organ located in the front of the neck below the larynx, and which regulates metabolism, growth and development, cardiovascular function, and temperature regulation of the body via the production of the hormones triiodothyronine (T3) and thyroxine (T4). Thyroid function is regulated by thyroid-stimulating hormone (TSH), which is secreted by the pituitary gland in response to thyrotropin-releasing hormone (TRH) released by the hypothalamus (Brent 2012). T3 and T4 play critical roles in almost every organ system, including the brain. Thyroid dysfunction can be classified as either the underproduction (hypothyroidism) or overproduction (hyperthyroidism) of thyroid hormones, both of which can result in significant neurological, cognitive, and neuropsychiatric symptoms (Smith et al. 2002). Clinical hypothyroidism is much more common than hyperthyroidism, with the former having a global prevalence of 3–8% compared with 0.5–2% for hyperthyroidism. While hypothyroidism has multiple possible causes, most cases of hyperthyroidism are due to Graves' disease or toxic multinodular goiter (Taylor et al. 2018).

Thyroid dysfunction can be further categorized into primary and secondary forms. Primary disease originates from the thyroid gland and includes both autoimmune etiologies (e.g., Hashimoto's thyroiditis, Graves' disease) as well as less common non-autoimmune pathology (e.g., iodine imbalance, thyroidectomy, radioactive iodine therapy). Secondary dysfunction results from disruption of the hypothalamic-pituitary-thyroid (HPT) axis and can be due to a wider array of etiologies (Patil et al. 2024; Ragusa et al. 2019). The severity of thyroid disease is often classified as overt (abnormalities in TSH and T4 levels) or subclinical (abnormalities in TSH with normal T4). Despite a typical lack of overt symptoms, subclinical thyroid dysfunction can lead to increased cardiovascular and neuropsychiatric risk and is thus important to consider and rule out (Patil et al. 2024).

Thyroid hormone abnormalities can lead to widespread effects on the central nervous system via multiple mechanisms. In hypothyroidism, lower levels of T3 and T4 result in impaired myelination, reduced synaptic efficiency, disrupted cerebral glucose metabolism, reduced hippocampal and prefrontal cortex activity, and negative effects on neurotransmitter systems including serotonin, dopamine, and norepinephrine (Zamwar and Muneshwar 2023). Hashimoto's thyroiditis specifically leads to progressive thyroid gland damage and chronic inflammation, further affecting brain function (Ragusa et al. 2019). In hyperthyroidism, excessive thyroid hormone levels affect sympathetic nervous system activity via enhanced oxidative stress and glutamatergic excitotoxicity, resulting in altered cerebral metabolism and symptoms of anxiety, agitation, restlessness, and cognitive fatigue (Mathew et al. 2023).

The onset of hypothyroidism is typically gradual, with early symptoms commonly misattributed to depression, aging, and chronic fatigue (Zamwar and Muneshwar 2023. Hashimoto's thyroiditis may fluctuate between periods of euthyroidism and hypothyroidism before progressing to permanent hormone deficiency (Ragusa et al. 2019). In contrast, hyperthyroidism typically has a subacute presentation, with symptoms developing over weeks to months.

The spectrum of neurobehavioral signs and symptoms in thyroid dysfunction is broad, and is contingent on course (rapid onset, transient, or longstanding), the degree of thyroid dysfunction, and whether a patient is experiencing hypothyroidism or hyperthyroidism. In overt hypothyroidism, patients commonly report memory problems, dysnomia, slowed processing speed, and reduced attention span (Khatiwada et al. 2016). Depression, lethargy, and psychomotor slowing are also common, which may take the form of a hypoactive delirium. More severe manifestations of hypothyroidism, typically accompanied by significant TSH and T4 derangements, include "myxedema madness" which is characterized by psychosis, hypothermia, and coma (Heinrich and Grahm 2003). In general, psychiatric symptoms are a more common mark of hyperthyroidism, and may include anxiety, agitation, distractibility, emotional lability, and sleep disturbances. Hyperthyroidism often presents as hyperactive delirium, although this is by no means always the case. Subclinical hypothyroidism presents more variably and is unlikely to be the precipitating cause of delirium but may be an important contribution to the burden of predisposing factors. Subclinical hyperthyroidism can also increase the risk for delirium but has additionally been associated with increased risk for dementia (Rieben et al. 2016). Finally, autoimmune hypothyroidism can progress to Hashimoto's encephalopathy, which is associated with seizures, psychosis, cognitive fluctuations, and altered consciousness (Ferracci and Carnevale 2006). EEG abnormalities and CSF changes may be seen, and symptoms typically respond to corticosteroids rather than thyroid hormone therapy.

Addressing Delirium in Endocrine Disorders

Delirium due to endocrine dysfunction is typically responsive to correction of the underlying abnormality, although as with all cases of delirium cognitive recovery can lag medical stability. As these syndromes are less likely to be the proximal precipitating factor for delirium, addressing more fulminant and potent deliriogenic factors (e.g., HE, hyponatremia, etc.) as targets to reverse delirium with the treatment team is recommended as an initial step. In cases of prolonged delirium or following correction of the identified precipitant, management of predisposing factors, including thyroid dysfunction, is recommended to further improve cognition and reduce the likelihood of future delirium.

Medical treatment for AI consists of physiologic glucocorticoid replacement, which may include hydrocortisone. Even in anatomically based etiologies, such as pituitary or adrenal tumors, perioperative glucocorticoids are used while HPA function is monitored closely (Husebye et al. 2018; Martin-Grace et al. 2024). All patients treated with glucocorticoids should be carefully monitored with individualized tapering schedules to reduce risk of TAI.

Treatment of thyroid dysfunction involves restoring and maintaining a euthyroid state. In hypothyroidism, replacement by levothyroxine is the standard of care. Management of Hashimoto's thyroiditis includes regular thyroid function testing and, in some cases, immunomodulatory support. Treatments for hyperthyroidism include antithyroid medications such as methimazole, consideration of radioactive iodine therapy, or thyroidectomy. Beta-blockers can also be used adjunctively to reduce sympathetic symptoms. Cognitive and psychiatric symptoms should be assessed as part of the standard care, especially in older adults and in those with premorbid dementia or other neuropsychiatric conditions.

POSTOPERATIVE COGNITIVE CHANGE

Postoperative Delirium

Postoperative delirium (POD) refers to alterations in mental status that occur after major surgical procedures. Incidence varies by population, type of surgery, and diagnostic criteria, but estimates range from 11% to 50% in general medical-surgical populations to >60% in high-risk subgroups, such as critical care patients and those undergoing cardiac or transplant procedures (Inouye et al. 2014; Xiang et al. 2023). POD may also be seen in ICU patients following minor surgeries, with rates as high as 80% in older adults with prolonged or complicated admissions (Wu et al. 2022).

Like all types of delirium, POD results from the synergistic interaction of predisposing and precipitating factors. Precipitating events in POD are typically linked to the surgical process, including the surgery itself, inflammation, metabolic disturbance, pain, and anesthesia exposure. Tissue disruption due to surgery triggers the release of neuroinflammatory peripheral cytokines, as well as general inflammatory proteins, that cross the BBB, activate microglia, and trigger neurotoxic cascades (Jin et al. 2020; Maldonado 2018). This results in oxidative stress, neurotransmitter dysregulation, and synaptic dysfunction.

Other pathophysiological considerations unique to POD include the use of anesthetic agents and procedures. Deeper levels of anesthesia, use of inhaled anesthetics, and volatile agents may increase the risk of POD through enhancement of neuroinflammatory and apoptotic pathways (Jin et al. 2020). In contrast, total intravenous anesthesia (TIVA) in which both induction and maintenance of anesthesia are delivered exclusively and continuously through infusion of intravenous agents, such as propofol, remifentanil, and other short-acting agents, may offer neuroprotective benefits in higher risk populations by limiting neurotoxicity and maintaining more consistent hemodynamic stability (Swarbrick and Partridge 2022).

POD is distinct from the acute confusional state observed shortly after a patient regains consciousness from general anesthesia. The latter, termed emergence delirium, is a benign, transient (~15–30 minutes) hyperactive confusional state that can be seen across the age spectrum. In contrast, POD typically lags surgery by ~1–3 days, often following a temporary post-surgical interval of lucidity and near-baseline mentation. However, the onset and course may vary depending on the type of surgery performed. POD may develop shortly following cardiac and orthopedic procedures but can be delayed following transplants due to prolonged ICU admissions, use of immunosuppressive drugs, and higher rate of complications (Fann et al. 2011; Fong et al. 2015). Regardless of when symptoms begin, most cases resolve within one week, though approximately 20%–30% of patients with preexisting cognitive or functional deficits will have a longer recovery. Like all causes of delirium, longer duration of symptoms portends a worse outcome (Inouye et al. 2014).

Postoperative Cognitive Decline

Regardless of a patient's premorbid baseline or medical status, POD increases the risk of chronic postoperative cognitive decline (POCD). In vulnerable populations, this risk is magnified, even when accounting for predisposing factors and after recovery from acute precipitants (de la Varga-Martínez et al. 2023; Maldonado 2017). For example, while POD is more frequent in those with pre-surgical cognitive dysfunction, it carries an independent risk for development of MCI or dementia in the months following a surgical procedure with general anesthesia (Sprung et al. 2017). For this reason, prompt identification of POD and targeted intervention, when possible, can help identify and mitigate risk for chronic symptoms. While POD is defined by acute and fluctuating mental status, POCD is characterized by persistent dysfunction or distal cognitive decline that is not attributable to reversible factors, and is akin to a dementia (de la Varga-Martínez et al. 2023; Sprung et al. 2017; Zimpfer et al. 2004).

While possible after any surgery, POCD appears to be most likely following heart procedures, particularly those, such as coronary artery bypass grafting (CABG), that involve extracorporeal membrane oxygenation (ECMO) (de la Varga-Martínez et al. 2023; Zimpfer et al. 2004). For example, Greaves et al. (2019) compared those undergoing CABG to those undergoing non-cardiac surgeries and found that patients who underwent CABG had elevated incidence of both POD and POCD, with up to 25% of cardiac patients experiencing persistent cognitive deficits 1-year post-surgery. The mechanism of permanent cognitive change following cardiac surgery is hypothesized to be due to mild hypoxia from ECMO, as well as inflammation and other diffuse effects of major surgery, in the context of patients who are likely at increased risk for vascular cognitive impairment at baseline.

Addressing Postoperative Delirium

The prevention and management of POD requires proactive, interdisciplinary collaboration to identify individuals at higher risk and to plan for behavioral and supportive care after surgery. This may include identifying a patient's risk for POD and POCD based on their medical history, considering the type of surgery proposed, and conducting pre-surgical neuropsychological evaluations for those deemed vulnerable to adverse cognitive outcomes. Neuropsychological workup offers high sensitivity to identify vulnerable patients preoperatively, even when consisting of briefer examinations focused on executive function, attention, and processing speed (Ghezzi et al. 2022). This has the additional benefit of rendering a post-surgical baseline for later comparison.

In addition to pre-surgical screening to detect at-risk individuals and to quantify their pre-surgical baselines, behavioral and supportive interventions to optimize health and cognitive functioning prior to surgery may also be helpful. Delirium prevention strategies that target modifiable risk factors, such as immobility, dehydration, sleep dysfunction and deprivation, as well as addressing hearing or vision impairments, can raise the threshold for delirium by ameliorating predisposing factors. Implementing strategies to address these risk factors well in advance of surgery has been shown to decrease the incidence of POD in hospitalized older adults by 30% (Siddiqi et al. 2016).

Intraoperatively, delirium risk can be reduced with conservative anesthesia practices, including minimizing the depth and duration of anesthesia, maintaining stable cerebral perfusion and hemodynamic stability throughout procedures, using gradual weaning of anesthesia at the end of the procedure to reduce risk of ED, and avoiding the use of anticholinergic medications (Jin et al. 2020; Swarbrick and Partridge 2022). Postoperatively, reducing environmental triggers for confusion and agitation is essential. Extraneous environmental stimuli should be minimized, such as bright lights, harsh sounds, and other factors that can influence physical comfort. Providers should speak softly, steadily, and calmly to reduce agitation and ensure comprehension. Once patients are medically cleared, physical movement should be encouraged as soon as possible when safe (Deiner and Silverstein 2009; Swarbrick and Partridge 2022).

SUMMARY

Delirium is a heterogenous, acute neurobehavioral syndrome with a multitude of potential etiologies that can serve as predisposing and/or precipitating factors. Common causes of delirium in inpatient settings include toxic-metabolic derangements, neuroendocrine dysfunction, and post-surgical alterations in mental status. Neuropsychologists possess unique skills in brain-behavior relationships, which lend themselves well to identifying and diagnosing delirium in inpatient settings. By familiarizing themselves with these common etiologies, as well as relevant lab values and other diagnostic markers, neuropsychologists are equipped to best serve patients and other team members by knowing what may be causing a patient's delirium, and what recommendations are the most helpful for their recovery.

REFERENCES

ACEP. American College of Emergency Physicians excited delirium task force. 2009. https://www.prisonlegalnewsr.org/media/publications/acep_report_on_excited_delirium_syndrome_sept_2009.pdf.

Adrogué HJ, Tucker BM, Madias NE. Diagnosis and Management of Hyponatremia: a review. JAMA. 2022;328(3):280–91. https://doi.org/10.1001/jama.2022.11176.

Allolio B. Extensive expertise in endocrinology. Adrenal crisis. European J Endocrinol. 2015;172(3):R115–24. https://doi.org/10.1530/EJE-14-0824.

Alagiakrishnan K, Wiens CA. An approach to drug induced delirium in the elderly. Postgrad Med J. 2004;80(945):388–93. https://doi.org/10.1136/pgmj.2003.017236.

Ameer M, Saadabadi A. 2023. Neuroleptic medications.

American Geriatrics Society. American Geriatrics Society 2019 updated AGS beers criteria® for potentially inappropriate medication use in older adults. J Am Geriatr Soc. 2019;67(4):674–94. https://doi.org/10.1111/jgs.15767.

Amodio P, Campagna F, Olianas S, Iannizzi P, Mapelli D, Penzo M, Angeli P, Gatta A. Detection of minimal hepatic encephalopathy: normalization and optimization of the psychometric hepatic encephalopathy score. A neuropsychological and quantified EEG study. J Hepatol. 2008;49(3):346–53. https://doi.org/10.1016/j.jhep.2008.04.022.

Amodio P. Current diagnosis and classification of hepatic encephalopathy. J Clin Exp Hepatol. 2018;8(4):432–7. https://doi.org/10.1016/j.jceh.2018.07.001.

Angel MJ, Young GB. Metabolic encephalopathies. Neurol Clin. 2011;29(4):837–82. https://doi.org/10.1016/j.ncl.2011.08.002.

Annweiler C, Montero-Odasso M, Llewellyn DJ, Richard-Devantoy S, Duque G, Beauchet O. Meta-analysis of memory and executive dysfunctions in relation to vitamin D. J Alzheimer's Dis. 2010;22(1):147–57.

Auer RN, Wieloch T, Olsson Y, Siesjö BK. The distribution of hypoglycemic brain damage. Acta Neuropathol. 1984;64(3):177–91. https://doi.org/10.1007/BF00688108.

Barbosa C, Cowell AJ, Dowd WN. Alcohol consumption in response to the COVID-19 pandemic in the United States. J Addict Med. 2021;15(4):341–4. https://doi.org/10.1097/ADM.0000000000000767.

Barthel A, Benker G, Berens K, Diederich S, Manfras B, Gruber M, Kanczkowski W, Kline G, Kamvissi-Lorenz V, Hahner S, Beuschlein F, Brennand A, Boehm BO, Torpy DJ, Bornstein SR. An update on Addison's disease. Exp Clin Endocrinol Diabetes : Off J German Soc Endocrinol German Diabetes Assoc. 2019;127(2–03):165–75. https://doi.org/10.1055/a-0804-2715.

Bass NM, Mullen KD, Sanyal A, Poordad F, Neff G, Leevy CB, Sigal S, Sheikh MY, Beavers K, Frederick T, Teperman L, Hillebrand D, Huang S, Merchant K, Shaw A, Bortey E, Forbes WP. Rifaximin treatment in hepatic encephalopathy. N Engl J Med. 2010;362(12):1071–81. https://doi.org/10.1056/NEJMoa0907893.

Belvederi Murri M, Guaglianone A, Bugliani M, Calcagno P, Respino M, Serafini G, Innamorati M, Pompili M, Amore M. Second-generation antipsychotics and neuroleptic malignant syndrome: systematic review and case report analysis. Drugs R&D. 2015;15(1):45–62. https://doi.org/10.1007/s40268-014-0078-0.

Benetos A, Rossignol P, Cherubini A, Joly L, Grodzicki T, Rajkumar C, Strandberg TE, Petrovic M. Polypharmacy in the aging patient: Management of Hypertension in octogenarians. JAMA. 2015;314(2):170. https://doi.org/10.1001/jama.2015.7517.

Berridge CW, Devilbiss DM. Psychostimulants as cognitive enhancers: the prefrontal cortex, catecholamines, and attention-deficit/hyperactivity disorder. Biol Psychiatry. 2011;69(12):e101–11. https://doi.org/10.1016/j.biopsych.2010.06.023.

Betterle C, Presotto F, Furmaniak J. Epidemiology, pathogenesis, and diagnosis of Addison's disease in adults. J Endocrinol Investig. 2019;42(12):1407–33. https://doi.org/10.1007/s40618-019-01079-6.

Bischoff-Ferrari HA, Borchers M, Gudat F, Dürmüller U, Stähelin HB, Dick W. Vitamin D receptor expression in human muscle tissue decreases with age. J Bone Miner Res. 2004;19(2):265–9.

Bivona G, Gambino CM, Iacolino G, Ciaccio M. Vitamin D and the nervous system. Neurol Res. 2019;41(9):827–35. https://doi.org/10.1080/01616412.2019.1622872.

Blair GJ, Mehmood T, Rudnick M, Kuschner WG, Barr J. Nonpharmacologic and medication minimization strategies for the prevention and treatment of ICU delirium: a narrative review. J Intensive Care Med. 2019;34(3):183–90. https://doi.org/10.1177/0885066618771528.

Blei AT, Larsen FS. Pathophysiology of cerebral edema in fulminant hepatic failure. J Hepatol. 1999;31(4):771–6. https://doi.org/10.1016/s0168-8278(99)80361-4.

Bleicken B, Hahner S, Ventz M, Quinkler M. Delayed diagnosis of adrenal insufficiency is common: a cross-sectional study in 216 patients. Am J Med Sci. 2010;339(6):525–31. https://doi.org/10.1097/MAJ.0b013e3181db6b7a.

Bornstein SR, Allolio B, Arlt W, Barthel A, Don-Wauchope A, Hammer GD, Husebye ES, Merke DP, Murad MH, Stratakis CA, Torpy DJ. Diagnosis and treatment of primary adrenal insufficiency: an Endocrine Society clinical practice guideline. J Clin Endocrinol Metab. 2016;101(2):364–89. https://doi.org/10.1210/jc.2015-1710.

Brent GA. Mechanisms of thyroid hormone action. J Clin Invest. 2012;122(9):3035–43. https://doi.org/10.1172/JCI60047.

Bugiani O. Why is delirium more frequent in the elderly? Neurol Sci. 2021;42(8):3491–503. https://doi.org/10.1007/s10072-021-05339-3.

Butterworth RF. Thiamin deficiency and brain disorders. Nutr Res Rev. 2009;22(1):49–59. https://doi.org/10.1017/S0954422409990036.

Calvo M, Gunnarsson T, Smith L, Hao M. Delirium tremens in an AUD patient after an intrathecal baclofen pump induced total alcohol abstinence. Eur Rev Med Pharmacol Sci. 2018;22:5371–6.

Carmel R. Current concepts in cobalamin deficiency. Annu Rev Med. 2000;51:357–75. https://doi.org/10.1146/annurev.med.51.1.357.

Campbell N, Boustani M, Limbil T, Ott C, Fox C, Maidment I, Schubert C, Munger S, Fick D, Miller D, Gulati R. The cognitive impact of anticholinergics: a clinical review. Clin Interv Aging. 2009;4:225–33. https://doi.org/10.2147/CIA.S5358.

Campbell N, Lane KA, Gao S, Boustani MA, Unverzagt F. Anticholinergics influence transition from Normal cognition to mild cognitive impairment in older adults in primary care. Pharm: the J Human Pharmacol Drug Therapy. 2018;38(5):511–9. https://doi.org/10.1002/phar.2106.

Chamorro AJ, Rosón-Hernández B, Medina-García J-A, Muga-Bustamante R, Fernández-Solá J, Martín-González M-C, Seco-Hernández E, Novo-Veleiro I, Suárez-Cuervo C, Mateos-Díaz AM, Monte-Secades R, Machado-Prieto B, Puerta-Louro R, Prada-González C, Fernández-Rial Á, Sabio-Repiso P, Vázquez-Vigo R, Antolí-Royo A-C, Gomila-Grange A, et al. Differences between alcoholic and nonalcoholic patients with Wernicke encephalopathy: a multicenter observational study. Mayo Clin Proc. 2017;92(6):899–907. https://doi.org/10.1016/j.mayocp.2017.02.019.

Chang L, Harrington DW, Milkotic A, Swerdloff RS, Wang C. Unusual occurrence of extrapontine myelinolysis associated with acute severe hypernatraemia caused by central diabetes insipidus. Clin Endocrinol. 2005;63(2):233–5. https://doi.org/10.1111/j.1365-2265.2005.02319.x.

Charlesworth CJ, Smit E, Lee DSH, Alramadhan F, Odden MC. Polypharmacy among adults aged 65 years and older in the United States: 1988–2010. J Gerontol Ser A Biol Med Sci. 2015;70(8):989–95. https://doi.org/10.1093/gerona/glv013.

Clegg A, Young JB. Which medications to avoid in people at risk of delirium: a systematic review. Age Ageing. 2011;40(1):23–9. https://doi.org/10.1093/ageing/afq140.

Coca SG, Singanamala S, Parikh CR. Chronic kidney disease after acute kidney injury: a systematic review and meta-analysis. Kidney Int. 2012;81(5):442–8. https://doi.org/10.1038/ki.2011.379.

Córdoba J. New assessment of hepatic encephalopathy. J Hepatol. 2011;54(5):1030–40. https://doi.org/10.1016/j.jhep.2010.11.015.

Cunningham C. Systemic inflammation and delirium: important co-factors in the progression of dementia. Biochem Soc Trans. 2011;39(4):945–53. https://doi.org/10.1042/BST0390945.

D'Amico G, Garcia-Tsao G, Pagliaro L. Natural history and prognostic indicators of survival in cirrhosis: a systematic review of 118 studies. J Hepatol. 2006;44(1):217–31. https://doi.org/10.1016/j.jhep.2005.10.013.

DeCarolis DD, Rice KL, Ho L, Willenbring ML, Cassaro S. Symptom-driven lorazepam protocol for treatment of severe alcohol withdrawal delirium in the intensive care unit. Pharmacotherapy. 2007;27(4):510–8. https://doi.org/10.1592/phco.27.4.510.

Deiner S, Silverstein J. Postoperative delirium and cognitive dysfunction. British journal of anaesthesia. 2009;103(suppl_1): i41–i46.

de la Varga-Martínez O, Gutiérrez-Bustillo R, Muñoz-Moreno MF, López-Herrero R, Gómez-Sánchez E, Tamayo E. Postoperative delirium: an independent risk factor for poorer quality of life with long-term cognitive and functional decline after cardiac surgery. J Clin Anesth. 2023;85:111030. https://doi.org/10.1016/j.jclinane.2022.111030.

Di Somma C, Scarano E, Barrea L, Zhukouskaya VV, Savastano S, Colao A. Vitamin D and neurological diseases: an endocrine view. Int J Mol Sci. 2017;18(11):2482.

Dineen R, Thompson CJ, Sherlock M. Hyponatraemia – presentations and management. Clin Med (Lond). 2017;17(3):263–9. https://doi.org/10.7861/clinmedicine.17-3-263.

Douaud G, Refsum H, de Jager CA, Jacoby R, Nichols TE, Smith SM, Smith AD. Preventing Alzheimer's disease–related gray matter atrophy by B-vitamin treatment. Proc Natl Acad Sci. 2013;110(23):9523–8. https://doi.org/10.1073/pnas.1301816110.

Dowell D, Ragan K, Jones C, Baldwin G, Chou R. CDC clinical practice guideline for prescribing opioids for pain—United States, vol. RR-3. Centers for Disease Control; 2022. p. 1–95. https://doi.org/10.15585/mmwr.rr7103a1.

Duffield JS. Cellular and molecular mechanisms in kidney fibrosis. J Clin Invest. 2014;124(6):2299–306. https://doi.org/10.1172/JCI72267.

Dunkley EJC, Isbister GK, Sibbritt D, Dawson AH, Whyte IM. The hunter serotonin toxicity criteria: simple and accurate diagnostic decision rules for serotonin toxicity. QJM. 2003;96(9):635–42. https://doi.org/10.1093/qjmed/hcg109.

Dutta C, Pasha K, Paul S, Abbas MS, Nassar ST, Tasha T, Desai A, Bajgain A, Ali A, Mohammed L. Urinary tract infection induced delirium in elderly patients: a systematic review. Cureus. 2022;14(12):e32321. https://doi.org/10.7759/cureus.32321.

Eckardt MJ, Martin PR. Clinical assessment of cognition in alcoholism. Alcohol Clin Exp Res. 1986;10(2):123–7. https://doi.org/10.1111/j.1530-0277.1986.tb05058.x.

Egberts A, van der Craats S, van Vijk M, Alkilabe S, van den bemt, P., & Mattace-Raso, F. Anticholinergic drug exposure is associated with delirium and postdischarge institutionalization in acutely ill hospitalized older patients. Pharmacol Res Perspect. 2017;5(3):e00310.

Eidelman LA, Putterman D, Putterman C, Sprung CL. The spectrum of septic encephalopathy. Definitions, etiologies, and mortalities. JAMA. 1996;275(6):470–3.

Elie M, Cole MG, Primeau FJ, Bellavance F. Delirium risk factors in elderly hospitalized patients. J Gen Intern Med. 1998;13(3):204–12. https://doi.org/10.1046/j.1525-1497.1998.00047.x.

Ellis SJ. Severe hyponatraemia: complications and treatment. QJM. 1995;88(12):905–9.

Elsaid MI, John T, Li Y, Pentakota SR, Rustgi VK. The health care burden of hepatic encephalopathy. Clin Liver Dis. 2020;24(2):263–75. https://doi.org/10.1016/j.cld.2020.01.006.

Ely EW, Shintani A, Truman B, Speroff T, Gordon SM, Harrell FE Jr, Inouye SK, Bernard GR, Dittus RS. Delirium as a predictor of mortality in mechanically ventilated patients in the intensive care unit. JAMA. 2004;291(14):1753–62. https://doi.org/10.1001/jama.291.14.1753.

Emmady PD, Murr NI. EEG Triphasic waves. In: StatPearls. StatPearls Publishing; 2023.

European Association for the Study of the Liver. EASL clinical practice guidelines on the management of hepatic encephalopathy. J Hepatol. 2022;77(3):807–24. https://doi.org/10.1016/j.jhep.2022.06.001.

Evans L, Rhodes A, Alhazzani W, Antonelli M, Coopersmith CM, French C, Machado FR, Mcintyre L, Ostermann M, Prescott HC, Schorr C, Simpson S, Wiersinga WJ, Alshamsi F, Angus DC, Arabi Y, Azevedo L, Beale R, Beilman G, Belley-Cote E, et al. Surviving sepsis campaign: international guidelines for Management of Sepsis and Septic Shock 2021. Crit Care Med. 2021;49(11):e1063–143. https://doi.org/10.1097/CCM.0000000000005337.

Eyles DW, Burne TH, McGrath JJ. Vitamin D, effects on brain development, adult brain function and the links between low levels of vitamin D and neuropsychiatric disease. Front Neuroendocrinol. 2013;34(1):47–64.

Fann JR, Hubbard RA, Alfano CM, Roth-Roemer S, Katon WJ, Syrjala KL. Pre- and post-transplantation risk factors for delirium onset and severity in patients undergoing hematopoietic stem-cell transplantation. J Clin Oncol. 2011;29(7):895–901. https://doi.org/10.1200/JCO.2010.28.4521.

Farrokh S, Roels C, Owusu KA, Nelson SE, Cook AM. Alcohol withdrawal syndrome in Neurocritical care unit: assessment and treatment challenges. Neurocrit Care. 2021;34(2):593–607. https://doi.org/10.1007/s12028-020-01061-8.

Ferenci P. Hepatic encephalopathy Gastroenterology report. 2017;5(2):138–47. https://doi.org/10.1093/gastro/gox013.

Ferracci F, Carnevale A. The neurological disorder associated with thyroid autoimmunity. J Neurol. 2006;253(8):975–84. https://doi.org/10.1007/s00415-006-0170-7.

Ferreira P, Ferreira AR, Barreto B, Fernandes L. Is there a link between the use of benzodiazepines and related drugs and dementia? A systematic review of reviews. European Geriatric Medicine. 2022;13(1):19–32. https://doi.org/10.1007/s41999-021-00553-.

Ferrer R, Martin-Loeches I, Phillips G, Osborn TM, Townsend S, Dellinger RP, Artigas A, Schorr C, Levy MM. Empiric antibiotic treatment reduces mortality in severe sepsis and septic shock from the first hour: results from a guideline-based performance improvement program. Crit Care Med. 2014;42(8):1749–55. https://doi.org/10.1097/CCM.0000000000000330.

Fong TG, Davis D, Growdon ME, Albuquerque A, Inouye SK. The interface between delirium and dementia in elderly adults. The Lancet Neurology. 2015;14(8):823–32. https://doi.org/10.1016/S1474-4422(15)00101-5.

Frier BM. Morbidity of hypoglycemia in type 1 diabetes. Diabetes Res Clin Pract. 2004;65(Suppl 1):S47–52. https://doi.org/10.1016/j.diabres.2004.07.008.

Gaieski DF, Mikkelsen ME, Band RA, Pines JM, Massone R, Furia FF, Shofer FS, Goyal M. Impact of time to antibiotics on survival in patients with severe sepsis or septic shock in whom early goal-directed therapy was initiated in the emergency department. Crit Care Med. 2010;38(4):1045–53. https://doi.org/10.1097/CCM.0b013e3181cc4824.

García-Martínez R, Diaz-Ruiz R, Poncela M. Management of Hepatic Encephalopathy Associated with advanced liver disease. Clin Drug Investig. 2022;42(Suppl 1):5–13. https://doi.org/10.1007/s40261-022-01146-6.

Garcion E, Wion-Barbot N, Montero-Menei CN, Berger F, Wion D. New clues about vitamin D functions in the nervous system. Trends in Endocrinology & Metabolism. 2002;13(3):100–5. https://doi.org/10.1016/S1043-2760(01)00547-1.

Ghezzi ES, Ross TJ, Sharman R, Davis D, Boord MS, Thanabalan T, Thomas J, Keage HAD. The neuropsychological profile of delirium vulnerability: a systematic review and meta-analysis. Neurosci Biobehav Rev. 2022;132:248–59. https://doi.org/10.1016/j.neubiorev.2021.11.046.

Gonçalves J, Baptista S, Silva AP. Psychostimulants and brain dysfunction: a review of the relevant neurotoxic effects. Neuropharmacology. 2014;87:135–49. https://doi.org/10.1016/j.neuropharm.2014.01.006.

Gonin P, Beysard N, Yersin B, Carron P. Excited delirium: a systematic review. Acad Emerg Med. 2018;25(5):552–65. https://doi.org/10.1111/acem.13330.

Gorelick DA. Cannabis-related disorders and toxic effects. N Engl J Med. 2023;389(24):2267–75. https://doi.org/10.1056/NEJMra2212152.

Greaves D, Psaltis PJ, Ross TJ, Davis D, Smith AE, Boord MS, Keage HAD. Cognitive outcomes following coronary artery bypass grafting: a systematic review and meta-analysis of 91,829 patients. Int J Cardiol. 2019;289:43–9. https://doi.org/10.1016/j.ijcard.2019.04.065.

Green R, Allen LH, Bjørke-Monsen AL, Brito A, Guéant JL, Miller JW, Molloy AM, Nexo E, Stabler S, Toh BH, Ueland PM, Yajnik C. Vitamin B_{12} deficiency. Nat Rev Dis Primers. 2017;3:17040. https://doi.org/10.1038/nrdp.2017.40.

Grossi CM, Richardson K, Fox C, Maidment I, Steel N, Loke YK, Arthur A, Myint PK, Campbell N, Boustani M, Robinson L, Brayne C, Matthews FE, Savva GM. Anticholinergic and benzodiazepine medication use and risk of incident dementia: a UK cohort study. BMC Geriatr. 2019;19(1):276. https://doi.org/10.1186/s12877-019-1280-2.

Grover S, Ghosh A. Delirium Tremens: Assessment and Management. J Clin Exp Hepatol. 2018;8(4):460–70. https://doi.org/10.1016/j.jceh.2018.04.012.

Grucza RA, Sher KJ, Kerr WC, Krauss MJ, Lui CK, McDowell YE, Hartz S, Virdi G, Bierut LJ. Trends in adult alcohol use and binge drinking in the early 21st-century United States: a meta-analysis of 6 National Survey Series. Alcohol Clin Exp Res. 2018;42(10):1939–50. https://doi.org/10.1111/acer.13859.

Gül F, Arslantaş MK, Cinel İ, Kumar A. Changing definitions of sepsis. Turkish J Anaesthesiol Reanima. 2017;45(3):129–38. https://doi.org/10.5152/TJAR.2017.93753.

Gurrera RJ, Caroff SN, Cohen A, Carroll BT, DeRoos F, Francis A, Frucht S, Gupta S, Levenson JL, Mahmood A, Mann SC, Policastro MA, Rosebush PI, Rosenberg H, Sachdev PS, Trollor JN, Velamoor VR, Watson CB, Wilkinson JR. An international consensus study of neuroleptic malignant syndrome diagnostic criteria using the Delphi method. J Clin Psychiatry. 2011;72(09):1222–8. https://doi.org/10.4088/JCP.10m06438.

Hanlon JT, Horner RD, Schmader KE, Fillenbaum GG, Lewis IK, Wall WE, Landerman LR, Pieper CF, Blazer DG, Cohen HJ. Benzodiazepine use and cognitive function among community-dwelling elderly*. Clin Pharmacol Therap. 1998;64(6):684–92. https://doi.org/10.1016/S0009-9236(98)90059-5.

Hanlon JT, Schmader KE. The medication appropriateness index at 20: where it started, where it has been, and where it may be going. Drugs Aging. 2013;30(11):893–900. https://doi.org/10.1007/s40266-013-0118-4.

Harbeck B, Kropp P, Mönig H. Effects of short-term nocturnal cortisol replacement on cognitive function and quality of life in patients with primary or secondary adrenal insufficiency: a pilot study. Appl Psychophysiol Biofeedback. 2009;34(2):113–9. https://doi.org/10.1007/s10484-009-9082-5.

Harper C. The incidence of Wernicke's encephalopathy in Australia–a neuropathological study of 131 cases. J Neurol Neurosurg Psychiatry. 1983;46(7):593–8. https://doi.org/10.1136/jnnp.46.7.593.

Häussinger D, Kircheis G, Fischer R, Schliess F, vom Dahl S. Hepatic encephalopathy in chronic liver disease: a clinical manifestation of astrocyte swelling and low-grade cerebral edema? J Hepatol. 2000;32(6):1035–8. https://doi.org/10.1016/s0168-8278(00)80110-5.

Healton EB, Savage DG, Brust JCM, Garrett TJ, Lindenbaum J. Neurologic aspects of cobalamin deficiency. N Engl J Med. 1991;326(14):1035–41. https://doi.org/10.1056/NEJM199204023261401.

Heinrich TW, Grahm G. Hypothyroidism presenting as psychosis: myxedema madness revisited. Primary Care Comp J Clin Psych. 2003;5(6):260–6. https://doi.org/10.4088/pcc.v05n0603.

Heinrich M, Nottbrock A, Borchers F, Mörgeli R, Kruppa J, Winterer G, Slooter AJC, Spies C, The BioCog Consortium. Preoperative medication use and development of postoperative delirium and cognitive dysfunction. Clin Transl Sci. 2021;14(5):1830–40. https://doi.org/10.1111/cts.13031.

Hill NR, Fatoba ST, Oke JL, Hirst JA, O'Callaghan CA, Lasserson DS, Hobbs FD. Global prevalence of chronic kidney disease – a systematic review and meta-analysis. PLoS One. 2016;11(7):e0158765. https://doi.org/10.1371/journal.pone.0158765.

Hirsch L, Yang J, Bresee L, Jette N, Patten S, Pringsheim T. Second-generation antipsychotics and metabolic side effects: a systematic review of population-based studies. Drug Saf. 2017;40:771–81.

Holick MF. Vitamin D deficiency. N Engl J Med. 2007;357(3):266–81.

Hooshmand B, Solomon A, Kåreholt I, Leiviskä J, Rusanen M, Ahtiluoto S, Winblad B, Laatikainen T, Soininen H, Kivipelto M. Homocysteine and holotranscobalamin and the risk of Alzheimer disease: a longitudinal study. Neurology. 2010;75(16):1408–14. https://doi.org/10.1212/WNL.0b013e3181f88162.

Hosokawa K, Gaspard N, Su F, Oddo M, Vincent JL, Taccone FS. Clinical neurophysiological assessment of sepsis-associated brain dysfunction: a systematic review. Crit Care. 2014;18(6):674. https://doi.org/10.1186/s13054-014-0674-y.

Hoste EA, Bagshaw SM, Bellomo R, Cely CM, Colman R, Cruz DN, Edipidis K, Forni LG, Gomersall CD, Govil D, Honoré PM, Joannes-Boyau O, Joannidis M, Korhonen AM, Lavrentieva A, Mehta RL, Palevsky P, Roessler E, Ronco C, Uchino S, et al. Epidemiology of acute kidney injury in critically ill patients: the multinational AKI-EPI study. Intensive Care Med. 2015;41(8):1411–23. https://doi.org/10.1007/s00134-015-3934-7.

Husebye ES, Anderson MS, Kämpe O. Autoimmune Polyendocrine Syndromes. N Engl J Med. 2018;378(12):1132–41. https://doi.org/10.1056/NEJMra1713301.

Iacobone E, Bailly-Salin J, Polito A, Friedman D, Stevens RD, Sharshar T. Sepsis-associated encephalopathy and its differential diagnosis. Crit Care Med. 2009;37(10 Suppl):S331–6. https://doi.org/10.1097/CCM.0b013e3181b6ed58.

Iba T, Levi M, Levy JH. Sepsis-induced coagulopathy and disseminated intravascular coagulation. Semin Thromb Hemost. 2020;46(1):89–95. https://doi.org/10.1055/s-0039-1694995.

Inouye SK, Westendorp RG, Saczynski JS. Delirium in elderly people. Lancet (London, England). 2014;383(9920):911–22. https://doi.org/10.1016/S0140-6736(13)60688-1.

Isenberg-Grzeda E, Kutner HE, Nicolson SE. Wernicke-Korsakoff-syndrome: under-recognized and under-treated. Psychosomatics. 2012;53(6):507–16.

Jin Z, Hu J, Ma D. Postoperative delirium: perioperative assessment, risk reduction, and management. Br J Anaesth. 2020;125(4):492–504. https://doi.org/10.1016/j.bja.2020.06.063.

Johansen KL, Chertow GM, Gilbertson DT, Herzog CA, Ishani A, Israni AK, Ku E, Li S, Li S, Liu J, Obrador GT, O'Hare AM, Peng Y, Powe NR, Roetker NS, St Peter WL, Saeed F, Snyder J, Solid C, Weinhandl ED, et al. US renal data system 2021 annual data report: epidemiology of kidney disease in the United States. American J Kidney Diseases : Off J Nat Kidney Found. 2022;79(4 Suppl 1):A8–A12. https://doi.org/10.1053/j.ajkd.2022.02.001.

Khatiwada S, Sah SK, Kc R, Baral N, Lamsal M. Thyroid dysfunction in metabolic syndrome patients and its relationship with components of metabolic syndrome. Clin Diabetes Endocrinol. 2016;2:3. https://doi.org/10.1186/s40842-016-0021-0.

Kohnke S, Meek CL. Don't seek, don't find: the diagnostic challenge of Wernicke's encephalopathy. Annal Clin Biochem: Int J Lab Med. 2021;58(1):38–46. https://doi.org/10.1177/0004563220939604.

Kok L, Slooter A, Hillegers M, van Dijk D, Veldhuijzen D. Benzodiazepine use and neuropsychiatric outcomes in the ICU: a systematic review. Crit Care Med. 2018;46(10):1673–80.

Koob GF, Le Moal M. Addiction and the brain Antireward system. Annu Rev Psychol. 2008;59(1):29–53. https://doi.org/10.1146/annurev.psych.59.103006.093548.

Kung CW, Chou YH. Acute kidney disease: an overview of the epidemiology, pathophysiology, and management. Kidney Res Clin Pract. 2023;42(6):686–99. https://doi.org/10.23876/j.krcp.23.001.

Kurisu K, Miyabe D, Furukawa Y, Shibayama O, Yoshiuchi K. Association between polypharmacy and the persistence of delirium: a retrospective cohort study. BioPsychoSocial Med. 2020;14(1):25. https://doi.org/10.1186/s13030-020-00199-3.

Lee B, Zhao W, Yang K-C, Ahn Y-Y, Perry BL. Systematic evaluation of state policy interventions targeting the US opioid epidemic, 2007-2018. JAMA Netw Open. 2021;4(2):e2036687. https://doi.org/10.1001/jamanetworkopen.2020.36687.

Leikin JB, Amusina O. Use of dexmedetomidine to treat delirium primarily caused by cannabis. Am J Emerg Med. 2017;35(3):524.e1–2. https://doi.org/10.1016/j.ajem.2016.10.027.

Levey AS, Eckardt KU, Tsukamoto Y, Levin A, Coresh J, Rossert J, De Zeeuw D, Hostetter TH, Lameire N, Eknoyan G. Definition and classification of chronic kidney disease: a position statement from kidney disease: improving global outcomes (KDIGO). Kidney Int. 2005;67(6):2089–100. https://doi.org/10.1111/j.1523-1755.2005.00365.x.

Levy MM, Artigas A, Phillips GS, Rhodes A, Beale R, Osborn T, Vincent JL, Townsend S, Lemeshow S, Dellinger RP. Outcomes of the surviving sepsis campaign in intensive care units in the USA and Europe: a prospective cohort study. Lancet Infect Dis. 2012;12(12):919–24. https://doi.org/10.1016/S1473-3099(12)70239-6.

Long D, Long B, Koyfman A. The emergency medicine management of severe alcohol withdrawal. Am J Emerg Med. 2017;35(7):1005–11. https://doi.org/10.1016/j.ajem.2017.02.002.

Lu H-C, Mackie K. Review of the endocannabinoid system. Biol Psych: Cogn Neurosci Neuroimag. 2021;6(6):607–15. https://doi.org/10.1016/j.bpsc.2020.07.016.

Lu X, Qin M, Walline JH, Gao Y, Yu S, Ge Z, Gong C, Zhu H, Annane D, Li Y. Clinical phenotypes of sepsis-associated encephalopathy: a retrospective cohort study. Shock (Augusta, Ga). 2023;59(4):583–90. https://doi.org/10.1097/SHK.0000000000002092.

Lupien SJ, Maheu F, Tu M, Fiocco A, Schramek TE. The effects of stress and stress hormones on human cognition: implications for the field of brain and cognition. Brain Cogn. 2007;65(3):209–37. https://doi.org/10.1016/j.bandc.2007.02.007.

Magistretti PJ, Allaman I. A cellular perspective on brain energy metabolism and functional imaging. Neuron. 2015;86(4):883–901. https://doi.org/10.1016/j.neuron.2015.03.035.

Mainerova B, Prasko J, Latalova K, Axmann K, Cerna M, Horacek R, Bradacova R. Alcohol withdrawal delirium—diagnosis, course and treatment. Biomedical Papers. 2015;159(1):044–52. https://doi.org/10.5507/bp.2013.089.

Maldonado JR. Delirium in the acute care setting: characteristics, diagnosis and treatment. Crit Care Clin. 2008;24(4):657–vii. https://doi.org/10.1016/j.ccc.2008.05.008.

Maldonado JR. Acute brain failure: pathophysiology, diagnosis, management, and sequelae of delirium. Crit Care Clin. 2017;33(3):461–519. https://doi.org/10.1016/j.ccc.2017.03.013.

Maldonado JR. Delirium pathophysiology: an updated hypothesis of the etiology of acute brain failure. Int J Geriatr Psychiatry. 2018;33(11):1428–57. https://doi.org/10.1002/gps.4823.

Manasco A, Chang S, Larriviere J, Hamm LL, Glass M. Alcohol withdrawal. South Med J. 2012;105(11):607–12. https://doi.org/10.1097/SMJ.0b013e31826efb2d.

Manchikanti L. Opioid epidemic in the United States. Pain Physician. 2012;3S;15(3S;7):ES9–ES38. https://doi.org/10.36076/ppj.2012/15/ES9.

Manzo G, De Gennaro A, Cozzolino A, Serino A, Fenza G, Manto A. MR imaging findings in alcoholic and nonalcoholic acute Wernicke's encephalopathy: a review. Biomed Res Int. 2014;2014:1–12. https://doi.org/10.1155/2014/503596.

Marcantonio ER. Delirium in Hospitalized Older Adults. N Engl J Med. 2017;377(15):1456–66. https://doi.org/10.1056/NEJMcp1605501.

Martin-Grace J, Tomkins M, O'Reilly MW, Sherlock M. Iatrogenic adrenal insufficiency in adults. Nat Rev Endocrinol. 2024;20(4):209–27. https://doi.org/10.1038/s41574-023-00929-x.

Martin GS, Mannino DM, Eaton S, Moss M. The epidemiology of sepsis in the United States from 1979 through 2000. N Engl J Med. 2003;348(16):1546–54. https://doi.org/10.1056/NEJMoa022139.

Mathew P, Kaur J, Rawla P. Hyperthyroidism. In: StatPearls. StatPearls Publishing; 2023.

Mattison ML. Delirium. Annals of internal medicine. 2020;173(7): ITC49-ITC64.

McCoy TH, Castro VM, Hart KL, Perlis RH. Stratified delirium risk using prescription medication data in a state-wide cohort. Gen Hosp Psychiatry. 2021;71:114–20. https://doi.org/10.1016/j.genhosppsych.2021.05.001.

McCreary AC, Müller CP, Filip M. Psychostimulants: basic and clinical pharmacology. Int Rev Neurobiol. 2015;120:41–83. https://doi.org/10.1016/bs.irn.2015.02.008.

Meagher D, Agar MR, Teodorczuk A. Debate article: antipsychotic medications are clinically useful for the treatment of delirium. Int J Geriatr Psychiatry. 2018;33(11):1420–7. https://doi.org/10.1002/gps.4759.

Mergenthaler P, Lindauer U, Dienel GA, Meisel A. Sugar for the brain: the role of glucose in physiological and pathological brain function. Trends Neurosci. 2013;36(10):587–97. https://doi.org/10.1016/j.tins.2013.07.001.

Midão L, Giardini A, Menditto E, Kardas P, Costa E. Polypharmacy prevalence among older adults based on the survey of health, ageing and retirement in Europe. Arch Gerontol Geriatr. 2018;78:213–20. https://doi.org/10.1016/j.archger.2018.06.018.

Moder KG, Hurley DL. Fatal hypernatremia from exogenous salt intake: report of a case and review of the literature. Mayo Clin Proc. 1990;65(12):1587–94. https://doi.org/10.1016/s0025-6196(12)62194-6.

Moritz ML, Ayus JC. The pathophysiology and treatment of hyponatraemic encephalopathy: an update. Nephrol Dialysis Trans Off Publ European Dial Trans Assoc – European Renal Assoc. 2003;18(12):2486–91. https://doi.org/10.1093/ndt/gfg394.

Morgan S, Weinsier R. Fundamentals of clinical nutrition. Mosby; 1998.

Morrison RS, Magaziner J, Gilbert M, Koval KJ, McLaughlin MA, Orosz G, Strauss E, Siu AL. Relationship between pain and opioid analgesics on the development of delirium following hip fracture. J Gerontol Ser A Biol Med Sci. 2003;58(1):M76–81. https://doi.org/10.1093/gerona/58.1.M76.

Moussa MK, Hall MAK, Akwe J. Cannabis-induced acute encephalopathy in a 94-year-old woman due to family Administration of Cannabidiol (CBD) products: case report. Cureus. 2023; https://doi.org/10.7759/cureus.37927.

Mueller A, Spies CD, Eckardt R, Weiss B, Pohrt A, Wernecke K-D, Schmidt M. Anticholinergic burden of long-term medication is an independent risk factor for the development of postoperative delirium: a clinical trial. J Clin Anesth. 2020;61:109632. https://doi.org/10.1016/j.jclinane.2019.109632.

Murray AM, Tupper DE, Knopman DS, Gilbertson DT, Pederson SL, Li S, Smith GE, Hochhalter AK, Collins AJ, Kane RL. Cognitive impairment in hemodialysis patients is common. Neurology. 2006;67(2):216–23. https://doi.org/10.1212/01.wnl.0000225182.15532.40.

Neufeld KJ, Yue J, Robinson TN, Inouye SK, Needham DM. Antipsychotic medication for prevention and treatment of delirium in hospitalized adults: a systematic review and meta-analysis. J Am Geriatr Soc. 2016;64(4):705–14. https://doi.org/10.1111/jgs.14076.

O'Donnell L, Gnjidic D, Nahas R, Bell JS, Hilmer SN. Anticholinergic burden: considerations for older adults. J Pharm Pract Res. 2017;47(1):67–77. https://doi.org/10.1002/jppr.1303.

O'Leary F, Samman S. Vitamin B12 in health and disease. Nutrients. 2010;2(3):299–316. https://doi.org/10.3390/nu2030299.

O'Mahony D, O'Sullivan D, Byrne S, O'Connor MN, Ryan C, Gallagher P. STOPP/START criteria for potentially inappropriate prescribing in older people: version 2. Age Ageing. 2014;44(2):213–8. https://doi.org/10.1093/ageing/afu145.

Ota Y, Capizzano AA, Moritani T, Naganawa S, Kurokawa R, Srinivasan A. Comprehensive review of Wernicke encephalopathy: pathophysiology, clinical symptoms and imaging findings. Jpn J Radiol. 2020;38(9):809–20. https://doi.org/10.1007/s11604-020-00989-3.

Ott M, Werneke U. Wernicke's encephalopathy — from basic science to clinical practice. Part 1: understanding the role of thiamine. Therap Adv Psychopharmacol. 2020;10:204512532097810. https://doi.org/10.1177/2045125320978106.

Oudman E, Wijnia JW, Oey MJ, Van Dam MJ, Postma A. Preventing Wernicke's encephalopathy in anorexia nervosa: a systematic review. Psychiatry Clin Neurosci. 2018;72(10):774–9. https://doi.org/10.1111/pcn.12735.

Oudman E, Wijnia JW, Oey M, Van Dam M, Painter RC, Postma A. Wernicke's encephalopathy in hyperemesis gravidarum: a systematic review. Eur J Obstet Gynecol Reprod Biol. 2019;236:84–93. https://doi.org/10.1016/j.ejogrb.2019.03.006.

Oudman E, Wijnia JW, Oey MJ, Van Dam M, Postma A. Wernicke's encephalopathy in Crohn's disease and ulcerative colitis. Nutrition. 2021;86:111182. https://doi.org/10.1016/j.nut.2021.111182.

Paice JA, Ferrell B. The management of cancer pain. CA Cancer J Clin. 2011;61(3):157–82. https://doi.org/10.3322/caac.20112.

Patidar KR, Thacker LR, Wade JB, Sterling RK, Sanyal AJ, Siddiqui MS, Matherly SC, Stravitz RT, Puri P, Luketic VA, Fuchs M, White MB, Noble NA, Unser AB, Gilles H, Heuman DM, Bajaj JS. Covert hepatic encephalopathy is independently associated with poor survival and increased risk of hospitalization. Am J Gastroenterol. 2014;109(11):1757–63. https://doi.org/10.1038/ajg.2014.264.

Patil N, Rehman A, Anastasopoulou C, Jialal I. Hypothyroidism. In: StatPearls. StatPearls Publishing; 2024.

Pertichetti M, Serioli S, Belotti F, Mattavelli D, Schreiber A, Cappelli C, Padovani A, Gasparotti R, Nicolai P, Fontanella MM, Doglietto F. Pituitary adenomas and neuropsychological status: a systematic literature review. Neurosurg Rev. 2020;43(4):1065–78. https://doi.org/10.1007/s10143-019-01134-z.

Pitel AL, Chételat G, Le Berre AP, Desgranges B, Eustache F, Beaunieux H. Macrostructural abnormalities in Korsakoff syndrome compared with uncomplicated alcoholism. Neurology. 2012;78(17):1330–3. https://doi.org/10.1212/WNL.0b013e318251834e.

Plauth M, Cabré E, Riggio O, Assis-Camilo M, Pirlich M, Kondrup J, Ferenci P, Holm E, Vom Dahl S, Müller MJ, Nolte W, DGEM (German Society for Nutritional Medicine), ESPEN (European Society for Parenteral and Enteral Nutrition). ESPEN guidelines on enteral nutrition: liver disease. Clinical Nutrition (Edinburgh, Scotland). 2006;25(2):285–94. https://doi.org/10.1016/j.clnu.2006.01.018.

Pow J, Baumeister A, Hawkins M, Cohen A, Garand J. Deinstitutionalization of American public hospitals for the mentally ill before and after the introduction of antipsychotic medications. Harv Rev Psychiatry. 2015;23(3):176–87.

Radomski JW, Dursun SM, Reveley MA, Kutcher SP. An exploratory approach to the serotonin syndrome: an update of clinical phenomenology and revised diagnostic criteria. Med Hypotheses. 2000;55(3):218–24. https://doi.org/10.1054/mehy.2000.1047.

Pribék IK, Kovács I, Kádár BK, Kovács CS, Richman MJ, Janka Z, Andó B, Lázár BA. Evaluation of the course and treatment of alcohol withdrawal syndrome with the clinical institute withdrawal assessment for alcohol – revised: a systematic review-based meta-analysis. Drug Alcohol Depend. 2021;220:108536. https://doi.org/10.1016/j.drugalcdep.2021.108536.

Pullen RG, DePasquale M, Cserr HF. Bulk flow of cerebrospinal fluid into brain in response to acute hyperosmolality. Am J Phys. 1987;253(3 Pt 2):F538–45. https://doi.org/10.1152/ajprenal.1987.253.3.F538.

Quesada-Gomez JM, Bouillon R, Heijboer AC. Cholecalciferol or calcifediol in the management of vitamin D deficiency. Nutrients. 2022;14(17):3608.

Randolph C, Hilsabeck R, Kato A, Kharbanda P, Li YY, Mapelli D, Ravdin LD, Romero-Gomez M, Stracciari A, Weissenborn K, International Society for Hepatic Encephalopathy and Nitrogen Metabolism (ISHEN). Neuropsychological assessment of hepatic encephalopathy: ISHEN practice guidelines. Liver Int : Off J Int Assoc Stud Liver. 2009;29(5):629–35. https://doi.org/10.1111/j.1478-3231.2009.02009.x.

Raabe W. Effects of hyperammonemia on neuronal function: NH4+, IPSP and cl(−)-extrusion. Adv Exp Med Biol. 1993;341:71–82. https://doi.org/10.1007/978-1-4615-2484-7_7.

Ragusa F, Fallahi P, Elia G, Gonnella D, Paparo SR, Giusti C, Churilov LP, Ferrari SM, Antonelli A. Hashimotos' thyroiditis: epidemiology, pathogenesis, clinic and therapy. Best Pract Res Clin Endocrinol Metab. 2019;33(6):101367. https://doi.org/10.1016/j.beem.2019.101367.

Rezende E, Silva JM Jr, Isola AM, Campos EV, Amendola CP, Almeida SL. Epidemiology of severe sepsis in the emergency department and difficulties in the initial assistance. Clinics (Sao Paulo, Brazil). 2008;63(4):457–64. https://doi.org/10.1590/s1807-59322008000400008.

Richards JR, Le JK. Cocaine toxicity. In: StatPearls. StatPearls Publishing; 2023.

Rickels MR, Schutta MH, Mueller R, Kapoor S, Markmann JF, Naji A, Teff KL. Glycemic thresholds for activation of counterregulatory hormone and symptom responses in islet transplant recipients. J Clin Endocrinol Metab. 2007;92(3):873–9. https://doi.org/10.1210/jc.2006-2426.

Rieben C, Segna D, da Costa BR, Collet TH, Chaker L, Aubert CE, Baumgartner C, Almeida OP, Hogervorst E, Trompet S, Masaki K, Mooijaart SP, Gussekloo J, Peeters RP, Bauer DC, Aujesky D, Rodondi N. Subclinical thyroid dysfunction and the risk of cognitive decline: a meta-analysis of prospective cohort studies. J Clin Endocrinol Metab. 2016;101(12):4945–54. https://doi.org/10.1210/jc.2016-2129.

Romero-Gómez M, Montagnese S, Jalan R. Hepatic encephalopathy in patients with acute decompensation of cirrhosis and acute-on-chronic liver failure. J Hepatol. 2015;62(2):437–47. https://doi.org/10.1016/j.jhep.2014.09.005.

Rudd KE, Johnson SC, Agesa KM, Shackelford KA, Tsoi D, Kievlan DR, Colombara DV, Ikuta KS, Kissoon N, Finfer S, Fleischmann-Struzek C, Machado FR, Reinhart KK, Rowan K, Seymour CW, Watson RS, West TE, Marinho F, Hay SI, Lozano R, et al. Global, regional, and national sepsis incidence and mortality, 1990-2017: analysis for the global burden of disease study. Lancet (London, England). 2020;395(10219):200–11. https://doi.org/10.1016/S0140-6736(19)32989-7.

Rudler M, Mallet M, Sultanik P, Bouzbib C, Thabut D. Optimal management of ascites. Liver Int : Off J Int Assoc Stud Liver. 2020;40(Suppl 1):128–35. https://doi.org/10.1111/liv.14361.

Ruxton K, Woodman RJ, Mangoni AA. Drugs with anticholinergic effects and cognitive impairment, falls and all-cause mortality in older adults: a systematic review and meta-analysis. Br J Clin Pharmacol. 2015;80(2):209–20. https://doi.org/10.1111/bcp.12617.

Saadi A, Naples-Mitchell J, Da Silva Bhatia B, Heisler M. End the use of "excited delirium" as a cause of death in police custody. Lancet. 2022;399(10329):1028–30. https://doi.org/10.1016/S0140-6736(22)00410-X.

Safdar M, Colosimo C, Khurshid MH, Spencer AL, Hejazi O, Castanon L, Hosseinpour H, Magnotti LJ, Bhogadi SK, Joseph B. Drugs, delirium, and trauma: substance use and incidence of delirium after traumatic brain injury. J Surg Res. 2024;301:45–53. https://doi.org/10.1016/j.jss.2024.05.042.

Salluh JI, Soares M, Teles JM, Ceraso D, Raimondi N, Nava VS, Blasquez P, Ugarte S, Ibanez-Guzman C, Centeno JV, Laca M, Grecco G, Jimenez E, Árias-Rivera S, Duenas C, Rocha MG, Delirium Epidemiology in Critical Care Study Group. Delirium epidemiology in critical care (DECCA): an international study. Crit Care. 2010;14(6):R210. https://doi.org/10.1186/cc9333.

Salmon DP, Butters N. The etiology and neuropathology of alcoholic Korsakoff's syndrome. Some evidence for the role of the basal forebrain. Recent Dev Alcohol : Off Publ American Med Soc Alcohol Res Soc Alcohol, Nat Council Alcohol. 1987;5:27–58.

Sanders AE, Zafar N, Sharma S. Myoclonus. [Updated 2024 Feb 26]. In: StatPearls [Internet]. Treasure Island (FL): StatPearls Publishing; 2025. Available from: https://www.ncbi.nlm.nih.gov/books/NBK537015/

Sanyal AJ, Van Natta ML, Clark J, Neuschwander-Tetri BA, Diehl A, Dasarathy S, Loomba R, Chalasani N, Kowdley K, Hameed B, Wilson LA, Yates KP, Belt P, Lazo M, Kleiner DE, Behling C, Tonascia J, NASH Clinical Research Network (CRN). Prospective study of outcomes in adults with nonalcoholic fatty liver disease. N Engl J Med. 2021;385(17):1559–69. https://doi.org/10.1056/NEJMoa2029349.

Scott IA, Hilmer SN, Reeve E, Potter K, Le Couteur D, Rigby D, Gnjidic D, Del Mar CB, Roughead EE, Page A, Jansen J, Martin JH. Reducing inappropriate polypharmacy: the process of Deprescribing. JAMA Intern Med. 2015;175(5):827. https://doi.org/10.1001/jamainternmed.2015.0324.

Scotton WJ, Hill LJ, Williams AC, Barnes NM. Serotonin syndrome: pathophysiology, clinical features, management, and potential future directions. Int J Tryptophan Res. 2019;12:117864691987392. https://doi.org/10.1177/1178646919873925.

Seay NW, Lehrich RW, Greenberg A. Diagnosis and Management of Disorders of body tonicity-hyponatremia and hypernatremia: Core curriculum 2020. American J Kidney Diseases : Off J Nat Kidney Found. 2020;75(2):272–86. https://doi.org/10.1053/j.ajkd.2019.07.014.

Sechi G, Serra A. Wernicke's encephalopathy: new clinical settings and recent advances in diagnosis and management. Lancet Neurol. 2007;6(5):442–55.

Sharma A, Nagalli S. Chronic liver disease. In: StatPearls. StatPearls Publishing; 2023.

Siddiqi N, House AO, Holmes JD. Occurrence and outcome of delirium in medical in-patients: a systematic literature review. Age Ageing. 2006;35(4):350–64. https://doi.org/10.1093/ageing/afl005.

Siddiqi N, Harrison JK, Clegg A, Teale EA, Young J, Taylor J, Simpkins SA. Interventions for preventing delirium in hospitalised non-ICU patients. Cochrane Database Syst Rev. 2016;3(3):CD005563. https://doi.org/10.1002/14651858.CD005563.pub3.

Singer M, Deutschman CS, Seymour CW, Shankar-Hari M, Annane D, Bauer M, Bellomo R, Bernard GR, Chiche JD, Coopersmith CM, Hotchkiss RS, Levy MM, Marshall JC, Martin GS, Opal SM, Rubenfeld GD, van der Poll T, Vincent JL, Angus DC. The third international consensus definitions for sepsis and septic shock (Sepsis-3). JAMA. 2016;315(8):801–10. https://doi.org/10.1001/jama.2016.0287.

Sinha S, Kataria A, Kolla BP, Thusius N, Loukianova LL. Wernicke encephalopathy—clinical pearls. Mayo Clin Proc. 2019;94(6):1065–72. https://doi.org/10.1016/j.mayocp.2019.02.018.

Slooter AJC, Otte WM, Devlin JW, Arora RC, Bleck TP, Claassen J, Duprey MS, Ely EW, Kaplan PW, Latronico N, Morandi A, Neufeld KJ, Sharshar T, MacLullich AMJ, Stevens RD. Updated nomenclature of delirium and acute encephalopathy: statement of ten societies. Intensive Care Med. 2020;46(5):1020–2. https://doi.org/10.1007/s00134-019-05907-4.

Smith JW, Evans AT, Costall B, Smythe JW. Thyroid hormones, brain function and cognition: a brief review. Neurosci Biobehav Rev. 2002;26(1):45–60. https://doi.org/10.1016/s0149-7634(01)00037-9.

Smith AD, Smith SM, de Jager CA, Whitbread P, Johnston C, Agacinski G, Oulhaj A, Bradley KM, Jacoby R, Refsum H. Homocysteine-lowering by B vitamins slows the rate of accelerated brain atrophy in mild cognitive impairment: a randomized controlled trial. PLoS One. 2010;5(9):e12244. https://doi.org/10.1371/journal.pone.0012244.

Smith H, McCoy M, Varughese K, Reinert JP. Thiamine dosing for the treatment of alcohol-induced Wernicke's encephalopathy: a review of the literature. J Pharm Technol. 2021;37(2):107–13. https://doi.org/10.1177/8755122520962859.

Sonneville R, de Montmollin E, Poujade J, Garrouste-Orgeas M, Souweine B, Darmon M, Mariotte E, Argaud L, Barbier F, Goldgran-Toledano D, Marcotte G, Dumenil AS, Jamali S, Lacave G, Ruckly S, Mourvillier B, Timsit JF. Potentially modifiable factors contributing to sepsis-associated encephalopathy. Intensive Care Med. 2017;43(8):1075–84. https://doi.org/10.1007/s00134-017-4807-z.

Spadaro A, Scott KR, Koyfman A, Long B. High risk and low prevalence diseases: serotonin syndrome. Am J Emerg Med. 2022;61:90–7. https://doi.org/10.1016/j.ajem.2022.08.030.

Spindle TR, Kuwabara H, Eversole A, Nandi A, Vandrey R, Antoine DG, Umbricht A, Guarda AS, Wong DF, Weerts EM. Brain imaging of cannabinoid type I (CB_1) receptors in women with cannabis use disorder and male and female healthy controls. Addict Biol. 2021;26(6):e13061. https://doi.org/10.1111/adb.13061.

Springer B. Hyperactive delirium with severe agitation. Psych Behav Emer Issue Emer Med Clin North America, E-Book. 2023;42(1):41.

Sprung CL, Peduzzi PN, Shatney CH, Schein RM, Wilson MF, Sheagren JN, Hinshaw LB. Impact of encephalopathy on mortality in the sepsis syndrome. The veterans Administration systemic sepsis cooperative study group. Crit Care Med. 1990;18(8):801–6. https://doi.org/10.1097/00003246-199008000-00001.

Sprung J, Roberts RO, Weingarten TN, Nunes Cavalcante A, Knopman DS, Petersen RC, Hanson AC, Schroeder DR, Warner DO. Postoperative delirium in elderly patients is associated with subsequent cognitive impairment. Br J Anaesth. 2017;119(2):316–23. https://doi.org/10.1093/bja/aex130.

Stabler SP. Clinical practice. Vitamin B12 deficiency. N Engl J Med. 2013;368(2):149–60. https://doi.org/10.1056/NEJMcp1113996.

Stahl S. Stahl's essential psychopharmacology: neuroscientific basis and practical applications. 5th ed. Cambridge University Press; 2021.

Steel TL, Giovanni SP, Katsandres SC, Cohen SM, Stephenson KB, Murray B, Sobeck H, Hough CL, Bradley KA, Williams EC. Should the CIWA-Ar be the standard monitoring strategy for alcohol withdrawal syndrome in the intensive care unit? Addict Sci Clin Pract. 2021;16(1):21. https://doi.org/10.1186/s13722-021-00226-w.

Steinman MA. Polypharmacy—time to get beyond numbers. JAMA Intern Med. 2016;176(4):482. https://doi.org/10.1001/jamainternmed.2015.8597.

Steinman MA, Landefeld CS. Overcoming inertia to improve medication use and Deprescribing. JAMA. 2018;320(18):1867. https://doi.org/10.1001/jama.2018.16473.

Sternbach H. The serotonin syndrome. Am J Psychiatry. 1991;148:705–13.

Sterns R. H. (2015). Disorders of plasma sodium--causes, consequences, and correction. N Engl J Med, 372(1), 55–65. https://doi.org/10.1056/NEJMra1404489.

Strawn JR, Keck PE, Caroff SN. Neuroleptic malignant syndrome. Am J Psychiatry. 2007;164:870.

Substance Abuse and Mental Health Services Administration. National survey on drug use and health: Detailed tables. 2021. https://www.samhsa.gov/data/report/2021-nsduh-detailed-tables.

Sullivan JT, Sykora K, Schneiderman J, Naranjo CA, Sellers EM. Assessment of alcohol withdrawal: the revised clinical institute withdrawal assessment for alcohol scale (CIWA-Ar). Br J Addict. 1989;84(11):1353–7. https://doi.org/10.1111/j.1360-0443.1989.tb00737.x.

Susantitaphong P, Cruz DN, Cerda J, Abulfaraj M, Alqahtani F, Koulouridis I, Jaber BL, Acute Kidney Injury Advisory Group of the American Society of Nephrology. World incidence of AKI: a meta-analysis. Clin J American Soc Nephrol: CJASN. 2013;8(9):1482–93. https://doi.org/10.2215/CJN.00710113.

Swarbrick CJ, Partridge JSL. Evidence-based strategies to reduce the incidence of postoperative delirium: a narrative review. Anaesthesia. 2022;77(Suppl 1):92–101. https://doi.org/10.1111/anae.15607.

Swart LM, Van Der Zanden V, Spies PE, De Rooij SE, Van Munster BC. The comparative risk of delirium with different opioids: a systematic review. Drugs Aging. 2017;34(6):437–43. https://doi.org/10.1007/s40266-017-0455-9.

Takeuchi A, Ahern T, Henderson S. Excited delirium. Western J Emer Med. 2011;12(1):77–83.

Talton C. Serotonin syndrome/serotonin toxicity. Fed Pract. 2020;37(10) https://doi.org/10.12788/fp.0042.

Tanaka S, Okusa MD. Crosstalk between the nervous system and the kidney. Kidney Int. 2020;97(3):466–76. https://doi.org/10.1016/j.kint.2019.10.032.

Tannenbaum C, Paquette A, Hilmer S, Holroyd-Leduc J, Carnahan R. A systematic review of amnestic and non-amnestic mild cognitive impairment induced by anticholinergic, antihistamine, GABAergic and opioid drugs. Drugs Aging. 2012;29(8):639–58.

Tapper EB. Predicting overt hepatic encephalopathy for the population with cirrhosis. Hepatology (Baltimore, Md). 2019;70(1):403–9. https://doi.org/10.1002/hep.30533.

Taylor PN, Albrecht D, Scholz A, Gutierrez-Buey G, Lazarus JH, Dayan CM, Okosieme OE. Global epidemiology of hyperthyroidism and hypothyroidism. Nat Rev Endocrinol. 2018;14(5):301–16. https://doi.org/10.1038/nrendo.2018.18.

Thompson AR, Kim CS, Kim GE, Keller MS, Marcum ZA, Brandt NJ. Challenges and successes of global Deprescribing networks: a qualitative key informant study. J Gerontol Nurs. 2022;48(1):7–14. https://doi.org/10.3928/00989134-20211206-02.

Torio CM, Moore BJ. National Inpatient Hospital Costs: the Most expensive conditions by payer, 2013. In: Healthcare cost and utilization project (HCUP) statistical briefs. Agency for Healthcare Research and Quality (US); 2016.

Torres RV, Elias MF, Seliger S, Davey A, Robbins MA. Risk for cognitive impairment across 22 measures of cognitive ability in early-stage chronic kidney disease. Nephrol Dialysis Trans Off Pub European Dialysis Trans Assoc – European Renal Assoc. 2017;32(2):299–306. https://doi.org/10.1093/ndt/gfw005.

United Nations Office on Drugs and Crime. World drug report 2022. (2022. https://www.unodc.org/unodc/en/data-and-analysis/world-drug-report-2022.html.

Upadhyay A, Jaber BL, Madias NE. Incidence and prevalence of hyponatremia. Am J Med. 2006;119(7 Suppl 1):S30–5. https://doi.org/10.1016/j.amjmed.2006.05.005.

Vaidya SR, Aeddula NR. Chronic kidney disease. In: StatPearls. StatPearls Publishing; 2024.

Van Velthuijsen EL, Zwakhalen SMG, Pijpers E, Van De Ven LI, Ambergen T, Mulder WJ, Verhey FRJ, Kempen GIJM. Effects of a medication review on delirium in older hospitalised patients: a comparative retrospective cohort study. Drugs Aging. 2018;35(2):153–61. https://doi.org/10.1007/s40266-018-0523-9.

Vanderlinden JA, Ross-White A, Holden R, Shamseddin MK, Day A, Boyd JG. Quantifying cognitive dysfunction across the spectrum of end-stage kidney disease: a systematic review and meta-analysis. Nephrology (Carlton). 2019;Vic.), 24(1):5–16. https://doi.org/10.1111/nep.13448.

Viggiano D, Wagner CA, Martino G, Nedergaard M, Zoccali C, Unwin R, Capasso G. Mechanisms of cognitive dysfunction in CKD. Nat Rev Nephrol. 2020;16(8):452–69. https://doi.org/10.1038/s41581-020-0266-9.

Vilstrup H, Amodio P, Bajaj J, Cordoba J, Ferenci P, Mullen KD, Weissenborn K, Wong P. Hepatic encephalopathy in chronic liver disease: 2014 practice guideline by the American Association for the Study of Liver Diseases and the European Association for the Study of the liver. Hepatology (Baltimore, Md.). 2014;60(2):715–35. https://doi.org/10.1002/hep.27210.

Vincent JL, de Mendonça A, Cantraine F, Moreno R, Takala J, Suter PM, Sprung CL, Colardyn F, Blecher S. Use of the SOFA score to assess the incidence of organ dysfunction/failure in intensive care units: results of a multicenter, prospective study. Working group on "sepsis-related problems" of the European Society of Intensive Care Medicine. Crit Care Med. 1998;26(11):1793–800. https://doi.org/10.1097/00003246-199811000-00016.

Walsh B, Agboola I, Agboola I, Coupet E, Rozel J, Wong A. Revisiting "excited delirium": does the diagnosis reflect and promote racial bias? Western Journal of Emergency Medicine. 2023;24(2):152–9. https://doi.org/10.5811/westjem.2022.10.56478.

Ware M, Feller D, Hall K. Neuroleptic malignant syndrome: diagnosis and management. Primary Care Comp for CNS Disorder. 2018;20(1):27030.

Wastesson JW, Morin L, Tan ECK, Johnell K. An update on the clinical consequences of polypharmacy in older adults: a narrative review. Expert Opin Drug Saf. 2018;17(12):1185–96. https://doi.org/10.1080/14740338.2018.1546841.

Webster AC, Nagler EV, Morton RL, Masson P. Chronic kidney disease. Lancet (London, England). 2017;389(10075):1238–52. https://doi.org/10.1016/S0140-6736(16)32064-5.

Weissenborn K. Hepatic encephalopathy: definition, clinical grading and diagnostic principles. Drugs. 2019;79(Suppl 1):5–9. https://doi.org/10.1007/s40265-018-1018-z.

Wilson JE, Mart MF, Cunningham C, Shehabi Y, Girard TD, MacLullich AMJ, Slooter AJC, Ely EW. Delirium. Nat Rev Disease Primer. 2020;6(1):90. https://doi.org/10.1038/s41572-020-00223-4.

Wood H, Dickman A, Star A, Boland JW. Updates in palliative care—overview and recent advancements in the pharmacological management of cancer pain. Clin Med. 2018;18(1):17–22. https://doi.org/10.7861/clinmedicine.18-1-17.

World Health Organization. (2018). Global status report on alcohol and health 2018.. [Licence: CC BY-NC-SA 3.0 IGO.]. World Health Organization.

Wu N, Zhang Y, Wang S, Zhao Y, Zhong X. Incidence, prevalence and risk factors of delirium in ICU patients: a systematic review and meta-analysis. Nurs Crit Care. 2022;28(5) https://doi.org/10.1111/nicc.12857.

Xiang PY, Boyle L, Short TG, Deng C, Campbell D. Incidence of postoperative delirium in surgical patients: an observational retrospective cohort study. Anaesth Intensive Care. 2023;51(4):260–7. https://doi.org/10.1177/0310057X231156459.

Yu AT, Gross A, Park K, Harvey EJ. Wernicke encephalopathy after bariatric surgery: a literature review. Obes Surg. 2023;33(11):3621–7. https://doi.org/10.1007/s11695-023-06840-8.

Zaal IJ, Devlin JW, Hazelbag M, Klein Klouwenberg PMC, Van Der Kooi AW, Ong DSY, Cremer OL, Groenwold RH, Slooter AJC. Benzodiazepine-associated delirium in critically ill adults. Intensive Care Med. 2015;41(12):2130–7. https://doi.org/10.1007/s00134-015-4063-z.

Zamwar UM, Muneshwar KN. Epidemiology, types, causes, clinical presentation, diagnosis, and treatment of hypothyroidism. Cureus. 2023;15(9):e46241. https://doi.org/10.7759/cureus.46241.

Zhang LN, Wang XT, Ai YH, Guo QL, Huang L, Liu ZY, Yao B. Epidemiological features and risk factors of sepsis-associated encephalopathy in intensive care unit patients: 2008-2011. Chin Med J. 2012;125(5):828–31.

Zhao L, Zhang J, Zhang Q, Liu Y. The role of vitamin D deficiency in hepatic encephalopathy: a review of pathophysiology, clinical outcomes, and therapeutic potential. Front Med. 2021;8:783195.

Zimpfer D, Czerny M, Vogt F, Schuch P, Kramer L, Wolner E, Grimm M. Neurocognitive deficit following coronary artery bypass grafting: a prospective study of surgical patients and nonsurgical controls. Ann Thorac Surg. 2004;78(2):513–9. https://doi.org/10.1016/j.athoracsur.2004.03.006.

Paraneoplastic Neurologic Disorders

8

Brittany Lang, Michelle Boston-Babicz, and Joel E. Kamper

ABBREVIATIONS

ACTH	Adrenocorticotropic hormone
ADH	Antidiuretic hormone
ALS	Amyotrophic lateral sclerosis
AMPA	α-amino-3-hydroxy-5-methyl-4-isoxazolepropionic
AMS	Altered mental status
ANNA	Antineuronal nuclear antibody
AQP4	aquaporin 4
BDUMP	Bilateral diffuse uveal melanocytic proliferation
CASPR2	Contactin-associated protein-like 2
CAR	Carcinoma-associated retinopathy
CCAS	Cerebellar cognitive affective syndrome
CNS	Central nervous system
CRMP5	Collapsin response-mediator protein 5
CSF	Cerebrospinal fluid
CT	Computed tomography
DNER	Delta/notch-like epidermal growth factor-related receptor
DPPX	Dipeptidyl peptidase-like protein
DWI	Diffusion weighted imaging
EEG	Electroencephalogram
FDG	Fludeoxyglucose-18
FTD	Frontotemporal dementia
GABA	Gamma-aminobutyric-acid
GAD	Glutamic acid decarboxylase
GFAP	Glial fibrillary acidic protein

B. Lang (✉)
Department of Neurology, University of Utah School of Medicine, Salt Lake City, UT, USA
e-mail: Brittany.Lang@hsc.utah.edu

M. Boston-Babicz
Madonna Ptak Memory Disorders Clinic, BayCare Medical Group, Clearwater, FL, USA
e-mail: Michelle.Babicz@baycare.org

J. E. Kamper
James A. Haley Veterans' Hospital, Mental Health and Behavioral Science Service (MS/BSS), Tampa, FL, USA
e-mail: Joel.Kamper@va.gov

E. K. Bailey et al. (eds.), *Inpatient Neuropsychology*, https://doi.org/10.1007/978-3-032-05880-5_8

GluR	Glutamate receptor
GlyR	Glycine receptor
HIV	Human immunodeficiency virus
IgLON	Immunoglobulin-like cell adhesion molecule
KCTD16	Potassium channel tetramerization domain containing
KLHL11	Kelch-like protein 11
LEMS	Lambert-Eaton myasthenic syndromes
LGI1	Leucine-rich glioma-inactivated protein 1
LUZP4	Leucine zipper protein 4
MAP 1B	Microtubule-associated protein 1B
MAR	Melanoma-associated retinopathy
mGluR	Metabotropic glutamate receptor
MMSE	Mini-Mental State Examination
MND	Motor neuron diseases
MoCA	Montreal cognitive assessment
MOG	Myelin oligodendrocyte glycoprotein
MRI	Magnetic resonance imaging
NMDA	N-methyl-D-aspartate
NSCLC	Non-small cell lung cancer
PCA-1	Purkinje cell antibody 1
PCD	Paraneoplastic cerebellar degeneration
PEM	Paraneoplastic encephalomyelitis
PERM	Progressive encephalomyelitis with rigidity and myoclonus
PET	Positive emission tomography
PLE	Paraneoplastic limbic encephalitis
PND	Paraneoplastic neurologic disorder
POM	Paraneoplastic opsoclonus-myoclonus
PON	Paraneoplastic optic neuropathy
SCLC	Small cell lung cancer
SE	Status epilepticus
SIADH	Syndrome of inappropriate antidiuretic hormone secretion
SOX	Sry-like high mobility group box
T2/FLAIR	T2-weighted-fluid-attenuated-inversion-recovery
TSH	Thyroid-stimulating hormone
VGCC	Voltage-gated calcium channel

INTRODUCTION

Paraneoplastic syndromes (PNSs) are a complex and heterogeneous group of disorders that typically arise from a triggered autoimmune response to an underlying malignancy (Darnell and Posner 2003). PNSs are not caused by a direct invasion of healthy tissue or by side effects of cancer treatment. Rather, malignant cells generate autoantibodies, cytokines, hormones, and/or peptides that can impact one or multiple organ systems irrespective of the presence of localized cancer cells. Although considered a negative oncological complication, symptoms associated with paraneoplastic syndromes often manifest before the underlying cancer is diagnosed. As such, proper identification of these clinical syndromes can help expedite the diagnosis of the occult malignancy, which can improve treatment outcomes for patients.

Recognition of the distal effects of cancer affecting organs and systems has been documented as early as 1865, when Armand Trousseau recorded migratory venous thrombi that occurred in relation to his own gastric carcinoma (Matharom et al. 2019). This discovery is now a recognized paraneoplastic syndrome called Trousseau's syndrome. The term "paraneoplastic" was coined in the 1940s when applied to polyradiculoneuritis (or inflammation of the ventral nerve roots and peripheral nerves) due to the systemic effects of cancer (Guichard and Vignon 1949) and is thought to come from the Greek *para* ("alongside"), *neo* ("new"), and *plasis* ("formation").

Paraneoplastic Neurologic Disorders

Paraneoplastic syndromes can be categorized into five main etiological groups that are determined by the affected systems (Henry 2019). These include neurologic, endocrine, hematologic, and musculocutaneous, with paraneoplastic disorders affecting systems outside of these categories specified as "other." This chapter will principally focus on paraneoplastic neurologic disorders (PNDs), the elucidation of which may be less familiar to many inpatient neuropsychologists. While some may view PNDs as the exclusive purview of oncology and other medical specialties, neuropsychologists and other providers familiar with neurobehavioral syndromes can be of added value. The benefits provided are twofold. First, as many PNDs appear before primary cancer symptoms, early identification of aberrant neurobehavioral functioning can alert other providers to the necessity of an oncology consult and additional medical workup. Second, many inpatient and critical care physicians may be unfamiliar with this topic, and the inpatient neuropsychologist can be helpful in raising consideration of a PND (versus other delirium) in a patient with known cancer. This is particularly important because early detection can lead to decreased morbidity and improved quality of life. For these reasons, discussion of these syndromes requires working familiarity with the antigens/antibodies, different types of cancer, and clinical phenotypes of PND.

In patients with PNDs, the malignant tumor cells aberrantly express antigens that are normally only found in the nervous system (McKeon et al. 2010). These antigens, referred to as onconeural antigens, can localize intracellularly on the neuronal cell-surface, or at the synapse, and are recognized as foreign by the immune system. The subsequent autoimmune response creates neuronal degeneration of both malignant and healthy tissue in areas where these antigens are endogenously expressed (Darnell 1996). This can sometimes be successful in suppressing tumor growth, which may contribute to the initial clinical difficulty in identifying the underlying cancer. Other PNS disorders, some of which are not PNDs, may indirectly affect cognition by way of causing delirium via direct hormonal expression, as will be further described below.

Epidemiology

Precise prevalence and incidence rates are difficult to quantify given the relative rarity of PNDs. Recent population-based epidemiology studies estimate prevalence rates ranging from 4.4 to 5.4 per 100,000 persons (Shah et al. 2021; Vogrig et al. 2020). Men and women appear to be affected at similar rates (Vogrig et al. 2020) though relative risk increases with age, likely due to the higher prevalence of cancer in older adults. There is minimal data available on prevalence or incidence across different racial and ethnic groups. The most common types of PND are limbic encephalitis, cerebellar degeneration, encephalomyelitis, and myasthenia gravis (Vogrig et al. 2020).

Paraneoplastic Antibodies

Antibodies that target onconeural antigens are referred to as paraneoplastic antibodies. Paraneoplastic antibodies are classified into three risk categories according to their frequency of being associated with cancer (Graus et al. 2021): (1) *high-risk antibodies* (associated with cancer in >70% of cases); (2) *intermediate-risk antibodies* (associated with cancer in 30–70% of cases), and *lower-risk antibodies* (associated with cancer in <30% of cases). Tables 8.1, 8.2, and 8.3 below identify high, intermediate-, and lower-risk antibodies and their associations with different neurologic phenotypes, tumor types, and clinical risk factors. While a direct pathogenic role of high-risk paraneoplastic antibodies has not been proven, the detection of high-risk paraneoplastic antibodies is highly associated with cancer and can be used in combination with clinical observations to assist in the targeted search for underlying malignancy (Graus et al. 2021).

In addition to risk classification, antibodies are also categorized based on the location of the targeted antigen: (1) antibodies directed against neuronal surface membrane proteins and (2) antibodies directed against intracellular neuronal proteins (Greenlee et al. 2022). Antibodies directed against neuronal surface membrane proteins can occur independent

Table 8.1 High-risk antibodies (>70% associated with *cancer)*

Antibody	Neurologic Phenotypes	Frequency of Cancer (%)	Usual Tumors	Clinical and Demographic Considerations
Hu (ANNA-1)	Sensory neuronopathy, chronic gastrointestinal pseudo-obstruction, encephalomyelitis, and limbic encephalitis	85	SCLC >> NSCLC, other neuroendocrine tumors, and neuroblastoma	Limbic encephalitis is usually nonparaneoplastic in patients aged <18y.
CV2/CRMP5	Encephalomyelitis and sensory neuronopathy	>80	SCLC and thymoma	Patients with an associated thymoma are younger in age and often present with myasthenia gravis and (less commonly) neuropathy.
SOX1	LEMS with and without rapidly progressive cerebellar syndrome	>90	SCLC	Stronger correlation with SCLC than with a particular neurologic presentation.
PCA2 (MAP1B)	Sensorimotor neuropathy, rapidly progressive cerebellar syndrome, and encephalomyelitis	80	SCLC, NSCLC, and breast cancer	N/A
Amphiphysin	Polyradiculoneuropathy, sensory neuronopathy, encephalomyelitis, stiff person syndrome	80	SCLC and breast cancer	Associated antibodies commonly coexist. Patients with isolated antiamphiphysin → women, with breast cancer and SPS.
Ri (ANNA-2)	Brainstem/cerebellar syndrome, opsoclonus-myoclonus syndrome	>70	Breast > lung (SCLC and NSCLC)	Breast cancer in women; lung cancer in men.
Yo (PCA-1)	Rapidly progressive cerebellar syndrome	>90	Ovary and breast cancers	Almost only in females; in men, antigen expression by tumor should be proven.

Table 8.1 (continued)

Ma2 and/or Ma	Limbic encephalitis, diencephalitis, and brainstem encephalitis	>75	Testicular cancer and NSCLC	Young men → testicular tumors and isolated Ma2 positivity; older patients → SCLC and both Ma1/2 positivity.
Tr (DNER)	Rapidly progressive cerebellar syndrome	90	Hodgkin lymphoma	N/A
KLH11	Brainstem/ cerebellar syndrome	80	Testicular cancer	Young men

DNER = delta/notch-like epidermal growth factor-related receptor; KLHL11 = Kelch-like protein 11; MAP1B = microtubule-associated protein 1B; NSCLS = non-small-cell lung cancer.
Reprinted with permission from Graus et al., 2021

of an underlying malignancy, with some specific antibodies conferring greater cancer risk. These antibodies appear to have a direct pathogenic effect by causing alterations in the function of the target antigen. Examples include antibodies against the N-methyl-D-aspartate (NMDA) receptor, α-amino-3-hydroxy-5-methyl-4-isoxazolepropionic acid (AMPA) receptor, gamma-aminobutyric-acid (GABA)-A, and GABA-B receptors (Greenlee et al. 2022). Antibodies directed against intracellular neuronal proteins are less common and are not directly pathogenic but rather indicative of T-cell-mediated toxicity (Greenlee et al. 2022). Some of these antibodies have been shown to target specific intracellular antigens (e.g., anti-glutamic acid decarboxylase 65 [GAD], anti-amphiphysin, and collapsin response-mediator protein 5 [CRMP5]), while the intracellular antigen targets of others are less understood (e.g., anti-Yo, anti-Hu, and anti-Ri). Clinical prognosis tends to be poorer in PNDs that are associated with antibodies against intracellular neuronal proteins compared to those associated with antibodies against neuronal surface membrane proteins (Graus and Dalmau 2019; Graus et al. 2021). In most cases, paraneoplastic antibodies can be detected in blood serum and/or cerebrospinal fluid (CSF; Li et al. 2023).

While at first glance it may seem as if there is a clear association between specific antibodies and respective PNDs, this is not always the case. Not all patients with PNDs are positive for a defined paraneoplastic antibody, and not all paraneoplastic antibodies are associated with the development of a PND. Further, the same paraneoplastic antibody may be present across multiple PNDs and associated with several different tumor types (Graus et al. 2021). However, several well-established associations between specific antibodies, neurologic syndromes, and underlying tumor types have been described (Tables 8.1, 8.2, and 8.3). Therefore, the presence of a paraneoplastic antibody can provide supportive evidence for a diagnosis, and in some cases, help identify the underlying tumor. As multiple paraneoplastic antibodies may cause similar clinical phenotypes, ordering phenotype-specific antibody panels allows for testing the appropriate antibodies based on any given patient's clinical presentation. Lastly, while some paraneoplastic antibodies are highly associated with specific types of cancer (e.g., Yo antibody with ovarian or breast cancer), reliable pathognomonic relationships have not been established. In cases of incongruency between antigen and cancer type, pathological examination of neoplastic tissue may be done to look for antigen expression and confirm the identified malignancy as causing the PND.

Table 8.2 Intermediate-risk antibodies (30%–70% associated with cancer)

Antibody	Neurologic Phenotypes	Frequency of Cancer (%)	Usual Tumors	Clinical and Demographic Considerations
AMPA	Limbic encephalitis	>50	SCLC and malignant thymoma	Paraneoplastic origin is more likely when other onconeuronal antibodies co-occur.
$GABA_B$ receptor	Limbic encephalitis	>50	SCLC	Paraneoplastic cases are more commonly observed in elderly men, smokers, with associated anti-KCTD16 antibodies. Most young patients are not paraneoplastic.
mGluR5	Encephalitis	~50	Hodkin lymphoma	N/A
P/Q VGCC	LEMS, rapidly progressive cerebellar syndrome	50 (LEMS, nearly 90 for rapidly progressive cerebellar syndrome)	SCLC	Co-occurrence with N-type VGCC antibodies might be slightly more common in paraneoplastic LEMS.
NMDA receptor	Anti-NMDAR encephalitis	38	Ovarian or extraovarian teratomas	Tumor (mostly ovarian teratomas) predominates in females aged 12 to 45 years (50%). Tumors in elderly patients are less frequent (<25%), but, when present, tend to be carcinomas. Paraneoplastic cases in children are very rare (<10%).
CASPR2	Morvan syndrome	50	Malignant thymoma	CASPR2 should be considered as an intermediate-risk antibody only in the setting of Morvan syndrome. When associated with other neurologic syndromes, the risk of cancer is very low.

CASPR2 = contactin-associated protein-like 2; mGluR5 = metabotropic glutamate receptor type 5

Reprinted with permission from Graus et al., 2021

Table 8.3 Lower-risk antibodies (<30% associated with cancer)

Antibody	Neurologic phenotypes	Frequency of Cancer (%)	Usual Tumors	Clinical and Demographic Considerations
mGluR1	Cerebellar ataxia	30	Mostly hematologic	N/A
$GABA_A$ Receptor	Encephalitis	<30	Malignant thymoma	Paraneoplastic origin is less frequent in children (10%) than in adults (60%).
CASPR2	Limbic encephalitis, acquired neuromyotonia (Isaac syndrome), and Morvan syndrome	<30	Malignant thymoma	Morvan syndrome is more associated (~50%) with malignant thymoma, whereas limbic encephalitis is almost always non-paraneoplastic.
GFAP	Menigoencephalitis	~20	Ovarian teratomas and adenocarcinomas	May occur as an immunologic accompaniment in anti-NMDA receptor encephalitis with ovarian teratomas.
GAD65	Limbic encephalitis, stiff person syndrome, and cerebellar ataxia	<15	SCLC, other neuroendocrine tumors, and malignant thymoma	Paraneoplastic patients are older, more frequently male, with associated neuronal antibodies, and atypical clinical presentations.
LGI1	Limbic encephalitis	<10	Malignant thymoma and neuroendocrine	Paraneoplastic cases are mainly observed in patients with Morvan syndrome and both serum LGI1 and CASPR2 antibodies.

Table 8.3 (continued)

DPPX	Encephalitis with CNS hyperexcitability and PERM	<10	B-cell neoplasms	N/A
GlyR	Limbic encephalitis and PERM	<10	Malignant thymoma and Hodgkin lymphoma	N/A
AQP4	Neuromyelitis optica spectrum disorder	<5	Adenocarcinomas	Older age, male, and severe nausea/vomiting at onset.
MOG	MOG antibody-associated disease	Only a handful of cases reported	Mostly ovarian teratomas	N/A

AQP4 = aquaporin 4; DPPX = dipeptidyl peptidase-like protein; CNS = central nervous system $GABA_AR$ = gamma-aminobutyric-acid-A receptor; GFAP = glial fibrillary acidic protein; GlyR = glycine receptor; LGI1 = leucine-rich glioma-inactivated protein 1; MOG = myelin oligodendrocyte glycoprotein; PERM = progressive encephalomyelitis with rigidity and myoclonus
Reprinted with permission from Graus et al., 2021

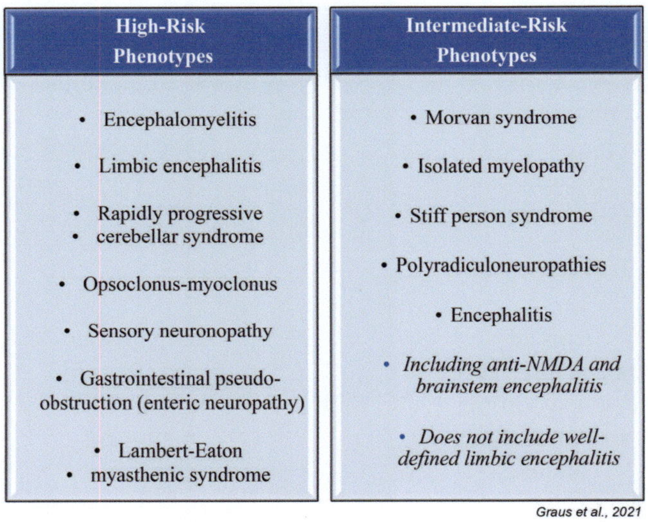

Fig. 8.1 High- and intermediate-risk phenotypes of PNDs

CLINICAL PHENOTYPES AND DIAGNOSIS

In the context of PNDs, the term *clinical phenotype* refers to a defined syndrome or set of symptoms that is associated with a paraneoplastic etiology. Whereas detection of autoantibodies can be thought of as a bottom-up approach to diagnosis, phenotype identification is a top-down process, in which clinical symptoms manifest prior to identification of underlying autoantibodies or other disease markers. Like antibody identification, however, clinical phenotypes are stratified into the likelihood that observed clinical phenomena reflect a PND. Certain clinical patterns have been associated with a higher risk of PND, as will be further discussed in the following sections (Fig. 8.1).

High-Risk Phenotypes

High-risk phenotypes (also called *classical PNDs*) that frequently have a paraneoplastic etiology include encephalomyelitis, limbic encephalitis, rapidly progressive cerebellar degeneration, opsoclonus-myoclonus, sensory neuronopathy, gastrointestinal pseudo-obstruction, and Lambert-Eaton myasthenic syndrome (Graus et al. 2021). High-risk phenotypes are frequently associated with cancer and, when these clinical syndromes are encountered, a work-up for underlying cancer should be undertaken. Identification of specific onconeural antibodies in these patients may help to focus the search for malignancy.

Intermediate-Risk Phenotypes

Intermediate-risk phenotypes (or *nonclassical PNDs*) include encephalitis (other than well-defined limbic encephalitis), Morvan syndrome, isolated myelopathy, stiff-person syndrome, and polyradiculoneuropathies (Graus et al. 2021). These syndromes are not limited to underlying malignant etiology and can occur idiopathically due to other autoimmune disruption. However, should these phenotypes be encountered, and no alternative explanation found, paraneoplastic etiology should be considered. Rapid progression of symptoms over a few months, inflammatory CSF findings, or magnetic resonance imaging (MRI) brain/spine findings concerning for an inflammatory process further increase the suspicion for a PND. Testing for paraneoplastic antibodies can help confirm a diagnosis, inform immunotherapy choices and prognostication, and guide a targeted search for underlying malignancy.

Diagnostic Criteria

As diagnostic criteria for PNDs have evolved over the past two decades, characterization of the clinical phenotype remains the starting point for any diagnosis of PND. This is especially important as extensive medical work-up including broad antibody panels and imaging studies have become readily available and are often ordered in the absence of a clinical history or presence of a high-risk phenotype. Common alternative causes of similar neurologic symptoms should not be overlooked, and careful consideration of infections, autoimmune non-paraneoplastic diseases, tumors, neurodegenerative disorders, and toxic metabolic disturbances should be included in the initial evaluation and re-considered whenever new data becomes available or symptoms.

The most recent diagnostic criteria define PNDs as neurologic disorders that: (1) can affect any part of the nervous system, often presenting with stereotyped clinical manifestations; (2) occur in association with cancer; and (3) have immune-mediated pathogenesis that is supported by the frequent presence of specific neuronal antibodies (Graus et al. 2021). These updated criteria are relatively conservative and emphasize specificity over sensitivity. From these criteria, a PNS-Care Score was developed to enable nuanced classification of PNDs for the purpose of evaluating the likelihood of an underlying cancer, predicting responsiveness to treatment, and informing prognosis (Graus et al. 2021). PNS-Care Scores are derived based on the clinical phenotype, antibody data, and cancer work-up to determine the level of certainty of a diagnosis of PND (Table 8.4). A patient's PNS-Care Score may change over time based on the length of follow-up (i.e., greater or less than two years). For example, a patient with a high-risk phenotype and an intermediate-risk antibody but no cancer at diagnosis would be determined to have a PNS-Care Score of 6 (probable). If a cancer, consistent with the clinical phenotype and antibody, is found 18 months later, the PNS-Care Score would increase to 9 (definite). Alternatively, cancer has not been detected within 24 months, the PNS-Care Score would be downgraded to 5 (possible).

Table 8.4 PNS-Care Score

PNS-Care Score Criteria	Score
Phenotype Risk Classification	**Points**
High-Risk Phenotype	3
Intermediate-Risk Phenotype	2
Other Phenotype $^{+}$	0
Antibody Risk Classification	**Points**
High-Risk Antibody	3
Intermediate-Risk Antibody	2
Lower-Risk Antibody	0
Underlying Cancer Detected	**Points**
Found, consistent with phenotype and (if present) antibody or not consistent but antigen expression demonstrated	4
Not found (or not consistent) but follow-up with within 2 years	1
Not found and follow-up ≥ 2 years	0
PNS-Care Score	**Total Points**
Definite	≥ 8
Probable	6-7
Possible	4-5
Non-PND	≤ 3

PARANEOPLASTIC DISORDERS OF THE CENTRAL NERVOUS SYSTEM

Limbic Encephalitis

Paraneoplastic limbic encephalitis (PLE) is characterized by dysfunction of the limbic system, including the hippocampus, amygdala, hypothalamus, and limbic cortex. PLE can occur alone or as part of a broader paraneoplastic syndrome (e.g., anti-Hu syndrome). PLE is a high-risk phenotype for an underlying cancer etiology.

Antibody Associations in PLE

Antibodies which recognize the GluR1 subunit of the NMDA receptor are the most common antibodies in PLE and are often termed “anti-NMDA encephalitis.” This genotype of PLE is more common in women, with the most frequent cancer association being ovarian teratomas (Marsili et al. 2023). PLE due to anti-NMDA receptor encephalitis can present with severe symptoms including coma, sleep dysfunction, autonomic dysfunction, and psychiatric disturbances including hallucinations, mania, and addictive or eating dysregulation (Casagrande et al. 2024; Dalmau et al. 2007). Fortunately, about 80% of individuals with anti-NMDA receptor encephalitis will have good recovery within 24 months if their clinical syndrome is recognized early and they receive prompt treatment with immunotherapy and teratoma removal (Titulaer et al. 2013). Other antibodies can also be implicated in PLE (see Tables 8.1, 8.2, and 8.3 above) with a perhaps less severe course.

Cancer Associations in PLE

Overall, small cell lung cancer (SCLC) and testicular cancer are the two most common causes of PLE, respectively (Gilligan et al. 2023). PLE in testicular cancer is associated with anti-Ma2 antibodies and widespread encephalopathy beyond the limbic system, with brainstem and hypothalamus involvement (Voltz et al. 1999; Rosenfield et al. 2001). As discussed above, PLE associated with ovarian teratomas causes severe cognitive disruptions. Other cancers associated with PLE include Hodgkin lymphoma, NSCLC, and thymoma (Chiu et al. 2023; Marsili et al. 2023).

Clinical Presentation of PLE

Onset of PLE is acute to subacute, with rapid development over several weeks to three months (Graus 2024). Clinical symptoms can include psychiatric disturbance, cognitive changes/memory loss, alterations of consciousness, and seizures. In many cases, these symptoms are present before cancer is discovered (Darnell and Posner 2011).

Psychiatric disturbance is often the initial symptom of PLE, developing rapidly over days to weeks. Psychiatric symptoms can be mild (e.g., irritability, agitation) to more severe in nature (e.g., hallucinations, psychosis, paranoia, delusions, obsessive compulsive behavior). Affective (e.g., depression and anxiety) and personality (e.g., disinhibition) changes, as well as psychomotor agitation, are common. Given the notable and abrupt change in psychiatric functioning, patients may often be misdiagnosed with a primary psychosis or other psychiatric disorder and initially admitted to inpatient psychiatric units prior to a PND, or underlying cancer, being identified. As experts in brain-behavior relationships, neuropsychologists can be particularly helpful in assisting differential diagnosis and guiding early detection of possible PNDs in these settings.

More so than other PNDs, PLE is also associated with acute, severe cognitive impairment. The hallmark cognitive profile in acute PLE is impaired learning and consolidation of new episodic memories with relatively spared remote memory (Graus et al. 2000). Given isolated amnesia in the absence of other impaired domains, early memory changes in acute PLE are sometimes referred to as *Ophelia syndrome* (Carr 1982). However, additional impairments in language and executive functioning commonly develop with the progression of PLE. Permanent cognitive impairments (i.e., dementia) are often demonstrated, especially without rapid treatment. Bedside assessment usually consists of cognitive screening measures along with a clinical interview with the patient, as well as collateral. The inclusion of tasks specifically targeting memory and attention is critical for capturing any fluctuations in mental status as well as archetypal amnestic deficits. For example, as memory impairment in limbic encephalitis could be missed in the context of an alteration of consciousness (AOC), adding a brief cognitive screener such as the Montreal Cognitive Assessment (MoCA; Nasreddine et al. 2005) or Mini-Mental State Examination (MMSE; Folstein et al. 1975) can be helpful in differentiating memory deficits due to poor initial encoding from true decrements in consolidation. As this cognitive profile can resemble other etiologies, such as Alzheimer's disease, neuropsychologists should carefully consider all sources of available data to avoid misdiagnoses.

Further obfuscating differential diagnosis is the risk of seizures associated with PLE. Ictal activity in PLE can occur with or without impaired awareness and tends to localize to the temporal lobes, which can exacerbate memory deficits. Seizures in PLE bear a poor prognosis (Altabakhi et al. 2024). Typically, refractory to anti-epileptic drugs (AEDs), patients with PLE are at high risk of secondary generalization of focal seizures and even status epilepticus (SE). In some cases, seizures have persisted years into cancer remission (Darnell and Posner 2011). Even without confirmed interictal discharges on EEG, the majority of PLE cases will have slow wave activity in one or both temporal lobes (Lawn et al. 2003).

Cerebellar Degeneration

Paraneoplastic cerebellar degeneration (PCD), also known as subacute cerebellar syndrome, is a PND that causes rapidly progressive cerebellar dysfunction. PCD can occur independently or in conjunction with other symptoms as part of an encephalomyelitis. In the former, Purkinje cells are the primary target of autoimmunity, with relative preservation of other types of cerebellar neurons (Dalmau and Rosenfeld 2008), leading to cerebellar degeneration and rapid development of pancerebellar symptoms (Chiu et al. 2023; Graus 2024). PCD is a high-risk phenotype for an underlying cancer etiology, and in about 60% of cases, PCD symptoms are present prior to tumor discovery (Dalmau and Rosenfeld 2008). Prognosis of PCD is poor as the associated neuronal degeneration typically results in permanent deficits (Chiu et al. 2023).

Antibody Associations in PCD

Anti-Yo/PCA1 is the most common antibody detected in PCD, and is relatively specific for adenocarcinomas of the ovary, uterus, fallopian tube, or breast in women. Anti-Tr/DNER is another antibody commonly associated with primary PCD, most seen in Hodgkin's Lymphoma. Other well-characterized antibodies implicated in PCD are described above (Tables 8.1, 8.2, and 8.3).

Cancer Associations in PCD

The most common cancers associated with PCD are breast, uterus, fallopian tube, and ovarian (Abbatemarco et al. 2024). SCLC and Hodgkin's lymphoma account for the next highest number of cases (Shams'ili et al. 2003). Melanoma, NSCLC, tonsillar cancer, gastrointestinal cancer, sarcomas, lymphoma, and testicular cancer have also been implicated in PCD (Abbatemarco et al. 2024).

Clinical Presentation of PCD

Onset of PCD is largely subacute in nature, developing over days to weeks. Although rare, a stroke-like, hyperacute onset related to the anti-Yo/PCA1 antibody is possible, accounting for approximately 8% of PCD cases (Vogrig et al. 2019). Initial symptoms of PCD include gait ataxia and upper extremity dysmetria, which can be identified via bedside behavioral tasks. While early motor and sensory signs in PCD may present unilaterally, bilateral symptoms typically develop over time (Abbatemarco et al. 2024). Other symptoms include truncal ataxia, bulbar symptoms (e.g., dysphagia), diplopia/double vision, and down-beating nystagmus. Patients may describe oscillopsia, or "the illusion of the world moving," with accompanied dizziness, vertigo, nausea, and vomiting.

In addition to motor and sensory symptoms, approximately 20% of patients with PCD experience cognitive changes early in the course (Abbatemarco et al. 2024). Cerebellar cognitive affective syndrome (CCAS), also called Schmahmann syndrome, can be seen in patients with PCD (Schmahmann et al. 1998) and is characterized by impairments in executive functioning, visuospatial skills, and language, as well as behavioral abnormalities (e.g., negativism, mutism, inattention, and affect dysregulation). While non-specific measures of mood and cognition can be utilized to capture symptoms, the Cerebellar Cognitive Affective/Schmahmann Syndrome scale (CCAS-Scale) is a 10-item screener developed specifically for assessment of CCAS (Hoche et al. 2018) and includes items that aid in differential diagnosis of CCAS (i.e., verbal fluency, digit span, set-switching, verbal recall, and abstract reasoning, as well as questions about mood and affect). Neuropsychologists can be helpful by characterizing symptoms and differentiating cerebellar dysfunction from other etiologies.

Paraneoplastic Opsoclonus-Myoclonus

Paraneoplastic opsoclonus-myoclonus (POM) has been known by multiple other names, including opsoclonus-myoclonus-ataxia, and Kinsbourne or Dancing Eyes-Dancing Feet syndrome. POM is a high-risk phenotype for an underlying cancer etiology, and symptoms often occur before cancer is diagnosed (Oh et al. 2019). Prognosis is poor as the degeneration of neurons is largely permanent.

Antibody Associations in POM

Anti-Ri/ANNA2 is the most common antibody associated with POM, largely found in women with breast cancer (Abbatemarco et al. 2024). Other antibodies that have been infrequently implicated in POM include anti-amphiphysin, anti-CV2/CRMP5, anti-GlyR, anti-Hu/ANNA1, anti-Ma1, anti-Ma2, anti-NMDA receptor, anti-Yo/PCA1, and antiZic2 (Chiu et al. 2023; Abbatemarco et al. 2024; Marsili et al. 2023; Oh et al. 2019).

Cancer Associations in POM

Besides breast cancer, the most common type of cancer associated with POM is SCLC (Chiu et al. 2023). Other associated cancer types include ovarian and testicular cancers (Chiu et al. 2023; Graus 2024; Marsili et al. 2023).

Clinical Presentation of POM

The onset of POM is typically rapid, over several days to a few weeks (Graus 2024). Diagnosis of POM can be difficult, as the presentation of its namesake features can be delayed. Proposed diagnostic criteria for opsoclonus-myoclonus include the presence of opsoclonus and myoclonus or ataxia, as well as behavioral changes or sleep disturbance (Matthay et al. 2005; Oh et al. 2019).

Physical symptoms associated with POM can be debilitating and cause a patient to become rapidly bed bound. Severe truncal ataxia, dizziness, and loss of balance can prevent patients from sitting or walking without substantial assistance. Limb ataxia is not uncommon and can add to ambulation difficulty. While not as severe as in PCD, limb ataxia can be difficult to assess for in POM given concurrent myoclonus (involuntary, sudden jerk-like movements). Myoclonus in POM largely occurs in the extremities and is non-epileptic in nature. Lastly, opsoclonus, a disorder of oculomotor utility characterized by continuous involuntary, arrhythmic, multidirectional, high-amplitude, and conjugate saccades, can cause patients to experience oscillopsia and nausea. Opsoclonus-related eye movements may be accompanied by synchronous blinking. This can occur when the patient is awake, but also at a slower rate when they are asleep (Sahu et al. 2011).

Behavioral abnormalities in POM include irritability, confusion, and impaired arousal. Cognitive sequela of POM are less researched in adult populations, though pediatric studies describe cognitive symptoms in as high as 80% of children with POM (Dalmau and Rosenfeld 2008; Gorman 2010). At minimum, decrements in attention are likely to occur across all age groups given the severity of physical symptoms and associated functional impact. Sleep disturbance associated with POM, such as poor sleep initiation and/or persistence, can also contribute to cognitive symptoms. Hypothesized mechanisms for sleep disturbance include neurological dysfunction secondary to the body's autoimmune response, as well as disruption of sleep associated with aberrant movements and low mood.

Brainstem Encephalitis

Paraneoplastic brainstem encephalitis most often occurs in the context of broader paraneoplastic encephalomyelitis (PEM) and rarely presents as an isolated condition (Abbatemarco et al. 2024). However, as a standalone syndrome, it is an intermediate-risk phenotype for an underlying cancer etiology. Prognosis tends to be poor due to the permanent nature of impairment following the loss of brainstem neurons, given the role these anatomical areas serve in maintaining vital functions.

Antibody Associations in Paraneoplastic Brainstem Encephalitis

Anti-Hu/ANNA-1 in the context of SCLC has been associated with brainstem encephalitis as a standalone syndrome (Blaes 2013). Otherwise, anti-Ri/ANNA-2 is largely found in women with breast cancer, while anti-Ma2 can look akin to progressive supranuclear palsy in symptom profile and is found in men with testicular cancer (Graus 2024; Marsili et al. 2023). Other antibodies implicated in paraneoplastic brainstem encephalitis include anti-NMDA, anti-GAD5, anti-Gly, anti-GQ1b, anti-immunoglobulin-like cell adhesion molecule (IgLON)5, anti-KLHL11, anti-LUZP4, and anti-Ma1 (Marsili et al. 2023; Graus 2024; Abbatemarco et al. 2024).

Cancer Associations in Paraneoplastic Brainstem Encephalitis

Paraneoplastic brainstem encephalitis is associated with both SCLC and NSCLC, as well as neuroendocrine, breast, testicular, and ovarian cancers (Abbatemarco et al. 2024).

Clinical Presentation of Paraneoplastic Brainstem Encephalitis

Isolated occurrence of paraneoplastic brainstem encephalitis tends to be subacute in nature, occurring over weeks to months (Graus 2024). Symptoms depend heavily on which areas of the brainstem are affected and which antibody is present. Sensory and motor abilities are impacted, as are sleep, breathing, and other important physiological processes.

Oculomotor and vision-based symptoms can include various types of paresis, palsy, and nystagmus. Additionally, patients may present with ptosis, impaired saccades, diplopia, and blepharospasms (involuntary and repetitive contractions of the eyelid muscles). Respiratory disorders such as hypoventilation and central sleep apnea are particularly dangerous, as low oxygen levels can lead to hypoxia. Hearing and vestibular dysfunction can include sensorineural hearing loss, tinnitus, and vestibular dysfunction, a syndrome named paraneoplastic cochleovestibulopathy.

Broader movement difficulties in the context of brainstem encephalitis can include general muscle weakness, rigidity, tremors, and spasms, as well as bradykinesia. Gait instability may mimic cerebellar dysfunction. Patients with brainstem encephalitis can also demonstrate cataplexy, or sudden onset of transient muscle paralysis. Dystonia, chorea, ataxia, myoclonus, and bulbar symptoms (e.g., dysphagia, dysarthria) can also occur. Jaw dystonia and trismus (i.e., limited opening of the mouth) are not uncommon. Laryngospasms, or involuntary contractions of the vocal cords that result in sudden closure of the airway, can further interfere with breathing, as well as impact speech.

Research is minimal on cognitive functioning in patients with paraneoplastic brainstem encephalitis, perhaps due to sensory, motor, speech, and arousal impairments limiting evaluation.

However, alterations of consciousness are likely to impact higher-order domains such as memory and executive functioning. Alertness training was cited to improve attention in a case study of an individual treated 5 years post-brainstem encephalitis, though the authors believed this success was at least in part due to the presence of intact cortical structures on brain imaging (Hauke et al. 2011). Overall, detailed behavioral observations are likely more useful than formal cognitive batteries in these scenarios given reduced attention and arousal, which can limit patient engagement.

Visual Syndromes

Paraneoplastic visual syndromes are characterized by vision concerns in the context of paraneoplastic involvement of the optic nerve (optic neuropathy) or the retina (retinopathy). Paraneoplastic optic neuropathy (PON) is homogeneous in nature, while the paraneoplastic retinopathies, including carcinoma-associated retinopathy (CAR), melanoma-associated retinopathy (MAR), and bilateral diffuse uveal melanocytic proliferation (BDUMP), tend to be more heterogenous in terms of cause, onset, and symptomology.

Antibody Associations in Paraneoplastic Visual Syndromes

According to a reviews by Tajfirouz et al. (2024), PON is most often affiliated with anti-CRMP5, but it has also been found in association with anti-Hu, anti-To, anti-Tr, anti-amphiphysin, and anti-Ma2. Anti-recoverin and anti-CRMP5 are the antibodies most associated with CAR, while anti-enolase and anti-transducin-a are less specifically reported. The most common antibody associated with MAR is anti-TRPM1. Current literature notes that antibodies associated with BDUMP are unknown.

Cancer Associations in Paraneoplastic Visual Syndromes

PON has been found in individuals with NSCLC, teratoma, and carcinoid tumors Tajfirouz et al. (2024). As summarized by Sarkar et al. (2021), CAR is most often associated with SCLC but has also been described in the context of many other cancers including ovarian, endometrial, breast, NSCLC, thymus, prostate, thyroid, colon, bladder, and pancreatic. MAR is associated with melanoma, most of the time cutaneous. BDUMP has been found in conjunction with carcinoma of the gynecological tract, lung cancer, and pancreatic cancer (Tajfirouz et al. 2024).

Clinical Presentation of Paraneoplastic Visual Syndromes

Paraneoplastic visual syndromes can result in visual field cuts, nystagmus, gaze paresis/paralysis, or color vision changes (dyschromatopsia), all of which can be observed at bedside. Understanding the cooccurrence of visual symptoms can help distinguish between each paraneoplastic visual syndrome.

CAR is the most common paraneoplastic visual syndrome, caused by cone, rod, and ganglion cell degeneration in the retina. About one-half of patients with CAR experience visual symptoms before a cancer diagnosis (Chan 2003). Night blindness (nyctalopia) and the experience of seeing shimmering or flickering lights (photopsia) can be initial signs of CAR-associated vision changes. Photosensitivity, dyschromatopsia, and ring scotoma, a ring-shaped blind spot, are also common. Vision loss develops acutely to subacutely, sometimes over the course of several months, and is painless. Initial symptoms usually present unilaterally, though bilateral loss of central or peripheral vision may progress over time (Jacobson 1990; Tajfirouz et al. 2024).

MAR presents acutely and differs from CAR in that it is often detected in patients who already know they have cancer, in most cases cutaneous melanoma (Singh 1995). In MAR, retinal rods and bipolar cells, interneurons that connect photoreceptors and ganglion cells, are impacted. According to reviews by Darnell and Posner (2011) and Tajfirouz et al. (2024), initial symptoms often include bilateral shimmering photopsia, nyctalopia, and mild peripheral vision loss. Other symptoms can include dyschromatopsia, crescent-shaped (arcuate) scotomas, and scotomas occurring in central or paracentral line of sight. Onset of MAR, like CAR, is cited to be painless.

PON is rarely seen outside the context of PEM (Tajfirouz et al. 2024). When PON does present independently, onset is subacute and painless, with bilateral vision loss. Bilateral internuclear ophthalmoplegia, dyschromatopsia, visual field deficits, and vertical gaze paresis can also occur.

BDUMP is the least common retinopathy. The mechanism of visual change is deposition of melanocytes, melanin-producing cells, in the eye's uveal tract, the middle layer of the eye between the white part of the eye and the retina that contains the iris, ciliary body, and choroid (Sarkar et al. 2021). Often, rapid development of cataracts and retinal detachment occur (Tajfirouz et al. 2024). Given these mechanisms, restriction or visual fields to complete vision loss are often swift and severe.

Paraneoplastic Encephalomyelitis

Paraneoplastic encephalomyelitis (PEM) is a broad-based inflammatory condition with multifocal nervous system involvement that can include combinations of the previously described PNDs. It is considered a high-risk phenotype for an underlying cancer etiology, and long-term prognosis is poor as it is generally refractory to treatment (Chiu et al. 2023).

Antibody Associations in PEM

As reviewed by Graus (2024), anti-Hu/ANNA1 is the antibody most often associated with PEM, while anti-CRMP5/CV2 is the second most detected. Other antibodies implicated in PEM include anti-amphiphysin, anti-Ma2, anti-PCA2, anti-Ri/ANNA2, and anti-SOX1.

Cancer Associations in PEM

The cancer type most associated with PEM is SCLC, particularly in those with anti-Hu/ANNA1 or anti-CRMP5/CV2 antibodies, although in those with the anti-CRMP5/CV2 antibody, thymoma is also common (Chiu et al. 2023; Honnorat et al. 2009; Yu et al. 2001). Breast cancer is often detected in those with the anti-amphiphysin antibody (Pittock et al. 2005). Other cancer types that have been documented in PEM include NSCLC, other neuroendocrine tumors, prostate cancer, gastrointestinal adenocarcinoma, and testicular germ cell tumors (Graus et al. 2001; Hoffmann et al. 2008; Vogrig et al. 2019).

Clinical Presentation of PEM

As summarized by Casagrande et al. (2024), the onset of PEM is generally subacute in nature and can include both central and peripheral nervous system disruption. Common brain regions implicated include the limbic system (temporal lobes), brain stem, and cerebellum; involvement of the spinal cord and autonomic nervous system is also possible. While PEM is characterized by multifocal dysfunction, symptoms may initially affect one brain region before spreading to others, making differentiation between PEM and more focal paraneoplastic syndromes difficult in early stages.

Adding to the complexity of bedside assessment and identification, advanced stages of PEM can present heterogeneously, depending on which areas of the nervous system are impacted and the implicated antibodies. As an example, PEM with anti-Hu/ANNA1 antibodies in the context of SCLC (also called anti-Hu syndrome) has a particularly poor prognosis due to involvement of the brainstem and cerebellum, disrupting life-sustaining functions and arousal (Graus et al. 2001). Limbic involvement is also implicated in anti-Hu syndrome, causing severe psychiatric disturbance, short-term memory impairment, and seizures. In contrast, PEM in the context of anti-CRMP5/CV2 is less prognostically severe, with common symptoms including chorea and optic neuropathy (Cohen et al. 2020).

Paraneoplastic Motor Neuron Disease

Most clinicians will be familiar with motor neuron diseases (MND), a group of syndromes that affect upper and/or lower motor neurons. However, a possible paraneoplastic etiology in a minority of cases may be less well understood.

Antibody Associations in Paraneoplastic MND

Early epidemiologic studies found no association between MND and malignancy. However, more recent studies have documented improvement or resolution of apparent MND after cancer treatment, and several autoantibodies have been found in patients with MND and cancer (e.g., Sanders et al. 1993; Waragai et al. 2006). Unfortunately, no antibodies have been consistently identified, and there have been no definitive links between paraneoplastic autoantibodies and MND. Clinical evidence of a hypothesized link between MND and cancer includes the apparent resolution or substantial improvement of an MND following cancer treatment (e.g., Brain et al. 1965; Forman et al. 1999).

Cancer Associations in Paraneoplastic MND

SCLC, germ cell tumors, lymphomas and blood cancers, and breast cancer have all been associated with autoantibodies and may be linked to paraneoplastic MND (Darnell and Posner 2011). While seemingly very uncommon, we recommend considering an oncology workup in patients with MND who exhibit atypical onset or course.

Clinical Presentation of Paraneoplastic MND

MNDs are a group of overlapping syndromes affecting upper and/or lower motor neurons. The classic presentation is of weakness, atrophy, fasciculations, spasticity, bulbar dysfunction, and other motor neuron signs. Cognitive symptoms are common, with up to 50% of amyotrophic lateral sclerosis (ALS) patients having some degree of meaningful cognitive impairment. A subset of patients will develop ALS-associated frontotemporal dementia (FTD-MND), usually accompanied by bulbar signs (Bersano et al. 2020; Ringholz et al. 2005). Most forms of MND are invariably progressive, although survival times differ across diagnoses, with loss of respiratory function a common endpoint. While drugs, physical therapy, and other interventions can treat symptoms and improve quality of life, most MNDs are terminal diagnoses for which no meaningful disease-modifying treatments currently exist. Paraneoplastic MND, while clinically similar to prototypic MND, may present with an atypical or altered course (e.g., rapid/acute onset or a more indolent progression; Brain et al. 1965).

Paraneoplastic Stiff Person Syndrome

Stiff person syndrome is an autoimmune neurologic disorder associated with fluctuating painful muscle spasms and rigidity, which typically worsen over time. The pathogenesis of classic stiff person syndrome is thought to involve GAD and GABA. GAD is an enzyme that catalyzes the decarboxylation of GABA, which is essential for maintaining inhibitory pathways in the central nervous system (Peng et al. 2023). Blockage of GAD and the resulting decline of GABA in the CNS cause a loss of neural inhibition and can lead to clinical manifestations of muscle stiffness and spasms. Stiff person syndrome is most common among individuals aged 20 to 50 years old, but in rare cases can occur in children and older adults. Estimated prevalence in the general population varies from 1–2 cases per 100,000 to 1–2 cases per million (Crane et al. 2024). Women are affected at rates 2-3x more than men (Crane et al. 2024).

Antibody Associations in Paraneoplastic Stiff Person Syndrome

Most patients with paraneoplastic stiff person syndrome harbor anti-amphiphysin antibodies, which synaptic proteins that play a role in vesicle endocytosis and are typically associated with small cell lung cancer and breast cancer (Dubey et al. 2019). Other implicated antibodies include anti-GAD68, anti-Gephyrin, and anti-$GABA_A$. Patients with the paraneoplastic stiff person syndrome variant are typically GAD antibody negative though antibodies targeting specific proteins such as amphiphysin have been detected.

Cancer Associations in Paraneoplastic Stiff Person Syndrome

Most patients with stiff person syndrome suffer from a primary autoimmune condition and do not have cancer. However, in the small subset of patients with paraneoplastic stiff person syndrome, breast cancer, followed by lung cancer and lymphoma, are the most common.

Clinical Presentation of Paraneoplastic Stiff Person Syndrome

There are three subtypes of stiff person syndrome based on the presence or absence of specific antibodies and other conditions (Ortiz et al. 2020; Buechner et al. 2015). *Classic Stiff Person Syndrome* is the most common subtype, comprising approximately 70 to 80% of cases, and presents as truncal stiffness, generalized rigidity, frequent muscle spasms, and wide-based gait. *Partial Stiff Person Syndrome*, comprising approximately 10 to 15% of cases, presents as localized stiffness, rigidity, and spasms. The most common form of partial stiff person syndrome is stiff-limb syndrome which generally affects one limb, typically one leg. *Paraneoplastic Stiff Person Syndrome Variant* is seen in less than 2% of patients with stiff person syndrome and is generally clinically indistinguishable from classic stiff person syndrome.

Rapid onset of severe pain, distal extremity involvement, and poor response to spasmolytic agents (e.g., diazepam, baclofen) should raise suspicion for the possibility of a PND. In the nonparaneoplastic form of stiff-person syndrome, there is often evidence of other autoimmune disorders, including diabetes, thyroiditis, pernicious anemia, and Graves' disease. These do not occur in the paraneoplastic form.

Treatment of paraneoplastic stiff person syndrome generally consists of control of the tumor and symptom management of spasms and rigidity with medication (e.g., benzodiazepines, baclofen, tizanidine) or immunosuppressive agents (e.g., glucocorticoids, intravenous immune globulin). Patients with GlyR antibodies tend to be more responsive to treatments than those with antibodies to GAD or amphiphysin or those without antibodies (Martinez-Hernandez et al. 2016).

ENDOCRINOPATHIES AND CANCER

All syndromes covered thus far have direct CNS effects. However, there are also indirect, non-immunologic methods by which cancer can affect cognition. While these are not PNDs in the strictest sense, they are cognitive syndromes that occur as an indirect effect of cancer, typically due to hormones directly expressed by cancers that then circulate in the body. Table 8.5 highlights these metabolic and endocrine disorders that can be paraneoplastic in origin, with specific syndromes discussed in more detail below. Note that in nearly all cases, these syndromes will present as delirium that is clinically indistinguishable from more typical causes. If focal neurologic deficits are seen, a true PND should instead be considered.

Syndrome of Inappropriate Antidiuretic Hormone Secretion

Syndrome of Inappropriate Antidiuretic Hormone Secretion (SIADH) is a relatively common cause of hyponatremia (low sodium). It is seen in up to 1% of all medical inpatients, with one recent study estimating that ~24% of SIADH cases may be paraneoplastic (Adrogué and Madias 2023). SIADH is a disorder of impaired water excretion. Excess antidiuretic hormone (ADH) acts on receptors in the kidneys causing abnormal water retention, low urine output, and low blood sodium. As hyponatremia is one of the most potent causes of (typically hypoactive) delirium, SIADH is thus a relatively frequent cause of delirium in hospital settings. While SIADH can arise from multiple causes, it is frequently associated with cancer. In neoplastic-induced SIADH, cancer will ectopically produce hormones typically released by the endocrine system, in this case ADH, leading to hyponatremia and, potentially, delirium.

SIADH is often a diagnosis of exclusion and further workup is recommended when patients present with seemingly idiopathic hyponatremia. Given that many paraneoplastic syndromes precede a diagnosis of cancer, consultation with oncology about a workup for occult malignancy may be helpful in cases where SIADH is suspected.

Pathophysiology of SIADH

Unlike many of the syndromes covered in this chapter, paraneoplastic SIADH is not the direct result of autoantibody activity but rather caused by ADH (also called vasopressin) produced directly by cancer. Laboratory tests will indicate high ADH and other values relatively specific for SIADH (e.g., euvolemia, high urine sodium and osmolality, low overall urine output)—as well as abnormalities common to patients with hyponatremia, such as low blood sodium.

Cancer Associations in SIADH

SIADH is most seen with SCLC, which frequently produces ADH ectopically, although only a minority of patients will develop clinical hyponatremia (Lokich 1982). Other common causes include so-called "functional" cancers of the adrenal glands or other endocrine origins that naturally produce hormones including ADH (Darnell and Posner 2011). Hematologic, breast, prostate, melanoma, gastrointestinal, and head and neck cancers are also possible causes. Finally, both primary and metastatic brain cancer may cause SIADH if there is hypothalamic involvement.

Paraneoplastic Cushing Syndrome

Cushing Syndrome is an endocrine disorder caused by chronic exposure to excess glucocorticoids, including cortisol. Estimates of the incidence and prevalence are imprecise, though Cushing Syndrome is relatively common—especially among those taking corticosteroids (Huscher et al. 2009). Paraneoplastic Cushing Syndrome comprises a subset of cases, with estimates ranging from nine to nearly 20% (Beuschlein and Hammer 2002).

Table 8.5 Endocrine disorders with possible paraneoplastic etiologies

Disorder	Features	Cancer Associations	Recommended Workup
SIADH	‣ Delirium/Altered Mental Status (AMS) ‣ Hyponatremia (low blood sodium) with low urine output ‣ High antidiuretic hormone (vasopressin) ‣ Seizures, myoclonus (in severe cases) ‣ Intracranial edema (in severe cases)	‣ SCLC ‣ Adrenal or other endocrine ‣ Brain (if hypothalamus is affected) ‣ Hematologic ‣ Melanoma ‣ Other solid tumors (e.g. prostate, breast, colon, bladder)	‣ Bedside neurobehavioral exam to diagnose delirium ‣ Full laboratory workup, incl electrolytes ‣ Rule out more common causes of hyponatremia (e.g., poor food/drink intake, beer potomania, hypothyroidism, etc.). syndrome of inappropriate antidiuretic hormone secretion (SIADH) is euvolemic, so less suspicion if high urine output. ‣ If SIADH is idiopathic, consider oncology consult for occult malignancy. ‣ If patient has known h/o cancer, recommend oncology consult for possible spread, recurrence, or other disease progression
Hypercalcemia	‣ Can be caused by hyperparathyroidism ‣ Was extremely common but less so now d/t prophylactic bisphosphonates ‣ Kidney stones ‣ Bone pain ‣ Abdominal discomfort ‣ Fatigue ‣ Headaches ‣ Excessive thirst and urination ‣ Constipation ‣ Delirium/AMS (in severe cases)	‣ Lung ‣ Breast ‣ Parathyroid ‣ Multiple Myeloma ‣ Other hematologic ‣ Other solid tumors ‣ Primary or secondary bone cancers	‣ Full laboratory workup, incl electrolytes ‣ Bedside neurobehavioral exam to diagnose delirium if symptoms are present. ‣ Rule out more common causes (e.g. benign parathyroid tumors, lithium use, Paget's disease, adrenal insufficiency, etc.). ‣ If idiopathic consider oncology consult for occult malignancy. ‣ In patients with known h/o cancer, unless obvious cause is identified consider workup to rule out bone metasets.

Table 8.5 (continued)

Hypoglycemia	• Delirium/AMS • Lightheadedness • Fatigue/obtundation • Hunger • Tachycardia • Palpitations • Seizures or permanent cognitive impairment with prolonged very low blood glucose (<20 mg/dl).	• Islet-cell tumor and other cancers of the pancreas • Liver • Sarcomas • Mesothelioma • Large retroperitoneal cancers	• Bedside neurobehavioral exam to diagnose delirium • Full laboratory workup, incl blood glucose and A1C. • Rule out more common causes (e.g. poorly managed diabetes incl. overuse of insulin, substance use/intoxication, hyperthyroidism, poor food/drink intake in the setting of diabetes, etc.). • If idiopathic and unresponsive to glucose infusion, consider neoplastic workup for occult malignancy. • If idiopathic and unresponsive to glucose infusion with known h/o cancer, recommend oncology consult for possible spread, recurrence, or other disease progression.
Hyperthroidism	• Typically, low thyroid-stimulating hormone (TSH) • Agitation/anxiety • Irritability • Poor sleep • Weakness • Tachycardia	• Testicular seminoma • Other germ cell tumors • Islet cell tumor of pancreas	• Full laboratory workup, incl TSH and other thyroid markers. • Rule out more common causes (e.g., Graves' disease, thyroid adenoma, accidental overconsumption of thyroid medications, etc.). • If idiopathic, consider oncology consult for possible occult malignancy.

Table 8.5 (continued)

Cushing Syndrome	▸ Long-term excess glucocorticoids (e.g., cortisol) ▸ Weight gain ▸ Round face ▸ Hirsutism ▸ Hypertension ▸ Diabetes Type II ▸ Neurobehavioral changes including emotional lability, mood disorder ▸ Impaired learning and memory	▸ Small-cell lung cancer ▸ Carcinoid tumor ▸ Islet cell tumor of pancreas ▸ Other neuroendocrine tumors ▸ Adrenal gland cancers	▸ Endocrine workup for Cushing Syndrome. ▸ Rule out Cushing's disease/pituitary adenoma, prescribed use of glucocorticoids, other adrenal dysfunction, etc. ▸ If idiopathic, consider oncology consult for possible occult malignancy.

TSH = Thyroid Stimulating Hormone, AMS = Altered Mental Status, SCLC = Small Cell Lung Cancer; Adapted from Darnell & Posner, 2011

Pathophysiology of Paraneoplastic Cushing Syndrome

Typical pathophysiology involves secretion of adrenocorticotropic hormone (ACTH), which is normally secreted by the pituitary gland and acts on the adrenal glands to produce cortisol, as part of the hypothalamic-pituitary-adrenal axis. Chronic excess cortisol causes the symptoms of Cushing Syndrome, including rapid weight gain, round face, hirsutism, hypertension, lethargy, diabetes, and abnormal fat deposits between the shoulder blades. Neuropsychiatric changes including poor learning and short-term memory (with corresponding hippocampal atrophy), emotional lability, depression, and anxiety (Bourdeau et al. 2002; Forget et al. 2000; Resmini et al. 2012; Starkman 2013).

Cancer Associations in Paraneoplastic Cushing Syndrome

Cushing Syndrome is most associated with neuroendocrine tumors including SCLC and islet cell tumors of the pancreas, as well as carcinoid tumors of the lung, thymus, or gastrointestinal tract, among others (Beuschlein and Hammer 2002). Such tumors ectopically produce ACTH, although only a minority of patients will develop Cushing symptoms (Beuschlein and Hammer 2002). The other common cause is adrenal cancers, which may directly produce cortisol.

OTHER CAUSES OF CANCER-RELATED DELIRIUM

While endocrine disorders represent an important method by which cancer can affect cognitive functioning, there are several more general pathophysiological considerations of note.

Metabolic Abnormalities

Cancer can lead to more general toxic-metabolic encephalopathies, either due to direct or indirect effects on organ systems or due to toxic effects from associated treatments. In our experience, these tend to present as typical cases of delirium with metabolic derangements and are more likely to be seen in solid tumors. While not necessarily associated with any specific neoplasms, such cases of delirium simply result from dysfunction of whichever organ system is affected. Thus, while not strictly paraneoplastic in nature, such cases are likely to improve with treatment of the underlying cancer, like in instances of true paraneoplastic syndromes. This is contrasted with more common delirium due to a primary metabolic disturbance, which should be a rule out prior to considering a cancer-related etiology. These rule-outs will be discussed further in the below sections.

MEDICAL WORKUP OF PARANEOPLASTIC NEUROLOGIC DISORDERS

Medical workup of potential PNDs and underlying oncological etiologies can be extensive, including blood serum and CSF studies, brain imaging such as MRI and fludeoxyglucose-18 (FDG)-PET, and electrodiagnostic tests like EEG or electromyography. Initial evaluation should be dictated by clinical history, as well as physical and neurologic examination findings. Infection or common toxic/metabolic causes of symptoms should be ruled out as early as possible, with additional iterative testing to expand the diagnostic workup based on new information.

While neuropsychologists are not directly responsible for medical treatment, the careful differential diagnosis and hypothesis testing services inherent to neuropsychological practice is not without merit. We have found that initial suspicion for paraneoplastic syndromes, as well as recommended first steps to initiate a workup, are often proposed by the neuropsychologist in non-specialized settings. In our experience this has come from a combination of critical problem solving (e.g., ruling out higher base rate syndromes), as well as the relative unfamiliarity many hospitalists have for PND. The following comprises a typical workup for suspected PND, which is helpful when providing recommended diagnostic steps to the medical team. Additional ways for inpatient neuropsychologists to be helpful will be discussed below.

Antibody Screening

Screening for antineuronal antibodies is an extremely important tool in the diagnosis of PNDs. In fact, the new diagnostic criteria for a definite (versus probable or possible) diagnosis of a paraneoplastic syndrome require the detection of high−/intermediate-risk neuronal antibodies (Graus et al. 2021). Testing for some antibodies may be more sensitive when performed in the blood serum, or more specific when performed in the CSF. Further, as the sensitivity and specificity of serum or CSF tests vary among different antibodies, it is generally recommended to perform antibody testing in both blood and CSF samples to increase diagnostic precision.

Commercial antibody testing is often offered as panels of antibodies grouped based on associated clinical phenotypes. Sending broad antibody panels that do not fit a specific clinical syndrome (low pre-test probability) decreases the positive predictive value of any given result, delaying the appropriate diagnosis and treatment of the patient. Similarly, as paraneoplastic and non-paraneoplastic antibody syndromes may present with identical acute to subacute neurologic symptoms, results of isolated paraneoplastic antibody testing are less likely to detect more common antibody-associated conditions. Panels that include both paraneoplastic and non-paraneoplastic antibodies boost the efficacy of this diagnostic tool.

Commercial antibody testing is a time- and cost-effective strategy but recognizing the limitations of this testing and when additional confirmatory testing at a research lab may be appropriate is important. While describing intricacies of these processes is beyond the scope of this chapter, it is notable that there may be up to a 10% false positive rate when using commercially manufactured antibody screening kits compared to "in-house" antibody testing in which individual laboratories create their own assays and testing procedures (Graus 2024). Additionally, studies show that up to 9% of negative findings from commercial antibody screening kits end up being positive when re-tested with in-house antibody screens (Gastaldi et al. 2020; Ruiz-Garcia et al. 2021). In general, it is recommended that any antibody results be confirmed by two different methods to increase the specificity of the testing. In general, we strongly recommend consulting with an autoimmune neurology specialist or contacting your clinical laboratory in cases of dubious test results.

Blood Serum

Initial evaluation of novel neurologic symptoms should include a complete blood cell count (CBC) and comprehensive metabolic panel (CMP), as certain markers in serum can indicate specific cancer types, which can aid in identifying and treating the underlying malignancy (e.g., elevated prostate-specific antigen suggesting possible prostate cancer, etc.) Serum testing can also rule out other potential causes of observed symptoms, including evidence for infectious and inflammatory processes, metabolic derangements, organ dysfunction, or nutritional deficiencies (Darnell and Posner 2011). For example, ruling out HSV as a potential cause of limbic encephalitis would raise the likelihood of a paraneoplastic process. Depending on the clinical history and presentation, additional serum studies can include evaluation for specific pathogens, endocrine abnormalities, inflammatory markers, toxins, heavy metals, and others.

Analysis of Cerebrospinal Fluid

While CSF antibody testing can detect the presence or absence of PND antibodies, routine CSF labs—including cell count, protein, glucose, oligoclonal bands, and infectious markers—can also yield clinical insight. For example, cerebellar ataxia can have a paraneoplastic etiology but can also be caused by non-paraneoplastic etiologies, such as Creutzfeldt-Jakob disease (Graus 2024), which will have a different pattern on CSF analysis. CSF examination is also particularly helpful in the identification of nonspecific inflammation, though additional clinical work-up will likely be needed to determine a paraneoplastic (versus other) etiology. For example, pleocytosis, or elevated white blood cells in the CSF, occurs in about 50% of individuals with PLE (Graus 2024), but can also be seen in immunocompromised individuals, such as those with human immunodeficiency virus (HIV; Graus 2024). Elevated protein and oligoclonal bands are also not specific to PND but are supportive of inflammatory processes of any cause.

Imaging

Imaging is advised in most cases when a paraneoplastic syndrome is suspected to provide additional clinical evidence in support of (or contrary to) a possible PND. Imaging is particularly helpful in cases when an identified antineuronal antibody is associated with cancer and matches the expected PND clinical phenotype. For example, if a patient presents with acute onset psychiatric symptoms, short-term memory loss, and seizures, and the anti-NMDA antibody is identified, T2-weighted or T2-weighted-fluid-attenuated-inversion-recovery (T2/FLAIR) MRI can be used to examine anticipated bilateral medial temporal lobe, hippocampal, amygdala, and insular abnormalities commonly seen in PLE. However, the absence of abnormal findings on imaging studies does not necessarily rule out PND. For example, in paraneoplastic brainstem encephalitis with anti-Hu/ANNA1 antibodies, MRI tends to be unremarkable despite clinical evidence of medulla involvement (Saiz et al. 2009).

Imaging can also aid in ruling out primary or metastatic brain disease, as well as non-cancer-related etiologies. For example, patients with temporal lobe gliomas may present with clinically similar symptoms of PLE, though the former would be easily appreciated on structural imaging. Similarly, posterior reversible encephalopathy syndrome (PRES) can resemble the onset of a paraneoplastic syndrome but can largely be differentiated from PNDs on MRI (Darnell and Posner 2011).

Functional imaging, such as FDG-PET, and computed tomography (CT), are also used to detect primary malignant and metastatic disease. These are the preferred screening tools for the tumors most often associated with paraneoplastic syndromes, with sensitivities of 87% and 89% for cancer detection when pooled (Garcia-Vicente et al. 2017; Sheikhbahaei et al. 2017). Targeted FDG-PET and CT can be used to detect cancer in suspected anatomical areas, should antibodies be identified that have strong association with a particular cancer type (e.g., anti-Hu/ANNA-1 antibodies and SCLC). In contrast, whole-body FDG-PET and CT may be ordered if a potential tumor location or tumor type is less clear. This can be particularly useful in cases when the paraneoplastic syndrome suspected does not always have a detectable neuronal antibody, such as those with POM (Armangué et al. 2016). FDG-PET of the brain can reveal metabolic changes due to a paraneoplastic syndrome before clinical symptoms are present, as well as before structural changes are detectable on MRI. If changes are evident, they are largely hypermetabolic (such as due to seizure activity or inflammation in PLE) but can also be hypometabolic in later stages of paraneoplastic syndromes if destruction of cells occurred.

Like structural imaging, the presence or absence of metabolic findings should not be used in isolation when determining a possible PND, as smaller tumors may be difficult to detect, and FDG-PET and CT had a false positive rate of 5%–20% (McKeon et al. 2010; Bresler et al. 2020). If tumor workup is negative, patients with a high-risk antibody, an intermediate risk antibody with a high-risk neurological syndrome, or a negative antibody screen with a high-risk neurological syndrome should be re-screened for a potential tumor every 4–6 months for 2 years (Graus 2024). Repeat tumor screening is not indicated in patients with low-risk antibody status and neurological syndrome, even if a paraneoplastic syndrome is suspected.

Electroencephalogram

Given cerebral electrical abnormalities in paraneoplastic syndromes, EEG can be a useful diagnostic tool, even detecting change before other clinical signs are present. However, EEG findings are usually nonspecific and additional work-up is required to refine etiologic contributors to EEG abnormalities.

ROLE OF THE INPATIENT NEUROPSYCHOLOGIST

Given the heterogeneity of PND presentations, fluency and nuance in detecting and distinguishing PNDs can be complex. While PND would be a likely explanation for new-onset neurological symptoms in a patient with recently diagnosed cancer, consideration of these oncological complications is less routine in approximately two-thirds of PND patients whose neurological symptoms precede identification of an underlying cancer. In these cases, PND is a diagnosis of exclusion and would be considered after ruling out higher-base rate etiologies. Below we present a case example to accompany discussion of the recommended procedure and workup.

CASE EXAMPLE

Below we present a case example to accompany discussion of the recommended procedure and workup.

History of Presenting Illness and Clinical Interview

One of the most developed tools in a neuropsychologist's skillset is the ability to integrate data from record reviews and a thorough clinical interview into a useful case history. When PND is suspected, this should include review of cancer history, as well as the course, severity, and pattern of cognitive or neurological symptoms. See Fig. 8.2 for initial background data on a relevant case example.

PARANEOPLASTIC CASE EXAMPLE:
Background and Presenting Symptoms

HPI/Background: Ms. N, a 65 y/o woman, was brought to the ED by her family with recurrent but recent-onset paranoia, auditory hallucinations, insomnia, agitation, and decreased appetite. Family was adamant there was no prior psychiatric history and noted that these changes began several months prior and were accompanied by memory loss and inattention. Family noted that similar symptoms had occurred ~1 year ago but abated with treatment for UTI.

Hospital Course and Referral: Initial workup was negative for metabolic abnormalities or urinary tract infection. Vitamin D was mildly low. Computed tomography (CT) of the head was negative for acute intracranial pathology. Ms. N was transferred to an inpatient psychiatry unit, where she was observed to be hyper-religious. Ambulatory electroencephalogram (EEG) was ordered but was negative for seizures. Inpatient neuropsychology was called for differential diagnosis. Ms. N was noted to be a poor historian.

Medical History: Diabetes mellitus type 2, herpes simplex virus, and cervical cancer 10 years prior (pathology unknown).

Psychiatric History: None prior to 1 year ago.

Substance Use History: Unremarkable.

Instrumental ADLs: Dependent on family for ~1 year since symptoms initially started.

Question 1. Besides the obvious concern for PND, what are some higher base rate differentials for this case?

Question 2. What is the likelihood of a primary psychotic disorder emerging later in life?

Fig. 8.2 Paraneoplastic case example, introduction

Cancer History

Unless the neuropsychologist is employed in a primary oncology setting, collecting a comprehensive cancer history may not typically be included in a routine clinical interview or record review. However, it is very important to gather detailed information about a patient's cancer history for multiple reasons. First, primary brain tumor should be a possible differential in any patient presenting with novel neurological symptoms. Second, while the presence of primary or metastatic cancer is the leading risk factor for PND development, treatments delivered for past cancer care can have acute-on-chronic CNS implications that can present like PND. We find it helpful to ask for as many details about cancer history as possible (e.g., known pathology, treatment modalities, prior chemotherapy and/or immunotherapy, and if so, which agents were used). Radiation necrosis is the often-latent degeneration of irradiated white matter and can occur as a side effect of ionizing radiation. If radiation was administered, particularly photonic radiation (X-rays or Gamma rays), to the head region for intracranial tumors or any other head/neck cancers, radiation necrosis and other treatment side effects should additionally be considered. Radiation necrosis appears as hyperintensities on T2/FLAIR sequences.

Symptom Course

In addition to having cancer history, the course of symptoms can also lend a clue to a potential PND, particularly if neurological symptoms began acutely and followed a subacute course, progressing over several weeks to months. It is also typical for symptoms to stabilize over time; this is unfortunately most often when functioning is already severely impaired. In rare instances, neurological symptoms of PND can onset very abruptly, within a few hours. In these instances, patients may present to the hospital thinking they are having a stroke, particularly if they have had prior strokes or have a family history of cardiovascular disease. Fortunately, stroke etiology can be quickly identified on brain imaging, easily clarifying the initial differential.

Symptom Severity

Information regarding symptom severity can be another good indicator of a potential PND versus other causes of neurological symptoms. Specifically, many PNDs cause severe neurological symptoms, with considerable associated decline in functional status. For example, there are many possible causes for nonspecific symptoms of dizziness and headaches; however, if a 70-year-old patient with no previous mental health history presents with subacute onset of psychosis accompanied by impaired memory, confusion, and seizures, possible diagnostic differentials are more focused and would certainly include PLE.

Symptom Pattern

Given widespread potential impact throughout the CNS, neurologic symptoms of PND subtypes are vastly heterogenous. However, as a rule of thumb, patients presenting with a clinical syndrome that affects a specific portion of the brain or nervous system (e.g., due to focal/proscribed involvement of cerebellar Purkinje cells, dorsal root ganglion) should raise suspicion for a potential PND. When there is diffuse CNS dysfunction, PND may still be considered, but many other processes (e.g., other autoimmune processes, infection, metabolic derangement) may be more parsimonious.

Cognitive Exam

As there is no single cognitive profile for PND, we recommend doing an initial, thorough chart review and identifying symptoms around which a battery can be designed. Since many patients being worked up for possible PND will be quite ill, we strongly encourage providers to thoughtfully select measures with testing burden in mind. In some cases, cognitive screening may be sufficient, although comprehensive testing is optimal for detecting focal defects. Neurobehavioral assessment is also strongly recommended. This should include motor sequencing tasks, rapid alternating movements, visual field testing, a sensory-perceptual exam, tasks of movement coordination (e.g., finger-to-nose for dysmetria), and perhaps a clock drawing test. Should patients be delirious, evaluations may show nonspecific inattention. While this is certainly not pathognomonic for PND, it should also not be immediately used to rule out such a syndrome either. Figure 8.3 outlines the results and discussion from the formal neuropsychological workup in our case example.

PARANEOPLASTIC CASE EXAMPLE:
Neuropsychological Findings

Findings from Exam:

- ***Behavioral*:** Neurobehavioral workup identified mild dysdiadochokinesia and motor sequencing impairments bilaterally, as well as mild ideomotor apraxia. There were no focal visual deficits or lateralizing signs. Grandiose hyperreligiosity was noted, as well as dyskinesia-like stereotyped motor behaviors.

- ***Formal Cognitive Evaluation*:** Testing included the California Verbal Learning Test Second Edition Short Form, Trail Making Test, elements of the Wechsler Adult Intelligence Scale-Fourth Edition, Delis-Kaplan Executive Function System Verbal Fluency subtest, Test of Practical Judgment, and the Rey Complex Figure Test. Results indicated impairments in visuospatial functions (particularly for complex tasks), executive functioning (including poor complex attention and problem solving), and memory. Executive dysfunction appeared to evidence a diffuse pattern (e.g., no evidence of cortical frontal lobe involvement). Memory profile implicated poor learning and recall with questionable consolidation deficits as well.

Differentials: Differential diagnoses from neuropsychology suggested atypical dementia vs. herpes simplex virus encephalitis vs. PND. Regarding the latter, an anti-NMDA receptor or other limbic encephalitis was thought most likely due to memory and executive decrements, as well as neuropsychiatric symptoms and course.

Recommendations: Given the strong possibility of limbic encephalitis, Neuropsychology recommended PND workup, including repeat EEG, MRI Brain, oncology consult, and neurology consult for lumbar puncture and CSF antibody panel. The inpatient psychiatry attending was unfamiliar with PNDs but agreed to act on Neuropsychology's recommendations.

Question 3. What additional follow-up testing would you recommend for this case?

Question 4. Which PND would you suspect? Which autoantibodies would most likely cause this?

Fig. 8.3 Paraneoplastic case example, findings and differentials

Drawing Conclusions and Providing Recommendations

Given the low base rates of these symptoms, it may be intimidating for the inpatient neuropsychologist to raise PND as part of a differential or to recommend additional neurologic workup. This is especially true when a PND has not already been suspected by the medical team. In cases where there is active cancer or a cancer history, it is reasonable to request specific assessment of high-risk areas (i.e., lymph nodes, breast, prostate, and testicles) during the routine physical exam. Alternatively, there are some symptom constellations that are nearly pathognomonic of a PND, making PND the leading differential even before exploring other etiological considerations. Such distinct clinical syndromes include PLE or PCD.

In situations where other differentials might be more likely than PND, or when PND is one of many etiological considerations, multiple recommendations to the treating team can be considered; these will vary depending on which PND is suspected. As reviewed above, recommendations for PND work-up include serum and CSF studies to assess for antibodies, brain imaging (CT, MRI, FDG-PET), and EEG. Generally, a strong case can be made for efficiency if one measure can be used to investigate multiple differential diagnoses. For example, brain imaging and EEG are useful to assess differentials for many clinical neurological presentations. Similarly, CSF and blood serum antibody tests can help rule out non-paraneoplastic autoimmune processes, which are considerably more common than paraneoplastic syndromes.

If findings of these tests suggest a possible PND, subsequent medical evaluation should focus on identification of an underlying cancer. While the presence of specific antibodies can guide such a search, should antibody testing be unavailable or unhelpful in this regard, whole-body FDG-PET can be performed to detect hypermetabolic lesions that can indicate malignancy. Combining FDG-PET with CT is encouraged to increase diagnostic sensitivity (Linke et al. 2008). However, these diagnostic tools may prove insufficient in detecting early stages of cancer (small size or low metabolic activity), in circumstance where the cancer is obscured by other organ systems (e.g., heart, kidney, and bladder), or in the event of spontaneous regression of cancer due to the immune response implicated in the PND. In these cases, it is reasonable to re-order CT and FDG-PET in about three months. Alternatively, if FDG-PET and CT are unremarkable, but there is strong evidence of PND due to a specific cancer type, other imaging studies or exam types may be considered. For example, if breast or gynecological cancers are suspected but CT and FDG-PET are negative, the next best steps may include ordering an ultrasound or mammogram. Figure 8.4 describes the next steps taken in our case example, highlighting possible referrals that may be placed.

PARANEOPLASTIC CASE EXAMPLE:
Additional Studies and Conclusions

MRI Brain: Mild temporal hyperintensities bilaterally on T2/FLAIR.

EEG: Repeat routine EEG was negative for seizures.

Lumbar Puncture: CSF was negative for herpes simplex virus but showed oligoclonal bands. CSF autoantibody testing was negative but not definitive.

Oncology Consult: Recurrent cervical cancer in situ.

Conclusions: Neurology felt the workup was suspicious enough for limbic encephalitis and immunomodulatory treatment was initiated. Ms. N underwent surgery to remove the cervical tumor. Psychiatric and behavioral symptoms gradually improved. However, follow-up neuropsychological evaluation 12 months later identified residual impairments in executive functioning (particularly for visuospatial functions) and memory.

Fig. 8.4 Paraneoplastic case example, additional workup and conclusions

DIFFERENTIAL DIAGNOSIS

PND remains a rare disorder, and more common differentials have been addressed throughout the chapter to this point. Here, we present a centralized discussion of the most common differentials to consider, with tips that we hope are useful in the inpatient clinical setting.

Primary Brain Tumor/Metastatic Disease

Primary brain tumors and metastatic disease are principal differentials to PNDs. Symptom onset can be similar, acute to subacute in nature for fast-growing, high-grade tumors; however, for low-grade or slower-growing tumors, onset can be prolonged, rendering easier differentiation from PNDs. Symptom profiles may look similar depending on the size and location of the tumor, including associated mass effect or edema. For example, widespread dysfunction caused by primary brain cancer can cause seizures, while symptoms associated with focal cerebellar tumors may resemble PCD. Given the significant overlap between cancer types most associated with PNDs and metastases (e.g., lung, breast, and skin cancers), workup for both primary and distal metastatic sites is strongly encouraged for patients with a history of cancer. Brain imaging (PET/CT, MRI) is by far the most useful tool in discerning brain tumors from PNDs when both are included on the differential list.

Lastly, cancer treatment effects should also be considered as part of the differential for PNDs. Understanding the typical cognitive effects of different cancer treatments can be useful in this case (Table 8.6). Treatment effects can be ruled out if symptoms begin before treatment started. The likelihood of a treatment effect etiology increases if symptoms wax and wane with treatment schedules (e.g., becoming worse when administered chemotherapy but better between treatments). Serum and CSF antibody testing will be very important for differentiation in scenarios when the symptoms started after treatment began and the patient cannot take a break in their treatment schedule.

Metabolic Derangement

Metabolic derangements can alter cognition—independent of, or secondary to cancer and associated oncological treatment—and should be considered as a potential etiology for acute neurological change. Symptoms are largely dependent on the nature of the disrupted metabolic process(es). For example, abnormal A1C and other causes of electrolyte imbalances, particularly those that impact sodium, are associated with impaired attention and may be top differentials for delirium (Fujisawa et al. 2016). Metabolic derangements are usually identified via serum analysis, and causative contribution can be determined if treatment of the underlying metabolic disruption results in symptom resolution (King 2024). However, should serum tests be delayed or unavailable, clinical history can also prove useful in identifying the contribution of metabolic disturbance to observed symptoms. For example, while both PLE and low thiamine can result in mem-

Table 8.6 Cancer treatment effects

Cancer Treatment	Effect	Resolution of Cognitive Symptoms
Chemotherapy	Slowed processing speed, inattention, fatigue	Cognitive symptoms usually resolve, though patients often experience residual fatigue.
Brain Radiation	Localized: Focal deficits Whole Brain: Diffuse deficits, radiation necrosis possible	Cognitive symptoms typically permanent.
Surgical Intervention	Focal deficits	Severity of initial cognitive symptoms may improve in early post-surgical recovery; however, milder chronic symptoms usually persist.
Hormone Therapy	Dependent on specific hormone(s) used	Cognitive symptoms usually resolve over time

ory impairment, the former is more likely to be accompanied by severe psychiatric disturbance and seizures, while the latter is most often associated with a history of nutritional deficiency or alcohol use (Arts et al. 2017).

Infectious Disease

Infectious diseases, especially those with neuroinvasive properties, can also cause acute neurologic change, although patients typically present with infection-specific symptoms such as fever, which are not usually present in PNDs. Nausea in the absence of motor or sensory concerns is another good indicator of an infectious process. Obtaining information about a patient's possible exposure to infections or recent travel history (e.g., foreign/domestic, regions traveled, types of food consumed, access to hygiene, lodging accommodations) can be supportive when an infectious disease is suspected or be used as justification for further medical workup (usually serum or CSF testing) for infectious etiology.

Dementia

In older adults, PND can largely be distinguished from possible dementia based on the protracted symptom course of most neurodegenerative disease, although rare and atypically rapid progressing dementias are possible (e.g., Creutzfeldt-Jakob disease). In addition to information gleaned from medical workup, any plateauing or resolution of symptoms (either spontaneous or in response to treatment) would raise concern for PND.

SUMMARY

Paraneoplastic disorders are complex and heterogeneous syndromes that arise from the indirect effects of a cancerous process, often mediated by autoimmune pathways. Although rare, PNDs have a high risk for mortality and permanent disability, related to both the PND syndrome, as well as the underlying cancer. As members of acute care or rehabilitation treatment teams, neuropsychologists are uniquely positioned to lend a critical diagnostic eye to patients with unexplained or treatment-resistant acute psychiatric, cognitive, and behavioral symptoms. Familiarity with these disorders, knowing when to refer or to recommend additional workup, and the ability to differentiate from more common neurocognitive and psychiatric syndromes will aid both patients and colleagues, and will ultimately serve to increase the value of neuropsychological services.

REFERENCES

Abbatemarco JR, Vedeler CA, Greenlee JE. Paraneoplastic cerebellar and brainstem disorders. Handb Clin Neurol. 2024;200:173–91. https://doi.org/10.1016/B978-0-12-823912-4.00030-X.

Adrogué HJ, Madias NE. The syndrome of inappropriate antidiuresis. N Engl J Med. 2023;389(16):1499–509. https://doi.org/10.1056/NEJMcp2210411.

Altabakhi IW, Babiker HM, Lui F. Paraneoplastic Limbic Encephalitis. [Updated 2024 Feb 12]. In: StatPearls [Internet]. Treasure Island (FL): StatPearls Publishing; 2025. Available from: https://www.ncbi.nlm.nih.gov/books/NBK519523/

Armangué T, Sabater L, Torres-Vega E, Martínez-Hernández E, Ariño H, Petit-Pedrol M, Planagumà J, Bataller L, Dalmau J, Graus F. Clinical and immunological features of opsoclonus-myoclonus syndrome in the era of neuronal cell surface antibodies. JAMA Neurol. 2016;73(4):417–24. https://doi.org/10.1001/jamaneurol.2015.4607.

Arts NJM, Walvoort SJW, Kessels RP, C. Korsakoff's syndrome: A critical review. Neuropsychiatr Dis Treat. 2017;13:2975–890. https://doi.org/10.2147/NDT.S130078.

Beuschlein F, Hammer GD. Ectopic pro-opiomelanocortin syndrome. Endocrinol Metab Clin N Am. 2002;31(1):191–234. https://doi.org/10.1016/s0889-8529(01)00025-1.

Bersano E, Sarnelli MF, Solara V, Iazzolino B, Peotta L, De Marchi F, Facchin A, Moglia C, Canosa A, Calvo A, Chiò A, Mazzini L. Decline of cognitive and behavioral functions in amyotrophic lateral sclerosis: a longitudinal study. Amyotrophic Lateral Sclerosis Frontotemp Degen. 2020;21(5–6):373–9. https://doi.org/10.1080/21678421.2020.1771732.

Blaes F. Paraneoplastic brain stem encephalitis. Curr Treat Options Neurol. 2013;15(2):201–9. https://doi.org/10.1007/s11940-013-0221-1.

Bourdeau I, Bard C, Noël B, Leclerc I, Cordeau MP, Bélair M, Lesage J, Lafontaine L, Lacroix A. Loss of brain volume in endogenous Cushing's syndrome and its reversibility after correction of hypercortisolism. J Clin Endocrinol Metab. 2002;87(5):1949–54. https://doi.org/10.1210/jcem.87.5.8493.

Brain L, Croft PB, Wilkinson M. Motor neurone disease as a manifestation of neoplasm (with a note on the course of classical motor neurone disease). Brain J Neurol. 1965;88(3):479–500. https://doi.org/10.1093/brain/88.3.479.

Bresler R, Schroeder HWI, Chow DZ, Lim R. 18F-fluorodeoxyglucose positron emission tomography/computed tomography in the diagnosis of suspected paraneoplastic syndromes: A retrospective analysis. World J Nucl Med. 2020;19(02):124–30. https://doi.org/10.4103/wjnm.WJNM_48_19.

Buechner S, Florio I, Capone L. Stiff person syndrome: A rare neurological disorder, heterogeneous in clinical presentation and not easy to treat. Case Rep Neurol Med. 2015; https://doi.org/10.1155/2015/278065.

Carr I. The Ophelia syndrome; memory loss in Hodgkin's disease. Lancet (London, England). 1982;1(8276):844–5. https://doi.org/10.1016/s0140-6736(82)91887-6.

Casagrande S, Zuliani L, Grisold W. Paraneoplastic encephalitis. Handb Clin Neurol. 2024;200(8):131–49. https://doi.org/10.1016/B978-0-12-823912-4.00019-0.

Chan JW. Paraneoplastic retinopathies and optic neuropathies. Surv Ophthalmol. 2003;48(1):12–38. https://doi.org/10.1016/s0039-6257(02)00416-2.

Chiu D, Rhee J, Gonzalez Castro LN. Diagnosis and treatment of paraneoplastic neurologic syndromes. Antibodies (Basel, Switzerland). 2023;12(3):50. https://doi.org/10.3390/antib12030050.

Cohen DA, Bhatti MT, Pulido JS, Lennon VA, Dubey D, Flanagan EP, Pittock SJ, Klein CJ, Chen JJ. Collapsin response-mediator protein 5-associated retinitis, Vitritis, and optic disc edema. Ophthalmology. 2020;127(2):221–9. https://doi.org/10.1016/j.ophtha.2019.09.012.

Crane PD, Sillau S, Dreher R, Fix R, Winters P, et al. Population-based study of the epidemiology of stiff person syndrome in a large Colorado-based health system. Neurology. 2024;103(12):e210078. https://doi.org/10.1212/WNL.0000000000210078.

Dalmau J, Rosenfeld MR. Paraneoplastic syndromes of the CNS. Lancet Neurol. 2008;7(4):327–40. https://doi.org/10.1016/S1474-4422(08)70060-7.

Dalmau J, Tuzun E, Wu HY, Masjuan J, Rossi JE, Voloschin A, Baehring JM, Shimazaki H, Koide R, King D, Mason W, Sansing LH, Dichter MA, Rosenfeld MR, Lynch DR. Paraneoplastic anti-N-methyl-D-aspartate receptor encephalitis associated with ovarian teratoma. Ann Neurol. 2007;61(1):25–36. https://doi.org/10.1002/ana.21050.

Darnell RB. Onconeural antigens and the paraneoplastic neurologic disorders: at the intersection of cancer, immunity, and the brain. Proc Natl Acad Sci. 1996;93(10):4529–36.

Darnell RB, Posner JB. Paraneoplastic syndromes involving the nervous system. N Engl J Med. 2003;349(16):1543–54.

Darnell RB, Posner JB. Paraneoplastic syndromes. Oxford University Press; 2011. https://doi.org/10.1093/med/9780199772735.001.1.

Dubey D, Jitprapaikulsan J, Bi H, Vazquez Do Campo R, McKeon A, Pittock SJ, Engelstad JK, Mills JR, Klein CJ. Amphiphysin-IgG autoimmune neuropathy: A recognizable clinicopathologic syndrome. Neurology. 2019;93(20):e1873–80. https://doi.org/10.1212/WNL.0000000000008472.

Folstein MF, Folstein SE, McHugh PR. "Mini-mental state". A practical method for grading the cognitive state of patients for the clinician. J Psychiatr Res. 1975;12(3):189–98. https://doi.org/10.1016/0022-3956(75)90026-6.

Forget H, Lacroix A, Somma M, Cohen H. Cognitive decline in patients with Cushing's syndrome. J Int Neuropsychol Soc JINS. 2000;6(1):20–9. https://doi.org/10.1017/s1355617700611037.

Forman D, Rae-Grant AD, Matchett SC, Cowen JS. A reversible cause of hypercapnic respiratory failure: lower motor neuronopathy associated with renal cell carcinoma. Chest. 1999;115(3):899–901. https://doi.org/10.1378/chest.115.3.899.

Fujisawa H, Sugimura Y, Takagi H, Mizoguchi H, Takeuchi H, Izumida H, Nakashima K, Ochiai H, Takeuchi S, Kiyota A, Fukumoto K, Iwama S, Takagishi Y, Hayashi Y, Arima H, Komatsu Y, Murata Y, Oiso Y. Chronic hyponatremia causes neurologic and psychologic impairments. J Am Soc Nephrol. 2016;27(3):766–80. https://doi.org/10.1681/ASN.2014121196.

Garcia-Vicente AM, Pérez-Beteta J, Pérez-García VM, Molina D, Jiménez-Londoño GA, Soriano-Castrejón A, Martínez-González A. Metabolic tumor burden assessed by dual time point [18 F] FDG PET/CT in locally advanced breast cancer: relation with tumor biology. Mol Imaging Biol. 2017;19:636–44. https://doi.org/10.1007/s11307-016-1034-x.

Gastaldi M, Zardini E, Scaranzin S, Uccelli A, Andreetta F, Baggi F, Franciotta D. Autoantibody diagnostics in neuroimmunology: experience from the 2018 Italian neuroimmunology association external quality assessment program. Front Neurol. 2020;10:1385. https://doi.org/10.3389/fneur.2019.01385.

Gilligan M, McGuigan C, McKeon A. Paraneoplastic Neurologic Disorders. Curr Neurol Neurosci Rep. 2023;23(3):67–82. https://doi.org/10.1007/s11910-023-01250-w. Epub 2023 Feb 13.

Gorman MP. Update on diagnosis, treatment, and prognosis in opsoclonusmyoclonus-ataxia syndrome. Curr Opin Pediatr. 2010;22(6):745–50. https://doi.org/10.1097/MOP.0b013e32833fde3f.

Graus F. Clinical approach to diagnosis of paraneoplastic neurologic syndromes. Handb Clin Neurol. 2024;200(5):79–96. https://doi.org/10.1016/B978-0-12-823912-4.00007-4.

Graus F, Dalmau J. Paraneoplastic neurological syndromes in the era of immune-checkpoint inhibitors. Nat Rev Clin Oncol. 2019;16(9):535–48. https://doi.org/10.1038/s41571-019-0194-4.

Graus F, Vogrig A, Muñiz-Castrillo S, Antoine JCG, Desestret V, Dubey D, Giometto B, Irani SR, Joubert B, Leypoldt F, McKeon A, Pruss H, Psimaras D, Thomas L, Titulaer MJ, Vedeler CA, Verschuuren JJ, Dalmau J, Honnorat J. Updated diagnostic criteria for paraneoplastic neurologic syndromes. Neurol Neuroimmunol Neuroinflamm. 2021;8(4):e1014. https://doi.org/10.1212/nxi.0000000000001014.

Graus F, Keime-Guibert F, Reñe R, Benyahia B, Ribalta T, Ascaso C, Escaramis G, Delattre JY. Anti-Hu-associated paraneoplastic encephalomyelitis; analysis of 200 patients. Brain J Neurol. 2001;214:1138–48. https://doi.org/10.1093/brain/124.6.1138.

Graus F, Gultekin SH, Rosenfeld MR, Voltz R, Eichen J, Posner JB, Dalmau J. Paraneoplastic limbic encephalitis: neurological symptoms, immunological findings and tumor association. Brain J Neurol. 2000;123(7):1481–94. https://doi.org/10.1093/brain/123.7.1481.

Greenlee JE, Carlson NG, Abbatemarco JR, Herdlevær I, Clardy SL, Vedeler CA. Paraneoplastic and Other Autoimmune Encephalitides: Antineuronal Antibodies, T Lymphocytes, and Questions of Pathogenesis. Front Neurol. 2022;12:744653. https://doi.org/10.3389/fneur.2021.744653.

Guichard MMA, Vignon G. La Polyradiculonéurite cancéreuse métastatique. Le J Médecine de Lyon. 1949;30(700):197–207

Hauke J, Fimm B, Sturm W. Efficacy of alertness training in a case of brainstem encephalitis: clinical and theoretical implications. Neuropsychol Rehabil. 2011;21(2):164–82. https://doi.org/10.1080/09602011.2010.541792.

Henry K. Paraneoplastic syndromes: definitions, classification, pathophysiology and principles of treatment. Semin Diagn Pathol. 2019;36:204–10. https://doi.org/10.1053/j.semdp.2019.01.002.

Hoche F, Guell X, Vangel MG, Sherman JC, Schmahmann JD. The cerebellar cognitive affective/Schmahmann syndrome scale. Brain. 2018;141(1):248–70. https://doi.org/10.1093/brain/awx317.

Hoffmann LA, Jarius S, Pellkofer HL, Schueller M, Krumbholz M, Koenig F, Johannis W, la Fougere C, Newman T, Vincent A, Voltz R. Anti-Ma and anti-Ta associated paraneoplastic neurological syndromes: 22 newly diagnosed patients and review of previous cases. J Neurol Neurosurg Psychiatry. 2008;79(7):767-73. https://doi.org/10.1136/jnnp.2007.118588. Epub 2008 Jan 25.

Honnorat J, Cartalat-Carel S, Ricard D, Camdessanche JP, Carpentier AF, Rogemond V, Chapuis F, Aguera M, Decullier E, Duchemin AM, Graus F, Antoine JC. Onco-neural antibodies and tumour type determine survival and neurological symptoms in paraneoplastic neurological syndromes with Hu or CV2/CRMP5 antibodies. J Neurol Neurosurg Psychiatry. 2009;80(4):412–6. https://doi.org/10.1136/jnnp.2007.138016. Epub 2008 Oct 17.

Huscher D, Thiele K, Gromnica-Ihle E, Hein G, Demary W, Dreher R, Zink A, Buttgereit F. Dose-related patterns of glucocorticoid-induced side effects. Ann Rheum Dis. 2009;68(7):1119–24. https://doi.org/10.1136/ard.2008.092163.

Jacobson DM, Thirkill CE, Tipping SJ. A clinical triad to diagnose paraneoplastic retinopathy. Ann Neurol. 1990;28(2):162–7. https://doi.org/10.1002/ana.410280208.

King LM. Comprehensive metabolic panel (CMP): an overview. WebMD; 2024. https://www.webmd.com/a-to-z-guides/comprehensive-metabolic-panel.

Lawn ND, Westmoreland BF, Kiely MJ, Lennon VA, Vernino S. Clinical, magnetic resonance imaging, and electroencephalographic findings in paraneoplastic limbic encephalitis. Mayo Clin Proc. 2003;78(11):1363–8. https://doi.org/10.4065/78.11.1363.

Li L, Guo Y, Wang J. Detection of paraneoplastic antibodies and their significance in paraneoplastic neurologic syndromes: a narrative review. Ann Transl Med. 2023;11(7):283. https://doi.org/10.21037/atm-21-2434. Epub 2022 Dec 19.

Linke R, Voltz R. FDG-pet in paraneoplastic syndromes. Recent Results Cancer Res. 2008;170:203–11. https://doi.org/10.1007/978-3-540-31203-1_16.

Lokich JJ. The frequency and clinical biology of the ectopic hormone syndromes of small cell carcinoma. Cancer. 1982;50(10):2111–4. https://doi.org/10.1002/1097-0142(19821115)50:10%3C2111::aid-cncr2820501023%3E3.0.co;2-o

Marsili L, Marcucci S, LaPorta J, Chirra M, Espay AJ, Colosimo C. Paraneoplastic neurological syndromes of the central nervous system: pathophysiology, diagnosis, and treatment. Biomedicine. 2023;11(5):1105–406. https://doi.org/10.3390/biomedicines11051406.

Martinez-Hernandez E, Ariño H, McKeon A, Iizuka T, Titulaer MJ, Simabukuro MM, et al. Clinical and immunologic investigations in patients with stiff-person spectrum disorder. JAMA Neurol. 2016;73(6):714–20. https://doi.org/10.1001/jamaneurol.2016.0133.

Matthay KK, Blaes F, Hero B, Plantaz D, De Alarcon P, Mitchell WG, Pike M, Pistoia V. Opsoclonus myoclonus syndrome in neuroblastoma a report from a workshop on the dancing eyes syndrome at the advances in neuroblastoma meeting in Genoa, Italy, 2004. Cancer Lett. 2005;228(1-2):275–82. https://doi.org/10.1016/j.canlet.2005.01.051.

McKeon A, Apiwattanakul M, Lachance DH, Lennon VA, Mandrekar JN, Boeve BF, Mullan B, Mokri B, Britton JW, Drubach DA, Pittock SJ. Positron emission tomography–computed tomography in paraneoplastic neurologic disorders: systematic analysis and review. Arch Neurol. 2010;67(3):322–9.

Nasreddine ZS, Phillips NA, Bédirian V, Charbonneau S, Whitehead V, Collin I, Cummings JL, Chertkow H. The Montreal cognitive assessment, MoCA: a brief screening tool for mild cognitive impairment. J Am Geriatr Soc. 2005;53(4):695–9. https://doi.org/10.1111/j.1532-5415.2005.53221.x.

Oh SY, Kim JS, Dieterich M. Update on opsoclonus-myoclonus syndrome in adults. J Neurol. 2019;266(6):1541–1548. https://doi.org/10.1007/s00415-018-9138-7. Epub 2018 Nov 27.

Ortiz JF, Ghani MR, Cox AM, Tambo W, Bashir F, Wirth W, Moya G. Cureus. 2020;12(12) https://doi.org/10.7759/cureus.11995.

Peng Y, Yang H, Xue Y, Chen Q, Jin H, Liu S, Yao S, Du M. An update on malignant tumor-related stiff person syndrome spectrum disorders: clinical mechanism, treatment, and outcomes. Front Neurol. 2023;14:1209302. https://doi.org/10.3389/2Ffneur.2023.1209302.

Pittock SJ, Lucchinetti CF, Parisi JE, Benarroch EE, Mokri B, Stephan CL, Kim KK, Kilimann MW, Lennon VA. Amphiphysin autoimmunity: paraneoplastic accompaniments. Ann Neurol. 2005;58(1):96–107. https://doi.org/10.1002/ana.20529.

Resmini E, Santos A, Gómez-Anson B, Vives Y, Pires P, Crespo I, Portella MJ, de Juan-Delago M, Barahona MJ, Webb SM. Verbal and visual memory performance and hippocampal volumes, measured by 3-tesla magnetic resonance imaging, in patients with Cushing's syndrome. J Clin Endocrinol Metab. 2012;97(2):663–71. https://doi.org/10.1210/jc.2011-2231.

Ringholz GM, Appel SH, Bradshaw M, Cooke NA, Mosnik DM, Schulz PE. Prevalence and patterns of cognitive impairment in sporadic ALS. Neurology. 2005;65(4):586–90. https://doi.org/10.1212/01.wnl.0000172911.39167.b6.

Rosenfeld MR, Eichen JG, Wade DF, Posner JB, Dalmau J. Molecular and clinical diversity in paraneoplastic immunity to Ma proteins. Ann Neurol. 2001;50(3):339–48.

Ruiz-Garcia R, Munoz-Sanchez G, Naranjo L, Guasp M, Sabater L, Saiz A, Dalmau J, Graus F, Martinez-Hernandez E. Limitations of a commercial assay as diagnostic test of autoimmune encephalitis. Front Immunol. 2021;12:691536. https://doi.org/10.3389/2Ffimmu.2021.691536.

Sanders KA, Rowland LP, Murphy PL, Younger DS, Latov N, Sherman WH, Pesce M, Lange DJ. Motor neuron diseases and amyotrophic lateral sclerosis: GM1 antibodies and paraproteinemia. Neurology. 1993;43(2):418–20. https://doi.org/10.1212/wnl.43.2.418.

Saiz A, Bruna J, Stourac P, Vigliani MC, Giometto B, Grisold W, et al. Anti-Hu-associated brainstem encephalitis. J Neurol Neurosurg Psychiatry. 2009;80(4):404–7. https://doi.org/10.1136/jnnp.2008.158246.

Sahu JK, Prasad K. The opsoclonus–myoclonus syndrome. Pract Neurol. 2011;11(3):160–6. https://doi.org/10.1136/practneurol-2011-000017.

Sarkar P, Mehtani A, Gandhi HC, Bhalla JS, Tapariya S. Paraneoplastic ocular syndrome: a pandora's box of underlying malignancies. Eye (Lond). 2022;36(7):1355-1367. https://doi.org/10.1038/s41433-021-01676-x. Epub 2021 Aug 3.

Schmahmann JD, Sherman JC. The cerebellar cognitive affective syndrome. Brain. 1998;121 (Pt 4):561–79. https://doi.org/10.1093/brain/121.4.561.

Shah S, Flanagan EP, Paul P, Smith CY, Bryant SC, Devine MF, Lennon VA, McKeon A, Pittock SJ, Dubey D. Population-based epidemiology study of paraneoplastic neurologic syndromes. Neurol Neuroimmunol Neuroinflamm. 2021;9(2):e1124. https://doi.org/10.1212/NXI.0000000000001124.

Shams'ili S, Grefkens J, de Leeuw B, van den Bent M, Hooijkaas H, van der Holt B, Vecht C, Sillevis Smitt P. Paraneoplastic cerebellar degeneration associated with antineuronal antibodies: analysis of 50 patients. Brain. 2003;126(Pt 6):1409–18. https://doi.org/10.1093/brain/awg133.

Sheikhbahaei S, Marcus CV, Fragomeni RS, Rowe SP, Javadi MS, Solnes LB. Whole-body 18F-FDG PET and 18F-FDG PET/CT in patients with suspected paraneoplastic syndrome: a systematic review and meta-analysis of diagnostic accuracy. J Nucl Med. 2017;58(7):1031–6. https://doi.org/10.2967/jnumed.116.183905.

Singh AD, Milam AH, Shields CL, De Potter P, Shields JA. Melanoma-associated retinopathy. Am J Ophthalmol. 1995;119(3):369-70. https://doi.org/10.1016/s0002-9394(14)71185-7.

Starkman MN. Neuropsychiatric findings in Cushing syndrome and exogenous glucocorticoid administration. Endocrinol Metab Clin N Am. 2013;42(3):477–88. https://doi.org/10.1016/j.ecl.2013.05.010.

Tajfirouz D, Chen JJ. Paraneoplastic vision loss. Handb Clin Neurol. 2024;200:203-210. https://doi.org/10.1016/B978-0-12-823912-4.00003-7.

Titulaer MJ, McCracken L, Gabilondo I, Armangué T, Glaser C, Iizuka T, Honig LS, Benseler SM, Kawachi I, Martinez-Hernandez E, Aguilar E, Gresa-Arribas N, Ryan-Florance N, Torrents A, Saiz A, Rosenfeld MR, Balice-Gordon R, Graus F, Dalmau J. Treatment and prognostic factors for long-term outcome in patients with anti-NMDA receptor encephalitis: an observational cohort study. Lancet Neurol. 2013;12(2):157–65. https://doi.org/10.1016/S1474-4422(12)70310-1.

Vogrig A, Gigli GL, Segatti S, Corazza E, Marini A, Bernardini A, Valent F, Fabris M, Curcio F, Brigo F, Iacono D, Passadore P, Rana M, Honnorat J, Valente M. Epidemiology of paraneoplastic neurological syndromes: a population-based study. J Neurol. 2020;267:26–35. https://doi.org/10.1007/s00415-019-09544-1.

Vogrig A, Fouret M, Joubert B, Picard G, Rogemond V, Pinto AL, Muñiz-Castrillo S, Roger M, Raimbourg J, Dayen C, Grignou L, Pallix-Guyot M, Lannoy J, Ducray F, Desestret V, Psimaras D, Honnorat J. Increased frequency of anti-Ma2 encephalitis associated with immune checkpoint inhibitors. Neurol Neuroimmun Neuroinflamm. 2019;6(6):e604. https://doi.org/10.1212/NXI.0000000000000604.

Vogrig A, Bernardini A, Gigli GL, Corazza E, Marini A, Segatti S, Fabris M, Honnorat J, Valente M. Stroke-Like Presentation of Paraneoplastic Cerebellar Degeneration: a Single-Center Experience and Review of the Literature. Cerebellum. 2019;18(5):976–982. https://doi.org/10.1007/s12311-019-01075-9.

Voltz R, Gultekin SH, Rosenfeld MR, Gerstner E, Eichen J, Posner JB, Dalmau J. A serologic marker of paraneoplastic limbic and brain-stem encephalitis in patients with testicular cancer. N Engl J Med. 1999;340(23):1788–95. https://doi.org/10.1056/NEJM199906103402303.

Waragai M, Chiba A, Uchibori A, Fukushima T, Anno M, Tanaka K. Anti-Ma2 associated paraneoplastic neurological syndrome presenting as encephalitis and progressive muscular atrophy. J Neurol Neurosurg Psychiatry. 2006;77(1):111–3. https://doi.org/10.1136/jnnp.2005.068775.

Yu Z, Kryzer TJ, Griesmann GE, Kim K, Benarroch EE, Lennon VA. CRMP-5 neuronal autoantibody: marker of lung cancer and thymoma-related autoimmunity. Ann Neurol. 2001;49(2):146–54.

Normal Pressure Hydrocephalus

9

Jason R. Soble, Devin M. Ulrich, Matthew S. Phillips, and Kyle J. Jennette

ABBREVIATIONS

ACS	Advanced Clinical Solutions
ACZ	Acetazolamide
AD	Alzheimer's Disease
AMS	Altered Mental Status
BVMT-R	Brief Visuospatial Memory Test-Revised
Cine MRI	Cine phase-contrast MRI
CNS	Central Nervous System
CSF	Cerebrospinal Fluid
CT	Computed Tomography
GDS	Geriatric Depression Scale
ETV	Endoscopic Third Ventriculostomy
GPT	Grooved Pegboard Test
H2O	Water
Hg	Mercury
LVLP	Large-Volume Lumbar Puncture
MCI	Mild Cognitive Impairment
mL	Milliliter
mm	Millimeter
MMSE	Mini-Mental State Examination
MoCA	Montreal Cognitive Assessment
MRI	Magnetic Resonance Imaging
NPH	Normal Pressure Hydrocephalus
RAVLT	Rey Auditory Verbal Learning Test
RCFT	Rey Complex Figure Test

J. R. Soble (✉) · K. J. Jennette
Department of Psychiatry, University of Illinois at Chicago – College of Medicine, Chicago, IL, USA

Department of Neurology and Rehabilitation, University of Illinois at Chicago – College of Medicine, Chicago, IL, USA
e-mail: Soble@uic.edu; Kjennett@uic.edu

D. M. Ulrich
Division of Movement Disorders, Department of Neurology, Rush University Medical Center, Chicago, IL, USA
e-mail: devin_ulrich@rush.edu

M. S. Phillips
Department of Psychiatry, University of Illinois at Chicago – College of Medicine, Chicago, IL, USA
e-mail: mphill8@uic.edu

E. K. Bailey et al. (eds.), *Inpatient Neuropsychology*, https://doi.org/10.1007/978-3-032-05880-5_9

RBANS	Repeatable Battery for the Assessment of Neuropsychological Status
RCIs	Reliable Change Indices
SDMT	Symbol-Digit Modalities Test
SPECT	Single-Photon Emission Computed Tomography
TMT	Trail Making Test
TOMM	Test of Memory Malingering
TOPF	Test of Premorbid Functioning
U.S.	United States
WAIS-V	Wechsler Adult Intelligence Scale-5th Edition
VPA	Ventriculoperitoneal Shunt

INTRODUCTION

Broadly defined, hydrocephalus refers to an abnormal accumulation of cerebrospinal fluid (CSF) in the ventricular system (Rekate 2009). Hydrocephalus can result from many etiologies, including congenital anomalies (e.g., Dandy-Walker syndrome), acquired conditions (e.g., subarachnoid hemorrhage, tumor), and/or unknown idiopathic causes that result in either inadequate CSF reabsorption or obstruction of CSF flow at some point within the ventricular system (Koleva and De Jesus 2023). Among adults, normal pressure hydrocephalus (NPH) is the most frequently encountered type of hydrocephalus, with an estimated prevalence of 1.5–10.2% among older adults overall and highest rates among those aged 80 and older (Constantinescu et al. 2024; Jaraj et al. 2014; Krauss and Halve 2004; Kuriyama et al. 2017). Despite this prevalence, NPH is often underappreciated in medical settings. As such, while other forms of hydrocephalus will be briefly reviewed, this chapter will primarily focus on NPH given its higher clinical relevance when evaluating adults in inpatient medical settings.

Pathophysiology

The ventricular system forms early in human development when the neural tube folds, creating several interconnected cavities deep within the brain (Fig. 9.1). Four ventricles comprise the ventricular system containing CSF that flows from the lateral ventricles to the third ventricle via the foramen of Monro, to the fourth ventricle via the cerebral aqueduct, and into the subarachnoid space around the brain and spinal cord, before ultimately being reabsorbed into the bloodstream via the arachnoid granulations (Fig. 9.2). The choroid plexus, located within the lateral, third, and fourth ventricles, produces CSF to cushion the brain from the inner skull and provide nourishment and waste removal. Healthy ventricles allow CSF to encase the brain and spinal cord without pushing the parenchyma against the inner skull or other periventricular structures and allow for unrestricted passage of CSF throughout the ventricular system. Disruption of this

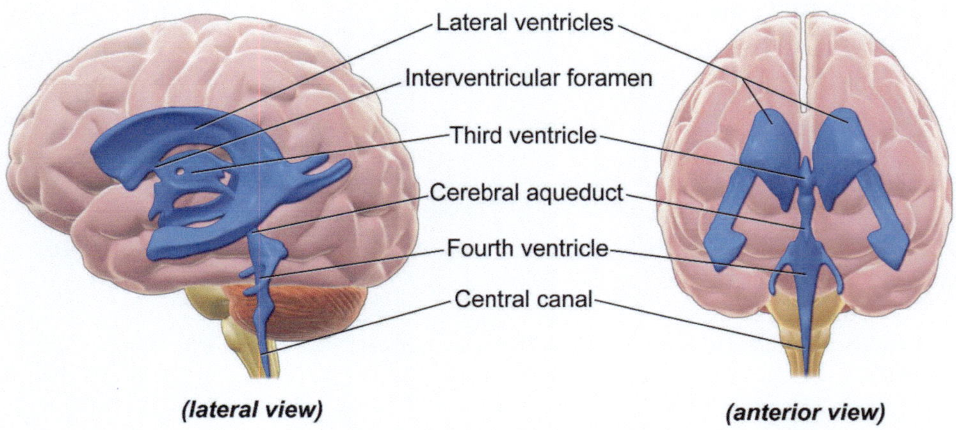

Fig. 9.1 Ventricles of the brain

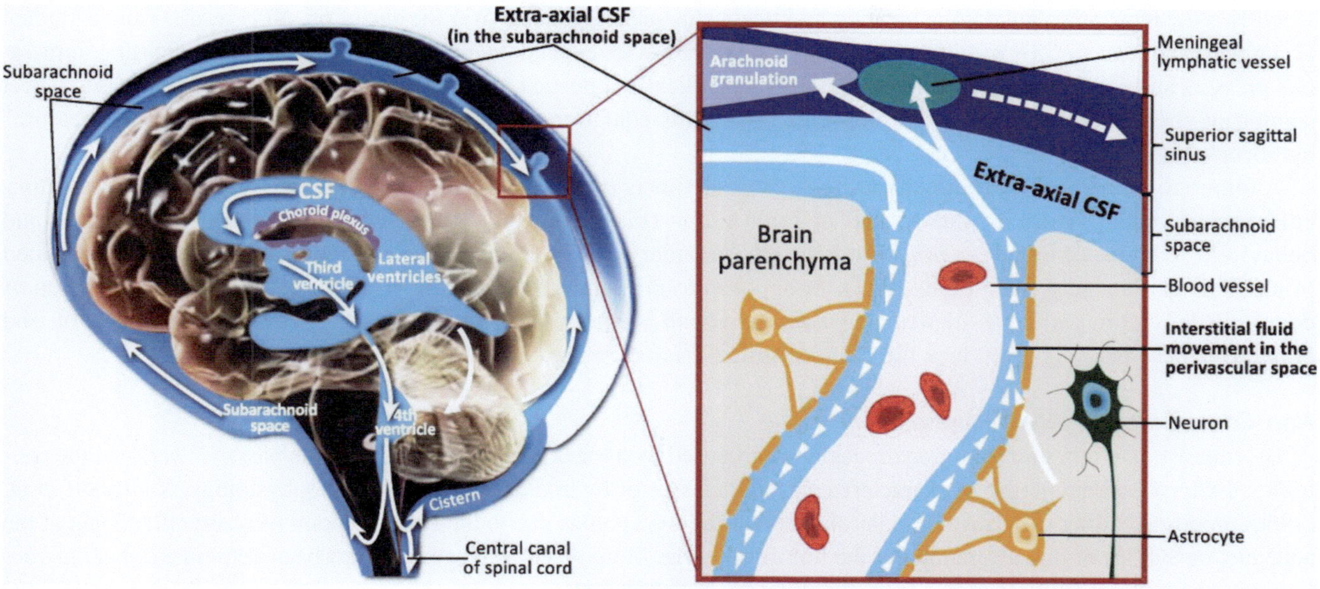

Shen, M.D. (2018), CC BY 4.0

Fig. 9.2 CSF circulation

healthy flow of CSF can lead to a condition known as hydrocephalus, in which excess CSF builds up in the central nervous system, which can damage brain tissue and cause various physical and neurobehavioral symptoms.

Excess CSF within the ventricular system can accumulate for several reasons, including general overproduction, a blockage or obstruction in flow, or a disruption in the reabsorption within the arachnoid granulations. Due to the constant production and resultant flow within the ventricular system, an increase of CSF in one compartment must be accompanied by a compensatory decrease in another. Otherwise, excessive fluid buildup in any given compartment can lead to increased intracranial pressure and a cascade of physical and cognitive symptoms. However, in clinical populations, obstruction and problems with reabsorption are more prevalent than overproduction of CSF.

Epidemiology

Congenital malformations in infancy are the most common cause of hydrocephalus (between 1 and 32 per 10,000 births), followed by idiopathic hydrocephalus in late life. Global prevalence of hydrocephalus is approximately 85 per 100,000, with higher rates seen in pediatric populations (i.e., 88 per 100,000) than in middle-aged adults (i.e., 11 per 100,000). Prevalence in older adult populations (ages 65+) equates to roughly 175 per 100,000, with patients 80 years and older showing rates of over 400 per 100,000. There are numerous risk factors associated with the development of hydrocephalus, the most prevalent being central nervus system (CNS) tumors, CNS infections, and any condition that leads to bleeding within the brain (e.g., stroke, brain injury; Isaacs et al. 2018; Koleva and De Jesus 2023).

Types of Hydrocephalus

Communicating Hydrocephalus

In clinical practice, hydrocephalus is divided into two types based on the specific mechanism of disruption: *communicating* and *non-communicating*. Communicating hydrocephalus results from either a disruption of CSF reabsorption within the arachnoid granulations or an obstruction of flow within the subarachnoid space. This class of hydrocephalus is referred to as "communicating" because the space between the ventricles remains open and CSF still maintains the ability to flow, and it is not until after the CSF reaches the end of the ventricular space that disruption occurs. Communicating hydrocephalus can occur because of injury (e.g., subarachnoid hemorrhage; traumatic brain injury) or acute inflammatory changes (e.g., meningitis) that thickens the subarachnoid space and restricts CSF flow between the brain and the inner skull (Bladowska and Sąsiadek 2018).

NPH is the most prevalent communicating hydrocephalus seen in patients over the age of 65. NPH results from a gradual accumulation of excess CSF within the ventricles. As the ventricles expand, brain parenchyma can shift, increasing intracranial pressure and causing clinical symptoms. However, medical diagnostic testing (e.g., lumbar puncture) often fails to detect significant changes in intracranial pressure, thus giving this condition the aptly named distinction of "normal pressure" hydrocephalus.

Another common type of communicating hydrocephalus is *hydrocephalus ex vacuo*, which refers to compensatory enlargement of the ventricles due to loss of brain tissue (Fig. 9.3). Commonly seen in older adults, cognitive and behavioral sequelae of *hydrocephalus ex vacuo* are secondary due to the presence of underlying pathology implicated in parenchymal atrophy (Kim et al. 2021), as opposed to primary effects of true NPH. In other words, the pattern of cognitive deficits in patients with *hydrocephalus ex vacuo* is contingent on the distal cause of neurodegeneration and clinically, these patients often lack the classic clinical triad (see Fig. 9.5) of NPH symptoms.

Non-Communicating Hydrocephalus

By contrast, non-communicating hydrocephalus is caused by a blockage or obstruction of the passages *between* the ventricles. Early on, when aiming to characterize specific classes of hydrocephalus and its pathophysiology, Ransohoff et al. (1960) proposed calling it "intraventricular obstructive" hydrocephalus due to the believed point of obstruction being at the aqueduct of Sylvius or outlet foramina of the fourth ventricle. More recently, however, it has been determined that obstruction can take place at multiple points throughout the ventricles (Rekate 2008).

Ventricular areas known to result in blockage most commonly are known as *obstacle sites* (Bladowska and Sąsiadek 2018). These sites include the foramen of Monro, the cerebral aqueduct, the fourth ventricle outlets, and the foramen magnum. Any malformation or restriction at each of these levels will result in the build-up of CSF ahead of the blockage and restriction of flow behind. Causes for flow restriction at each of these sites are widespread and include cysts, tumors, and adhesions, among others. Other common conditions that can contribute to noncommunicating hydrocephalus include acquired or congenital aqueductal stenosis, cerebellar infarcts, or intraventricular hemorrhages (Fig. 9.4).

In sum, hydrocephalus occurs when a disruption or obstruction within the brain's ventricular system leads to excessive accumulation of CSF that can result in many physical and neurocognitive symptoms. Hydrocephalus can result from several different conditions and is typically classified as communicating or non-communicating depending on the mechanism of disruption. Hydrocephalus can occur across the lifespan with a bimodal prevalence in infancy and later adulthood. NPH is the most prevalent form of communicating hydrocephalus in older adults that onsets insidiously, progresses gradually, and can cause physical and neurobehavioral symptoms without increasing intracranial pressure. Due to NPH's relatively high occurrence in hospital settings (as opposed to other forms of hydrocephalus)—and, given its association with older adult populations who are more likely be referred for inpatient neuropsychological consultation—a comprehensive understanding of pathophysiology, clinical presentation, and treatment trajectory of NPH is critical for clinical neuropsychologists working in inpatient medical settings. NPH is difficult to diagnose without formal cognitive and neurobehavioral testing, thus making neuropsychology an integral part of the evaluation and treatment process.

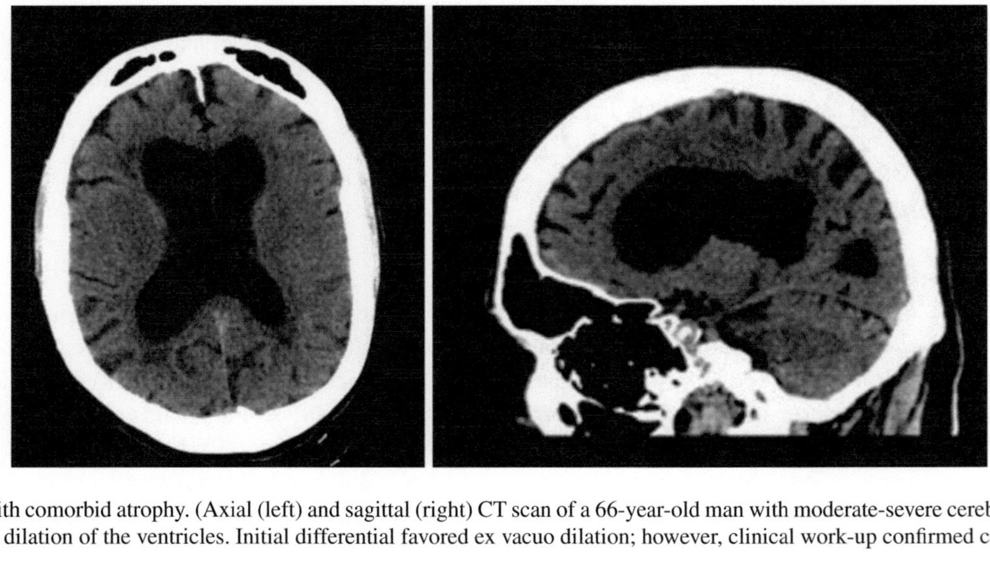

Fig. 9.3 NPH with comorbid atrophy. (Axial (left) and sagittal (right) CT scan of a 66-year-old man with moderate-severe cerebral atrophy, sulcal enlargement, and dilation of the ventricles. Initial differential favored ex vacuo dilation; however, clinical work-up confirmed concomitant NPH)

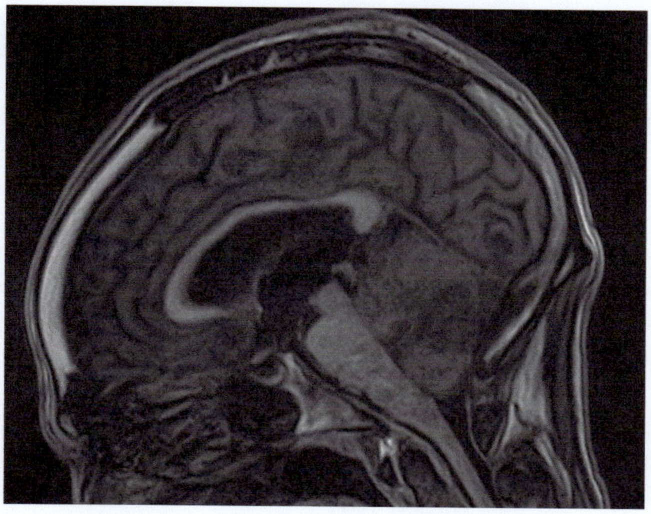

Fig. 9.4 Obstructive hydrocephalus due to metastatic disease. (Sagittal T1 showing obstructive hydrocephalus, including dilation of the third ventricle, due to compression of the cerebral aqueduct from metastatic lesion in the left cerebellum of a 55-year-old man)

Fig. 9.5 Classic NPH symptom triad

CLINICAL PRESENTATION

Clinically, NPH may present with a broad array of physical, cognitive, and neuropsychiatric symptoms, although it is classically associated with a symptom triad consisting of progressively impaired ambulation, urinary incontinence, and altered mental status (AMS; Blumenfeld 2021; Passos-Neto et al. 2022). Sometimes colloquially referred to as "wet, wobbly, and wacky" (Patel et al. 2023), this hallmark triad commonly differentiates NPH from other causes of hydrocephalus (Fig. 9.5).

Although the presence of the full symptom triad aids in diagnostic confidence, it is not present in all cases. According to Damasceno (2009), the complete triad is seen in 50–75% of cases, with gait and cognitive disturbances being the most common (seen in 80–95% of cases) followed by urinary incontinence (seen in 45–90% of cases). An additional clinical challenge for diagnosing and managing NPH in inpatient medical settings is that, individually, these core physical, cognitive, and neuropsychiatric symptoms are nonspecific and can occur for reasons unrelated to NPH (e.g., general debility, malnourishment, exacerbation of chronic baseline medical conditions, delirium, other underlying neurodegenerative dementia processes), particularly among older adults. These relevant clinical considerations germane to inpatient medical settings are addressed below with each respective symptom type.

Physical Symptoms

Physically, gait disturbance is the most common symptom of NPH. The primary gait abnormality associated with NPH often is referred to as "magnetic gait" or "apraxic gait" and is characterized as slow and wide-based with outwardly rotated feet, a diminished height of steps, and *en bloc* turning requiring several steps to change direction (Passos-Neto et al. 2022; Stolze et al. 2001). In addition to being the most common NPH symptom, gait disturbance is the most frequently improved symptom following neurosurgical intervention with improvement seen in upwards of 90% of NPH patients (Marmarou et al. 2005; McGrath et al. 2024). In fact, immediate gait improvement is the best prognostic indicator of post-shunt positive outcome (Marmarou et al. 2005). Other motor signs can be seen in individuals with pre-existing gait dysfunction and NPH (Damasceno 2009). These can include postural instability, spastic movements, hyperreflexia, rigidity, and certain primitive reflexes (e.g., snout, Babinski sign, palmomental, sucking). Bladder dyscontrol/urinary dysfunction is another core symptom found among 45–90% of patients with NPH and occurs on a spectrum ranging from increased urgency/frequency to full incontinence (Malm et al. 2013; Rendtorff et al. 2012; Sakakibara et al. 2008; Verrees and Selman 2004). Urinary incontinence also has been shown to improve post-shunting, albeit significantly less frequently than gait (McGirt et al. 2005).

While onset of incontinence and/or gait disturbance in a patient with no prior history of these symptoms would raise clinical suspicion for NPH, these symptoms are ultimately not specific to NPH and can overlap with other acute conditions commonly seen in inpatient medical settings. For instance, estimated rates of urinary incontinence among community-dwelling adults aged 60 and older range from 8% to 18% (Goode et al. 2008). Similarly, gait problems have been found to affect approximately 15% of individuals in their 60s, 35% in their 70s, and 80% in their 80s (Jahn et al. 2010). To aid in differential diagnosis of NPH vs other etiology, we recommend obtaining a thorough clinical history, ideally with independent collateral, to characterize the onset and course of symptoms. Even in cases with some degree of baseline gait disturbance and/or bladder dysfunction, obtaining relevant temporal information (e.g., a patient's spouse clarified that the patient has had mildly unsteady gait for the past 7 years, but this has appreciably worsened over the past 4–6 months in the absence of any other known cause) can clarify a clinical diagnosis of NPH. Appreciation for any baseline gait or urinary symptoms also provides a relevant comparison benchmark when assessing for meaningful changes in physical symptoms following neurosurgical intervention (e.g., shunting; large volume lumbar puncture; discussed below).

Unfortunately, for a minority of patients who are significantly immobile or medically compromised at baseline, improvements in gait and/or urinary symptoms may not be as effective as an objective indicator of favorable response to intervention. Rather, in these cases, changes in mental status from baseline may function as a more useful objective improvement metric.

Cognitive Symptoms

Though not universal, cognitive impairment often accompanies NPH, such that patients with NPH typically perform worse on standard neuropsychological testing relative to healthy controls (Hellström et al. 2007; Langheinrich et al. 2022). Unfortunately, no pathognomonic neuropsychological profile can be used in isolation to identify and differentiate NPH from other causes of cognitive dysfunction. Rather, cognitive deficits in NPH are heterogenous and are primarily attributed to dysfunction in frontal-subcortical circuitry with secondary widespread disruption of the entorhinal-hippocampal circuits and posterior cortical function (Saito et al. 2011; Xiao et al. 2022). On testing, individuals can demonstrate impairments in global cognition, attention/working memory, executive functions (e.g., abstraction, planning, decision making/reasoning), memory, visuospatial/visuoperceptual abilities, and/or psychomotor speed (Passos-Neto et al. 2022; Picascia et al. 2015; Saito et al. 2011; Xiao et al. 2022).

Further confounding identification of NPH-related cognitive impatient is the elevated prevalence rates of mild cognitive impairment (MCI; ~3–42%) and dementia (~3%) among adults aged 65 and older (U.S. Preventive Services Task Force, 2020). Consequently, many older adults (both with and without NPH) evaluated in inpatient medical settings will have some degree of baseline cognitive impairment, which further exacerbate NPH-related cognitive deficits. Hellström et al. (2007) found that patients with NPH and vascular risk factors (e.g., hypertension, diabetes) performed worse on objective neuropsychological testing relative to patients with NPH without vascular comorbidity. The nonspecific cognitive deficits associated with NPH can also be seen in other underlying dementia processes, such as Alzheimer's disease (Langheinrich et al. 2022), though memory dysfunction in NPH tends to be milder than the dense amnestic memory deficit hallmark of Alzheimer's disease (AD; Picascia et al. 2015; Iddon et al. 1999).

In sum, for many older adults with NPH-related cognitive impairment, impairment may not be solely attributable to hydrocephalus. Delirium is a very relevant consideration, occurring in 30–55% of hospitalized older patients, though it is often unrecognized and can obscure accurate assessment of any NPH-related cognitive dysfunction (Al Farsi et al. 2023; Saxena and Lawley 2009). Notwithstanding, NPH should remain on the differential for diagnostic rule-out when a patient presenting with cognitive impairment and complaints of co-occurring gait disturbance and urinary incontinence (Langheinrich et al. 2022). Therefore, it is essential that neuropsychologists in inpatient medical settings fully consider all potential etiological contributors when cognitive impairment is detected on testing.

Unlike many other dementia processes, which involve progressive cognitive deterioration, NPH is typically referred to as a "reversible dementia" or "treatable cause of dementia" given that mental status can markedly improve following shunting, particularly in executive and motor performance (Hellström et al. 2008; Lenfeldt et al. 2008; Patel et al. 2023; Pierson et al. 1982; Saito et al. 2011). Notably, a meta-analysis found that despite heterogeneity among individual neuropsychological tests of various cognitive abilities, shunting results in significant improvements in several domains of neurocognitive functioning among patients with NPH, including global cognition, learning and memory, psychomotor processing speed, and select executive functions (Peterson et al. 2016). Given this, serial neuropsychological testing is critical to objectively monitor cognition over time, in order to assess for symptom change following neurosurgical intervention. This can occur both in inpatient settings as well as longitudinally on an outpatient basis after discharge. If neuropsychological testing is conducted as part of a pre/post-shunting work-up, patient performance should be compared to their own baseline, as well as appropriate demographically corrected normative data.

As previously discussed, given the prevalence rates of cognitive impairment among older adults, it is important to remember that many patients with NPH likely will have cognitive impairment due to multiple etiologies such that even if the NPH is well-managed following neurosurgical intervention, a return to normal cognition is less probable. However, even if cognitive impairment does not fully remit, objective cognitive improvement often is seen pre- to post-shunting and can have significant positive implications for patients' functioning and quality of life post-discharge (Hülser et al. 2022).

Neuropsychiatric Symptoms

Although physical/motoric and cognitive dysfunction are classic symptoms of NPH, the disorder has also been associated with neuropsychiatric and behavioral sequelae. Individuals with NPH commonly present with various constellations of psychomotor retardation, AMS/slowed mentation, depression, lethargy/drowsiness, diminished volition/amotivation, apathy, and affective blunting (Malm et al. 2013; Mathew et al. 2018; Passos-Neto et al. 2022; Price and Tucker 1977). Differentiating symptoms of depression from neurogenic apathy can be challenging in this population. Rarer symptoms, including delusions and aggressive/hostile behavior, also have been described in association with NPH, albeit typically among small numbers of patients in case reports (Crowell et al. 1973; Srivastava et al. 2015).

DIAGNOSIS OF NPH

NPH lacks a universally accepted set of standardized diagnostic criteria (Andersson et al. 2017). Two sets of international guidelines have been developed and are widely used: American-European guidelines (Relkin et al. 2005) and Japanese guidelines (Ishikawa et al. 2008; Mori et al. 2012; Nakajima et al. 2021). Discussed in detail below, we recommend familiarization with both diagnostic guidelines (outlined in Tables 9.1 and 9.2) as NPH may be differentially diagnosed based on which set of criteria are used by the medical team.

Under both sets of guidelines, individuals can be diagnosed as having "probable," "possible," or "unlikely" NPH based on clinical history, brain imaging, physical/cognitive symptoms, and physiological criteria. Under the Japanese guidelines, individuals can also meet criteria for "definite" NPH.

NPH is considered unlikely when there is:

- NO evidence of ventriculomegaly.
- NO signs of increased intracranial pressure.
- NO clinical features of NPH present.
- And/or symptoms are better explained by other causes.

Clinical History

According to the American-European guidelines, *Probable* NPH is characterized by an insidious onset after the age of 40 with a minimum duration of at least 3–6 months, a progressive course, no evidence of a precipitating event (e.g., traumatic brain injury, intracerebral hemorrhage, meningitis, and other causes of secondary hydrocephalus), and no evidence of other neurological, psychiatric, or medical explanation for presenting symptoms.

Table 9.1 American-European criteria for normal pressure hydrocephalus

Domain	Unlikely	Possible NPH	Probable NPH
Clinical History	No criteria met	• Subacute or indeterminate mode of onset. • Begin any time after childhood. • Less than 3 month or indeterminate duration. • Nonprogressive or not clearly progressive course. • May follow events (e.g., TBI, meningitis, intracerebral hemorrhage) that in the judgment of the clinician are not likely to be causally related. • Coexist with other neurological, psychiatric, or general medical disorders but symptoms not entirely attributable to these conditions.	• Insidious onset • Onset after age 40 years • Minimum 3–6-month duration • Progressive course • No evidence of antecedent events • No other neurological, psychiatric, or general medical conditions sufficient to explain symptoms
Brain Imaging	No evidence of ventriculomegaly	Ventricular dilation (Evans's index > 0.3)	Ventricular dilation (Evans index > 0.3) AND At least one of the following: • Narrow callosal angle • Enlargement of the temporal horns • Periventricular signal changes not attributable to ischemic changes or demyelination

Table 9.1 (continued)

Clinical Features	No clinical features present OR Symptoms better explained by other causes	EITHER • Incontinence and/or cognitive impairment in the absence of gait/balance disturbance OR • Gait disturbance alone	At least one symptom of the clinical triad: • Gait disturbance • Cognitive impairment • Urinary incontinence
Physiological Findings	None	Opening pressure unavailable OR Outside the range required for probable	CSF opening pressure in the range of 5-18 mm Hg (or 70-245 mmH_2O)

Evans Index: ratio of the maximum width of the frontal horns of the lateral ventricles and the maximal internal diameter of the skull at the same level on neuroimaging; Venkatesh et al., 2013.
Narrow callosal angle: results from enlarged widening of the Sylvian fissures and cisterns with crowding of the gyri at the vertex on neuroimaging; Gaillard et al. (2015)

In contrast, a diagnosis of *possible* NPH is rendered when the above criteria are not definitively met. For example, a diagnosis of *possible* NPH may be appropriate when there is evidence of subacute or indeterminate onset, onset occurs any time after childhood but not necessarily after the age of 40, duration is fewer than 3 months or indeterminate, and/or there is a nonprogressive or not clearly progressive course. A diagnosis of *possible* NPH may also be appropriate when patients have a history of injury or illness that could produce ventriculomegaly but are deemed unlikely to be clinically related (e.g., mild traumatic brain injury, remote history of intracerebral hemorrhage, childhood/adolescent meningitis) and/or a history of other neurological, psychiatric, or other medical disorders that could explain some features of the clinical presentation, but are not considered to be causally-related.

The Japanese guidelines, in contrast, state that individuals must develop symptoms in their 60s or older, and the course may be progressive or may fluctuate over time. These guidelines share the requirement that clinical symptoms cannot be completely explained by other neurological or medical diseases and there must not be any obvious preceding diseases that cause ventricular dilation. Other neurological diseases (e.g., Parkinson's disease, Alzheimer's disease, cerebrovascular diseases) *may* be present, but must be mild.

Since all three components of the NPH "classic triad" are not required for a diagnosis of NPH, a thorough review of a patient's medical and surgical history should be conducted to rule out disorders which are clinically similar (e.g., neurodegenerative, neurological, infectious, urological, or other disorders). Under both sets of guidelines, other etiologies of neurocognitive dysfunction (e.g., Alzheimer's disease, dementia with Lewy bodies, and vascular dementia), diseases affecting both gait and cognition (e.g., Parkinson's disease, progressive supranuclear palsy), and other conditions affecting gait specifically (e.g., spinal stenosis, orthopedic conditions) should all be ruled out as primary causes of clinical symptoms. Additionally, urinary tract infections, which can cause altered mental status and urinary symptoms, should be excluded before a diagnosis of NPH is rendered, as should other urological disorders and causes of incontinence in older adults (e.g., bladder/prostate cancer, benign prostatic enlargement). As many patients with suspected NPH present with AMS and may not be able to provide reliable self-report, independent collateral information from family member or trusted friend/caregiver is strongly recommended.

Table 9.2 Japanese diagnostic criteria for normal pressure hydrocephalus

Domain	Unlikely	Possible NPH	Probable NPH	Definite NPH
Clinical History	No criteria met	• Individuals must be in their 60s or older. • Clinical symptoms cannot be better explained by other neurological or medical diseases. • No obvious preceding disease possibly causing ventricular dilation.	All criteria for possible met	All criteria for probable met
Brain Imaging	No evidence of ventriculomegaly	Ventricular dilation (Evans index > 0.3)	All criteria for possible met AND: • Narrowing of the sulci over the high convexity/ disproportionately enlarged subarachnoid space • Sylvian fissures and basal cistern enlargement	All criteria for probable met
Clinical Features	No clinical features present	At least ONE of the clinical triad: • Gait disturbance • Cognitive impairment • Urinary incontinence	All criteria for possible met AND: • Ataxic/apraxic gait	All criteria for probable met
Physiological Findings	None	CSF examination unavailable	Both of the following: • CSF pressure of $200 mmH_2O$ or less and normal CSF content • Improvement following lumbar puncture	Improvement following shunting procedure

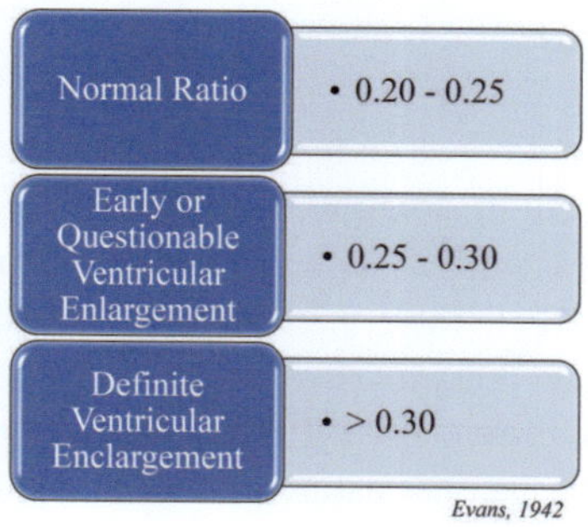

Fig. 9.6 Evans index

Brain Imaging

Brain imaging studies, such as a CT or MRI, constitute another essential piece of evidence to provide diagnostic clarity in cases of suspected NPH. Imaging is useful in determining the *callosal angle*, or the angle between the superior walls of the lateral ventricles on a coronal plane. In NPH, smaller callosal angles are seen due to enlargement of the lateral ventricles relative to the cortex (Damasceno 2015). Per the American-European guidelines, a diagnosis of *probable* NPH is rendered when imaging is remarkable for (1) enlargement of the temporal horns of the lateral ventricles not entirely attributable to hippocampal atrophy, (2) callosal angle of 40–90°, (3) evidence of periventricular signal changes on CT and MRI that is not attributable to microvascular ischemic changes or demyelination, (4) an aqueductal or fourth ventricular flow void on MRI, and (5) ventricular enlargement (Evans Index >0.3) that is not entirely attributable to cerebral atrophy or congenital enlargement. The Evans Index is defined as the ratio of the maximal width of the frontal horns to the maximum inner skull diameter (Evans 1942; Fig. 9.6). When there is ventricular enlargement consistent with hydrocephalus but there is evidence of sufficient cerebral atrophy to potentially explain the ventricular size or structural lesions that may impact ventricular size, a diagnosis of *possible* NPH is indicated.

Like the American-European guidelines, the Japanese guidelines require evidence of ventricular dilation (Evans index >0.3). Supportive features include enlargement of the Sylvian fissures and basal cistern and narrowing of the sulci and subarachnoid spaces of the high convexity/midline surface. Although less commonly used, additional radiographic support for a diagnosis of NPH would include a neuroimaging study demonstrating interval change in ventricle size from a prior examination, radionuclide cisternogram showing delayed clearance of a radiotracer over the cerebral convexities after 48–72 h, increased ventricular flow rate as seen on Cine MRI or other technique, or decreased periventricular perfusion that is not altered by an intravenous injection of acetazolamide (ACZ) on a SPECT-ACZ challenge. Periventricular changes, however, are not essential according to these guidelines.

Physiological Findings

In cases for which NPH is strongly suspected due to clinical history, neuroimaging, and symptom presentation, procedures such as a large-volume lumbar puncture (LVLP) or lumbar drain can be performed to provide additional diagnostic clarification. In the case of an LVLP, a neuroradiologist performs a lumbar puncture under fluoroscopy, and opening and closing pressures are recorded. Ideally, a total of 50 mL of CSF is collected and laboratory examinations are obtained on the collected CSF (Oliveira et al. 2019). Opening pressure in the range of 5–18 mm Hg (or 70–245 mm H2O) is consistent with a diagnosis of *probable* NPH according to the American-European guidelines. When opening pressure measurement is not available or is outside of this range, a diagnosis of *possible* NPH can still be rendered. In some cases, repeated lumbar punctures or a continuous lumbar drain procedure is preferred. In the latter, CSF is continually drained over several days with assessments of gait and mental status before the procedure and again 24–72 h later to determine whether the patient experienced any measurable improvement.

Per the Japanese guidelines, CSF examination via lumbar puncture or lumbar drain is the primary distinguishing feature between *possible* and *probable* NPH. That is, individuals with *probable* NPH should show symptoms improvement after a lumbar puncture, in addition meeting all prerequisite requirements for *possible* NPH based on clinical history, neuroimaging, and clinical presentation. *Definite* NPH is diagnosed when patients experience improvement of symptoms following neurosurgical intervention (i.e., shunt procedure). A LVLP performed with the specific goal of assessing change in gait and cognition among patients with suspected NPH is sometimes referred to as a Miller Fisher Test.

Clinical Signs/Symptoms and Evaluation

Per the American-European guidelines, incontinence and/or AMS in the absence of observable gait or balance difficulties, or when gait disturbance or dementia are present in isolation, is suggestive of a diagnosis of *possible* NPH. Evaluations by physical therapy and neuropsychology to assess gait/balance and mental status, respectively, can be essential pieces of the diagnostic puzzle in the context of suspected NPH. Table 9.3 describes the clinical features characteristic of each symptom of the triad according to the American-European guidelines.

Per the Japanese guidelines, individuals with *possible* NPH should have more than one symptom of the clinical triad, while individuals with *probable* NPH have at least one feature of the clinical triad including gait disturbance characterized by an unstable, slow, magnetic, wide-based gait with a small stride and small, unstable steps when turning.

Table 9.3 Clinical features of the NPH according to the American-European guidelines

Gait/Balance Disturbance (*at least 2 needed*)	Cognitive Impairment (*at least 2 needed*)	Urinary Incontinence (*at least 1 needed*)
• Decreased step height. • Decreased step length. • Decreased cadence (speed) of walking. • Increased truck sway while walking. • Toes turned out while walking. • Widened base when standing. • Retropulsion (i.e., an inability to maintain balance following a backward pull on the shoulders.) • Requiring three or more steps to turn 180 degrees. (i.e., *en bloc* turning) • Impaired balance while walking. (i.e., two or more corrections out of eight steps during tandem gait testing.)	• Psychomotor slowing. (i.e., response latency) • Decreased fine motor speed. • Decreased fine motor accuracy. • Difficulty dividing or maintaining attention. • Impaired recall, especially for recent events. • Executive dysfunction • Behavioral/personality changes	• Episodic or persistent urinary incontinence in the absence of a primary urological disorder. • Persistent urinary incontinence • Urinary and fecal incontinence

COGNITIVE EVALUATION

Pre/post-LVLP neuropsychological evaluation in the inpatient setting allows for rapid assessment of patients with suspected NPH to monitor measurable post-procedure changes, aid in the overall NPH diagnostic process, and predict whether a patient is likely to benefit cognitively from shunting. Below, we summarize relevant points of consideration identified in the scientific literature to best guide evidence-based assessment when selecting tests and performing these evaluations.

First, the timing of testing is critical when performing pre/post-LVLP evaluations. For instance, in their proposed standardized NPH workup protocol, Acosta et al. (2021) used a 1–3-h window after the LVLP was performed. As such, the neuropsychologist must remain cognizant that there is a circumscribed timeframe following LVLP in which repeated evaluation must occur to detect meaningful improvements in cognition. Some research also suggests that repeat assessment following serial LVLP may yield more detectable cognitive improvement. For example, da Rocha et al. (2021) found that measuring cognition following a series of two LVLPs separated by at least 24 h (repeat neuropsychological evaluation was completed 6–8 h following the second LVLP) resulted in improvements across a full battery of neuropsychological tests, whereas improvement was only seen on a cognitive screener and one executive test after the first LVLP. Thus, if possible, serial evaluation following serial LVLP may result in more detectable post-procedure cognitive changes.

Second, in this context, the pre/post-LVLP neuropsychological evaluation ideally includes tests which assess multiple domains of cognitive and psychological functioning (e.g., apathy, depression). Careful neuropsychological test selection can go a long way, particularly when factors related to the inpatient setting significantly limit the number of tests that can be administered. In general, tests of complex attention/executive functions, verbal fluency, memory, complex visuoconstructional skills, and psychomotor speed have shown as sensitive and consistent predictors of pre/post-intervention outcomes (Katzen et al. 2011; Laidet et al. 2015; Mayorga et al. 2024; McGovern et al. 2019; Peterson et al. 2016; Solana et al. 2012). Improvement in executive functioning test scores may be especially useful for measuring the effect of intervention for improving NPH-related cognitive impairment (Laidet et al. 2015).

Several proposed batteries have been developed, albeit with limited empirical support (DeVito et al. 2005; Micchia et al. 2022). A sample neuropsychological battery is provided in Table 9.4.

However, while a comprehensive battery naturally provides robust characterize of a patient's cognitive status and more baseline datapoints for measuring change post-LVLP, a "one-size-fits-all" approach to pre/post-LVLP neuropsychological evaluation is not practical (and arguably infeasible) when assessing patients with known or suspected NPH in inpatient settings. One very important consideration for test selection is severity of cognitive dysfunction and patients' ability to tolerate lengthier evaluations beyond brief cognitive screening (e.g., Montreal Cognitive Assessment [MoCA]; Nasreddine et al. 2005). Among cognitive screeners, the Mini-Mental State Exam (MMSE) has demonstrated evidence of good sensitivity for measuring pre-post LP cognitive changes (e.g., de Rocha et al., 2021; Solana et al. 2012; Schmidt et al. 2014). Therefore, if battery length is a concern, the neuropsychologist should judiciously prioritize tests that target these cognitive domains to construct a brief, focused pre/post LP neuropsychological test battery. Unfortunately, with few exceptions (e.g., McGovern et al. (2019) specifically recommended use of the RAVLT for measuring pre/post LP memory function), there is a paucity of direct comparison research to inform which specific test(s) of each of the cognitive domains listed above are most sensitive for detecting post-intervention changes. Irrespective of the test battery selected, neuropsychological test and norm selection should consider individual patient factors including patient demographics, including age, linguistic preference, educational attainment, and other relevant social determinants of health. Some objective assessment of performance validity also should be included, per current practice standards (Sweet et al. 2021). Another critical consideration when performing repeat neuropsychological testing, particularly over short time intervals as is the case with pre/post-LVLP evaluations, is that test scores often will improve as a function of prior test exposure (i.e., practice effects) which can yield inaccurate conclusions of cognitive improvement if not appropriately considered. In clinical practice, there are two primary approaches to mitigating practice effects. First, using tests with psychometrically equivalent parallel/alternate forms (e.g., RAVLT, BVMT-R) minimizes exposure to identical test stimuli. Second, calculation of reliable change indices (RCIs) allows for objective determination of whether any improvement(s) in pre- to post-test scores reflect statistically significant improvement versus nonsignificant/artifactual test score increases due to factors such as practice effects and/or regression to the mean (Chelune et al. 1993; Gonzalez et al. 2022a; Maassen et al. 2009). The former approach is typically more pragmatic in inpatient settings, as pre/post-LVLP evaluations often utilize cognitive screening measures which may offer multiple forms or options for item substitution (e.g., MoCA, MMSE, Repeatable Battery for the Assessment of Neuropsychological Status [RBANS]).

Table 9.4 Sample NPH neuropsychological battery

Sample NPH Neuropsychological Battery (Pre/Post-Lumbar Puncture)	
Domain(s) Assessed	**Test(s)**
Performance Validity	• Test of Memory Malingering (TOMM)-Trial 1
Estimated Premorbid Intellectual Functioning	• ACS: Test of Premorbid Functioning (TOPF)
Learning & Memory	• Rey Auditory Verbal Learning Test (RAVLT) • Brief Visuospatial Memory Test-Revised (BVMT-R)
Attention, Processing Speed & Executive Functions	• Wechsler Adult Intelligence Scale-V (WAIS-V) Digit Span • Phonemic and Semantic Verbal Fluency • Trail Making Test A & B (TMT) • Symbol-Digit Modalities Test (SDMT) • Stroop Color and Word Test
Visuoperceptual & Visuoconstructional Abilities	• Rey Complex Figure Test - Copy (RCFT-Copy)
Psychomotor Testing	• Grooved Pegboard Test (GPT)
Psychological Functioning	• Geriatric Depression Scale (GDS)

Finally, the neuropsychologist should adjust expectations for post-intervention cognitive improvement based on the individual patient's full clinical history. As previously discussed, many older patients will have some baseline cognitive impairment from etiologies other than NPH, which can temper the degree to which cognitive improvement is detectable on repeat, post-intervention testing. Figure 9.7 includes a clinical case example that summarizes an inpatient neuropsychological workup of a patient with suspected NPH.

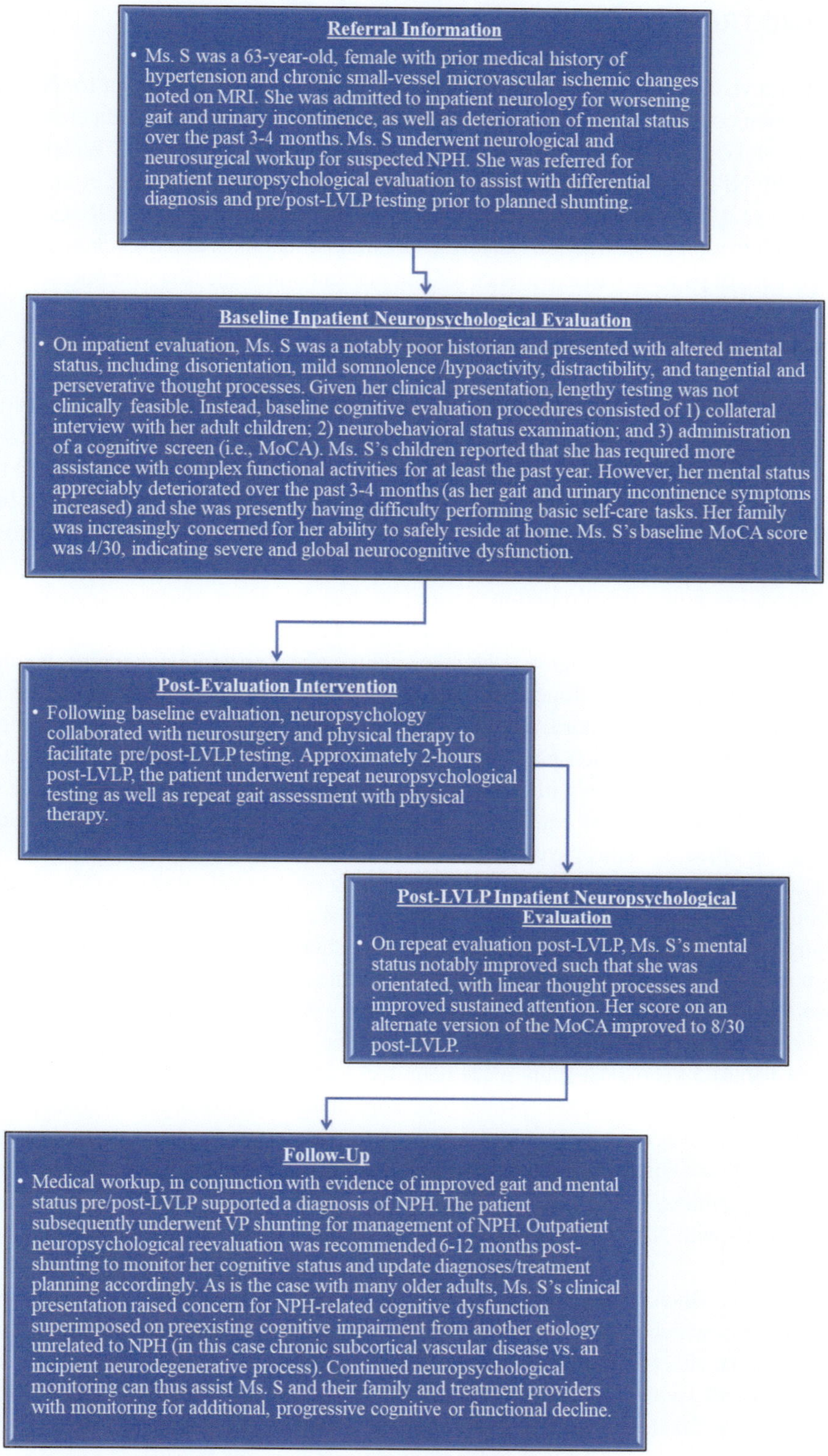

Fig. 9.7 Inpatient neuropsychological workup of an older adult with suspected NPH

INTERVENTION AND TREATMENT

Surgical placement of a ventriculoperitoneal shunt (VPS) is the most common treatment for NPH (Oliveira et al. 2019). A ventriculoperitoneal shunt consists of a ventricular catheter that is typically placed in the lateral ventricle and is connected to a valve and then to a distal catheter which is placed into the peritoneal cavity (abdomen). Widely accepted as the first-line treatment for patients with NPH, VPS provides continuous drainage of excess CSF (Soler et al. 2018) with relatively few absolute contraindications (e.g., infection in the CSF or at the entry site; Fowler et al. 2023). Risks of VPS placement include shunt malfunction due to infection or obstruction, seizures, subdural hematomas, and over-drainage. VPS placement can be up to 90% effective, 11–25% of patients will experience a shunt malfunction in the first year after shunt placement (Paff et al. 2018).

Typically, surgical treatment with the insertion of a VPS is only indicated for those patients who experience clear improvement in gait, usually with concurrent improvement in urinary symptoms, following an LVLP or lumbar drain procedure. This clinical indication is based on research showing that gait and balance difficulties are the most likely to improve following intervention and have over 90% accuracy in predicting positive post-shunt outcome (Marmarou et al. 2005). By contrast, studies examining cognitive outcomes show greater variability in improvement following LVLP, lumbar drain, or shunting, with around 50% of patients demonstrating improvements in mental status following shunt procedure (Solana et al. 2012; Thomas et al. 2005). Inconsistencies in measurement of cognitive functioning across studies may account for discrepancies in the literature, resulting in a call for greater protocol standardization (Acosta et al. 2021; da Rocha et al. 2021). Notwithstanding this limitation, several studies document evidence of improvement in at least one cognitive domain post-shunting (e.g., Katzen et al. 2011; McGovern et al. 2019; Peterson et al. 2016). Furthermore, Nimni et al. (2021) found that due to subtlety in cognitive change occurring across various domains, a combination of different tests—namely tests of memory, executive functions, and naming—is most sensitive in detecting meaningful cognitive outcomes following intervention. Furthermore, some evidence suggests that global baseline cognition as measured by the MMSE may have utility in predicting which patients cognitively improve post-shunting, as patients with lower baseline global cognitive functioning, such as those with MCI or dementia, are less likely to demonstrate cognitive improvement (Solana et al. 2012; Thomas et al. 2005). By contrast, CSF biomarkers were not found to be a significant predictor of post-shunt cognitive changes on objective neuropsychological testing (McGovern et al. 2019).

Early intervention also appears particularly important in maximizing post-shunting cognitive outcomes (Kambara et al. 2021). That said, although a preponderance of the literature suggests shunting results in varying degrees of immediate cognitive improvement, long-term cognitive outcomes associated with shunt-responsive NPH are more elusive. For instance, Kambara and colleagues found that 77% of NPH patients demonstrate postsurgical cognitive improvement at 3 months and 1 year and maintained these gains at 2 years post-shunting (Kambara et al. 2021). Younger age and milder NPH symptoms were predictive of a better cognitive prognosis. By contrast, another study examining a large sample of shunt-responsive NPH patients with a median of 5 years post shunt placement documented that 80% had declined cognitively and 46% met criteria for clinical dementia with age, male sex, primary memory deficits, and time duration between shunting and follow up emerging as significant predictors of cognitive decline (Koivisto et al. 2018). Interestingly, a small subset of patients who declined (i.e., 5%) had no other clinical, radiological, or histopathological indicators of a concomitant neurodegenerative process. Considering these findings, longitudinal monitoring and serial neuropsychological reevaluation on an outpatient basis is warranted for patients with NPH who have undergone shunting, even if the initial post-shunt cognitive outcome was favorable.

Another alternative to VPS, which is continuing to gain empirical support, is an endoscopic third ventriculostomy (ETV). ETV is a surgical procedure in which a stoma, or hole, is created in the floor of the third ventricle to allow for the drainage of CSF into the interpeduncular cistern. Overall success rate has been estimated to be as high as 61%, with average complication rate estimated to be 4.86%, with a higher risk for initial failure than shunting but lower risk of later complications (Tasiou et al. 2016; Isaacs et al. 2017; Yadav et al. 2021). Finally, a medication called acetazolamide, a carbonic anhydrase inhibitor, can also be prescribed to inhibit CSF production in patients who are considered non-surgical candidates or who decline surgery (Del Bigio and Di Curzio 2015).

SUMMARY

NPH is a reasonably prevalent, though often underdiagnosed cause of cognitive and physical morbidity in older adults resulting from gradual accumulation of excess CSF within the ventricles, leading to progressive ventricular enlargement and shifting brain parenchyma. Although multiple diagnostic criteria/guidelines exist, co-occurrence of any combination of the classic triad of progressively impaired ambulation (particularly magnetic gait), urinary incontinence, and cognitive dysfunction/neuropsychiatric changes should prompt further evaluation for NPH. Unlike other progressive neurodegenerative conditions, the physical and cognitive sequelae associated with NPH can be treated, to varying degrees, with neurosurgical intervention/shunting, thereby reducing morbidity and increasing patient quality of life.

Although neuropsychological evaluation plays a critical role in differential diagnosis and pre/post-intervention test in patients with known or suspected NPH, several key areas of future research remain. First, more head-to-head comparison studies using a uniform cognitive test battery are needed to inform evidence-based practice by elucidating which test(s) offer the greatest sensitivity for reliably detecting pre/post-treatment changes as well as development of NPH-specific RCIs. Relatedly, the diagnostic accuracy of cognitive screeners can be enhanced when combined with other independent data (e.g., functional measures; Gonzalez et al. 2022b) while simultaneously minimizing patient burden. Finally, improved identification of risk factors for post-shunt, long-term cognitive decline will refine the prediction of surgical outcomes and maximize patient (and caregiver) quality of life.

REFERENCES

Acosta LMY, Stubblefield K, Conwell T, Espaillat K, Koons H, Konrad P, Fang J, Kirshner H, Davis T. Protocolizing the workup for idiopathic normal pressure hydrocephalus improves outcomes. Neurol Clin Pract. 2021;11(4):e447–53.

Al Farsi RS, Al Alawi AM, Al Huraizi AR, Al-Saadi T, Al-Hamadani N, Al Zeedy K, Al-Maqbali JS. Delirium in medically hospitalized patients: prevalence, recognition and risk factors: a prospective cohort study. J Clin Med. 2023;12(12):3897.

Andersson J, Rosell M, Kockum K, Söderström L, Laurell K. Challenges in diagnosing normal pressure hydrocephalus: evaluation of the diagnostic guidelines. ENeurologicalSci. 2017;7:27–31.

Bladowska J, Sąsiadek MJ. Obstructive hydrocephalus in adults. In: Barkhof F, Jager R, Thurnher M, Rovira Cañellas A, editors. Clinical neuroradiology: the ESNR textbook. Springer; 2018. p. 1–23.

Blausen B. Ventricles of the brain. In Medical gallery of Blausen Medical 2014. WikiJ Med. 2014;1(2) https://doi.org/10.15347/wjm/2014.010. ISSN 2002-4436. CC BY 3.0

Blumenfeld H. Neuroanatomy through clinical cases. 3rd ed. Sinauer Associates/Oxford University Press; 2021.

Chelune GJ, Naugle RI, Luders H, Sedlak J, Awad IA. Individual change after epilepsy surgery: practice effects and base-rate information. Neuropsychology. 1993;7:41–52.

Constantinescu C, Wikkelsø C, Westman E, Ziegelitz D, Jaraj D, Rydén L, Skoog I, Tullberg M. Prevalence of possible idiopathic normal pressure hydrocephalus in Sweden: a population-based MRI study in 79,170 70-year-old participants. Neurology. 2024;102(2):e208037.

Crowell RM, Tew JM Jr, Mark VH. Aggressive dementia associated with normal pressure hydrocephalus: report of two unusual cases. Neurology. 1973;23(5):461–4.

da Rocha SFB, Kowacs PA, de Souza RKM, Pedro MKF, Ramina R, Teive HAG. Serial tap test of patients with idiopathic normal pressure hydrocephalus: impact on cognitive function and its meaning. Fluids Barriers CNS. 2021;18(1):22.

Damasceno BP. Normal pressure hydrocephalus: diagnostic and predictive evaluation. Dement Neuropsychol. 2009;3(1):8–15.

Damasceno BP. Neuroimaging in normal pressure hydrocephalus. Dement Neuropsychol. 2015;9(4):350–5.

Del Bigio MR, Di Curzio DL. Nonsurgical therapy for hydrocephalus: a comprehensive and critical review. Fluids Barriers CNS. 2015;13:1–20.

DeVito EE, Pickard JD, Salmond CH, Iddon JL, Loveday C, Sahakian BJ. The neuropsychology of normal pressure hydrocephalus (NPH). Br J Neurosurg. 2005;19(3):217–24.

Evans WA. An encephalographic ratio for estimating ventricular enlargement and cerebral atrophy. Arch Neurol Psychiatr. 1942;47(6):931–7.

Fowler JB, De Jesus O, Mesfin FB. Ventriculoperitoneal shunt. StatPearls; 2023.

Gaillard F, Murphy A, Lukies M, et al. Callosal angle. rID: 34874. 2015. Radiopaedia.org.

Gonzalez DA, Gonzales MM, Jennette KJ, Soble JR, Fongang B. Cognitive screening with functional assessment improves diagnostic accuracy and attenuates bias. Alzheimers Dement (Amst). 2022a;13(1):e12250.

Gonzalez DA, Resch ZJ, Gonzales MM, Soble JR. A novel method for establishing functional change in older adults with cognitive impairment. Alzheimer Dis Assoc Disord. 2022b;36:238–43.

Goode PS, Burgio KL, Redden DT, Markland A, Richter HE, Sawyer P, Allman RM. Population based study of incidence and predictors of urinary incontinence in black and white older adults. J Urol. 2008;179(4):1449–54.

Hellström P, Edsbagge M, Archer T, Tisell M, Tullberg M, Wikkelsø C. The neuropsychology of patients with clinically diagnosed idiopathic normal pressure hydrocephalus. Neurosurgery. 2007;61(6):1219–28.

Hellström P, Edsbagge M, Blomsterwall E, Archer T, Tisell M, Tullberg M, Wikkelsø C. Neuropsychological effects of shunt treatment in idiopathic normal pressure hydrocephalus. Neurosurgery. 2008;63(3):527–36.

Hülser M, Spielmann H, Oertel J, Sippl C. Motor skills, cognitive impairment, and quality of life in normal pressure hydrocephalus: early effects of shunt placement. Acta Neurochir. 2022;164(7):1765–75.

Iddon JL, Pickard JD, Cross JJ, Griffiths PD, Czosnyka M, Sahakian BJ. Specific patterns of cognitive impairment in patients with idiopathic normal pressure hydrocephalus and Alzheimer's disease: a pilot study. J Neurol Neurosurg Psychiatry. 1999;67(6):723–32. https://doi.org/10.1136/jnnp.67.6.723.

Isaacs AM, Williams MA, Hamilton MG. Hydrocephalus in the elderly: Surgical management of idiopathic normal pressure hydrocephalus. In: Berhouma M, Krolak-Salmon P, editors. Brain and spine surgery in the elderly. Springer; 2017. p. 211–26.

Isaacs AM, Riva-Cambrin J, Yavin D, Hockley A, Pringsheim TM, Jette N, Lethebe BC, Lowerison M, Dronyk J, Hamilton MG. Age-specific global epidemiology of hydrocephalus: systematic review, metanalysis and global birth surveillance. PLoS One. 2018;13(10):e0204926. https://doi.org/10.1371/journal.pone.0204926.

Ishikawa M, Hashimoto M, Kuwana N, Mori E, Miyake H, Wachi A, Takeuchi T, Kazui H, Koyama H. Guidelines for management of idiopathic normal pressure hydrocephalus. Neurol Med Chir (Tokyo). 2008;48(Suppl):S1–S23.

Jahn K, Zwergal A, Schniepp R. Gait disturbances in old age: classification, diagnosis, and treatment from a neurological perspective. Deutsches Arzteblatt Int. 2010;107(17):306–16. https://doi.org/10.3238/arztebl.2010.0306.

Jaraj D, Rabiei K, Marlow T, Jensen C, Skoog I, Wikkelsø C. Prevalence of idiopathic normal-pressure hydrocephalus. Neurology. 2014;82(16):1449–54.

Kambara A, Kajimoto Y, Yagi R, Ikeda N, Furuse M, Nonoguchi N, Kawabata S, Kuroiwa T, Kuroda K, Tsuji S, Saura R, Wanibuchi M. Long-term prognosis of cognitive function in patients with idiopathic normal pressure hydrocephalus after shunt surgery. Front Aging Neurosci. 2021;12:617150.

Katzen H, Ravdin LD, Assuras S, Heros R, Kaplitt M, Schwartz TH, Fink M, Levin E, Relkin NR. Postshunt cognitive and functional improvement in idiopathic normal pressure hydrocephalus. Neurosurgery. 2011;68(2):416–9.

Kim M, Park SW, Lee JY, Kim H, Rhim JH, Park S, Lee JY, Son H, Kim YK, Lee SH. Differences in brain morphology between hydrocephalus ex vacuo and idiopathic normal pressure hydrocephalus. Psychiatry Investig. 2021;18(7):628–35.

Koivisto AM, Alafuzoff I, Savolainen S, Sutela A, Rummukainen J, Kurki M, Jääskeläinen JE, Soininen H, Rinne J, Leinonen V, Kuopio NPH Registry. Cognitive and histopathological changes in idiopathic normal pressure hydrocephalus. Acta Neuropathol Commun. 2018;6(1):1–9.

Koleva M, De Jesus O. Hydrocephalus. National Library of Medicine; 2023. https://www.ncbi.nlm.nih.gov/pmc/articles/PMC6731849/

Krauss JK, Halve B. Normal pressure hydrocephalus: survey on contemporary diagnostic algorithms and therapeutic decision-making in clinical practice. Acta Neurochir. 2004;146(4):379–88.

Kuriyama N, Miyajima M, Nakajima M, Kurosawa M, Fukushima W, Watanabe Y, Ozaki E, Hirota Y, Tamakoshi A, Mori E, Kato T, Tokuda T, Urae A, Arai H. Nationwide hospital-based survey of idiopathic normal pressure hydrocephalus in Japan: epidemiological and clinical characteristics. Brain Behav. 2017;7(3):e00635.

Laidet M, Herrmann FR, Momjian S, Assal F, Allali G. Improvement in executive subfunctions following cerebrospinal fluid tap test identifies idiopathic normal pressure hydrocephalus from its mimics. Eur J Neurol. 2015;22(12):1533–9.

Langheinrich T, Chen C, Thomas O. Update on the cognitive presentations of iNPH for clinicians. Front Neurol. 2022;13:894617.

Lenfeldt N, Larsson A, Nyberg L, Andersson M, Birgander R, Eklund A, Malm J. Idiopathic normal pressure hydrocephalus: increased supplementary motor activity accounts for improvement after CSF drainage. Brain. 2008;131(Pt 11):2904–12.

Maassen GH, Bossema E, Brand N. Reliable change and practice effects: outcomes of various indices compared. J Clin Exp Neuropsychol. 2009;31:339–52.

Malm J, Graff-Radford NR, Ishikawa M, Kristensen B, Leinonen V, Mori E, Owler BK, Tullberg M, Williams MA, Relkin NR. Influence of comorbidities in idiopathic normal pressure hydrocephalus - research and clinical care. A report of the ISHCSF task force on comorbidities in INPH. Fluids Barriers CNS. 2013;10(1):22.

Marmarou A, Young HF, Aygok GA, Sawauchi S, Tsuji O, Yamamoto T, Dunbar J. Diagnosis and management of idiopathic normal-pressure hydrocephalus: a prospective study in 151 patients. J Neurosurg. 2005;102(6):987–97.

Mathew R, Pavithran S, Byju P. Neuropsychiatric manifestations of cognitively advanced idiopathic normal pressure hydrocephalus. Dement Geriatr Cogn Disord Extra. 2018;8(3):467–75.

Mayorga P, Hurtado C, Gempeler A, et al. Cognitive changes after tap test in patients with normal pressure hydrocephalus: a before-and-after study. Discov Psychol. 2024;4:21.

McGirt MJ, Woodworth G, Coon AL, Thomas G, Williams MA, Rigamonti D. Diagnosis, treatment, and analysis of long-term outcomes in idiopathic normal-pressure hydrocephalus. Neurosurgery. 2005;57(4):699–705.

McGrath K, Laurent D, Otero O, Hey G, Tomdio M, Sorrentino Z, Riklan J, Chowdhury MAB, Isom E, Schreffler A, Musalo M, Rahman M. An Interdisciplinary Protocol for Ventriculoperitoneal Shunt Patient Selection in Normal Pressure Hydrocephalus. World neurosurgery. 2024;187:e1–e11. https://doi.org/10.1016/j.wneu.2024.04.124.

McGovern RA, Nelp TB, Kelly KM, Chan AK, Mazzoni P, Sheth SA, Honig LS, Teich AF, McKhann GM. Predicting cognitive improvement in normal pressure hydrocephalus patients using preoperative neuropsychological testing and cerebrospinal fluid biomarkers. Neurosurgery. 2019;85(4):E662–9.

Micchia K, Formica C, De Salvo S, Muscarà N, Bramanti P, Caminiti F, Marino S, Corallo F. Normal pressure hydrocephalus: neurophysiological and neuropsychological aspects: a narrative review. Medicine. 2022;101(9):e28922.

Mori E, Ishikawa M, Kato T, Kazui H, Miyake H, Miyajima M, Nakajima M, Hashimoto M, Kuriyama N, Tokuda T, Ishii K, Kaijima M, Hirata Y, Saito M, Arai H, Japanese Society of Normal Pressure Hydrocephalus. Guidelines for management of idiopathic normal pressure hydrocephalus: Second edition. Neurol Med Chir (Tokyo). 2012;52(11):775–809.

Nakajima M, Yamada S, Miyajima M, Ishii K, Kuriyama N, Kazui H, et al. Guidelines for management of idiopathic normal pressure hydrocephalus: endorsed by the Japanese society of normal pressure hydrocephalus. Neurol Med Chir. 2021;61(2):63–97.

Nasreddine ZS, Phillips NA, Bédirian V, Charbonneau S, Whitehead V, Collin I, et al. The Montreal Cognitive Assessment, MoCA: a brief screening tool for mild cognitive impairment. J Am Geriatr Soc. 2005;53(4):695–9.

Nimni M, Weiss P, Cohen C, Laviv Y. Neuropsychological assessments and cognitive profile mostly associated with shunt surgery in idiopathic normal pressure hydrocephalus patients: diagnostic and predictive parameters and practical implications. Acta Neurochir. 2021;163(12):3373–86.

Oliveira LM, Nitrini R, Román GC. Normal-pressure hydrocephalus: a critical review. Dement Neuropsychol. 2019;13(2):133–43.

Paff M, Alexandru-Abrams D, Muhonen M, Loudon W. Ventriculoperitoneal shunt complications: a review. Interdiscip Neurosurg. 2018;13:66–70.

Passos-Neto CEB, Lopes CCB, Teixeira MS, Studart Neto A, Spera RR. Normal pressure hydrocephalus: an update. Arq Neuropsiquiatr. 2022;80(5 Suppl 1):42–52.

Patel S, Ditamo M, Mangal R, Gould M, Ganti L. Normal pressure hydrocephalus. Cureus. 2023;15(2):e35131.

Peterson KA, Savulich G, Jackson D, Killikelly C, Pickard JD, Sahakian BJ. The effect of shunt surgery on neuropsychological performance in normal pressure hydrocephalus: a systematic review and meta-analysis. J Neurol. 2016;263(8):1669–77.

Picascia M, Zangaglia R, Bernini S, Minafra B, Sinforiani E, Pacchetti C. A review of cognitive impairment and differential diagnosis in idiopathic normal pressure hydrocephalus. Funct Neurol. 2015;30(4):217–28.

Pierson EW, Thompson JR, Wolcott GJ. Normal pressure hydrocephalus: a treatable cause of dementia. Nebr Med J. 1982;67(1):6.

Price TR, Tucker GJ. Psychiatric and behavioral manifestations of normal pressure hydrocephalus: a case report and brief review. J Nerv Ment Dis. 1977;164(1):51–5.

Ransohoff J, Shulman K, Fishman RA. Hydrocephalus: a review of etiology and treatment. J Pediatr. 1960;56:399–411.
Rekate H. The definition and classification of hydrocephalus: a personal recommendation to stimulate debate. Cerebrospinal Fluid Res. 2008;5(2)
Rekate HL. A contemporary definition and classification of hydrocephalus. Semin Pediatr Neurol. 2009;16(1):9–15.
Relkin N, Marmarou A, Klinge P, Bergsneider M, Black P. Diagnosing idiopathic normal-pressure hydrocephalus. Neurosurgery. 2005;57(3):S2–S16.
Rendtorff R, Novak A, Tunn R. Normal pressure hydrocephalus as cause of urinary incontinence: a shunt for incontinence. Geburtshilfe Frauenheilkd. 2012;72(12):1130–1.
Saito M, Nishio Y, Kanno S, Uchiyama M, Hayashi A, Takagi M, Kikuchi H, Yamasaki H, Shimomura T, Iizuka O, Mori E. Cognitive profile of idiopathic normal pressure hydrocephalus. Dement Geriatr Cogn Disord Extra. 2011;1(1):202–11.
Sakakibara R, Kanda T, Sekido T, Uchiyama T, Awa Y, Ito T, Liu Z, Yamamoto T, Yamanishi T, Yuasa T, Shirai K, Hattori T. Mechanism of bladder dysfunction in idiopathic normal pressure hydrocephalus. Neurourol Urodyn. 2008;27(6):507–10.
Saxena S, Lawley D. Delirium in the elderly: a clinical review. Postgrad Med J. 2009;85(1006):405–13.
Schmidt H, Elster J, Eckert I, Wiefek J, Paulus W, von Steinbuechel N, Abatih EN, Blocher J. Cognitive functions after spinal tap in patients with normal pressure hydrocephalus. J Neurol. 2014;261(12):2344–50.
Shen MD. CSF Circulation. In cerebrospinal fluid and the early brain development of autism. J Neurodev Disord. 2018;10(1):39. https://doi.org/10.1186/s11689-018-9256-7. CC BY 4.0
Solana E, Sahuquillo J, Junqué C, Quintana M, Poca MA. Cognitive disturbances and neuropsychological changes after surgical treatment in a cohort of 185 patients with idiopathic normal pressure hydrocephalus. Arch Clin Neuropsychol. 2012;27(3):304–17.
Soler GJ, Bao M, Jaiswal D, Zaveri HP, DiLuna ML, Grant RA, Hoshino K. A review of cerebral shunts, current technologies, and future endeavors. Yale J Biol Med. 2018;91(3):313–21.
Srivastava S, Bhatia MS, Gautam P. Normal pressure hydrocephalus presenting as delusional disorder. Ind Psychiatry J. 2015;24(1):94–6.
Stolze H, Kuhtz-Buschbeck JP, Drücke H, Jöhnk K, Illert M, Deuschl G. Comparative analysis of the gait disorder of normal pressure hydrocephalus and Parkinson's disease. J Neurol Neurosurg Psychiatry. 2001;70(3):289–97.
Sweet JJ, Heilbronner RL, Morgan JE, Larrabee GJ, Rohling ML, Boone KB, Kirkwood MW, Schroeder RW, Suhr JA, Conference Participants. American Academy of Clinical Neuropsychology (AACN) 2021 consensus statement on validity assessment: update of the 2009 AACN consensus conference statement on neuropsychological assessment of effort, response bias, and malingering. Clin Neuropsychol. 2021;35(6):1053–106.
Tasiou A, Brotis AG, Esposito F, Paterakis KN. Endoscopic third ventriculostomy in the treatment of idiopathic normal pressure hydrocephalus: a review study. Neurosurg Rev. 2016;39(4):557–63.
Thomas G, McGirt MJ, Woodworth G, Heidler J, Rigamonti D, Hillis AE, Williams MA. Baseline neuropsychological profile and cognitive response to cerebrospinal fluid shunting for idiopathic normal pressure hydrocephalus. Dement Geriatr Cogn Disord. 2005;20(2–3):163–8.
US Preventive Services Task Force. Screening for cognitive impairment in older adults: US preventive services task force recommendation statement. JAMA. 2020;323(8):757–63.
Venkatesh M, Yap J, Gaillard F, et al. Evans index. rID:26125. 2013. Radiopaedia.org.
Verrees M, Selman WR. Management of normal pressure hydrocephalus. Am Fam Physician. 2004;70(6):1071–8.
Xiao H, Hu F, Ding J, Ye Z. Cognitive impairment in idiopathic normal pressure hydrocephalus. Neurosci Bull. 2022;38(9):1085–96.
Yadav YR, Bajaj J, Ratre S, Yadav N, Parihar V, Swamy N, et al. Endoscopic third ventriculostomy: a review. Neurol India. 2021;69(Suppl 2):S502–13.

Rare and Rapidly Progressive Syndromes

10

Emily N. Vanderbleek, Daniel S. Weitzner, and Joseph J. Boscarino

ACRONYMS AND ABBREVIATIONS

Aβ	Amyloid Beta
ACR	American College of Rheumatologists
AD	Alzheimer's Disease
ADC	Apparent Diffusion Coefficient
ADEM	Acute Disseminated Encephalomyelitis
AHLE	Acute Hemorrhagic Leukoencephalitis
AKI	Acute Kidney Injury
AMA	Against Medical Advice
ANAM	Automated Neuropsychological Assessment Metrics
APOE	Apolipoprotein E
APP	Amyloid Precursor Protein
BSE	Bovine Spongiform Encephalopathy
CA	Cerebral Catheter Angiography
CAA	Cerebral Amyloid Angiopathy
CAA-RI	CAA-related Inflammation
CADASIL	Cerebral Autosomal Dominant Arteriopathy with Subcortical Infarcts and Leukoencephalopathy
CARASIL	Cerebral Autosomal Recessive Arteriopathy with Subcortical Infarcts and Leukoencephalopathy
CDT	Clock Drawing Test
CJD	Creutzfeldt-Jakob Disease
CNS	Central Nervous System
CPM	Central Pontine Myelinolysis
CSF	Cerebrospinal Fluid
CVLT3	California Verbal Learning Test-3
DLB	Dementia with Lewy Bodies
DSA	Digital Subtraction Angiography
DWI	Diffusion Weighted Imaging
EBV	Epstein-Barr Virus
ED	Emergency Department
EEEV	Eastern Equine Encephalitis Virus

E. N. Vanderbleek (✉) · J. J. Boscarino
Mental Health and Behavioral Science Service (MH/BSS), James A. Haley Veterans' Hospital, Tampa, FL, USA
e-mail: Joseph.Boscarino@va.gov

D. S. Weitzner
Madonna Ptak Memory Disorders Clinic, BayCare Medical Group, Clearwater, FL, USA
e-mail: Daniel.Weitzner@baycare.org

E. K. Bailey et al. (eds.), *Inpatient Neuropsychology*, https://doi.org/10.1007/978-3-032-05880-5_10

EEG	Electroencephalogram
EPM	Extrapontine Myelinolysis
FLAIR	Fluid Attenuated Inversion Recovery
FSIQ	Full-Scale Intelligence Quotient
GRE	Gradient Recalled Echo
HE	Hashimoto's Encephalopathy
HIV	Human Immunodeficiency Virus
HSE	Herpes Simplex Encephalitis
HSV	Herpes Simplex Virus
ICH	Intracerebral Hemorrhage
IVIG	Intravenous Immunoglobulin
JCV	John Cunningham Virus
LDH	Lactate Dehydrogenase
LP	Lumbar Puncture
MDEM	Multiphasic Disseminated Encephalomyelitis
MoCA	Montreal Cognitive Assessment
MRI	Magnetic Resonance Imaging
MS	Multiple Sclerosis
NBD	Neuro-Behçet Disease
NPSLE	Neuropsychiatric Systemic Lupus Erythematosus
NTZ	Natalizumab-Treated
ODS	Osmotic Demyelination Syndrome
PCR	Polymerase Chain Reaction
PET	Positron Emission Tomography
PML	Progressive Multifocal Leukoencephalopathy
PRES	Posterior Reversible Encephalopathy Syndrome
PRNP	Prion Protein Gene
RAVLT	Rey Auditory Verbal Learning Test
RBANS	Repeatable Battery for the Assessment of Neuropsychological Status
RCFT	Rey Osterrieth Complex Figure Test
RTQuIC	Real-time Quaking-Induced Conversion
PrPSc	Prion Protein Scrapie
SCWT	Stroop Color and Word Test
SDMT	Symbol Digit Modalities Test
SLE	Systemic Lupus Erythematosus
SWI	Susceptibility-Weighted Imaging
TIA	Transient Ischemic Attack
TMT	Trail Making Test
TSEs	Transmissible Spongiform Encephalopathies
US	United States
vCJD	Variant CJD
VCI	Vascular Cognitive Impairment
WAIS-IV	Wechsler Abbreviated Scale of Intelligence-IV- 4th Edition
WBC	White Blood Cell
WMS-IV	Wechsler Memory Scale—4th Edition
WNV	West Nile Virus
WhD	Whipple's Disease
WD	Wilson's Disease

INTRODUCTION

Rare and rapidly progressive syndromes prove to be some of the most perplexing diagnostic puzzles, and many clinicians do not receive training in their identification. With these syndromes, the role of the neuropsychologist may be to assist with diagnostic clarification and provide insight on prognosis. Due to symptom acuity, comprehensive cognitive testing is often of limited utility in elucidating diagnosis. Rather, establishing a clear history with a well-demarcated symptom onset, course progression, and temporal association with other medical illnesses, recent stressors, or medication changes is paramount.

Though comprehensive assessment is usually infeasible, the inpatient setting does provide access to beneficial contextual data that would not be as readily accessible in outpatient settings. Recent medical labs, neuroimaging, and electroencephalogram (EEG) results are more frequently available in the inpatient setting and can help to refine diagnosis, as well as identify what has likely already been ruled out from the current work-up. Highly detailed clinical interviews and obtaining thorough history from collateral informants are also critical aspects of inpatient neuropsychology. To this end, the neuropsychologist is uniquely poised to offer extended time for gathering background information that may have been missed in routine care.

It is important to keep in mind that when assessing patients with rapid decline, the well-known clinical adage still applies: "*when you hear hoofbeats, think horses, not zebras*." Using the clinical presentation in conjunction with a hypothesis-driven approach can help rule out subacute worsening of gradually progressive disorders and reduce misdiagnosis. For this reason—in addition to comprehensive work-up—population-based rates should also be considered when weighing potential etiologies (Fig. 10.1).

Common disorders with atypical manifestations, such as variants of neurodegenerative diseases (Alzheimer's disease [AD], dementia with Lewy bodies [DLB], etc.), comprise a considerable proportion of "rapid dementia" referrals and should not be ruled out simply because they do not exactly follow the prototypical presentation. Concomitant pathologies are frequent and individuals with previous neurological compromise may be more prone to acute changes in mental status. Polypharmacy should also be considered. Borrowing from a recently developed framework (Johnson et al. 2021), clinicians should be mindful of these potential diagnostic mimics, or "chameleons", of rare and rapidly progressive syndromes. Awareness of the incipient signs and symptoms of these syndromes, or the "canaries", will also improve diagnosis. While not exhaustive in scope, this chapter will cover many rare and rapidly progressive syndromes, across vascular, metabolic, infectious, and neurodegenerative etiologies, that may be encountered in inpatient settings.

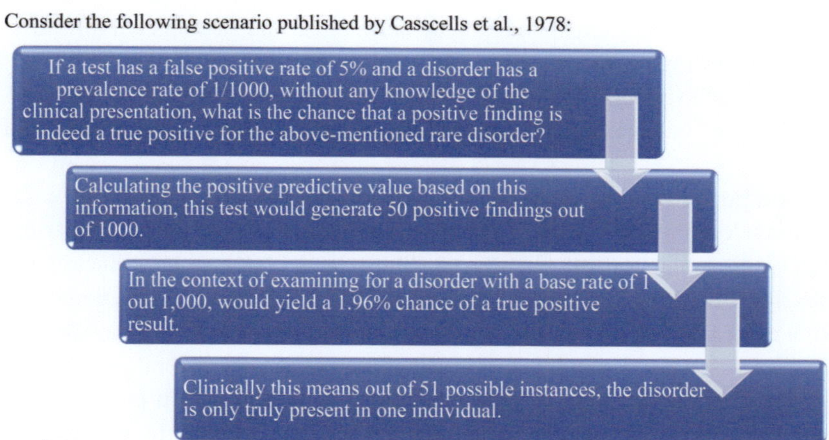

Fig. 10.1 Base rates and type 1 error

RARE AND RAPIDLY PROGRESSIVE SYNDROMES

Prion Diseases

Prion diseases, or transmissible spongiform encephalopathies (TSEs), are a group of fatal neurodegenerative diseases characterized by aberrant, misfolded proteins that recruit other proteins to misfold, cascading intra- and/or extra-cellularly in the brain and causing increasing atrophy and neuronal dysfunction. Unlike other proteinopathies, prions are infectious and transmittable. Pathognomonic of prion disease is its neuronal aggregation of misfolded scrapie prion protein (PrPSc) leading to sponge-like cerebral atrophy and cell death (Prusiner 1998). Post-mortem autopsy remains the gold standard for definitive confirmation but as will be discussed, there are other reliable and valid tests that allow for a probabilistic antemortem diagnosis to be made.

Regarded as the archetypal rapid dementia, these fatal proteinopathies typically have an expected disease duration of around 6 months after initial symptom onset, though course varies depending on the phenotype (Zerr and Poser 2002). Despite their exceptionally low incidence rate, ranging from 1 to 2 cases per million yearly worldwide or about 350 newly diagnosed cases annually in the United States (Chen and Dong 2016), prion diseases represent a sizable proportion (~13–54%) of rapid dementia referrals to tertiary clinics (Day 2022; Geschwind et al. 2008; Papageorgiou et al. 2009; Sala et al. 2012), and detection and surveillance remain important to public health given the potential for iatrogenic transmission.

Creutzfeldt-Jakob Disease

The most prevalent type of prion disease is Creutzfeldt-Jakob disease (CJD), which can be further divided into three subtypes ordered from most to least common: sporadic (~85% of CJD), genetic (~10–15% of CJD) and acquired or secondary (<1% of CJD). Other prion diseases which are also characterized by PrP misfolding include Gerstmann-Straussler-Scheinker disease (Ghetti et al. 1995; Kim et al. 2018) typified by prominent ataxia and dementia; and fatal familial insomnia, characterized by progressive insomnia, panic attacks, and hallucinations (Wu et al. 2018).

Epidemiology: Incident rates of CJD vary among subtypes. Sporadic CJD more often occurs in older adults (mean age of onset = 67 years; Uttley et al. 2020), while mean age of onset for genetic CJD is more variable (Minikel et al. 2019) but tends to be slightly younger, between 40 and 50 years. Exceptions to both above patterns have been documented (Tam et al. 2023; Yang et al. 2010). Family history of CJD would also be expected in genetic forms of CJD, most often due to autosomal-dominant inheritance of a prion protein (*PRNP*) mutation. Acquired cases of CJD are overall quite rare but more likely to occur in younger adult populations with mean age of onset being 43 years old for iatrogenic causes (e.g., neurosurgical, corneal implant, or blood transfusion procedures) and ranging between 25 and 32 years old for cause related to consumption of cattle infected by bovine spongiform encephalopathy (BSE) or "mad cow disease" (Appleby et al. 2007). Cases due to the latter have been alternatively termed variant CJD (vCJD), which was discovered formally in 1996 and has been mainly attributed to an outbreak of BSE in the United Kingdom between the 1980s and 1990s. No sex vulnerability has been noted for any subtypes of CJD.

Clinical Work-Up: Clinical guidelines have been published to aid in ante-mortem diagnosis of CJD (Zerr et al. 2009) and probable diagnosis may be rendered when expected clinical presentation and positive laboratory tests align. Neuroimaging findings supportive of CJD include the cortical ribbon sign (Shiga et al. 2004), or curvilinear (ribbon-like) appearance of increased or hyperintense signal on diffusion weighted imaging (DWI) and fluid attenuated inversion recovery (FLAIR) in the cerebral cortical gyri of at least two cortical areas, with corresponding hypointensities in the same areas on apparent diffusion coefficient (ADC) sequences (Fig. 10.2). Imaging findings may be present several years before initial symptom manifestation (Shiga et al. 2004). The pulvinar sign is highly suggestive of vCJD, which presents on FLAIR with symmetric hyperintensities in the pulvinar and/or dorsomedial nucleus of the thalamus resembling the shape of a hockey stick (Collie et al. 2003). Supportive EEG findings are periodic sharp wave complexes (Wieser et al. 2006). Cerebrospinal fluid (CSF) analysis for elevations of 14–3-3 protein have demonstrated good sensitivity but variable specificity for suspected cases of CJD (Muayqil et al. 2012), whereas real-time quaking-induced conversion (RT-QuIC) assay has exhibited stronger diagnostic accuracy for CJD across the clinical spectrum (Atarashi et al. 2011).

Symptom Course: Some combination of ataxia, myoclonus, behavioral/cognitive symptoms, pyramidal/extrapyramidal symptoms, akinetic mutism (often late-stage), and visual dysfunction with a rapid course are commonly noted, typically leading to death in 6–12 months (Zerr et al. 2009). Importantly, while considerable clinical heterogeneity exists for CJD, cognitive and affective symptoms often serve as early warning signs, with motor symptoms following in the weeks to months after initial onset of the disease (Nakatani et al. 2016). As such, atypical manifestations of other neurodegenerative diseases (e.g., Lewy body dementia, Alzheimer's disease, frontotemporal dementia, vascular dementia) represent important diagnostic considerations.

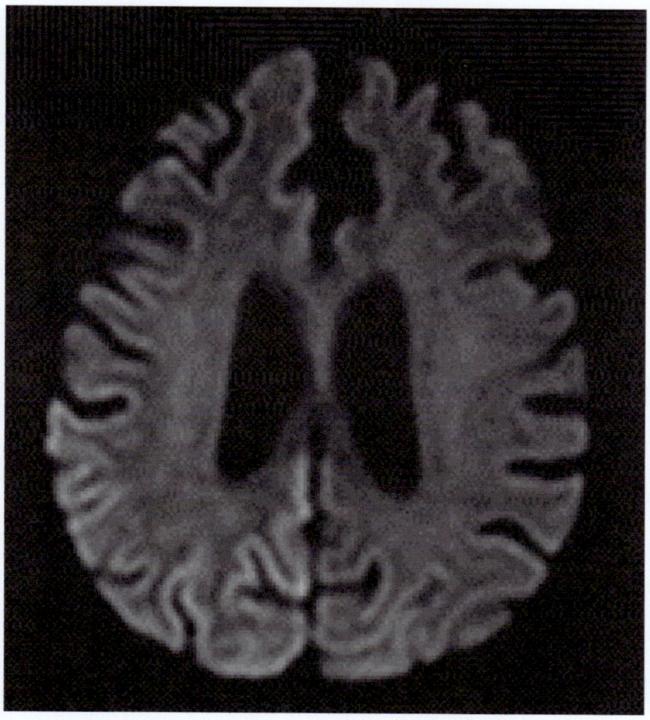

Fig. 10.2 CJD on MRI. Axial DWI showing "cortical ribboning" or abnormal signal hyperintensity of the posterior cortical gyri in a 71-year-old patient. Creutzfeldt-Jakob disease was later confirmed on autopsy

Neuropsychological Assessment: Diffuse/non-specific cognitive deficits are expected with ongoing progression; however, early on the most commonly affected domain is executive functioning with recurrent perseveration and intrusion errors (Caine et al. 2015) followed by memory (Cordery et al. 2005). Preliminary evidence suggests that visuospatial deficits may provide prognostic value in CJD, with worse performance associated with faster decline (Sundaram et al. 2020). Assessment of attention, in conjunction with history and course, may be helpful in differentiating presentations of delirium and acute mental status change, as while patients with CJD may appear unresponsive at times there are less likely to be fluctuations in arousal (Snowden et al. 2002). Clinical signs of cortical dysfunction on bedside exam, such as apraxia, finger agnosia, hemispatial neglect, right-left disorientation, alexia, and aphasia are not uncommon in CJD. Most common behavioral changes in patients with CJD include symptoms of depression and anxiety, as psychotic symptoms are less likely to be initially encountered but do not constitute a definitive rule out. In later stages, akinetic mutism typically precludes formal neuropsychological testing, but represents an important data point in differentially diagnosing CJD from other similar syndromes, such as psychiatric disorders (Fig. 10.3).

Differential Diagnosis: Ultimately, differential considerations for CJD are vast and should include a wide array of diagnostic possibilities prior to increasing suspicion for this rare etiology. Signs and symptoms that are uncommon in CJD and should prompt alternative work-up include fever, unremarkable neurological examination, generalized seizures, facial movement disorder, pleocytosis, aberrant contrast-enhancement, and T2 hyperintense lesions outside the thalamus, cortex, and striatum (Mead and Rudge 2017). Brief serial evaluations across multiple time points on successive days can also differentiate persistent fulminant decline in CJD versus fluctuating mental status and varying levels of consciousness associated with delirium, more commonly seen in prolonged hospitalizations among elderly, medically complex patients.

After ruling out other etiologies such as those listed above, sudden periods of unresponsiveness but maintained arousal may represent an incipient stage of akinetic mutism and raise suspicion for CJD (Snowden et al. 2002). Syndromes that should be considered rule-outs when suspecting akinetic mutism include mood symptoms in severely vegetative depression, respiratory difficulties as well as lack of brain stem reflexes and paralyzed voluntary movements in locked-in syndrome, and echolalia and posturing in catatonia. All the aforesaid conditions may also present with limited to no spontaneous verbal output and movements and thus are key rule outs when evaluating patients, especially when formal cognitive testing is impractical (Fig. 10.4).

Treatment: Treatment for CJD is typically symptomatic or supportive with no disease-modifying options available at present. Timely discussion of suspected CJD cases with inpatient medical teams is important to prevent undue medical testing

Mr. A was a 64-year-old man who initially presented to primary care for subjective memory complaints, difficulty managing work responsibilities, and irritability, with reported onset in the past few months. One month after the primary care visit, he was brought to the ED by his wife, who reported increasing confusion, agitation, and new gait disturbance.

- CT scan was read as showing no acute intracranial abnormalities and routine labs suggested an acute kidney injury.
- Inpatient neuropsychology was consulted to evaluate for delirium versus a neurocognitive disorder, and initial evaluation revealed prominent executive dysfunction and memory impairment. Neuropsychological differential raised initial concern for contribution from a toxic-metabolic encephalopathy, but deficits persisted even after AKI resolved.
- Follow-up outpatient assessment and MRI Brain were recommended to rule possible early-onset neurodegenerative considerations. However, Mr. A left AMA before further workup was completed, and he was lost to ambulatory follow-up.

Two months later, Mr. A was again brought to the ED due to severe cognitive decline. His wife described him as bed-bound and minimally responsive to his environment. Laboratory data was negative for any contributory abnormalities.

- Inpatient neuropsychology was consulted to provide further diagnostic clarification with MRI ordered and results pending. Formal cognitive evaluation was discontinued due to prominent akinetic mutism and immediate discussion with medical team was initiated by the inpatient neuropsychologist for rapid dementia work-up.
- Neurological exam demonstrated spastic contractions of upper and lower extremities and diffuse myoclonic twitching.
- MRI showed cortical ribboning and high signal in the caudate and putamen; positive 14-3-3 CSF assay led to a diagnosis of probable sporadic CJD and transition to palliative care unit.

Mr. A passed away several weeks later and pathology confirmed diagnosis.

Fig. 10.3 CJD case example

Creutzfeldt-Jakob disease (CJD) is the most prevalent prion disease.

- **Common signs:** Behavioral changes (depression, anxiety), ataxia, myoclonus, akinetic mutism (late-stage)
- **Supportive features:** Cortical ribboning, pulvinar sign
- **Practice guidance:** The neuropsychologist can be helpful through serial monitoring and assisting in differential diagnosis. Formal testing is usually limited.

Fig. 10.4 Summary of CJD

and iatrogenic transmission, as well as to facilitate prompt reporting to the public health authorities. Providing psychoeducation to the family about CJD will also help to inform their understanding, guide appropriate future planning, and manage expectations for the future.

Vascular Syndromes

Incidence of vascular cognitive impairment (VCI) is common. Traditionally described as having stepwise progression, it is more frequently characterized by slow, gradual decline due to increasing cerebrovascular burden. However, some uncommon vascular pathologies show more rapid onset and course, such as those outlined below.

Cerebral Amyloid Angiopathy

Cerebral amyloid angiopathy (CAA) is a type of cerebrovascular disorder characterized by the accumulation of isoforms of amyloid precursor protein (APP) into characteristic insoluble plaques known as amyloid beta (Aβ). While this pathophysiology is identical to that seen in Alzheimer's disease, CAA describes vascular Aβ deposition in the leptomeninges and cortical small blood vessels. This deposition weakens vessel walls, increasing the likelihood of microhemorrhages. CAA is described as an important cause of normotensive intracerebral hemorrhage in older adults (Yamada 2015), preferentially affecting cortical-subcortical (lobar) regions, particularly the occipital and posterior temporal lobes including white/grey matter junctions (Rosand et al. 2005).

Epidemiology: Based on pathology, the prevalence rate of CAA is estimated at ~6% in the older adult population (Jäkel et al. 2022), although not all patients will exhibit clinical symptoms. Prevalence rates of CAA are similar in East Asian and Western countries, although the prevalence of deep and mixed microbleeds is higher in East Asian populations. The reason for this increased susceptibility is unknown (De Kort et al. 2024).

Genetics: Given their shared pathologies, clinical CAA and Alzheimer's disease (AD) often overlap, with moderate-to-severe CAA pathology observed in approximately 48% of patients with AD (Greenberg et al. 2020). Like AD, the risk of CAA is associated with apolipoprotein E (APoE), although the risk is mediated depending on the specific alleles expressed. For example, those with APoE ε4 are more likely to develop CAA in capillaries (Yu et al. 2015). Alternatively, APoE ε2 is demonstrated as protective against AD pathology and thus, may moderate the clinical and pathological effects of CAA (Rannikmäe et al. 2014).

Clinical Work-Up: Like AD, the definitive diagnosis of CAA is only achieved postmortem. Clinically, gradient recalled echo (GRE) MRI can be used to visualize microhemorrhages characteristic of CAA (Figs. 10.5 and 10.6). The updated Boston criteria v2.0 offers radiographic guidelines for a diagnosis of probable CAA based on 1) the presence of strictly lobar hemorrhagic lesions (intracerebral hemorrhage [ICH], cerebral microbleeds, cortical superficial siderosis/convexity sub-

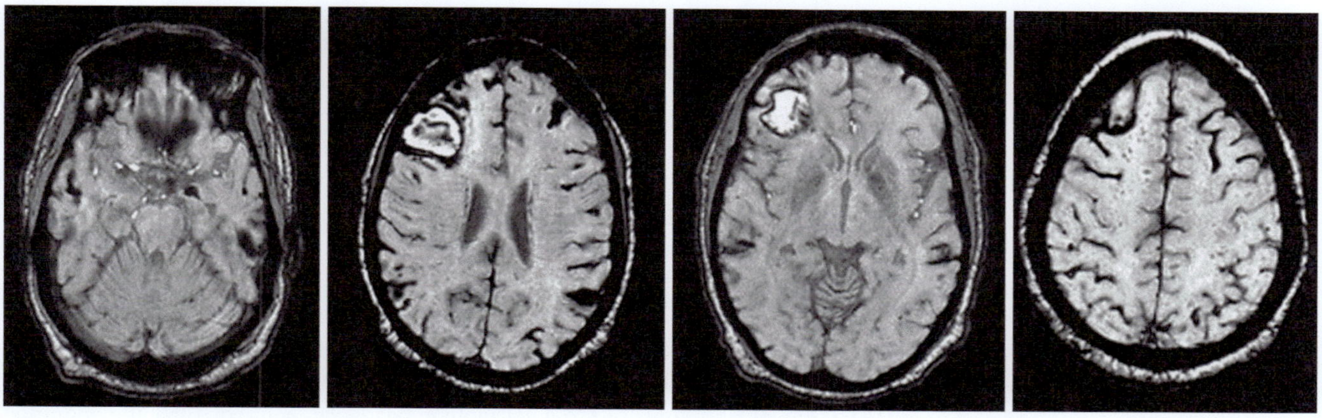

Fig. 10.5 CAA on SWI. MRI of a 62-year-old patient with suspected cerebral amyloid angiopathy. Axial SWI shows a large subacute right frontal hematoma, as well as multiple lobar-distributed microhemorrhages, remote hematomas throughout the bilateral cerebellar (center-left) and cerebral (center-right) hemispheres with encephalomalacia and hemosiderin deposition in the left inferior parietal lobe, and prominent perivascular space. Multifocal supratentorial superficial siderosis can also be appreciated on SWI (right)

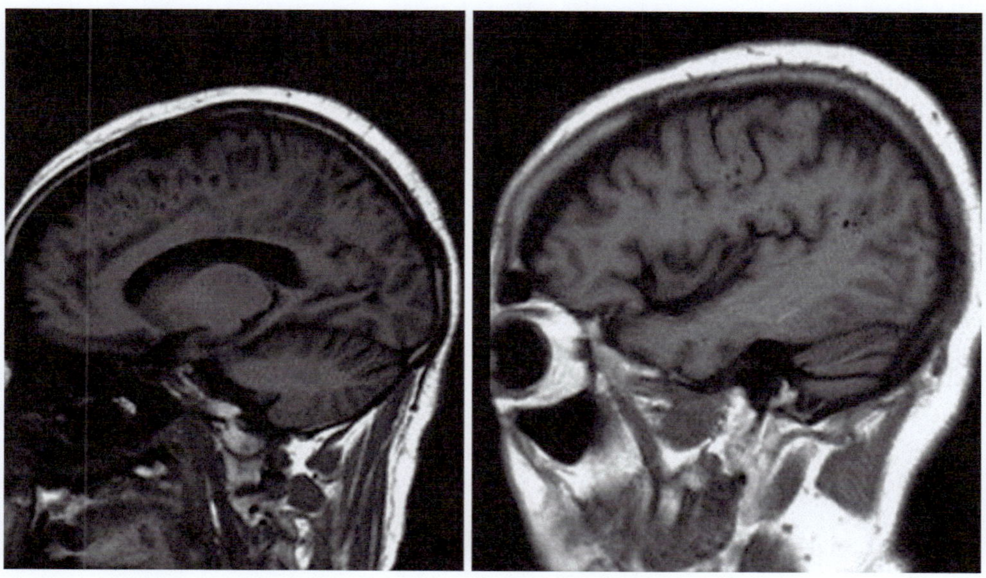

Fig. 10.6 CAA on sagittal T1. Sagittal TI of the above 62-year-old patient with suspected cerebral amyloid angiopathy

CAA is caused by amyloid beta deposition in the cerebrovasculature, weaking the artery walls and increasing risk of hemorrhage.

- **Common signs:** Headache, altered mental status, personality change, focal neurologic signs.
- **Supportive features:** Asymmetric T2-hyperintense white matter lesions, along with typical findings in CAA including the presence of multiple cerebral microbleeds.
- **Practice guidance:** CAA can be asymptomatic or show subcortical dysfunction, but rapidly progressive decline can be seen with CAA-related inflammation. Evidence of CAA on neuroimaging should not be assumed to be a definitive cause of clinical cognitive decline and should not preclude further workup.

Fig. 10.7 Summary of CAA

arachnoid hemorrhage) or white matter abnormalities (severe visible perivascular spaces in centrum semiovale or 2) white matter hyperintensities) in a multi-spot pattern (Charidimou et al. 2022).

Neuropsychological Assessment: Cognitive changes in CAA vary due to the co-occurrence with other disorders (e.g., AD, ICH) and heterogeneity of localized pathology. In the absence of focal vascular events (Chan et al. 2023), multiple studies have demonstrated a diffuse, subcortical cognitive profile in CAA (Case et al. 2016), with impairments in processing speed, executive functioning, and aspects of memory (Arvanitakis et al. 2011; Boyle et al. 2015; Durrani et al. 2023). In evaluating patients with CAA, the neuropsychologist can help localize lesions, characterize focal and diffuse patterns of cognitive impairment, differentially diagnose (e.g., CAA vs. small-vessel VCI), and track progression of symptoms over time (Fig. 10.7).

Symptom Course: Neurological sequelae in CAA range from stable and asymptomatic (Jäkel et al. 2022) to progressive and severe. Trajectories of decline depend on the nature of CAA-associated cerebrovascular disease. Small microhemorrhages, lobar lacunes, and microinfarcts present as clinically insidious, while large distribution, recurrent hemorrhages appear in more of a stepwise pattern. Rapid, progressive decline is usually attributed to an immune response to vascular amyloid-beta deposition (termed CAA-related inflammation [CAA-RI]).

CAA-Related Inflammation: CAA-RI is exceedingly rare. Few epidemiological studies suggest a prevalence rate of 0.13 per 100,000 population (Sakai et al. 2019), a slight male predominance, and an average age of 67 years at diagnosis (Corovic et al. 2018). Neuroimaging is characterized by asymmetric T2-hyperintense white matter lesions, along with cerebral microbleeds typical of CAA (Ronsin et al. 2016). Clinical symptoms of CAA-RI include headache, acute or subacute cognitive decline (e.g., memory decline, impaired concentration), confusion, personality/behavioral changes, focal neurological deficits (e.g., dysphagia, visual field defects, hemiparesis), and seizures (Chung et al. 2011; Grasso et al. 2021).

Some evidence suggests that the inflammatory encephalopathy in CAA-RI is responsive to high-dose glucocorticoids (Wu et al. 2021) or other immunosuppressive agents (Eng et al. 2004). However, early recognition and treatment of symptoms remains critical (Martucci et al. 2014). Patients are at risk for decline over the ensuing months following immunosuppressant treatment due to infectious complications or progressive ICHs (Grasso et al. 2021). A review of outcome data in patients treated for CAA-RI found that approximately half had some degree of residual disability and approximately one quarter had died at a 2-year follow-up interval (Corovic et al. 2018).

Posterior Reversible Encephalopathy Syndrome

Posterior reversible encephalopathy syndrome (PRES; also known as reversible posterior leukoencephalopathy syndrome) is characterized by a range of neurological deficits and distinctive neuroimaging findings. Despite this nomenclature, PRES is a hypertensive encephalopathy that is neither confined to the posterior distribution nor fully reversible.

Epidemiology: In a nationwide sample of general population medical inpatients, the frequency of PRES was estimated at 0.03% out of 21 million, though the condition is likely underdiagnosed. PRES has been observed in all age groups, though some evidence supports greater incidence in female and white populations (Buchhanolla et al. 2022; Liman et al. 2012).

Pathophysiology: PRES is attributed to a rapid increase in blood pressure (hypertensive crisis) that disrupts cerebral homeostasis, leading to cerebral hyperperfusion, vascular leakage, vasogenic edema, and blood-brain barrier dysfunction—of which the posterior circulation is more susceptible. Acute arterial hypertension and/or blood pressure fluctuations are typical. However, 20–30% of patients do not show significant hypertension, and PRES is often associated with pre-eclampsia, sepsis, immunosuppression, or cytotoxic agents (Fischer and Schmutzhard 2017).

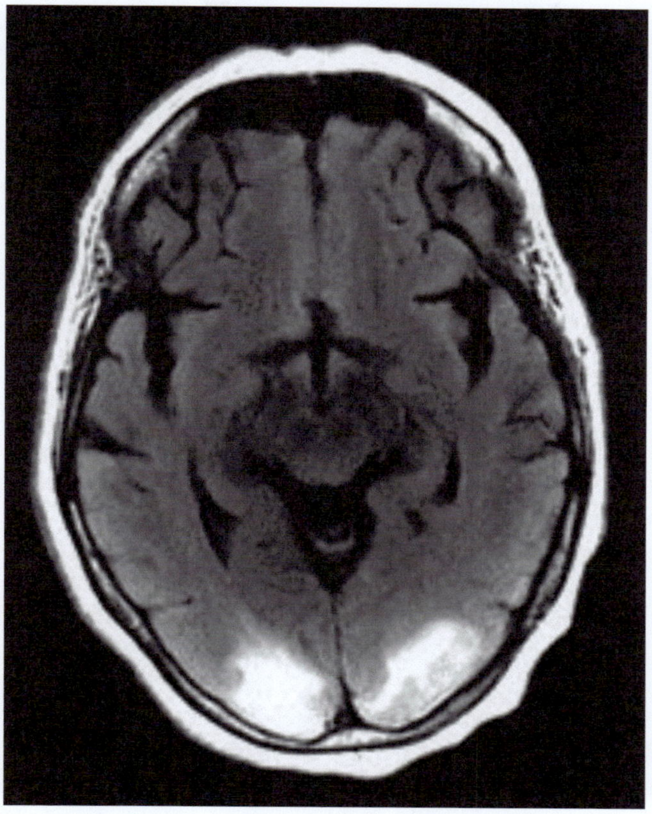

Fig. 10.8 PRES on MRI. Axial T2 FLAIR showing confluent areas of hyperintense signal within the bilateral posterior occipital lobes of a 57-year-old woman, consistent with PRES

Clinical Work-Up: While there is no gold standard diagnostic test, MRI is considered an important tool. Classically, imaging shows vasogenic edema involving the parieto-occipital regions, nearly always bilateral though may be asymmetric, and predominately affecting the white matter (Fig. 10.8). Less commonly, PRES may affect the watershed area, superior frontal sulcus, brainstem, or cerebellum. Vasoconstriction and vasospasm are frequently seen. Lab findings are highly varied depending on underlying etiology and medical comorbidities. In general, many patients demonstrate elevated lactate dehydrogenase levels (LDH), low serum albumin, and elevated CSF protein levels (Hinduja 2020).

Symptom Course: Symptoms typically manifest over several hours or days. Most patients present with delirium, ranging from mild confusion to coma, and seizures are possible early in the disease course. In rare cases, seizures may result in status epilepticus. Visual disturbances such as visual field deficits, reduced acuity, hallucinations, cortical blindness, and color vision abnormalities are common. Headache (described as dull and diffuse) and focal neurological symptoms can also occur. Unusual visual disturbances in the context of altered mental status and lab abnormalities raise suspicion for PRES. Prompt diagnosis is essential. Though PRES is termed as "reversible," persistent deficits and residual lesions have been observed in approximately 30–40% of patients. Mortality rates are higher in patients with PRES and associated cerebral infarctions (Buchhanolla et al. 2022).

Neuropsychological Assessment: In acute settings, the neuropsychologist can be helpful in identifying early symptoms, recommending additional diagnostic work-up (e.g., MRI), tracking symptom course, and providing education to the patient, family, and treatment team (Fig. 10.9). For example, in a patient with PRES-associated visual disturbance, providers may wish to place important objects in locations that are easily visible to the patient and offer verbal cueing. Few studies address post-acute neurocognitive outcomes, but declines in visuo-spatial, attention, and executive functions—as well as distribution-specific focal deficits—have been documented (Hill et al. 2023; Stroescu et al. 2011).

Treatment: Management consists of supportive care, medical treatment of the underlying pathology leading to PRES, and management of complications (e.g., anticonvulsive treatment for seizures).

PRES is commonly associated with rapid, hypertensive crisis, though other causes possible. Historically considered reversible, persistent deficits are demonstrated clinically.

- **Common signs:** Altered mental status, seizures (early in the course), visual disturbance.
- **Supportive features:** Edema in the parieto-occipital regions
- **Practice guidance:** Prompt diagnosis is essential to treat the underlying cause and complications of PRES. Visual disturbances may be noted on your neurobehavioral exam and have implications for patient care. Post-acute visuospatial deficits and attentional/executive dysfunction are common and facilitating outpatient follow-up is beneficial.

Fig. 10.9 Summary of PRES

Moyamoya

Moyamoya disease is a rare, chronic, cerebrovascular disease characterized by stenosis and occlusion of the terminal internal carotid arteries, with an unknown etiology. It has a characteristic "puff of smoke" appearance due to the tiny, wispy blood vessels that form to compensate for the reduced blood flow. Moyamoya disease is associated with acute infarcts, which may occur in relatively quick succession and thus can present with stepwise or progressive decline. In children, the first symptoms often include stroke or recurrent transient ischemic attacks (TIAs), which can cause weakness or hemiparesis. Adults are at increased risk for hemorrhagic stroke. Additional symptoms can include headaches, seizures, altered mental status, or vision impairments.

Epidemiology: Moyamoya is more common in people living in East Asian countries, specifically South Korea and Japan (Kim 2016), and more commonly seen in women (Ghaffari-Rafi et al. 2020). Incidence rates in the United States from 2008 to 2015 are estimated at 0.293 per 100,000, with South Korea having the largest incidence globally at 2.3 per 100,000 (Ghaffari-Rafi et al. 2020). The overall hospitalization rate is estimated at 10.4 cases per million, with a substantial increase in the number of patients diagnosed or hospitalized due to Moyamoya thought to be related to improvements in awareness and more advanced imaging modalities (Patel et al. 2020). In the United States, Patel et al. (2020) showed that African Americans had the highest prevalence of Moyamoya hospitalizations among all racial groups, and that Hispanic Americans evidence an increase in hospitalizations related to Moyamoya (2.8 cases per 1 million in 2010 to 11.8 cases per million in 2016). The reason for the growing proportion of hospitalized patients with Moyamoya who belong to lower socioeconomic status is unknown but is likely related to increased environmental risk factors and/or untreated medical comorbidities.

Clinical Work-Up: Moyamoya disease is associated with both ischemic and hemorrhagic events. Symptomatic patients treated conservatively have a high risk of experiencing recurrent cerebrovascular events (Gonzalez et al. 2023). Ischemia typically involves the anterior circulation; hemorrhage most commonly presents with intracranial or intraventricular patterns. Pathological involvement of the Circle of Willis is a complicating factor and may include aneurysm formation and rupture. Digital subtraction angiography (DSA), also called cerebral catheter angiography (CA), is the gold standard for diagnosis of Moyamoya disease (Zhang et al. 2022).

Neuropsychological Assessment: Cognitive impairment is common in Moyamoya, with difficulties particularly in executive functioning, processing speed, performance IQ, and visual memory (Chan et al. 2023). Specific patterns and/or focal deficits are contingent upon locality of associated vascular pathology (Fig. 10.10).

CADASIL/CARASIL

Cerebral autosomal dominant arteriopathy with subcortical infarcts and leukoencephalopathy (CADASIL) is the most prevalent inherited cause of cerebral small-vessel disease, caused by mutations in NOTCH3. The mean age of onset of CADASIL is between 40 and 50 years, with a mean disease duration of approximately 14 years (Mizuno et al. 2020). However, certain conditions (e.g., acute hypotension in the setting of impaired vasoreactivity) can accelerate CADASIL pathology (Pettersen et al. 2017). CADASIL can induce cerebral infarction, white matter disease, and microbleeds. Clinical features include migraine with aura, strokes, and psychiatric symptoms. CADASIL is associated with vascular cognitive impairment in patients as young as 50 years of age (Jolly et al. 2022).

Although exceedingly rare, an autosomal recessive variant (CARASIL) has also been identified, caused by *HTRA1* mutation (Fig. 10.11). Compared to CADASIL, patients with genetically confirmed CARASIL present with a more attenuated course often accompanied by spasticity, non-hypertensive cerebrovascular disease, mood changes, spondylosis, disc degeneration/herniation, and alopecia (Khandelwal et al. 2021; Nozaki et al. 2015).

Moyamoya disease is a chronic cerebrovascular disease with an unknown etiology.

- **Common signs:** Stroke or recurrent TIAs in children; hemorrhagic stroke in adults; headaches, seizures, altered mental status.
- **Supportive features:** Progressive narrowing of internal carotid arteries and circle of Willis vessels, resulting in appearance of tiny abnormal intracranial collateral vessel networks ("puff of smoke" sign) seen on angiogram.
- **Practice guidance:** Prompt diagnosis is necessary to reduce risk of recurrent stroke. Given increased risk of recurrent stroke, cognitive impairments can be progressive, necessitating serial monitoring while on an inpatient unit.

Fig. 10.10 Summary of Moyamoya

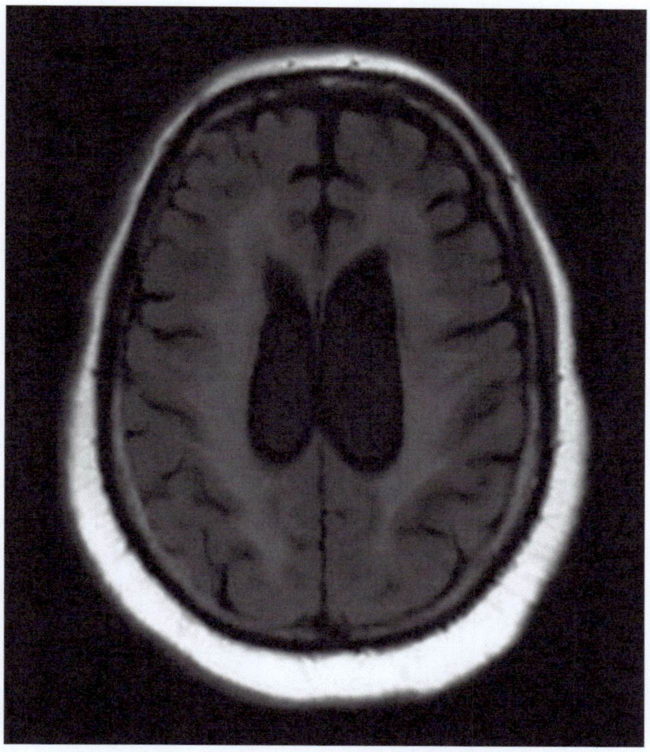

Fig. 10.11 Suspected CARASIL on MRI. Axial T2 FLAIR sequence of a 52-year-old woman showing diffuse volume loss with confluent hyperintensities in the periventricular and subcortical white matter. Multiple sclerosis (MS) was initially considered as a differential, though clinical work-up was more suspicious for CARASIL given protracted course, medical comorbidities, and absence of symptoms classically associated with MS. Genetic testing was recommended

Neuroinvasive Infections

Infections can alter normal brain function through direct invasion of the nervous system, or indirectly via systemic inflammatory or autoimmune responses. Neuroinvasive infections cause a broad spectrum of cognitive, psychiatric, and physical sequelae. Symptoms may become evident during acute infection or may arise long after initial infection, in cases where the virus is dormant, and thus detailed history taking is essential. In most cases, symptom onset and course will be rapid. The following section covers some of the more common infections associated with known neurological manifestations.

Herpes Simplex Encephalitis (HSE)

Human herpesviruses are comprised of eight known viruses (Whitley 1996), including herpes simplex virus-1 and -2 (HSV-1 and HSV-2, respectively). HSV infection is very widespread, with an estimated 40–63% of people in the United States seropositive for HSV-1 (Fatahzadeh and Schwartz 2007). Though relatively rare, herpes simplex encephalitis (HSE) is the most common infectious cause of sporadic encephalitis, with HSV-1 accounting for over 90% of the cases.

Epidemiology: Worldwide incidence of HSE is estimated at 1 in 250,000–500,000 and there is a bimodal age distribution, with most cases occurring in patients 6 months to 20 years old or in patients older than 50 years. Rates are the same for males and females (Gnann Jr. and Whitley 2017; Modi et al. 2017).

Pathophysiology: HSV is typically spread via mucous membranes or damaged skin. The exact mechanism through which HSV gains access to the CNS is not fully understood, but it is thought to be transmitted via the olfactory nerve, which passes directly to the frontal and mesial temporal lobes, or trigeminal nerve, which innervates the meninges (Bradshaw and Venkatesan 2016). With HSE, the inflammatory response to infection leads to necrosis involving all cell types. This necrotizing encephalitis tends to involve the temporal lobe, frontal lobe, and/or the insular cortex. Although usually unilateral in early stages, HSE can spread to the contralateral temporal lobe via the anterior commissure (Jennische et al. 2015; Taylor et al. 2005). About a third of cases are related to initial HSV infection (more commonly in children and adolescents) and the rest are related to HSV reactivation (Gnann Jr. and Whitley 2017).

Clinical Work-Up: When clinical suspicion is high, lumbar puncture should be obtained. Polymerase chain reaction (PCR) assay is the diagnostic test of choice, though false negatives can occur early in the disease course. Moderate CSF pleocytosis is also typical (Gnann Jr. and Whitley 2017). MRI typically shows abnormalities in the mesial temporal lobe structures, orbitofrontal lobe, insular cortex, and/or the thalamus (Fig. 10.12). EEG is abnormal in most patients, but findings are nonspecific (Singh et al. 2016).

Symptom Course: The clinical features of HSE have poor specificity. Patients may present with prodromal symptoms including headache, malaise, signs of systemic infection, or behavioral changes, such as hypomania or Kluver-Bucy syndrome (Fig. 10.13). The most common manifestations include fever, altered mental status, seizures (occurring in about half of cases), headaches, focal neurological deficits, and eventually impaired consciousness. Olfactory hallucinations and feelings of déjà vu are rare but if present, raise suspicion for HSE due to temporal lobe involvement (Rayan et al. 2022). Symptoms usually progress over a few days. Patients receiving immunosuppressive therapy may be at increased risk for HSE, and these patients often have atypical presentations such as more widespread CNS lesions. Compared with their immunocompetent peers, the mortality rate of HSE is up to 6 times higher in immunocompromised groups (Tan et al. 2012).

Neuropsychological Assessment: Cognitive deficits usually include memory impairment (commonly anterograde memory), executive dysfunction, slowed processing speed, and language impairment (Riancho et al. 2013; Sili et al. 2014).

Treatment: Early initiation of antiviral therapy is essential and it is recommended to start intravenous acyclovir therapy in patients with clinically suspected HSE regardless of laboratory test results (Hughes and Jackson 2012). The most important negative prognostic indicators are delay in initiation of antiviral treatment, older age, and lower level of consciousness at pre-

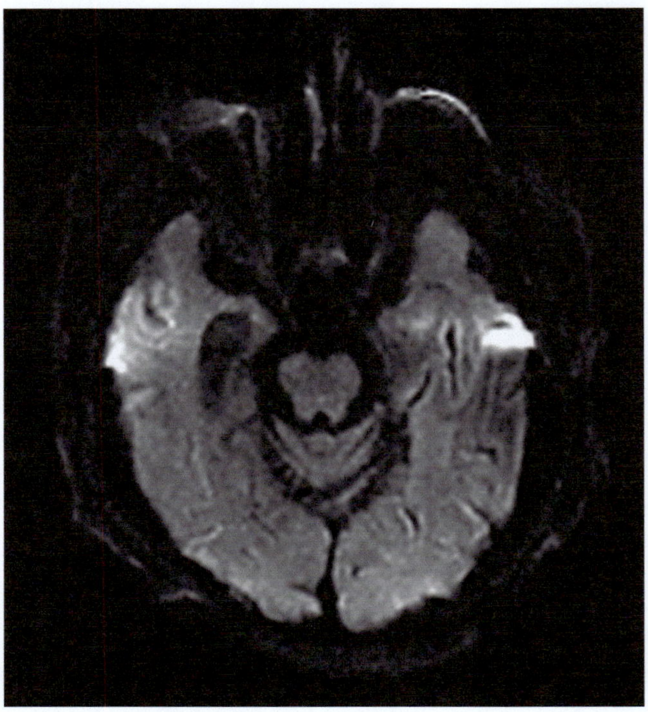

Fig. 10.12 Bitemporal viral encephalitis on MRI. Axial DWI showing bitemporal hyperintensities associated with viral encephalitis in a 58-year-old man

HSV is the most common infectious cause of sporadic encephalitis (HSE).

- **Common signs:** Behavioral changes, altered mental status, seizures (~50%), headache, focal neurological signs.
- **Supportive features:** Abnormalities in the mesial temporal lobe, orbitofrontal lobe, insular cortex, and/or thalamus; abnormal EEG; olfactory hallucinations and déjà vu.
- **Practice guidance:** A patient with a history of HSV and signs of encephalitis should receive antiviral therapy regardless of test results, particularly if they are receiving immunosuppressive treatment or are in a higher-risk age range (6 months-20 years or >50 years).

Fig. 10.13 Summary of HSE

sentation (Bradshaw and Venkatesan 2016; Gnann Jr. and Whitley 2017). If left untreated, the mortality rate of HSE is around 70%, but with advances in treatment, outcomes have improved dramatically. Current 1-year mortality rate is estimated at 5–15% (Bradshaw and Venkatesan 2016). Even with treatment, most patients experience persistent neurological sequelae.

Epstein-Barr Virus

As noted above, there are nine known human herpesviruses, most of which can result in neurological syndromes. One of the most widespread is the Epstein-Barr virus (EBV) which, in adolescence and adulthood, causes infectious mononucleosis. Acutely, EBV is associated with aseptic meningitis, encephalitis, and cranial nerve palsies. Alice in Wonderland syndrome, an uncommon perceptual phenomenon characterized by discrete episodes of visual distortions in the proportions of objects in the environment, body image disturbance, or tendency for linear objects to appear curved, is also associated with EBV infection (Carneiro et al. 2022; Soldan and Lieberman 2020).

Beyond the acute stage, lymphoproliferation driven by EBV can cause neurological disease, and a history of EBV or positive lab findings may raise suspicion for these conditions. This includes primary CNS lymphoma (typically in immunocompromised patients) and posttransplant lymphoproliferative disease, which presents as a rapidly progressive severe form of mononucleosis and sepsis-like symptoms and is associated with CNS lesions in advanced stages (Tselis et al. 2020).

Varicella Zoster Virus

Another common human herpesvirus is varicella zoster, the virus causing chickenpox (i.e., varicella). After primary infection, the virus becomes latent in dorsal root ganglia and can reactivate, causing shingles (i.e., herpes zoster or zoster). Acute varicella infection causes a distinctive rash, fever, and possible acute CNS involvement, such as encephalitis, acute cerebellar ataxia, myelitis, and Reye's syndrome, a severe and often fatal syndrome involving rapid brain swelling and liver toxicity (Amlie-Lefond and Jubelt 2009; Kennedy and Gershon 2018). Risk of reactivation increases with aging, rising sharply over age 50, and in immunosuppressed patients. Immunosuppression also increases the likelihood of CNS manifestations and infection.

Clinically, zoster typically presents as a mild, segmental rash related to ganglionitis, most commonly in the thoracic region and trigeminal nerve. Rash is sometimes accompanied by severe pain which persists following recovery for 10–20% of patients, thought to be due to persistent infection and chronic inflammation in the ganglia (Amlie-Lefond and Jubelt 2009; Skripuletz et al. 2018). Cranial nerve palsies are also common. Due to zoster's ability to replicate in both large and small cerebral arteries, vasculopathy is a significant complication of zoster and can lead to stroke, TIA, aneurysm, arterial dissection, progressive stenosis, or thrombosis. Note that for patients presenting with neurological or vascular symptoms, diagnosis of zoster is often delayed or missed because rash may be absent or may have occurred weeks or months prior to presenting symptoms. Early diagnosis is crucial, as antiviral therapy may lead to stabilization or improvement of disease.

Neurosyphilis

Syphilis is an infection caused by the *treponema pallidum* bacterium. Transmission is most common via sexual contact, though transmission can occur in infants born to infected mothers.

Epidemiology: Once ubiquitous, rates of syphilis have declined dramatically since the introduction of penicillin. However, worldwide incidence remains high, and cases have been steadily increasing in the United States since 2000. Rates of syphilis are much higher in the male population compared to the female population (estimated at 26.8 versus 8.7 per 100,000, respectively), particularly among men who have sex with men (Centers for Disease Control and Prevention [CDC] 2023; Schmidt et al. 2019). Neurosyphilis is rare, though specific data on prevalence is lacking due to variable case reporting, as well as changes in surveillance and treatment. There is an increased risk of developing neurosyphilis associated with HIV coinfection and with inadequate treatment of early syphilis (de Voux et al. 2018; Tuddenham and Ghanem 2018).

Clinical Work-Up: Accurate diagnosis is based on serum and CSF antibody tests and on elevations in CSF white blood count, but the sensitivity and specificity of these tests is suboptimal, and no agreed-upon benchmarks exist (Hamill et al. 2023). Neuroimaging may be useful in detecting brain infarcts associated with meningovascular neurosyphilis but findings are generally nonspecific, and may include diffuse or focal cortical atrophy, subcortical and periventricular white matter changes, or hyperintensities in the frontotemporal lobes and/or hippocampal region (Hamill et al. 2023; Stefani et al. 2013).

Symptom Course: CNS invasion can occur within days of primary infection. Symptoms may manifest at any stage of the disease, even up to decades after the initial infection (Ghanem 2010). Early neurosyphilis is typically asymptomatic, detectable only through serological or CSF abnormalities, but may present as acute syphilitic meningitis with headaches, nuchal rigidity, photophobia, nausea, vomiting, cranial nerve palsies (including optic or auditory neuropathies), ocular abnormalities, altered mental status, and occasionally seizures (Hamill et al. 2023; Ropper 2019). Meningovascular syphilis, often occurring 5–10 years after initial infection, involves vasculitis of small- and medium-sized arteries in the CNS resulting in thrombosis and infarction. Focal neurological signs and meningomyelitis are the most common manifestations. In the late stages, decades after initial infection, the two archetypal syndromes are *general paresis* and *tabes dorsalis*. *General paresis* is characterized by progressive dementia (may be insidious or sudden), personality changes, psychiatric syndromes, halting speech, tremors, and grandiose delusions. *Tabes dorsalis* is less common and characterized by ataxic gait, Romberg's sign, paresthesia, impaired vibratory sense and proprioception, and Charcot joints (i.e., neuropathic arthropathy). Argyll Robertson pupils—small bilateral pupils that do not constrict to bright light but do constrict when focusing on a close object—are strongly associated with late stage neurosyphilis, as is "the candy sign," a type of orofacial dyskinesia resembling sucking on a candy (Martinelli et al. 2013; Thompson and Kardon 2006).

Neuropsychological Assessment: Due to its diverse presentation, syphilis is often called "the great imitator." Symptoms of neurosyphilis can easily be misattributed to primary psychiatric disorders and should be considered as a differential when psychotic symptoms initially present in older adulthood. For example, *general paresis*, in particular, is often misdiagnosed as schizophrenia. Other common misdiagnoses include frontotemporal dementia, movement disorders, Creutzfeldt-Jakob disease, or toxic-metabolic encephalopathy (Kaur and Khanna 2023; Stefani et al. 2013). Often presenting in acute settings, the inpatient neuropsychologist can be helpful in assisting with differential diagnosis, thus expediting treatment, and providing prognostic guidance to families. See Fig. 10.14 for a case example.

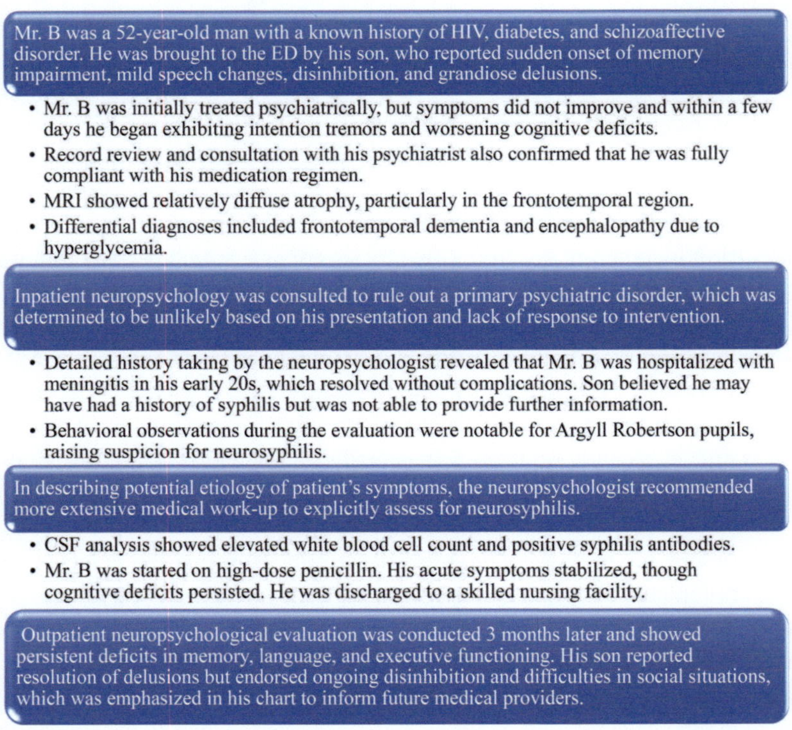

Fig. 10.14 Neurosyphilis case example

Treatment: Treatment consists of high-dose penicillin, and guidelines recommend intensive treatment of symptomatic neurosyphilis. In early neurosyphilis, symptoms are expected to resolve quickly, though recovery may be prolonged in immunocompromised individuals, such as those with HIV. Full recovery is unlikely in later stages and treatment is not guaranteed to stop progressive decline. Persistent cognitive deficits are common, especially in memory, attention, and executive functioning, with many patients demonstrating a diffuse pattern of impairment (Fig. 10.15). Medial temporal lobe atrophy is associated with poorer cognitive outcomes. Psychiatric disturbances and deficits in social functioning are also common (Beauchemin and Laforce Jr. 2014; Gao et al. 2021; Stefani et al. 2013).

John Cunningham Virus and PML

Progressive multifocal leukoencephalopathy (PML) is a progressive, multifocal disease of the central nervous system, initially described in the late 1950s (Astrom et al. 1958). It is caused by the John Cunningham polyomavirus (JCV), which was first isolated in 1971 (Padgett et al. 1971).

Pathophysiology: Although JCV is widespread throughout the world, infection is typically asymptomatic except in those who are immunocompromised (Fig. 10.16). For these individuals, JCV becomes pathological with suppression of cell-mediated immunity, causing lysis of oligodendrocytes and rapid demyelination of axons in the CNS. Although classically distributed throughout subcortical white matter, intracortical lesions at gray-white junction areas and in cortical grey matter (Wüthrich and Koralnik 2012) can also be seen. In neuroimaging, widespread asymmetrical hypointense lesions on T1 sequences and hyperintense lesions on T2 and FLAIR sequences with ill-defined borders without contrast enhancements are seen (Shah et al. 2010).

Epidemiology: Although PML is rare overall (incidence from 2010 to 2017 in France of 0.11 per 100,000 person-years; Joly et al. 2023), specific patient populations tend to be at particular risk. PML is most common among those infected with human immunodeficiency virus (HIV+), with an incidence rate of 1.3 cases per 1000 person-years from 2000–2006 (Engsig et al. 2009). Within the HIV+ population, patients with low CD4 T-cell count (<200 cells/μL) are at particularly high risk due to immunosuppression. Improvements in the treatment of HIV have reduced the incidence of PML in this population, although HIV+ patients continue to have poor outcomes.

Individuals with cancer (particularly hematological malignancies) are also at higher risk for PML, with incidence rates of 8.3 and 11.1 per 100,000 person-years in non-Hodgkin lymphoma and chronic lymphocytic leukemia, respectively (Amend et al. 2010). Other groups at risk for PML include individuals with autoimmune conditions (e.g., multiple sclerosis, lupus, rheumatoid arthritis) and individuals undergoing chronic corticosteroid/immunosuppressive therapy for organ transplant.

Neurosyphilis may occur at any disease stage, even decades after infection.

- **Common signs:** Early stage neurosyphilis is typically asymptomatic or presents as syphilitic meningitis. Patients may later present with *general paresis* or *tabes dorsalis*.
- **Supportive features:** Argyll-Robertson pupils and the candy sign are strongly associated with neurosyphilis.
- **Practice guidance:** Symptoms are often attributed to primary psychiatric disorders or other causes of neurocognitive impairment. Progressive damage may occur even with treatment, and the neuropsychologist can assist with psychoeducation and future planning.

Fig. 10.15 Summary of neurosyphilis

Although typically asymptomatic, JCV is an opportunistic infection associated with PML in individuals with compromised immune systems.

- **Common signs:** Visual disturbances e.g., (homonymous hemianopsia), progressive weakness, gait disturbance, ataxia, headache, executive dysfunction
- **Supportive features:** Widespread asymmetrical lesions on neuroimaging, immunocompromising medical condition/treatment
- **Practice guidance:** Neuropsychological profiles differ depending on course and severity.

Fig. 10.16 Summary of JCV and PML

PML incidence in these high-risk populations appears to be increasing, mostly due to the introduction of new therapies that extend life expectancy but increase the risk of PML (Pavlovic et al. 2015). For example, although monoclonal antibody therapies (e.g., natalizumab) have improved treatment of chronic inflammatory disease, some of these therapies suppress the immune system for long periods of time, predisposing patients to opportunistic infections such as JCV (Kartau et al. 2019). Indeed, patients with diseases such as MS are routinely monitored for JCV, with some otherwise promising immunosuppressant treatments being contraindicated due to the risk of PML if a patient is JCV positive.

Symptom Course: The clinical symptoms of PML can vary depending on the extent of demyelination and the brain structures affected. Visual disturbances are common, related to damage to cerebral visual pathways, with homonymous hemianopia being the most common ophthalmic manifestation (Sudhakar et al. 2015). Other commonly reported symptoms include progressive weakness, gait disturbance, ataxia, and headache. Neurologic deficits associated with PML typically progress over days to weeks following JCV infection and decline accelerates as the disease progresses. The clinical course has historically spanned over weeks or months but has improved with advancement in treatments for co-occurring medical disease, coinciding with evidence that early detection of PML can improve clinical outcomes (Lindå and von Heijne 2013; Phan-Ba et al. 2012; Vermersch et al. 2011). Although survival rates have increased, many individuals with PML continue to experience neurological symptoms resulting in disability (Cortese et al. 2021).

Neuropsychological Assessment: Neuropsychological deficits have previously been identified to include subcortical impairments, apraxia, visual deficits, and motor abnormalities (Berger and Major 1999; Krupp et al. 1985). More recently, researchers have attempted to identify specific neurocognitive profiles in PML across different clinical cohorts. For example, diffuse cognitive impairments are seen in rapidly progressive PML, while those with a more stable PML course evidenced a primary dysexecutive pattern (Zucchella et al. 2011). When assessing patients with PML, Kinner et al. (2021) recommend including measures of visuospatial abilities (e.g., Clock Drawing Test [CDT]), attention/working memory (e.g., digit span tests), processing speed (e.g., Symbol-Digit Modalities Test [SDMT]), verbal and visual memory (e.g., Rey Auditory Verbal Learning Test [RAVLT] and Wechsler Memory Scale-IV [WMS-IV] visual reproduction), and fine motor speed (e.g., Grooved Pegboard).

Other Infectious Etiologies

While it is beyond the scope of this chapter to describe the full range of neuroinvasive diseases, the section below includes four infections that have rare, but notable, neurocognitive sequalae.

- *West Nile Virus:* West Nile Virus (WNV) is a flavivirus typically transmitted by mosquitoes. Since 2013, incidence in the United States has ranged from 700 to 3000 annually, though rates of mosquito-borne illnesses are expected to increase with climate change (CDC 2023). WNV is usually asymptomatic, but about 1% of people develop neuroinvasive disease including meningitis, encephalitis, or a poliomyelitis-like syndrome with acute flaccid limb paralysis. Ocular findings are common (i.e., choriotetinitis, vitritis). Neuroimaging is often normal, though may show abnormalities in the basal ganglia, thalami, cerebellum, and brainstem. CSF invariably shows pleocytosis and WNV antibodies (Davis et al. 2006; Sejvar 2016). On neuropsychological testing, patients may demonstrate impairments in fine motor dexterity, verbal learning and memory, and executive functioning, although cognitive findings can be highly variable (Carson et al. 2006). Fatality rates of neuroinvasive infection range from 10 to 20%. Individuals who survive usually show persistent cognitive and physical (e.g., residual fatigue, headache, and/or movement disorders) deficits.

- *Eastern Equine Encephalitis:* Eastern equine encephalitis virus (EEEV) is an alphavirus also typically transmitted through mosquito bites that affects both mammals and birds. Human infections are extremely rare, with national average annual incidence of 0.03 cases per million from 2003 to 2016, although there has been an increase in virus activity and several outbreaks in the past decade (Corrin et al. 2021; Lindsey et al. 2018), with 38 cases in 2019 alone. While initial symptoms are indistinguishable from more common viral illnesses (e.g., fever, chills, vomiting), atypical neurological signs (e.g., headache, confusion, focal weakness) usually develop within days followed by a rapid decline in arousal (Morens et al. 2019). Older adults and children are at greater risk for severe neurological manifestations. Diffuse cerebral edema as well as basal ganglia and brainstem hemorrhage can be seen early in the disease course on MRI (Deresiewicz et al. 1997; Ludlow et al. 2016). Fatality rate is ~30–40% and survivors often show global cognitive impairment.

- *Rabies*: Rabies lyssavirus is a neurotropic virus that causes rabies in a wide host of animals and is typically transmitted via bite from an infected organism. Clinical cases of human rabies are very rare in the United States with <10 cases per year, mostly from bat bites and scratches. Worldwide, there are about 60,000 deaths from human rabies per year with dogs being the most common vector; nearly all cases are in Asia (primarily India) and Africa (Pieracci et al. 2019). Clinically, the disease may manifest as encephalitic rabies (~80%) or paralytic rabies. Prodromal symptoms include fatigue, insomnia, irritability, fever, and paresthesias or pain at the site of entry. Encephalitic rabies is characterized by discrete periods of arousal/hyperexcitability, autonomic dysfunction, and hydrophobia or aerophobia, progressing to impaired consciousness and quadriparesis. Paralytic rabies is characterized by weakness in the infected limb, which progresses to quadriparesis and facial weakness. Bat variants show atypical features such as Horner's syndrome, focal brainstem signs, and myoclonus (Hemachudha et al. 2013; Jackson 2016). On MRI, T2 hyperintensities can be seen in the spinal cord, brainstem, thalamus, limbic structures, and white matter, and widespread lesions are seen as symptoms progress. Antibodies and/or viral RNA can be detected in CSF, saliva, nuchal skin biopsy, or hair follicles.

- Postexposure prophylaxis immediately after exposure prevents disease but is not widely available in all countries. Once symptoms appear, rabies is always fatal and there is an average of 5–7 days from clinical onset to death for encephalitic rabies. With intensive care support, survival may be extended to ~1 month. Given the rapid progression of this virus, comprehensive cognitive testing is not typically performed. However, inpatient neuropsychologists can still be helpful by managing behavior and providing support and recommendations to families and the treatment team.

- *Whipple's Disease:* While the bacterium *T. whipplei* is common in the general population, clinical incidence of Whipple's disease (WhD) is rare and estimated at about 1–3 cases per million. It predominantly affects men (~85%) and White individuals (~98%) and is most common in middle age. There appears to be a genetic predisposition to developing WhD (Puéchal 2013). Arthritis and arthralgia are usually present and may precede diagnosis by several years; intestinal signs (i.e., weight loss, diarrhea) are classic. Neurological signs occur in 20–40% of patients including dementia, supranuclear gaze palsy, myoclonus (comprising the classic triad of WhD), altered consciousness, and psychiatric disturbance. Oculomasticatory myorhythmia—slow, pendular convergent-divergent oscillations of the eye synchronous with jaw movements—is a rare, but highly specific, symptom of WhD. Most patients with CNS involvement will also have detectable involvement of the small intestine on biopsy and positive PCR tests in stool or saliva samples. On neuroimaging, focal/multifocal brain lesions are the most common finding and are typically supratentorial. Cognitive impairment is the most common neurological manifestation of WhD, though presentation is variable. Observed deficits include impaired learning and memory, attention, processing speed, and temporal and spatial disorientation. Overall outcomes have improved with proper antibiotic treatment, but neurological symptoms often persist even with recovery of other clinical symptoms. Disease recurrence is more common when there is neurological involvement and preferentially affects the CNS (Manini et al. 2022; Mecklenburg et al. 2023).

Metabolic/Autoimmune-Related Disorders

Acute Disseminated Encephalomyelitis

In addition to neuroinvasive infections, there are several rapidly progressive syndromes that tend to develop post-infectiously, though causal relationships have not been definitively established. One example is acute disseminated encephalomyelitis (ADEM), a rare demyelinating CNS disorder typically preceded by a viral infection or immunization. Course is rapid with maximal deficits within 2–5 days. The incidence is estimated at 0.3–0.6 per 100,000 per year and it is much more frequent in children, with median age of 5–8 at presentation and a slight male predominance (Filippi and Rocca 2020; Pohl et al. 2016).

Clinically, ADEM is characterized as an acute encephalopathy with multifocal neurological deficits. Although clinical presentation depends on the location of lesions, common symptoms include fever, meningeal signs, ataxia, cognitive impairment. Cranial nerve abnormalities (e.g., palsy), motor plegia, and sensory changes. MRI shows multifocal lesions of varying sizes, commonly localized to the brainstem, midbrain, supratentorial white matter, cortical gray matter, thalami, basal ganglia, and cerebellum (Schwarz et al. 2001). Lymphocytic pleocytosis, or elevated white blood cells (WBCs) in the CSF, has also been identified through lumbar puncture (LP) analyses in patients with ADEM (Sonneville et al. 2008).

ADEM is usually monophasic, but 10–30% of patients experience a relapse of symptoms months to years after the first episode (Berzero et al. 2016). The term multiphasic disseminated encephalomyelitis (MDEM) is used to describe patients who have experienced a relapse of symptoms in the weeks to months proximal to the initial episode (Dale and Branson 2005). However, a subset of patients (~15–32%) may continue to experience longitudinal relapses and will eventually be diagnosed with multiple sclerosis (MS; De Seze et al. 2007; Schwarz et al. 2001). Differential diagnosis is challenging in this regard due to the heterogeneity of symptoms and lack of specific biomarkers that differentiate ADEM from MS. Apart from course, a polysymptomatic initial presentation, the presence of grey matter lesions with relative sparing of periventricular areas, and the presence of symptoms not traditionally associated with MS (e.g., seizure, hypersomnia, alterations in consciousness, hemi/tetraplegia, bilateral optic neuritis, and language changes) can support a clinical diagnosis of ADEM (De Seze et al. 2007). Seizures, in particular, are associated with a higher incidence of mortality among hospitalized patients with ADEM (Sonneville et al. 2008).

Treatment of ADEM includes high-dose corticosteroids and intravenous immunoglobulin (IVIG). Prognosis—especially for adults—is usually positive, and most individuals will return to baseline within weeks. However, residual symptoms and increased risk of mortality are possible especially in those with severe presentations. Chronic impairments include deficits in attention and executive functioning, as well as behavioral changes (Kanmaz et al. 2020; Pohl et al. 2016).

Acute hemorrhagic leukoencephalitis (AHLE), also called Hurst disease, is often regarded as a severe form of ADEM and occurs most frequently in adults aged 20–40 (Alromaihi 2022). AHLE has a high mortality rate due to herniation and brain swelling and most survivors experience severe neurological deficits.

Behçet Syndrome

Behçet syndrome is a systemic, immune-mediated vasculitis with a wide range of clinical manifestations, which can affect blood vessels of all sizes (small, medium, large) in the venous and arterial systems (Yazici et al. 2021).

Epidemiology: It is most common in the regions extending from eastern Asia to the Mediterranean, particularly in individuals of Turkish ancestry, and estimates of prevalence in the United States range from 0.12 to 7.5 per 100,000 people. It occurs most frequently in individuals between the ages of 20 and 40 (Calamia et al. 2009).

Symptom Course: Nearly all patients with Behçet syndrome present with recurrent oral aphthous lesions (i.e., canker sores) which are often the first manifestation of the disease, and most patients also present with similar-appearing genital lesions. The clinical presentation is otherwise quite diverse and may include cutaneous lesions, ocular disease (uveitis, retinal vasculitis), joint issues (arthralgia, nonerosive arthritis, ankylosing spondylitis), gastrointestinal manifestations, and vascular disease. Around 10% of patients have neurologic involvement (Neuro-Behçet disease; NBD) which is more common in males, typically several years after initial symptom onset (Borhani-Haghighi et al. 2020; Peine et al. 2022). Neurologic symptoms may be acute or progressive; acute symptoms are often due to focal lesions associated with vascular thrombosis and are highly dependent on the location of the lesions. Motor symptoms, encephalopathy, personality change, and psychiatric symptoms are common.

Clinical Work-Up: MRI typically shows acute/subacute focal lesions which are multiple and appear to preferentially impact the brainstem, mesencephalon, and subcortical white matter structures (Peine et al. 2022). NBD is associated with a worse prognosis than Behçet syndrome without neurologic involvement and there is high risk of relapse. Disease activity fluctuates, resulting in waxing and waning course, and the cumulative burden of damage increases over time.

Neuropsychological Assessment: Diagnostic clarification is not usually the presenting question for a neuropsychologist, as the symptoms of Behçet syndrome are typically present for years before neurologic manifestations are observed, and the oral and genital lesions are near-universal and relatively easy to identify. Neurocognitive testing is helpful to identify areas of deficit that may interfere with the patient's ability to manage the multiple other symptoms they may be experiencing. The inpatient neuropsychologist may also be consulted to assess personality changes and psychiatric symptoms, which can be challenging to disentangle from non-neurological factors given that NBD is associated with both acute and chronic progressive change. Capacity determinations, behavioral management, and psychoeducation to the family and treatment team are often beneficial. Outpatient monitoring of NBD patients is critical given waxing and waning course. Permanent neurocognitive dysfunction is seen in most NBD patients, specifically in visuospatial ability, memory, and processing speed (Fisher and Bernard 2019).

Wilson's Disease

Wilson's disease (WD), or hepatolenticular degeneration, is a rare, genetic, and progressive disorder typified by excess copper deposition in various organs, including the liver and brain. It follows an autosomal recessive pattern of inheritance caused by ATP7B mutations on chromosome 13.

Clinical Work-Up: In most cases, age of onset tends to occur between 5 and 35 years old (Członkowska et al. 2018). Common neurological symptoms include dysarthria, ataxia, proximal/wing-beating tremor (high amplitude tremor when extending upper extremities), and parkinsonism, while pyramidal symptoms and seizures are far less likely (Członkowska et al. 2018; Dusek et al. 2019; Machado et al. 2006). Kayser-Fleischer rings, or copper deposition in the cornea of the eyes, are a strong indicator of WD and have been found in 95% of cases of WD with neurologic symptom presentation (Lorincz 2010). Neuroimaging findings supportive of WD include T2 FLAIR hyperintensities of the basal ganglia, thalamus, brain stem, and cerebellum (Lorincz 2010).

Neuropsychological Assessment: Frontal-subcortical dysfunction is the most common pattern of cognitive impairment exhibited by patients with WD, which is typically mild to moderate in severity but can present as a rapid dementia in untreated patients with greater organ damage (Frota et al. 2009). Neuropsychiatric symptoms, present in about 30–40% of cases when formally diagnosed, may include anxiety, agitation, depression, mania, disorganized thought processes, and delusions (Zimbrean and Schilsky 2014). Prognosis depends upon the degree and extent of multi-systemic dysfunction, as earlier intervention with copper chelator agents, zinc, and dietary changes may yield normal lifespans (Schilsky et al. 2023). While improvements in neuropsychological symptoms are expected with aggressive treatment, patients are often left with some degree of residual cognitive impairment. Like other rare diseases, cognitive sequelae of WD are nonspecific, and can even be conflated with other autosomal recessive disorders (Hogarth 2015).

Heavy Metal Toxicity

Acute and chronic neurotoxic effects of heavy metal exposure represent another relatively rare, yet germane, cause of cognitive decline. Inpatient neuropsychologists play an important role in helping to differentiate cognitive decline related to heavy metal exposure from other, more common conditions with similar presentations. The heavy metals most associated with cognitive decline include lead, mercury, manganese, arsenic, cadmium, and chromium (Althomali et al. 2024)—the latter, albeit exceptionally rare, can present following surgical implantation of metallic joint replacement (Green et al. 2017).

Clinical Work-Up: Heavy metal levels may be assessed via urine, hair, fingernails, or blood, with tests with the latter being the most common first step. In addition, a comprehensive clinical history and focused bedside cognitive testing are paramount to informing differential diagnosis. Elevated concentrations of heavy metals are dose-dependent, and often associated with cumulative occupational (e.g., mining, factory, welding, or construction work) or residential (e.g., lead-based paint) exposure (Jacobs et al. 2002). Indeed, the phrase "mad as a hatter" was colloquially coined to describe the cognitive and behavioral changes associated with mercury poisoning amongst hatters in the eighteenth and nineteenth centuries (Huang et al. 2014). To help determine an individual's risk of heavy metal exposure, we recommend obtaining a thorough employment history, as well as inquiring about any known manifestations of similar symptoms in coworkers or family members. Symptoms that fluctuate throughout the week or appear temporally associated with one setting can raise clinical suspicion for context-dependent heavy metal exposure. Some alternative medicines and herbal supplements may also unwittingly expose individuals to high levels of heavy metals (Luo et al. 2020; Mikulski et al. 2017; Saper et al. 2008). For this reason, we advise expanding the discussion of patients' medication regimen to include inquiry of any over-the-counter (OTC) supplements or alternative treatments.

Alongside cognitive decline, other systemic symptoms may also be present and help inform differential diagnosis, including ataxia and other cerebellar signs in mercury poisoning, tremors in manganese poisoning, and seizures in acutely high levels of any metal exposure. Behavioral changes associated with heavy metal toxicity can include insomnia, agitation/irritability, depression, anxiety, hallucinations, and delusions. The most acute systemic signs and symptoms may help indicate the route of exposure, depending on the severity (e.g., respiratory problems in inhalation). In addition, unexplained symptoms impacting the nervous (neuropathy, headaches, weakness), cardiovascular (hypertension, arrhythmia), renal, and gastrointestinal systems (nausea, vomiting, diarrhea) are commonly found to co-occur and are important to inquire about on clinical interview. Symptoms of heavy metal poisoning are often non-specific, but new-onset multi-systemic illness alongside recent encephalopathy may prompt taking a more in-depth history of possible exposures.

Neuropsychological Assessment: Excess levels of manganese have been associated with parkinsonism, cognitive dysfunction (attention, executive, processing speed, letter fluency, and visuospatial/constructional functions), and frequent affective symptoms (Singer 2011). Elevated lead exposure has been correlated with lower FSIQ and higher rates of learning delays in children, as well as impairments in sustained attention, fine motor, memory encoding and recall, and executive functions (Singer 2011). Mercury toxicity has been negatively associated with performance on tasks of working memory, visuospatial, motor, and executive functions alongside ataxia likely due to the metal's tendency to target the cerebellum, though overall cognitive impairment has been described as relatively mild (Althomali et al. 2024; Haut et al. 2017). Quality research on the relationship between cadmium toxicity and cognitive functioning is scant, but one observational study revealed worsening cognitive performance alongside increasing cadmium levels in older adults (Peng et al. 2020). Furthermore, in a case series of 10 subjects who underwent failed metal-on-metal arthroplasty with suspected cobalt and chromium toxicity, the majority of patients exhibited short-term memory deficits and elevated depressive symptoms (Green et al. 2017).

Treatment: Chelation therapy has been shown to be helpful for some cases of heavy metal poisoning, namely in acute exposures, but is not fully curative (Kim et al. 2019). Chelates, or agents that bind to metal ions, have demonstrated efficacy in reducing concentrations of these toxins in the body typically by excretion. Chelation therapy is controversial for chromium, and while chelates and supplements to help induce excretion may be trialed alongside symptomatic and supportive treatment for this type of heavy metal exposure, acute poisoning typically confers a poor prognosis (Balali-Mood et al. 2025). Lead and mercury poisoning may partially respond to chelation therapy, but residual central nervous system effects are not generally modifiable (Haut et al. 2017). Furthermore, although intuitive, removal from possible heavy metal exposure sources is another key component of treatment.

Osmotic Demyelination Syndrome

Osmotic demyelination syndrome (ODS) comprises central pontine myelinolysis (CPM) and extrapontine myelinolysis (EPM). It is a secondary disease associated with rapid correction of hyponatremia, chronic alcohol use and/or alcohol withdrawal, liver transplantation, and general electrolyte imbalances. ODS is considered rare, with estimated prevalence of 0.25 to 0.5% in the general population, and higher rates in at-risk populations (i.e., alcoholism, liver transplant recipients; King and Rosner 2010). In brief, the brain has the potential for limited homeostatic adaptation to metabolic disturbances and does not adapt well to rapid change. Thus, overly rapid correction of hyponatremia in clinical settings risks exhausting the brain's ability to adapt, which can lead to demyelination.

Most patients have CPM or both CPM and EPM, with an estimated 6–13% showing only EPM (Fitts et al. 2021; Singh et al. 2014). Clinical presentation is quite variable depending on location of pathology. CPM is associated most frequently with acute encephalopathy, disorders of consciousness, rapidly progressive paresis, dysphagia, dysarthria, and ocular abnormalities. Locked-in syndrome may be seen in severe cases of CPM. EPM usually occurs in the basal ganglia, cerebellar white matter, thalamus, or hippocampus, and is associated with extrapyramidal motor symptoms and ataxia. Both types of ODS are associated with seizures, neurocognitive dysfunction, and psychiatric symptoms (e.g., catatonia, emotional lability, apathy). Neuroimaging classically shows demyelinating lesions which can be seen even in milder courses of disease. However, neuroimaging may appear unremarkable for up to a week after symptom onset and normal CT or MRI early in the symptom course does not necessarily rule out ODS. Initial electrolyte imbalance, the correction of which does not improve cognition, and presence of notable risk factors (e.g., history of liver transplant, underlying cirrhosis) may raise suspicion for ODS (Lambeck et al. 2019; Singh et al. 2014). However, more common causes of cognitive dysfunction (e.g., metabolic encephalopathy) should be considered prior to pursuing this rare etiology. Mortality rates vary widely, ranging from 5.8% to 48%, likely due to heterogeneous populations studied. About half of patients achieve good recovery, though neurocognitive sequelae are frequent and usually show a pattern of frontal/subcortical dysfunction (Fitts et al. 2021; Odier et al. 2010).

Hashimoto's Encephalopathy

Hashimoto's encephalopathy (HE), also known as steroid-responsive encephalopathy, is a rare syndrome associated with elevated thyroid antibodies, independent of thyroid functioning (Fig. 10.17; Ferracci et al. 2003).

Epidemiology: Due to a wide range of reported symptoms and an unclear pathophysiology, there is controversy whether it represents a distinct clinical entity (Waliszewska-Prosół and Ejma 2022). It has an estimated prevalence of 2.1 per 100,000 (Ferracci et al. 2004), a mean age of approximately 44 years, and predominantly affects women (Chong et al. 2003).

Hashimoto's encephalopathy is associated with elevated thyroid antibodies, independent of thyroid functioning.

- **Common signs:** Progressive cognitive impairment with sudden onset of psychosis (older adults), stroke-like focal symptoms (younger adults), seizures and myoclonus common in both groups
- **Supportive features:** Responsive to steroid treatment
- **Practice guidance:** Symptom presentation is quite variable, but steroid treatment is beneficial even for psychiatric symptoms and seizures. Serial testing in the outpatient setting is beneficial given varied recovery trajectories, and shorter batteries can be given as an inpatient to establish baseline.

Fig. 10.17 Summary of Hashimoto's encephalopathy

Clinical Work-Up: Neuroimaging findings are nonspecific and have included cerebral atrophy, nonspecific cortical focal white matter abnormality, diffuse subcortical abnormality, and focal cortical abnormality; most have an abnormal EEG, typically showing diffuse slowing. Lab work demonstrates that HE patients can be euthyroid, hypothyroid, or hyperthyroid (Chong et al. 2003). Diagnosis is made via identification of antithyroid antibodies (TPOAb and/or TgAb), although false positives may occur. CSF abnormalities are also typically seen, with high CSF protein the most common finding.

Symptom Course: Apart from acute or subacute encephalopathy, varied symptoms have been reported. There is some evidence for two patterns of symptom clusters: progressive cognitive impairment with sudden onset of hallucinations and delusions tends to occur in older individuals, whereas stroke-like and focal neurologic symptoms appear more common in younger individuals (Chaudhuri et al. 2023). Seizures and myoclonus are common in both groups. HE has also been found to affect children (mean age of approximately 10 years of age), with symptoms including epilepsy, behavioral problems, decline in school performance, and hallucinations (Boelen and de Vries 2021).

Neuropsychological Assessment: The inpatient neuropsychologist can help identify acute symptoms (e.g., hallucinations and delusions) that may occur in older adults with HE. Cognitive impairment tends to be global prior to treatment. Regarding outpatient evaluation, four months after tapering of steroid treatment, there is evidence for improvements in many areas, but ongoing difficulties with executive functioning, confrontation naming, semantic fluency, and working memory are common (Cummings et al. 2007). Testing at 5 months (Lagström et al. 2019), 1 year (Mazzù et al. 2012), and 3 years (Sabbah-Talasazan and Piryatinsky 2018) posttreatment suggests ongoing executive dysfunction. However, differences in trajectories and the specific domains of cognitive functioning affected have been identified, likely related to different presentations and co-occurring conditions. For example, HE with stroke-like presentation initially demonstrated executive dysfunction and memory impairment, with resolution of symptoms following treatment (Fukunaga et al. 2006). In addition, there is evidence for ongoing memory impairments in HE with hippocampal changes on MRI (McCabe et al. 2000; Wang et al. 2013).

Treatment: Patients often respond to steroid treatment, leading to the inclusion of improvement in neurologic status with high-dose glucocorticoid therapy as part of the diagnostic criteria for HE (Zhou et al. 2017). Such treatment has beneficial impacts on associated symptoms of HE as well. For example, psychotic symptoms, which appear more frequently in those with later onset, are effectively treated by steroid therapy (Lin and Liao 2009). Similarly, seizures associated with HE are more effectively controlled via immunotherapy as compared to antiepileptic medications (Li and Li 2019).

Neuropsychiatric Systemic Lupus Erythematosus

Systemic lupus erythematosus (SLE) is a chronic, clinically heterogenous autoimmune disorder that can impact multiple organ systems (Fig. 10.18). Incidence is 9x higher among women than men, with Black females having the highest prevalence rate (Izmirly et al. 2021).

Clinical Work-Up: SLE with CNS involvement is known as neuropsychiatric SLE (NPSLE), which can include both focal (e.g., seizures, movement disorders) and diffuse (e.g., headache, mood disorder, acute confusional state; Manca 2022). MRI findings of NPSLE most commonly show small punctate focal white matter lesions (Souirti et al. 2013), cortical atrophy, ventricular dilation, and large distribution infarctions (Popescu et al. 2011).

Symptom Course: Symptoms of SLE are heterogenous and may develop at any stage of the disease. Typical symptoms observed in the early stages of SLE include fever, fatigue, joint pain, and a "butterfly" rash (Kuhn et al. 2015). Patients with SLE are at increased risk for progressive multifocal leukoencephalopathy (PML), with an incidence rate of 2 to 4 cases per 100,000 patients, and prevalence of 13–27 per 100,000 discharges following SLE-related hospitalization (Kapoor et al. 2020). Such cases can lead to rapid onset of neurological deficit and mortality. The cause of JCV activation in SLE is unclear, but is likely due to a combination of genetic factors and drug-induced immunosuppression (Emmanouilidou et al. 2023). Although very rare and limited to case study examples, there is also evidence of NPSLE vasculitis presenting with symptoms suggestive of Lewy body dementia (e.g., visual hallucinations, progressive memory decline), with resolution of symptoms following administration of prednisone (Abraham et al. 2021).

Neuropsychiatric systemic lupus erythematosus (NPSLE) is an autoimmune disorder associated with severe CNS involvement and cognitive dysfunction.

- **Common signs:** May show focal or diffuse neurologic symptoms, rarely presents similarly to Lewy body dementia.
- **Supportive features:** Fatigue, joint pain, "butterfly" rash
- **Practice guidance:** The MoCA is a sensitive screening tool, with recommendation for follow-up on an outpatient basis.

Fig. 10.18 Summary of neuropsychiatric SLE

Neuropsychological Assessment: Cognitive dysfunction frequently occurs in patients with SLE even without detectable CNS dysfunction, with impairments in visual attention, immediate visual memory, and visual reasoning (Leslie and Crowe 2018). However, patients with NPSLE display more severe cognitive impairments compared to those without overt CNS involvement (Ho et al. 2018). Specifically, patients with NPSLE demonstrate impairments in complex attention, delayed verbal memory, language, and verbal reasoning (Leslie and Crowe 2018). In inpatient settings, use of the MoCA is recommended due to its utility as a screening tool for detecting cognitive impairments in SLE (Adhikari et al. 2011). Following discharge, more comprehensive evaluation is often warranted.

The American College of Rheumatologists (ACR) recommended outpatient SLE neuropsychological battery that consists of:

- Wechsler Abbreviated Scale of Intelligence-IV (WAIS-IV; Wechsler 2009).
- Stroop Color and Word Test (SCWT; Golden and Freshwater 2002).
- Trail Making Test (TMT; Reitan 1992).
- California Verbal Learning Test-3 (CVLT-3; Delis et al. 2017).
- Rey-Osterrieth Complex Figure Test (RCFT; Osterrieth 1944; Rey 1941).

Given the time-consuming nature of traditional neuropsychological testing and limited norms for non-English speaking patients, computerized assessment (e.g., Automated Neuropsychological Assessment Metrics; ANAM) represents a potential cost-effective alternative for ethnically diverse populations (Holliday et al. 2003). In time-limited inpatient contexts, provision of briefer first-line testing (e.g., MoCA) may also be utilized to determine if cognitive impairment is indeed present, before progressing a longer and more comprehensive battery.

SUMMARY

Inpatient neuropsychologists play many important roles in the evaluation and treatment of patients with rare and rapidly progressive syndromes, including aiding in differential diagnosis, synthesizing data, determining a pattern of neuropsychological strengths and weaknesses, and communicating findings to patients, family members, and/or providers who may have little familiarity with these conditions. Prompt diagnosis is essential to help tailor clinical decisions, reduce uncertainty, and enhance prognosis. Opportunities for comprehensive cognitive evaluation are limited given the severity and rapid course; however, serial monitoring using brief screening measures is useful for documenting the rate of change. Given changes in mental status and cognition, determining capacity for medical decisions is often an essential role for inpatient neuropsychologists working with these populations. While these cases can initially appear overwhelmingly complex and daunting, taking an approach focused on clinical history can elucidate a clearer understanding of symptom course and provide a basis for further work-up. Recommendations can then be offered, with the inpatient neuropsychologist uniquely suited to provide education to the treatment team and families and assist with behavioral management and environmental modifications.

REFERENCES

Abraham P, Neel I, Bishay S, Sewell DD. Central nervous system systemic lupus erythematosus (CNS-SLE) Vasculitis mimicking Lewy body dementia: A Case report emphasizing the role of imaging with an analysis of 33 comparable cases from the scientific literature. J Geriatr Psychiatry Neurol. 2021;34(2):128–41. https://doi.org/10.1177/0891988720901788.

Adhikari T, Piatti A, Luggen M. Cognitive dysfunction in SLE: development of a screening tool. Lupus. 2011;20(11):1142–6. https://doi.org/10.1177/0961203311405374.

Alromaihi M. Acute hemorrhagic Leukoencephalitis (AHLE): A comprehensive review on causes, symptoms, link with COVID-19, diagnosis, and treatment. Biomed Res Int. 2022;2022:6008375. https://doi.org/10.1155/2022/6008375.

Althomali RH, Abbood MA, Saleh EAM, Djuraeva L, Abdullaeva BS, Habash RT, et al. Exposure to heavy metals and neurocognitive function in adults: a systematic review. Environ Sci Eur. 2024;36(1):18. https://doi.org/10.1186/s12302-024-00843-7.

Amend KL, Turnbull B, Foskett N, Napalkov P, Kurth T, Seeger J. Incidence of progressive multifocal leukoencephalopathy in patients without HIV. Neurology. 2010;75(15):1326–32. https://doi.org/10.1212/WNL.0b013e3181f73600.

Amlie-Lefond C, Jubelt B. Neurologic manifestations of varicella zoster virus infections. Curr Neurol Neurosci Rep. 2009;9(6):430–4. https://doi.org/10.1007/s11910-009-0064-z.

Appleby BS, Appleby KK, Rabins PV. Does the presentation of Creutzfeldt-Jakob disease vary by age or presumed etiology? A meta-analysis of the past 10 years. J Neuropsychiatry Clin Neurosci. 2007;19(4):428–35. https://doi.org/10.1176/jnp.2007.19.4.428.

Arvanitakis Z, Leurgans SE, Wang Z, Wilson RS, Bennett DA, Schneider JA. Cerebral amyloid angiopathy pathology and cognitive domains in older persons. Ann Neurol. 2011;69(2):320–7. https://doi.org/10.1002/ana.22112.

Astrom KE, Mancall EL, Richardson EP Jr. Progressive multifocal leuko-encephalopathy; a hitherto unrecognized complication of chronic lymphatic leukaemia and Hodgkin's disease. Brain. 1958;81(1):93–111. https://doi.org/10.1093/brain/81.1.93.

Atarashi R, Sano K, Satoh K, Nishida N. Real-time quaking-induced conversion: a highly sensitive assay for prion detection. Prion. 2011;5(3):150–3. https://doi.org/10.4161/pri.5.3.16893.

Balali-Mood M, Eizadi-Mood N, Hassanian-Moghaddam H, Etemad L, Moshiri M, Vahabzadeh M, Sadeghi M. Recent advances in the clinical management of intoxication by five heavy metals: mercury, lead, chromium, cadmium and arsenic. Heliyon. 2025;11(4):e42696. https://doi.org/10.1016/j.heliyon.2025.e42696.

Beauchemin P, Laforce R Jr. Neurocognitive changes in tertiary neurosyphilis: a retrospective chart review. Can J Neurol Sci. 2014;41(4):452–8. https://doi.org/10.1017/s0317167100018485.

Berger JR, Major EO. Progressive multifocal leukoencephalopathy. Semin Neurol. 1999;19(2):193–200. https://doi.org/10.1055/s-2008-1040837.

Berzero G, Cortese A, Ravaglia S, Marchioni E. Diagnosis and therapy of acute disseminated encephalomyelitis and its variants. Expert Rev Neurother. 2016;16(1):83–101. https://doi.org/10.1586/14737175.2015.1126510.

Boelen R, de Vries T. Clinical characteristics of paediatric Hashimoto's encephalopathy. Eur J Paediatr Neurol. 2021;32:122–7. https://doi.org/10.1016/j.ejpn.2021.04.006.

Borhani-Haghighi A, Kardeh B, Banerjee S, Yadollahikhales G, Safari A, Sahraian MA, Shapiro L. Neuro-Behcet's disease: an update on diagnosis, differential diagnoses, and treatment. Mult Scler Relat Disord. 2020;39:101906. https://doi.org/10.1016/j.msard.2019.101906.

Boyle PA, Yu L, Nag S, Leurgans S, Wilson RS, Bennett DA, Schneider JA. Cerebral amyloid angiopathy and cognitive outcomes in community-based older persons. Neurology. 2015;85(22):1930–6. https://doi.org/10.1212/wnl.0000000000002175.

Bradshaw MJ, Venkatesan A. Herpes simplex Virus-1 encephalitis in adults: pathophysiology, diagnosis, and management. Neurotherapeutics. 2016;13(3):493–508. https://doi.org/10.1007/s13311-016-0433-7.

Buchhanolla P, Bir S, Angelette A, Lewis A, Kandregula S, Guthikonda B, et al. Determination of prevalence of posterior reversible encephalopathy syndrome (PRES) and its association with cerebral infarction, and outcome in the Nationwide inpatient sample, 2016–2018 (P16-10.002). Neurology. 2022;98(18_supplement):3555.

Caine D, Tinelli RJ, Hyare H, De Vita E, Lowe J, Lukic A, et al. The cognitive profile of prion disease: a prospective clinical and imaging study. Ann Clin Transl Neurol. 2015;2(5):548–58. https://doi.org/10.1002/acn3.195.

Calamia KT, Wilson FC, Icen M, Crowson CS, Gabriel SE, Kremers HM. Epidemiology and clinical characteristics of Behçet's disease in the US: a population-based study. Arthritis Rheum. 2009;61(5):600–4. https://doi.org/10.1002/art.24423.

Carneiro VCS, Pereira JG, de Paula VS. Family Herpesviridae and neuroinfections: current status and research in progress. Mem Inst Oswaldo Cruz. 2022;117:e220200. https://doi.org/10.1590/0074-02760220200.

Carson PJ, Konewko P, Wold KS, Mariani P, Goli S, Bergloff P, Crosby RD. Long-term clinical and neuropsychological outcomes of West Nile virus infection. Clin Infect Dis. 2006;43(6):723–30. https://doi.org/10.1086/506939.

Case NF, Charlton A, Zwiers A, Batool S, McCreary CR, Hogan DB, et al. Cerebral amyloid Angiopathy is associated with executive dysfunction and mild cognitive impairment. Stroke. 2016;47(8):2010–6. https://doi.org/10.1161/strokeaha.116.012999.

Centers for Disease Control and Prevention. U.S. Department of Health and Human Services. Retrieved from https://www.cdc.gov/westnile/statsmaps/data-and-maps.html (2023).

Chan E, Gal AM, Van Harskamp N, Adams ME, Brown MM, Werring DJ, et al. Long-term study of the cognitive profile of Moyamoya disease in adults. J Stroke Cerebrovasc Dis. 2023;32(6):107064. https://doi.org/10.1016/j.jstrokecerebrovasdis.2023.107064.

Charidimou A, Boulouis G, Frosch MP, Baron JC, Pasi M, Albucher JF, et al. The Boston criteria version 2.0 for cerebral amyloid angiopathy: a multicentre, retrospective, MRI-neuropathology diagnostic accuracy study. Lancet Neurol. 2022;21(8):714–25. https://doi.org/10.1016/s1474-4422(22)00208-3.

Chaudhuri J, Mukherjee A, Chakravarty A. Hashimoto's encephalopathy: case series and literature review. Curr Neurol Neurosci Rep. 2023;23(4):167–75. https://doi.org/10.1007/s11910-023-01255-5.

Chen C, Dong XP. Epidemiological characteristics of human prion diseases. Infect Dis Poverty. 2016;5(1):47. https://doi.org/10.1186/s40249-016-0143-8.

Chong JY, Rowland LP, Utiger RD. Hashimoto encephalopathy: syndrome or myth? Arch Neurol. 2003;60(2):164–71. https://doi.org/10.1001/archneur.60.2.164.

Chung KK, Anderson NE, Hutchinson D, Synek B, Barber PA. Cerebral amyloid angiopathy related inflammation: three case reports and a review. J Neurol Neurosur Psychiatry. 2011;82(1):20–6. https://doi.org/10.1136/jnnp.2009.204180.

Collie DA, Summers DM, Sellar RJ, Ironside JW, Cooper S, Zeidler M, et al. Diagnosing variant Creutzfeldt-Jakob disease with the pulvinar sign: MR imaging findings in 86 neuropathologically confirmed cases. AJNR Am J Neuroradiol. 2003;24(8):1560–9.

Cordery RJ, Alner K, Cipolotti L, Ron M, Kennedy A, Collinge J, Rossor MN. The neuropsychology of variant CJD: a comparative study with inherited and sporadic forms of prion disease. J Neurol Neurosurg Psychiatry. 2005;76(3):330–6. https://doi.org/10.1136/jnnp.2003.030320.

Corovic A, Kelly S, Markus HS. Cerebral amyloid angiopathy associated with inflammation: a systematic review of clinical and imaging features and outcome. Int J Stroke. 2018;13(3):257–67. https://doi.org/10.1177/1747493017741569.

Corrin T, Ackford R, Mascarenhas M, Greig J, Waddell LA. Eastern equine encephalitis virus: a scoping review of the global evidence. Vector Borne Zoonotic Dis. 2021;21(5):305–20. https://doi.org/10.1089/vbz.2020.2671.

Cortese I, Reich DS, Nath A. Progressive multifocal leukoencephalopathy and the spectrum of JC virus-related disease. Nat Rev Neurol. 2021;17(1):37–51. https://doi.org/10.1038/s41582-020-00427-y.

Cummings RR, Bagley SC, Syed S, Dong N, Lin A. Hashimoto's encephalopathy: a case report with neuropsychological testing. Gen Hosp Psychiatry. 2007;29(3):267–9. https://doi.org/10.1016/j.genhosppsych.2007.01.003.

Członkowska A, Litwin T, Dusek P, Ferenci P, Lutsenko S, Medici V, et al. Wilson disease. Nat Rev Dis Primers. 2018;4(1):21. https://doi.org/10.1038/s41572-018-0018-3.

Davis LE, DeBiasi R, Goade DE, Haaland KY, Harrington JA, Harnar JB, et al. West Nile virus neuroinvasive disease. Ann Neurol. 2006;60(3):286–300. https://doi.org/10.1002/ana.20959.

Day GS. Rapidly progressive dementia. Continuum (Minneap Minn). 2022;28(3):901–36. https://doi.org/10.1212/con.0000000000001089.

De Kort AM, Verbeek MM, Schreuder F, Klijn CJM, Jäkel L. Prevalence of cerebral amyloid Angiopathy pathology and strictly lobar microbleeds in east-Asian versus Western populations: a systematic review and meta-analysis. J Stroke. 2024;26(2):179–89. https://doi.org/10.5853/jos.2023.04287.

De Seze J, Debouverie M, Zephir H, Lebrun C, Blanc F, Bourg V, et al. Acute fulminant demyelinating disease: a descriptive study of 60 patients. Arch Neurol. 2007;64(10):1426–32.

de Voux A, Kidd S, Torrone EA. Reported cases of Neurosyphilis among early syphilis cases-United States, 2009 to 2015. Sex Transm Dis. 2018;45(1):39–41. https://doi.org/10.1097/olq.0000000000000687.

Dale RC, Branson JA. Acute disseminated encephalomyelitis or multiple sclerosis: can the initial presentation help in establishing a correct diagnosis?. Arch Dis Child. 2005;90(6):636-9. https://doi.org/10.1136/adc.2004.062935.

Delis DC, Kramer JH, Kaplan E, Ober BA. California verbal learning test. 3rd ed. The Psychological Corporation; 2017.

Deresiewicz RL, Thaler SJ, Hsu L, Zamani AA. Clinical and neuroradiographic manifestations of eastern equine encephalitis. N Engl J Med. 1997;336(26):1867–74. https://doi.org/10.1056/nejm199706263362604.

Durrani R, Wang M, Cox E, Irving E, Saad F, McCreary CR, et al. Mediators of cognitive impairment in cerebral amyloid angiopathy. Int J Stroke. 2023;18(1):78–84. https://doi.org/10.1177/17474930221099352.

Dusek P, Litwin T, Członkowska A. Neurologic impairment in Wilson disease. Ann Transl Med. 2019;7(Suppl 2):S64. https://doi.org/10.21037/atm.2019.02.43.

Emmanouilidou E, Kosmara D, Papadaki E, Mastorodemos V, Constantoulakis P, Repa A, et al. Progressive multifocal leukoencephalopathy in systemic lupus erythematosus: A consequence of patient-intrinsic or -extrinsic factors? J Clin Med. 2023;12(21):6945. https://www.mdpi.com/2077-0383/12/21/6945

Eng JA, Frosch MP, Choi K, Rebeck GW, Greenberg SM. Clinical manifestations of cerebral amyloid angiopathy–related inflammation. Ann Neurol. 2004;55(2):250–6. https://doi.org/10.1002/ana.10810.

Engsig FN, Hansen AB, Omland LH, Kronborg G, Gerstoft J, Laursen AL, et al. Incidence, clinical presentation, and outcome of progressive multifocal leukoencephalopathy in HIV-infected patients during the highly active antiretroviral therapy era: a nationwide cohort study. J Infect Dis. 2009;199(1):77–83. https://doi.org/10.1086/595299.

Fatahzadeh M, Schwartz RA. Human herpes simplex virus infections: epidemiology, pathogenesis, symptomatology, diagnosis, and management. J Am Acad Dermatol. 2007;57(5):737–63. https://doi.org/10.1016/j.jaad.2007.06.027.

Ferracci F, Moretto G, Candeago RM, Cimini N, Conte F, Gentile M, et al. Antithyroid antibodies in the CSF: their role in the pathogenesis of Hashimoto's encephalopathy. Neurology. 2003;60(4):712–4. https://doi.org/10.1212/01.wnl.0000048660.71390.c6.

Ferracci F, Bertiato G, Moretto G. Hashimoto's encephalopathy: epidemiologic data and pathogenetic considerations. J Neurol Sci. 2004;217(2):165–8. https://doi.org/10.1016/j.jns.2003.09.007.

Filippi M, Rocca MA. Acute disseminated encephalomyelitis. White Matter Diseases. 2020;109–125 https://doi.org/10.1007/978-3-030-38621-4_5.

Fischer M, Schmutzhard E. Posterior reversible encephalopathy syndrome. J Neurol. 2017;264(8):1608–16. https://doi.org/10.1007/s00415-016-8377-8.

Fisher CA, Bernard C. A systematic review of neurocognitive functioning in Behçet's disease. Neuropsychol Rev. 2019;29(4):498–521. https://doi.org/10.1007/s11065-019-09416-5.

Fitts W, Vogel AC, Mateen FJ. The changing face of osmotic demyelination syndrome: a retrospective, observational cohort study. Neurol Clin Pract. 2021;11(4):304–10. https://doi.org/10.1212/cpj.0000000000000932.

Frota NAF, Caramelli P, Barbosa ER. Cognitive impairment in Wilson's disease. Dementia Neuropsychol. 2009;3(1):16–21. https://doi.org/10.1590/s1980-57642009dn30100004.

Fukunaga M, Murai H, Minohara M, Kikuchi H, Ohyagi Y, Kira J. Hashimoto's encephalopathy with bilateral pallidal lesions presenting memory disturbance and executive dysfunction. Rinsho Shinkeigaku. 2006;46(8):568–71.

Gao J-H, Li W-R, Xu D-M, Zheng B-W, Huang Y-M, Wu W-Q, Zhang W. Clinical manifestations, fluid changes and neuroimaging alterations in patients with general paresis of the insane. Neuropsychiatr Dis Treat. 2021;17(null):69–78. https://doi.org/10.2147/NDT.S279265.

Geschwind MD, Shu H, Haman A, Sejvar JJ, Miller BL. Rapidly progressive dementia. Ann Neurol. 2008;64(1):97–108. https://doi.org/10.1002/ana.21430.

Ghaffari-Rafi A, Ghaffari-Rafi S, Leon-Rojas J. Socioeconomic and demographic disparities of moyamoya disease in the United States. Clin Neurol Neurosurg. 2020;192:105719. https://doi.org/10.1016/j.clineuro.2020.105719.

Ghanem KG. REVIEW: Neurosyphilis: a historical perspective and review. CNS Neurosci Ther. 2010;16(5):e157–68. https://doi.org/10.1111/j.1755-5949.2010.00183.x.

Ghetti B, Dlouhy SR, Giaccone G, Bugiani O, Frangione B, Farlow MR, Tagliavini F. Gerstmann-Sträussler-Scheinker disease and the Indiana kindred. Brain Pathol. 1995;5(1):61–75. https://doi.org/10.1111/j.1750-3639.1995.tb00578.x.

Gnann JW Jr, Whitley RJ. Herpes simplex encephalitis: an update. Curr Infect Dis Rep. 2017;19(3):13. https://doi.org/10.1007/s11908-017-0568-7.

Golden C, Freshwater S. The Stroop color and word test: a manual for clinical and experimental uses. Stoelting Co.; 2002.

Gonzalez NR, Amin-Hanjani S, Bang OY, Coffey C, Du R, Fierstra J, et al. Adult Moyamoya disease and syndrome: current perspectives and future directions: a scientific statement from the American Heart Association/American Stroke Association. Stroke. 2023;54(10):e465–79. https://doi.org/10.1161/STR.0000000000000443.

Grasso D, Castorani G, Borreggine C, Simeone A, De Blasi R. Cerebral amyloid angiopathy related inflammation: a little known but not to be underestimated disease. Radiol Case Rep. 2021;16(9):2514–21. https://doi.org/10.1016/j.radcr.2021.05.080.

Green B, Griffiths E, Almond S. Neuropsychiatric symptoms following metal-on-metal implant failure with cobalt and chromium toxicity. BMC Psychiatry. 2017;17(1):33. https://doi.org/10.1186/s12888-016-1174-1.

Greenberg SM, Bacskai BJ, Hernandez-Guillamon M, Pruzin J, Sperling R, van Veluw SJ. Cerebral amyloid angiopathy and Alzheimer disease – one peptide, two pathways. Nat Rev Neurol. 2020;16(1):30–42. https://doi.org/10.1038/s41582-019-0281-2.

Hamill MM, Ghanem KG, Tuddenham S. State-of-the-art review: neurosyphilis. Clin Infect Dis. 2023; https://doi.org/10.1093/cid/ciad437.

Haut MW, Hartzell JW, Moran MT. Toxins in the central nervous system. Taylor & Francis; 2017.

Hemachudha T, Ugolini G, Wacharapluesadee S, Sungkarat W, Shuangshoti S, Laothamatas J. Human rabies: neuropathogenesis, diagnosis, and management. Lancet Neurol. 2013;12(5):498–513. https://doi.org/10.1016/s1474-4422(13)70038-3.

Hill G, Archibald N, Larkin P, Farquhar J, Evans J. Foreign accent syndrome and other neuropsychological sequelae of a parieto-occipital lesion following COVID-19 associated posterior reversible encephalopathy syndrome. Clin Neuropsychol. 2023;37(6):1321–35. https://doi.org/10.1080/13854046.2022.2108903.

Hinduja A. Posterior reversible encephalopathy syndrome: clinical features and outcome. Front Neurol. 2020;11:71. https://doi.org/10.3389/fneur.2020.00071.

Ho RC, Husain SF, Ho CS. Cognitive dysfunction in patients with systemic lupus erythematosus: the challenge in diagnosis and management. Rheumatol Practice Res. 2018;3:2059902118792434. https://doi.org/10.1177/2059902118792434.

Hogarth P. Neurodegeneration with brain iron accumulation: diagnosis and management. J Mov Disord. 2015;8(1):1–13. https://doi.org/10.14802/jmd.14034.

Holliday SL, Navarrete MG, Hermosillo-Romo D, Valdez CR, Saklad AR, Escalante A, Brey RL. Validating a computerized neuropsychological test battery for mixed ethnic lupus patients. Lupus. 2003;12(9):697–703. https://doi.org/10.1191/0961203303lu442oa.

Huang X, Law S, Li D, Yu X, Li B. Mercury poisoning: a case of a complex neuropsychiatric illness. Am J Psychiatry. 2014;171(12):1253–6. https://doi.org/10.1176/appi.ajp.2013.12101266.

Hughes PS, Jackson AC. Delays in initiation of acyclovir therapy in herpes simplex encephalitis. Can J Neurol Sci. 2012;39(5):644–8. https://doi.org/10.1017/s0317167100015390.

Izmirly PM, Parton H, Wang L, McCune WJ, Lim SS, Drenkard C, et al. Prevalence of systemic lupus erythematosus in the United States: estimates from a meta-analysis of the Centers for Disease Control and Prevention National Lupus Registries. Arthritis Rheumatol. 2021;73(6):991–6. https://doi.org/10.1002/art.41632.

Jackson AC. Human rabies: a 2016 update. Curr Infect Dis Rep. 2016;18(11):38. https://doi.org/10.1007/s11908-016-0540-y.

Jacobs DE, Clickner RP, Zhou JY, Viet SM, Marker DA, Rogers JW, et al. The prevalence of lead-based paint hazards in U.S. housing. Environ Health Perspect. 2002;110(10):A599–606. https://doi.org/10.1289/ehp.021100599.

Jäkel L, De Kort AM, Klijn CJM, Schreuder FHBM, Verbeek MM. Prevalence of cerebral amyloid angiopathy: a systematic review and meta-analysis. Alzheimers Dement. 2022;18(1):10–28. https://doi.org/10.1002/alz.12366.

Jennische E, Eriksson CE, Lange S, Trybala E, Bergström T. The anterior commissure is a pathway for contralateral spread of herpes simplex virus type 1 after olfactory tract infection. J Neurovirol. 2015;21(2):129–47. https://doi.org/10.1007/s13365-014-0312-0.

Johnson JCS, McWhirter L, Hardy CJD, Crutch SJ, Marshall CR, Mummery CJ, et al. Suspecting dementia: canaries, chameleons and zebras. Pract Neurol. 2021; https://doi.org/10.1136/practneurol-2021-003019.

Jolly AA, Nannoni S, Edwards H, Morris RG, Markus HS. Prevalence and predictors of vascular cognitive impairment in patients with CADASIL. Neurology. 2022;99(5):e453–61. https://doi.org/10.1212/WNL.0000000000200607.

Joly M, Conte C, Cazanave C, Le Moing V, Tattevin P, Delobel P, et al. Progressive multifocal leukoencephalopathy: epidemiology and spectrum of predisposing conditions. Brain. 2023;146(1):349–58. https://doi.org/10.1093/brain/awac237.

Kanmaz S, Köse S, Eraslan C, Şimşek E, Serin HM, Yılmaz S, et al. Neuropsychological outcome in cases with acute disseminated encephalomyelitis. Turk J Pediatr. 2020;62(4):594–605. https://doi.org/10.24953/turkjped.2020.04.009.

Kapoor T, Mahadeshwar P, Hui-Yuen J, Quinnies K, Tatonetti N, Gartshteyn Y, et al. Prevalence of progressive multifocal leukoencephalopathy (PML) in adults and children with systemic lupus erythematosus. Lupus Sci Med. 2020;7(1) https://doi.org/10.1136/lupus-2020-000388.

Kartau M, Sipilä JO, Auvinen E, Palomäki M, Verkkoniemi-Ahola A. Progressive multifocal leukoencephalopathy: current insights. Degener Neurol Neuromuscul Dis. 2019;9:109–21. https://doi.org/10.2147/dnnd.S203405.

Kaur B, Khanna D. A narrative review of the many psychiatric manifestations of Neurosyphilis: the great imitator. Cureus. 2023;15(9):e44866. https://doi.org/10.7759/cureus.44866.

Kennedy PGE, Gershon AA. Clinical features of varicella-zoster virus infection. Viruses. 2018;10(11) https://doi.org/10.3390/v10110609.

Khandelwal D, Mathur V, Vyas A, Ghunawat J, Bagaria AK. CARASIL–A review of patients from India. Neurol India. 2021;69(5):1359–62. https://doi.org/10.4103/0028-3886.329544.

Kim JS. Moyamoya disease: epidemiology, clinical features, and diagnosis. J Stroke. 2016;18(1):2–11. https://doi.org/10.5853/jos.2015.01627.

Kim MO, Takada LT, Wong K, Forner SA, Geschwind MD. Genetic PrP prion diseases. Cold Spring Harb Perspect Biol. 2018;10(5) https://doi.org/10.1101/cshperspect.a033134.

Kim JJ, Kim YS, Kumar V. Heavy metal toxicity: an update of chelating therapeutic strategies. J Trace Elem Med Biol. 2019;54:226–31. https://doi.org/10.1016/j.jtemb.2019.05.003.

King JD, Rosner MH. Osmotic demyelination syndrome. Am J Med Sci. 2010;339(6):561–7. https://doi.org/10.1097/MAJ.0b013e3181d3cd78.

Kinner M, Prehn C, Schneider R, Schroeder C, Kolb E, Gold R, Hoepner R, Chan A. Course of neuropsychological impairment during natalizumab-associated progressive multifocal leukoencephalopathy. Eur J Neurol. 2021;28(3):921–7. https://doi.org/10.1111/ene.14604.

Krupp LB, Lipton RB, Swerdlow ML, Leeds NE, Llena J. Progressive multifocal leukoencephalopathy: clinical and radiographic features. Ann Neurol. 1985;17(4):344–9. https://doi.org/10.1002/ana.410170407.

Kuhn A, Bonsmann G, Anders HJ, Herzer P, Tenbrock K, Schneider M. The diagnosis and treatment of systemic lupus erythematosus. Dtsch Arztebl Int. 2015;112(25):423–32. https://doi.org/10.3238/arztebl.2015.0423.

Lagström RMB, Østerbye NN, Henriksen OM, Høgh P. Hashimoto's encephalopathy: follow-up data from neuropsychology, lumbar puncture, and FDG-PET. Clin Case Rep. 2019;7(9):1750–3. https://doi.org/10.1002/ccr3.2367.

Lambeck J, Hieber M, Dreßing A, Niesen WD. Central pontine Myelinosis and osmotic demyelination syndrome. Dtsch Arztebl Int. 2019;116(35–36):600–6. https://doi.org/10.3238/arztebl.2019.0600.

Leslie B, Crowe SF. Cognitive functioning in systemic lupus erythematosus: a meta-analysis. Lupus. 2018;27(6):920–9. https://doi.org/10.1177/0961203317751859.

Li J, Li F. Hashimoto's encephalopathy and seizure disorders. Front Neurol. 2019;10:440. https://doi.org/10.3389/fneur.2019.00440.

Liman TG, Bohner G, Heuschmann PU, Endres M, Siebert E. The clinical and radiological spectrum of posterior reversible encephalopathy syndrome: the retrospective Berlin PRES study. J Neurol. 2012;259(1):155–64. https://doi.org/10.1007/s00415-011-6152-4.

Lin YT, Liao SC. Hashimoto encephalopathy presenting as schizophrenia-like disorder. Cogn Behav Neurol. 2009;22(3):197–201. https://doi.org/10.1097/WNN.0b013e318197926e.

Lindå H, von Heijne A. Presymptomatic diagnosis with MRI and adequate treatment ameliorate the outcome after Natalizumab-associated progressive multifocal leukoencephalopathy [clinical case study]. Front Neurol. 2013;4 https://doi.org/10.3389/fneur.2013.00011.

Lindsey NP, Staples JE, Fischer M. Eastern equine encephalitis virus in the United States, 2003-2016. Am J Trop Med Hyg. 2018;98(5):1472–7. https://doi.org/10.4269/ajtmh.17-0927.

Lorincz MT. Neurologic Wilson's disease. Ann N Y Acad Sci. 2010;1184:173–87. https://doi.org/10.1111/j.1749-6632.2009.05109.x.

Ludlow M, Kortekaas J, Herden C, Hoffmann B, Tappe D, Trebst C, et al. Neurotropic virus infections as the cause of immediate and delayed neuropathology. Acta Neuropathol. 2016;131(2):159–84. https://doi.org/10.1007/s00401-015-1511-3.

Luo L, Wang B, Jiang J, Fitzgerald M, Huang Q, Yu Z, et al. Heavy metal contaminations in herbal medicines: determination, comprehensive risk assessments, and solutions. Front Pharmacol. 2020;11:595335. https://doi.org/10.3389/fphar.2020.595335.

Machado A, Chien HF, Deguti MM, Cançado E, Azevedo RS, Scaff M, Barbosa ER. Neurological manifestations in Wilson's disease: report of 119 cases. Mov Disord. 2006;21(12):2192–6. https://doi.org/10.1002/mds.21170.

Manca E. Autoantibodies in neuropsychiatric systemic lupus erythematosus (NPSLE): can they be used as biomarkers for the differential diagnosis of this disease? Clin Rev Allergy Immunol. 2022;63(2):194–209. https://doi.org/10.1007/s12016-021-08865-2.

Manini A, Querzola G, Lovati C, Pantoni L. Rapidly progressive dementia and intractable diarrhea: a teaching case report and a systematic review of cognitive impairment in Whipple's disease. Neurol Sci. 2022;43(2):907–26. https://doi.org/10.1007/s10072-021-05844-5.

Martinelli P, Rizzo G, Scaglione C, Capellari S. Neurosyphilis orofacial dyskinesia: the candy sign. Mov Disord. 2013;28(2):246–7. https://doi.org/10.1002/mds.25281.

Martucci M, Sarria S, Toledo M, Coscojuela P, Vert C, Siurana S, et al. Cerebral amyloid angiopathy-related inflammation: imaging findings and clinical outcome. Neuroradiology. 2014;56(4):283–9. https://doi.org/10.1007/s00234-014-1330-6.

Mazzù I, Mosti S, Caltagirone C, Carlesimo GA. Hashimoto's encephalopathy: neuropsychological findings. Neurol Sci. 2012;33(3):653–6. https://doi.org/10.1007/s10072-011-0813-z.

McCabe DJ, Burke T, Connolly S, Hutchinson M. Amnesic syndrome with bilateral mesial temporal lobe involvement in Hashimoto's encephalopathy. Neurology. 2000;54(3):737–9. https://doi.org/10.1212/wnl.54.3.737.

Mead S, Rudge P. CJD mimics and chameleons. Pract Neurol. 2017;17(2):113–21. https://doi.org/10.1136/practneurol-2016-001571.

Mecklenburg J, Moos V, Moter A, Siebert E, Nave AH, Schneider T, et al. The spectrum of central nervous system involvement in Whipple's disease. Eur J Neurol. 2023;30(11):3417–29. https://doi.org/10.1111/ene.15511.

Mikulski MA, Wichman MD, Simmons DL, Pham AN, Clottey V, Fuortes LJ. Toxic metals in ayurvedic preparations from a public health lead poisoning cluster investigation. Int J Occup Environ Health. 2017;23(3):187–92. https://doi.org/10.1080/10773525.2018.1447880.

Minikel EV, Vallabh SM, Orseth MC, Brandel JP, Haïk S, Laplanche JL, et al. Age at onset in genetic prion disease and the design of preventive clinical trials. Neurology. 2019;93(2):e125–34. https://doi.org/10.1212/wnl.0000000000007745.

Mizuno T, Mizuta I, Watanabe-Hosomi A, Mukai M, Koizumi T. Clinical and genetic aspects of CADASIL [review]. Front Aging Neurosci. 2020;12 https://doi.org/10.3389/fnagi.2020.00091.

Modi S, Mahajan A, Dharaiya D, Varelas P, Mitsias P. Burden of herpes simplex virus encephalitis in the United States. J Neurol. 2017;264(6):1204–8. https://doi.org/10.1007/s00415-017-8516-x.

Morens DM, Folkers GK, Fauci AS. Eastern equine encephalitis virus—another emergent arbovirus in the United States. N Engl J Med. 2019;381(21):1989–92. https://doi.org/10.1056/NEJMp1914328.

Muayqil T, Gronseth G, Camicioli R. Evidence-based guideline: diagnostic accuracy of CSF 14-3-3 protein in sporadic Creutzfeldt-Jakob disease: report of the guideline development subcommittee of the American Academy of Neurology. Neurology. 2012;79(14):1499–506. https://doi.org/10.1212/WNL.0b013e31826d5fc3.

Nakatani E, Kanatani Y, Kaneda H, Nagai Y, Teramukai S, Nishimura T, et al. Specific clinical signs and symptoms are predictive of clinical course in sporadic Creutzfeldt-Jakob disease. Eur J Neurol. 2016;23(9):1455–62. https://doi.org/10.1111/ene.13057.

Nozaki H, Sekine Y, Fukutake T, Nishimoto Y, Shimoe Y, Shirata A, Yanagawa S, Hirayama M, Tamura M, Nishizawa M, Onodera O. Characteristic features and progression of abnormalities on MRI for CARASIL. Neurology. 2015;85(5):459–63. https://doi.org/10.1212/WNL.0000000000001803

Odier C, Nguyen DK, Panisset M. Central pontine and extrapontine myelinolysis: from epileptic and other manifestations to cognitive prognosis. J Neurol. 2010;257(7):1176–80. https://doi.org/10.1007/s00415-010-5486-7.

Osterrieth PA. Le test de copie d'une figure complexe; contribution à l'étude de la perception et de la mémoire. Test of copying a complex figure; contribution to the study of perception and memory. Arch Sci Psychol. 1944;30:206–356.

Padgett BL, Walker DL, ZuRhein GM, Eckroade RJ, Dessel BH. Cultivation of papova-like virus from human brain with progressive multifocal leucoencephalopathy. Lancet. 1971;1(7712):1257–60. https://doi.org/10.1016/s0140-6736(71)91777-6.

Papageorgiou SG, Kontaxis T, Bonakis A, Karahalios G, Kalfakis N, Vassilopoulos D. Rapidly progressive dementia: causes found in a Greek tertiary referral center in Athens. Alzheimer Dis Assoc Disord. 2009;23(4):337–46. https://doi.org/10.1097/WAD.0b013e31819e099b.

Patel SD, Desai N, Rane S, Patel N, Desai R, Mehta T, et al. Trends in hospitalizations and epidemiological characteristics of adults Moyamoya disorder in the United States. J Neurol Sci. 2020;419:117165. https://doi.org/10.1016/j.jns.2020.117165.

Pavlovic D, Patera AC, Nyberg F, Gerber M, Liu M. Progressive multifocal leukoencephalopathy: current treatment options and future perspectives. Ther Adv Neurol Disord. 2015;8(6):255–73. https://doi.org/10.1177/1756285615602832.

Peine B, Figueroa C, Robinette N. Neuro-Behcet's syndrome: Case report and literature review. Radiol Case Rep. 2022;17(9):3064–70. https://doi.org/10.1016/j.radcr.2022.05.070.

Peng Y, Li Z, Yang X, Yang L, He M, Zhang H, et al. Relation between cadmium body burden and cognitive function in older men: a cross-sectional study in China. Chemosphere. 2020;250:126535. https://doi.org/10.1016/j.chemosphere.2020.126535.

Pettersen JA, Keith J, Gao F, Spence JD, Black SE. CADASIL accelerated by acute hypotension. Neurology. 2017;88(11):1077–80. https://doi.org/10.1212/WNL.0000000000003717.

Phan-Ba R, Lommers E, Tshibanda L, Calay P, Dubois B, Moonen G, et al. MRI preclinical detection and asymptomatic course of a progressive multifocal leucoencephalopathy (PML) under natalizumab therapy. J Neurol Neurosurg Psychiatry. 2012;83(2):224–6. https://doi.org/10.1136/jnnp-2011-300511.

Pieracci EG, Pearson CM, Wallace RM, Blanton JD, Whitehouse ER, Ma X, et al. Vital signs: trends in human rabies deaths and exposures – United States, 1938-2018. MMWR Morb Mortal Wkly Rep. 2019;68(23):524–8. https://doi.org/10.15585/mmwr.mm6823e1.

Pohl D, Alper G, Van Haren K, Kornberg AJ, Lucchinetti CF, Tenembaum S, Belman AL. Acute disseminated encephalomyelitis: updates on an inflammatory CNS syndrome. Neurology. 2016;87(9 Suppl 2):S38–45. https://doi.org/10.1212/wnl.0000000000002825.

Popescu A, Kao H, A. Neuropsychiatric systemic lupus erythematosus. Curr Neuropharmacol. 2011;9(3):449–57. https://doi.org/10.2174/157015911796557984.

Prusiner SB. Prions. Proc Natl Acad Sci USA. 1998;95(23):13363–83. https://doi.org/10.1073/pnas.95.23.13363.

Puéchal X. Whipple's disease. Ann Rheum Dis. 2013;72(6):797–803. https://doi.org/10.1136/annrheumdis-2012-202684.

Rannikmäe K, Kalaria RN, Greenberg SM, Chui HC, Schmitt FA, Samarasekera N, et al. APOE associations with severe CAA-associated vasculopathic changes: collaborative meta-analysis. J Neurol Neurosurg Psychiatry. 2014;85(3):300–5. https://doi.org/10.1136/jnnp-2013-306485.

Rayan MN, Bassi R, Khazem M, Pozo DA, Abduljaber W, Burtis DB. Herpes simplex encephalitis: detection, management, and outcomes. Cureus. 2022;14(11):e31962. https://doi.org/10.7759/cureus.31962.

Reitan RM Trail making test: manual for administration and scoring; 1992.

Rey A. L'examen psychologique dans les cas d'encéphalopathie traumatique. Arch Sci Psychol. 1941;28:215–85.

Riancho J, Delgado-Alvarado M, Sedano MJ, Polo JM, Berciano J. Herpes simplex encephalitis: clinical presentation, neurological sequelae and new prognostic factors. Ten years of experience. Neurol Sci. 2013;34(10):1879–81. https://doi.org/10.1007/s10072-013-1475-9.

Ronsin S, Deiana G, Geraldo AF, Durand-Dubief F, Thomas-Maisonneuve L, Formaglio M, et al. Pseudotumoral presentation of cerebral amyloid angiopathy-related inflammation. Neurology. 2016;86(10):912–9. https://doi.org/10.1212/wnl.0000000000002444.

Ropper AH. Neurosyphilis. N Engl J Med. 2019;381(14):1358–63. https://doi.org/10.1056/NEJMra1906228.

Rosand J, Muzikansky A, Kumar A, Wisco JJ, Smith EE, Betensky RA, Greenberg SM. Spatial clustering of hemorrhages in probable cerebral amyloid angiopathy. Ann Neurol. 2005;58(3):459–62. https://doi.org/10.1002/ana.20596.

Sabbah-Talasazan L, Piryatinsky I. Neuropsychological impairment in Hashimoto's encephalopathy: a case report and literature review. Appl Neuropsychol Adult. 2018;25(6):572–80. https://doi.org/10.1080/23279095.2017.1326048.

Sakai K, Ueda M, Fukushima W, Tamaoka A, Shoji M, Ando Y, Yamada M. Nationwide survey on cerebral amyloid angiopathy in Japan. Eur J Neurol. 2019;26(12):1487–93. https://doi.org/10.1111/ene.14031.

Sala I, Marquié M, Sánchez-Saudinós MB, Sánchez-Valle R, Alcolea D, Gómez-Ansón B, et al. Rapidly progressive dementia: experience in a tertiary care medical center. Alzheimer Dis Assoc Disord. 2012;26(3):267–71. https://doi.org/10.1097/WAD.0b013e3182368ed4.

Saper RB, Phillips RS, Sehgal A, Khouri N, Davis RB, Paquin J, et al. Lead, mercury, and arsenic in US- and Indian-manufactured Ayurvedic medicines sold via the internet. JAMA. 2008;300(8):915–23. https://doi.org/10.1001/jama.300.8.915.

Schilsky ML, Roberts EA, Bronstein JM, Dhawan A, Hamilton JP, Rivard AM, et al. A multidisciplinary approach to the diagnosis and management of Wilson disease: executive summary of the 2022 practice guidance on Wilson disease from the American Association for the Study of Liver Diseases. Hepatology. 2023;77(4):1428–55. https://doi.org/10.1002/hep.32805.

Schmidt R, Carson PJ, Jansen RJ. Resurgence of syphilis in the United States: an assessment of contributing factors. Infect Dis (Auckl). 2019;12:1178633719883282. https://doi.org/10.1177/1178633719883282.

Schwarz S, Mohr A, Knauth M, Wildemann B, Storch-Hagenlocher B. Acute disseminated encephalomyelitis: a follow-up study of 40 adult patients. Neurology. 2001;56(10):1313–8.

Sejvar JJ. West Nile virus infection. Microbiol Spectr. 2016;4(3) https://doi.org/10.1128/microbiolspec.EI10-0021-2016.

Shah R, Bag AK, Chapman PR, Curé JK. Imaging manifestations of progressive multifocal leukoencephalopathy. Clin Radiol. 2010;65(6):431–9. https://doi.org/10.1016/j.crad.2010.03.001.

Shiga Y, Miyazawa K, Sato S, Fukushima R, Shibuya S, Sato Y, et al. Diffusion-weighted MRI abnormalities as an early diagnostic marker for Creutzfeldt-Jakob disease. Neurology. 2004;63(3):443–9. https://doi.org/10.1212/01.wnl.0000134555.59460.5d.

Sili U, Kaya A, Mert A. Herpes simplex virus encephalitis: clinical manifestations, diagnosis and outcome in 106 adult patients. J Clin Virol. 2014;60(2):112–8. https://doi.org/10.1016/j.jcv.2014.03.010.

Singer R. Neurotoxicity in neuropsychology. In: Schoenberg MR, Scott JG, editors. The little Black book of neuropsychology: a syndrome-based approach. US: Springer; 2011. p. 813–38.

Singh TD, Fugate JE, Rabinstein AA. Central pontine and extrapontine myelinolysis: a systematic review. Eur J Neurol. 2014;21(12):1443–50. https://doi.org/10.1111/ene.12571.

Singh TD, Fugate JE, Hocker S, Wijdicks EFM, Aksamit AJ Jr, Rabinstein AA. Predictors of outcome in HSV encephalitis. J Neurol. 2016;263(2):277–89. https://doi.org/10.1007/s00415-015-7960-8.

Skripuletz T, Pars K, Schulte A, Schwenkenbecher P, Yildiz Ö, Ganzenmueller T, et al. Varicella zoster virus infections in neurological patients: a clinical study. BMC Infect Dis. 2018;18(1):238. https://doi.org/10.1186/s12879-018-3137-2.

Snowden JS, Mann DM, Neary D. Distinct neuropsychological characteristics in Creutzfeldt-Jakob disease. J Neurol Neurosurg Psychiatry. 2002;73(6):686–94. https://doi.org/10.1136/jnnp.73.6.686.

Soldan SS, Lieberman PM. Epstein-Barr virus infection in the development of neurological disorders. Drug Discov Today Dis Models. 2020;32(Pt A):35–52. https://doi.org/10.1016/j.ddmod.2020.01.001.

Sonneville R, Demeret S, Klein I, Bouadma L, Mourvillier B, Audibert J, et al. Acute disseminated encephalomyelitisin the intensive care unit: clinical features and outcome of 20 adults. Intensive Care Med. 2008;34:528–32.

Souirti Z, Lahlou M, Ouali OE, Chtaou N, Aarab C, Ghazouani FE, et al. Neuropsychiatric systemic lupus erythematosus. Open J Rheumatol Autoimm Dis. 2013;03(02):31388. https://doi.org/10.4236/ojra.2013.32013.

Stefani A, Riello M, Rossini F, Mariotto S, Fenzi F, Gambina G, et al. Neurosyphilis manifesting with rapidly progressive dementia: report of three cases. Neurol Sci. 2013;34(11):2027–30. https://doi.org/10.1007/s10072-013-1531-5.

Stroescu I, Salinas CM, Nahab FB, Stringer AY. Long-term neurocognitive and neuroimaging outcomes in posterior reversible encephalopathy syndrome: two case reports and implications. Clin Neuropsychol. 2011;25(8):1386–402. https://doi.org/10.1080/13854046.2011.628339.

Sudhakar P, Bachman DM, Mark AS, Berger JR, Kedar S. Progressive multifocal leukoencephalopathy: recent advances and a neuro-ophthalmological review. J Neuroophthalmol. 2015;35(3):296–305. https://doi.org/10.1097/wno.0000000000000271.

Sundaram SE, Staffaroni AM, Walker NC, Casaletto KB, Casey M, Golubjatnikov A, et al. Baseline neuropsychological profiles in prion disease predict survival time. Ann Clin Transl Neurol. 2020;7(9):1535–45. https://doi.org/10.1002/acn3.51115.

Tam J, Centola J, Kurudzhu H, Watson N, MacKenzie J, Leitch M, et al. Sporadic Creutzfeldt-Jakob disease in the young (50 and below): 10-year review of United Kingdom surveillance. J Neurol. 2023;270(2):1036–46. https://doi.org/10.1007/s00415-022-11467-3.

Tan IL, McArthur JC, Venkatesan A, Nath A. Atypical manifestations and poor outcome of herpes simplex encephalitis in the immunocompromised. Neurology. 2012;79(21):2125–32. https://doi.org/10.1212/WNL.0b013e3182752ceb.

Taylor SW, Smith RM, Pari G, Wobeser W, Rossiter JP, Jackson AC. Herpes simplex encephalitis. Can J Neurol Sci. 2005;32(2):246–7. https://doi.org/10.1017/s0317167100004054.

Thompson HS, Kardon RH. The Argyll Robertson pupil. J Neuroophthalmol. 2006;26(2):134–8. https://doi.org/10.1097/01.wno.0000222971.09745.91.

Tselis AC, Rajamani K, Bhattacharya P. Epstein–Barr virus and the nervous system. In: Nath A, Berger JR, editors. Clinical neurovirology. CRC Press; 2020. p. 73–87.

Tuddenham S, Ghanem KG. Neurosyphilis: knowledge gaps and controversies. Sex Transm Dis. 2018;45(3):147–51. https://doi.org/10.1097/olq.0000000000000723.

Uttley L, Carroll C, Wong R, Hilton DA, Stevenson M. Creutzfeldt-Jakob disease: a systematic review of global incidence, prevalence, infectivity, and incubation. Lancet Infect Dis. 2020;20(1):e2–e10. https://doi.org/10.1016/s1473-3099(19)30615-2.

Vermersch P, Kappos L, Gold R, Foley JF, Olsson T, Cadavid D, et al. Clinical outcomes of natalizumab-associated progressive multifocal leukoencephalopathy. Neurology. 2011;76(20):1697–704. https://doi.org/10.1212/WNL.0b013e31821a446b.

Waliszewska-Prosół M, Ejma M. Hashimoto encephalopathy-still more questions than answers. Cells. 2022;11(18) https://doi.org/10.3390/cells11182873.

Wang J, Zhang J, Xu L, Shi Y, Wu X, Guo Q. Cognitive impairments in Hashimoto's encephalopathy: a case-control study. PLoS One. 2013;8(2):e55758. https://doi.org/10.1371/journal.pone.0055758.

Wechsler D. WMS-IV Administration and scoring manual. Pearson; 2009.

Whitley RJ. Chapter 68: herpesviruses. In: Medical microbiology. Galveston: University of Texas Medical Branch at Galveston; 1996.

Wieser HG, Schindler K, Zumsteg D. EEG in Creutzfeldt-Jakob disease. Clin Neurophysiol. 2006;117(5):935–51. https://doi.org/10.1016/j.clinph.2005.12.007.

Wu LY, Zhan SQ, Huang ZY, Zhang B, Wang T, Liu CF, et al. Expert consensus on clinical diagnostic criteria for fatal familial insomnia. Chin Med J. 2018;131(13):1613–7. https://doi.org/10.4103/0366-6999.235115.

Wu JJ, Yao M, Ni J. Cerebral amyloid angiopathy-related inflammation: current status and future implications. Chin Med J. 2021;134(6):646–54. https://doi.org/10.1097/cm9.0000000000001427.

Wüthrich C, Koralnik IJ. Frequent infection of cortical neurons by JC virus in patients with progressive multifocal leukoencephalopathy. J Neuropathol Exp Neurol. 2012;71(1):54–65. https://doi.org/10.1097/NEN.0b013e31823ede59.

Yamada M. Cerebral amyloid Angiopathy: emerging concepts. J Stroke. 2015;17(1):17–30. https://doi.org/10.5853/jos.2015.17.1.17.

Yang TI, Jung DS, Ahn BY, Jeong BH, Cho HJ, Kim YS, et al. Familial Creutzfeldt-Jakob disease with V180I mutation. J Korean Med Sci. 2010;25(7):1097–100. https://doi.org/10.3346/jkms.2010.25.7.1097.

Yazici Y, Hatemi G, Bodaghi B, Cheon JH, Suzuki N, Ambrose N, Yazici H. Behçet syndrome. Nat Rev Dis Primers. 2021;7(1):67. https://doi.org/10.1038/s41572-021-00301-1.

Yu L, Boyle PA, Nag S, Leurgans S, Buchman AS, Wilson RS, et al. APOE and cerebral amyloid angiopathy in community-dwelling older persons. Neurobiol Aging. 2015;36(11):2946–53. https://doi.org/10.1016/j.neurobiolaging.2015.08.008.

Zerr I, Poser S. Clinical diagnosis and differential diagnosis of CJD and vCJD. With special emphasis on laboratory tests. APMIS. 2002;110(1):88–98. https://doi.org/10.1034/j.1600-0463.2002.100111.x.

Zerr I, Kallenberg K, Summers DM, Romero C, Taratuto A, Heinemann U, et al. Updated clinical diagnostic criteria for sporadic Creutzfeldt-Jakob disease. Brain. 2009;132(Pt 10):2659–68. https://doi.org/10.1093/brain/awp191.

Zhang X, Xiao W, Zhang Q, Xia D, Gao P, Su J, et al. Progression in Moyamoya disease: clinical features, neuroimaging evaluation, and treatment. Curr Neuropharmacol. 2022;20(2):292–308. https://doi.org/10.2174/1570159x19666210716114016.

Zhou JY, Xu B, Lopes J, Blamoun J, Li L. Hashimoto encephalopathy: literature review. Acta Neurol Scand. 2017;135(3):285–90. https://doi.org/10.1111/ane.12618.

Zimbrean PC, Schilsky ML. Psychiatric aspects of Wilson disease: a review. Gen Hosp Psychiatry. 2014;36(1):53–62. https://doi.org/10.1016/j.genhosppsych.2013.08.007.

Zucchella C, Sinforiani E, Tavazzi E, Del Bue S, Novati S, Maserati R, et al. Neuropsychological evaluation and follow up in jcv- and non-jcv-related leukoencephalopathies in HIV infection. Neurol Sci. 2011;32(5):833–9. https://doi.org/10.1007/s10072-011-0626-0.

Acquired Brain Injury

11

Rachel Keelan, Amanda Zwilling, and Robin Hanks

ABBREVIATIONS

ABS	Agitated Behavior Scale
ACRM	American Congress of Rehabilitation Medicine
ADL	Activities of Daily Living
AED	Antiepileptic Drug
AGS	American Geriatrics Society
AOC	Alteration of Consciousness
BIAA	Brain Injury Association of America
BSI-18	Brief Symptom Inventory
CACE	Communication Aid to Capacity Evaluation
CBT	Cognitive Behavioral Therapy
CDC	Center for Disease Control
CES-D	Center of Epidemiological Studies—Depression Scale
COMBI	Center for Outcome Measurement in Brain Injury
3D-CAM	Confusional Assessment Method-Defined Delirium
Cog-Log	Cognitive Log
CRS-R JFK	Coma Recovery Scale—Revised
CRSR-FAST	Coma Recovery Scale-Revised for Accelerated Standardized Testing
CT	Computed Tomography
DAI	Diffuse Axonal Injury
DME	Durable Medical Equipment
DOC	Disorders of Consciousness
DoD/DOD	Department of Defense
DSM-V-TR	Diagnostic Statistical Manual—5th Edition Text Revision
DTI	Diffuse Tensor Imaging
EEG	Electroencephalogram
FLAIR	Fluid-Attenuated Inversion Recovery
FSMC	Fatigue Scale for Motor and Cognitive Functions
FSS	Fatigue Severity Scale
GAD-7	Generalized Anxiety Disorder 7-item
GCS	Glasgow Coma Scale

R. Keelan (✉) · A. Zwilling · R. Hanks (✉)
Department of Rehabilitation Psychology and Neuropsychology, Detroit Medical Center, Detroit, MI, USA

Department of Physical Medicine and Rehabilitation, Wayne State University, Detroit, MI, USA
e-mail: Rkeelan@dmc.org; Azwillin@dmc.org; Rhanks@med.wayne.edu

E. K. Bailey et al. (eds.), *Inpatient Neuropsychology*, https://doi.org/10.1007/978-3-032-05880-5_11

GOAT	Galveston Orientation and Amnesia Test
GRE	Gradient Echo
HAM-D	Hamilton Depression Rating-Scale
HPA	Hypothalamic-Pituitary-Adrenal (*axis*)
ICH	Intracerebral Hemorrhages
ICP	Intracranial Pressure
IPH	Intraparenchymal Hemorrhage
LOC	Loss of Consciousness
MAPI	Multivariable Apnea Prediction Index
MAST	Mississippi Aphasia Screening Test
MBSR	Mindfulness-Based Stress Reduction
MCS	Minimally Conscious State
MDD	Major Depressive Disorder
MFIS	Modified Fatigue Impact Scale
MRI	Magnetic Resonance Imaging
mTBI	Mild Traumatic Brain Injury
NASEM	National Academy of Sciences, Engineering, and Medicine
NIH	National Institute of Health
NIHSS	National Institute of Health Stroke Scale
NPO	Nothing by mouth (Latin: *nil* per os)
OEF	Operation Enduring Freedom
OIF	Operation Iraqi Freedom
O-Log	Orientation Log
OSA	Obstructive Sleep Apnea
PCS	Posttraumatic Confusional State
PHQ-9	Patient Health Questionnaire-9 Item
PSD	Post-Stroke Depression
PTA	Posttraumatic Amnesia
PTC	Posttraumatic Confusion
SAH	Subarachnoid Hemorrhage
SE	Status Epilepticus
SDH	Subdural Hematoma
SLP	Speech-Language Pathologist
SSRI	Selective Serotonin Reuptake Inhibitor
SWI	Susceptibility-Weighted Imaging
TBI	Traumatic Brain Injury
TIA	Transient Ischemic Attack
TNK	Tenecteplase
tPA	Tissue Plasminogen Activator
QoL	Quality of Life
UWS	Unresponsive Wakefulness Syndrome
VA	Veterans Affairs, Department of
VS	Vegetative State
WAB	Western Aphasia Battery
WAB-R	Western Aphasia Battery—Revised
WHO	World Health Organization

INTRODUCTION

The role of the clinical neuropsychologist in acute care and inpatient rehabilitation for patients with acquired brain injury (ABI) is unique and specialized. Patients with acute ABI typically present with a complex interplay of medical, cognitive, behavioral, and psychiatric systems. Given our expertise in brain-behavior relationships, neuropsychologists are uniquely suited to identify post-injury symptom sequalae, assess recovery trajectory, opine on prognosis, and address barriers to care that arise in the inpatient setting. In this role, neuropsychologists provide strategies for behavioral management, work to optimize patient engagement in rehabilitation therapies, and help facilitate meaningful communication and recommendations between the inpatient team, patients, and families. As many patients with ABI will be subsequently referred for outpatient neuropsychological services, inpatient neuropsychological consultation enables a seamless continuity of care across inpatient and outpatient settings and better prepares patients for services they may not have previously considered.

To function effectively in these settings, the inpatient neuropsychologist must possess a strong knowledge of the etiologies, pathophysiology, and sequelae of these syndromes to best serve patients and the treatment team. The current chapter outlines fundamental knowledge, processes, and applications of neuropsychological services in hospital and rehabilitation settings following ABI. While the primary patient populations treated may vary setting by setting, inpatient neuropsychologists working with acute ABI are most involved in the care of traumatic brain injuries and strokes. As such, these etiologies will be the primary focus of the current chapter, though much of the presented information may be relevant and easily applied to other ABIs encountered in the acute and inpatient rehabilitation hospital settings.

ACQUIRED BRAIN INJURIES

Acquired brain injury is an inclusive category for any injury to the brain that occurs after birth and impacts its functioning. Brain injuries are often initially dichotomized into two groups—traumatic or nontraumatic—then subclassified based on various injury characteristics and mechanisms, as discussed later in this chapter.

Traumatic Versus Non-traumatic Brain Injuries

Broadly speaking, a traumatic brain injury (TBI) is defined as an injury or alteration to the brain caused by an external force, although specific criteria vary across practice organizations (National Academy of Sciences, Engineering, and Medicine [NASEM] 2019). Common mechanisms of TBI include falls, strike to the head by an object, motor vehicle collisions, intentional self-harm, and assault or domestic violence (Peterson et al. 2019). Blast injuries also are a common mechanism of TBI associated with veterans—especially those who served during select conflicts, such as veterans from the Operation Iraqi Freedom/Operation Enduring Freedom campaigns (OIF/OEF; Jackson et al. 2008). In contrast to traumatic injuries, nontraumatic brain injuries are those that do not result from external trauma. Most commonly, nontraumatic brain injuries in acute medical and inpatient rehabilitation settings include stroke, though may also include other neurological conditions (e.g., neoplasms, autoimmune lesions). Stroke mechanisms are classically characterized as *hemorrhagic* and *ischemic* in nature and will be outlined in a later section.

Understanding the mechanism of ABI and being able to synthesize this information with other associated data (e.g., neuroimaging findings, cognitive impairment) is vitally important, as these injury characteristics may differentially impact recovery and prognosis. For example, TBI sustained from high velocity impact are more susceptible to diffuse axonal injury (DAI), which is associated with greater diffuse cognitive impairment and correspondingly poorer outcomes (Macruz et al. 2022; Chen et al. 2023; Humble et al. 2018; Lohani et al. 2020). The current chapter will focus primarily on detailing the mechanisms of TBI and stroke, though much of the applied portion of the current chapter outlines transferable assessment and intervention guidance for diverse ABI etiologies.

Focal and Diffuse Injuries

Focal Contusions, Hemorrhages, and Hematomas

Focal contusions are local abrasions to the surface of the brain resulting from impact of parenchyma to the skull following a linear and/or rotation force (Georges 2023). Inferior frontal orbital regions are particularly vulnerable to contusion due to the proximity of the orbital plate of the frontal bone and cribriform plate of the ethmoid bone. Just anterior to the temporal lobes, the raised sphenoid wing of the skull also poses risk of contusion to inferior anterior temporal areas. As with focal penetrating injuries, the neurocognitive sequela of contusions are usually localized to the site of injury (Georges 2023; Maas et al. 2008).

Cerebral hemorrhages (active bleeding within the skull cavity) and hematomas (pooled, clotted blood) can also result in deficits localized to the region of injury, although sequelae are also contingent on the anatomical boundaries of the bleed itself. Intracranial bleeds are subdivided into four types: Epidural hematomas (EDH) which form between the skull and the dura mater, subdural hematomas (SDH) which occur between the dura mater and leptomeninges, subarachnoid hemorrhages (SAH) which form between the pia mater and arachnoid mater, and lastly, intraparenchymal and/or intracerebral hemorrhages (IPH/ICH) which refers to bleeding within the brain tissue. Each of these bleed types has unique pathological and clinical features. Epidural bleeds are particularly associated with high mortality, as patients will often experience a period of immediate post-injury lucidity (lucid interval) prior to symptom onset that can cause delayed intervention (Khairat and Waseem 2023). SAHs develop rapidly, with sudden onset of severe head pain (thunderclap headache), with complaints of "*the worst headache of my life*." Symptoms of both SAH and IPH include headache, nausea, vomiting, sensory/motor changes, and loss of consciousness (LOC), and both are typically associated with arterial hemorrhages which tend to be higher pressure/volume and rapidly evolving. In contrast, SDHs often evolve more slowly due to the venous source of hemorrhaging blood.

Familiarity with the type and location of hemorrhages can contextualize a patient's early signs and symptoms, point to potential mechanisms of injury, and guide the determination of appropriate assessment methods. For example, a neuropsychologist who learns that a patient sustained a left frontal SAH would be prepared to assess and accommodate potential expressive aphasia in their intake, whereas knowledge of a right parietal IPH may prompt evaluation of visual disturbance and/or sensory inattention. Understanding the location and type of bleed also affords the neuropsychologist an opportunity to help family members or loved ones make sense of the events leading up to hospitalization (e.g., "*he was fine but then got worse and stopped responding*" is a common description of a lucid interval in a patient with an EDH). This knowledge can also inform prognosis, manage expectations for recovery, and guide post-discharge recommendations.

Diffuse Axonal Injury

Although stroke can cause widespread secondary disruptions to brain functioning (e.g., edema, mass effect), most primary diffuse brain injuries are traumatic in nature. DAI is caused by rapid acceleration-deceleration and rotational forces inside the skull, disrupting both the shape and density of cellular structures and ultimately resulting in *shearing* (or pulling) of the neuronal axons and disturbance of cell bodies (Georges 2023). Although DAI is often seen in more severe injuries, neuropathology may be initially undiagnosed as emergent neuroimaging, often computed tomography (CT), is unlikely to detect the associated white matter changes. Instead, magnetic resonance imaging (MRI)—particularly susceptibility-weighted and diffusion tensor sequences (SWI and DTI, respectively), can be used to detect DAI when CT scan is negative (Davis 2007; Mittl et al. 1994).

Multiple neuroimaging classification symptoms exist to appreciate DAI, including the Adams Diffuse Axonal Injury Classification System (Fig. 11.1; Adams et al. 1989), which characterizes the extent of injury into Grades 1–3 based on the anatomical distribution and severity of axonal injury (Mesfin et al. 2023). While there appears to be a relationship between the severity of DAI and functional prognosis (van Eijck et al. 2018), the utility of using specific grading systems to predict outcomes has shown mixed efficacy in the literature.

Areas of high axonal volume are particularly vulnerable to DAI, including the splenium of the corpus callosum or rostral brainstem and cortical areas adjacent to skull cavities, such as the frontal and temporal lobes of the cerebrum (Slazinski and

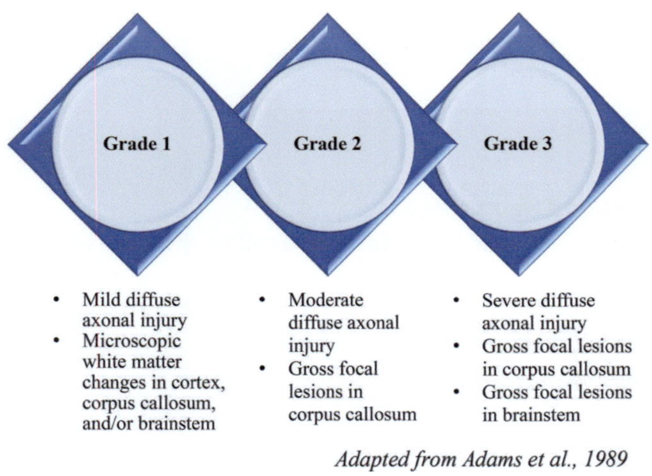

Fig. 11.1 Adams diffuse axonal injury classification system

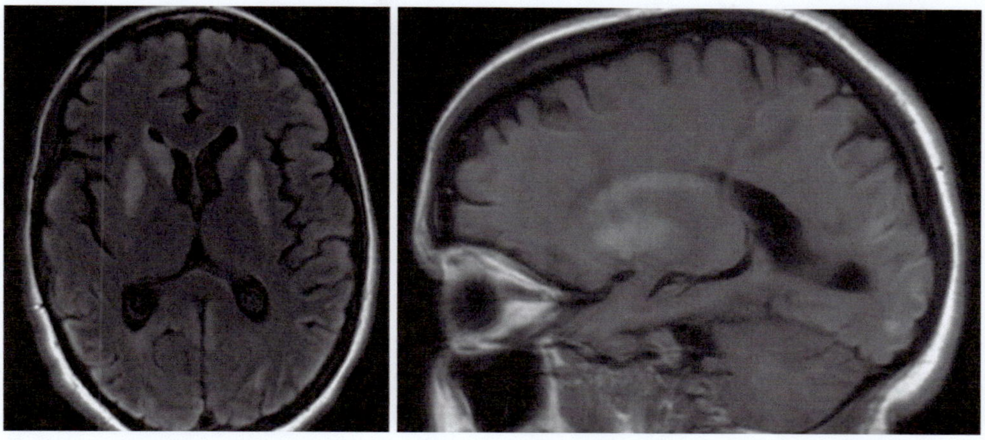

Fig. 11.2 Anoxic brain injury on MRI. MRI of a 27-year-old female patient with severe anoxic brain injury s/p accidental carbon monoxide poisoning. T1 (right) and T2 Flair (left) sequences show multiple areas of hyperintense signal through the cerebral cortices, as well as the caudate and putamina compatible with laminar necrosis

Johnson 1994). Mild DAI may only affect scattered regions of parasagittal white matter resulting in motor impairment (often more severe in legs than arms), though severe DAI can cause widespread axonal disruption across both hemispheres and focal softening in the corpus callosum and the dorsolateral tegmentum of midbrain (Slazinski and Johnson 1994).

Although patients with DAI typically have less severe focal contusions, and are less likely to have skull fractures, increased intracerebral pressure, or intracranial hematoma, DAI is usually associated with a severe TBI and carries a higher likelihood for a disorder of consciousness (DOC) such as coma (Hilton 1995; van Eijck et al. 2018). More specifically, the degree of injury to the brainstem is predictive of persisting DOC, in contrast to neurological and neuropsychiatric symptoms characteristic of injury to supratentorial regions (Medana and Esiri 2003). DAI poses increased risk for secondary complications, including hypoxia and intracranial hypertension, among others (discussed below).

Anoxic Brain Injury

Cerebral anoxia is another diffuse ABI caused by a complete lack of oxygen delivery to the brain, critically disrupting metabolic homeostasis (Lacerte et al. 2023). While prolonged reductions in oxygen flow (hypoxia) can eventually lead to chronic neural damage (most frequently in the territories of the bilateral posterior cerebral arteries), anoxia results in permanent cell death within minutes and widespread ischemia (e.g., laminar necrosis), particularly in vulnerable arterial border zones, termed "watershed areas," basal ganglia, and cerebellum (Fig. 11.2; Vella et al. 2017; Salasky and Chang 2023). Mechanisms of anoxic brain injury include cardiac arrest, strangulation, drowning, smoke inhalation, carbon monoxide poisoning, drug overdose, severe pulmonary disruptions, and hemorrhage. Anoxic brain injury is associated with high mortality and poor functional prognosis, particularly among patients who experience post-injury status epilepticus (SE), disrupted cranial nerve function and reflexes, or patterns of so-called "malignant" suppressions on electroencephalogram (EEG; Fugate 2017).

TRAUMATIC BRAIN INJURY

Penetrating Versus Non-penetrating Injuries

Traumatic brain injuries can be subdivided into penetrating and non-penetrating injuries. Penetrating TBIs result from external mechanisms (e.g., bullet, bone fragment, shrapnel) that puncture through skull into the brain. Associated damage to brain tissue is typically proximal to the narrow path of the projectile or the area directly under a depressed fracture (Alao et al. 2024). With increased momentum (*mass x velocity*) of the projectile, there is a higher likelihood of additional neuroanatomical consequences occurring distally to the primary penetrating injury. Penetrating injuries are less likely to cause severe alterations of consciousness (AOC), such as coma, because of the reduced likelihood of diffuse injury (e.g., DAI). However, there is a greater risk of complications, such as infection (due to the invasive nature of injury) or posttraumatic seizures, compared to non-penetrating injuries. It is not uncommon for antiepileptic drugs (AEDs) to be prophylactically administered as part of acute management of penetrating injuries (Alao et al. 2024).

Non-penetrating TBIs have historically been termed closed head injuries due to the inherent preservation of skull integrity despite focal or diffuse damage to brain tissue. While differing cognitive profiles have been found between penetrating and non-penetrating TBIs, the latter are often associated with more diffuse cognitive sequalae compared to the heterogeneity of focal deficits associated with penetrating injury. That said, the rate and degree of cognitive improvement appears to be generally similar among these groups—proceeding asymptotically, with the most prominent recovery occurring within the first year and gradually stabilizing around two years post injury (Ylioja et al. 2010).

Defining TBI Severity

There are several guidelines for classifying TBI severity, all of which utilize acute injury characteristics such as the duration of LOC or AOC, duration of post-traumatic confusion, the earliest available post-injury Glasgow Coma Scale (GCS) score, and neuroimaging findings (Silverberg et al. 2023; VA/DoD 2021; Stuss et al. 1999; Teasdale and Jennett 1976; Teasdale et al. 2014).

The Glasgow Coma Scale (GCS; Teasdale and Jennett 1974; Jain and Iverson 2023) is a commonly used measure that enables assessment of patient arousal and responsiveness through three subdomains: eye-opening (4), verbal response (5), and motor response (6) with a total possible score range from 3 to 15 (Fig. 11.3). If a patient is intubated, this can be indicated by adding the letter T after the score (e.g., GCS5T) and the potential highest possible score then becomes 10 due to the expected absence of verbal responses.

Historically, duration of posttraumatic amnesia (PTA)—or the period of posttraumatic anterograde memory loss—was used as an AOC severity indicator, although nomenclature has changed in recent years given evolution in the understanding of this state as broadly confusional, as opposed to only amnestic. Presently, the term posttraumatic confusional state (PTC/PTCS or PCS) is preferred and describes the duration of time a patient experiences fluctuating derangements in orientation, attention, memory, sleep-wake cycles, and behavior following a TBI (Stuss et al. 1999; Sherer et al. 2020).

Broadly, TBI severity falls into three classes: mild, moderate, and severe—with the former often being subdivided into complicated and uncomplicated mild TBI (mTBI) based on the presence or absence of unambiguous, trauma-related intracranial injury on structural neuroimaging, respectively (Table 11.1; Silverberg et al. 2023). Commonly used TBI severity criteria are outlined in Table 11.2.

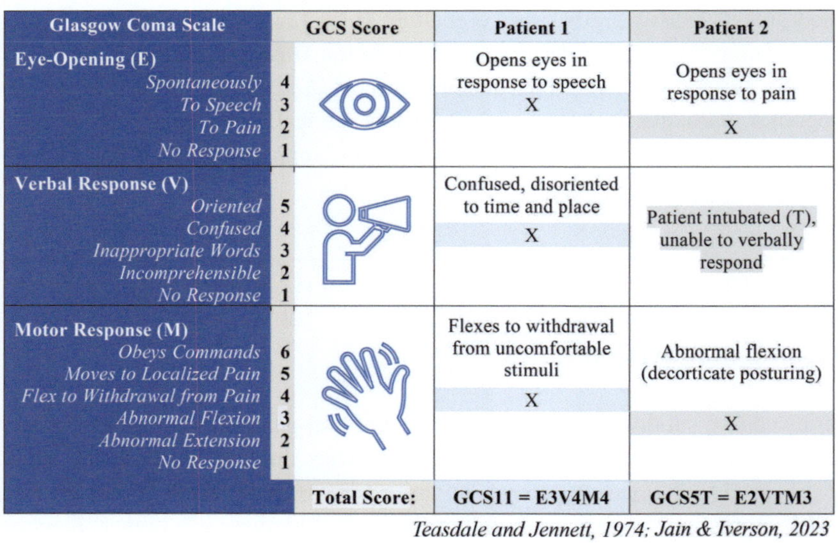

Fig. 11.3 GCS in practice

Table 11.1 ACRM classification of mild TBI severity

	GCS Score (*30 minutes post-injury*)	Duration of LOC	Duration of PTC	Structural Imaging
Uncomplicated* Mild TBI	13-15	$\leq$ 30 minutes	$\leq$ 24 hours	Normal
Complicated+ Mild TBI	13-15	$\leq$ 30 minutes	$\leq$ 24 hours	Abnormal

**Also referred to as "mild TBI without neuroimaging evidence of structural intracranial injury"*
+Also referred to as "mild TBI with neuroimaging evidence of structural intracranial injury"
If neuroimaging is not completed, no qualifier is used.
GCS = Glasgow coma scale; LOC = loss of consciousness; PTC = posttraumatic confusion
Adapted from Silverberg et al., 2023

Table 11.2 Classification of TBI severity

	GCS Score (*best available in first 24 hours*)	Duration of LOC	Duration of AOC+	Duration of PTC/PTA	Structural Imaging
Mild	13-15	0-30 min	Up to 24 hours	0-1 day	Normal*
Moderate	9-12	>30 min and <24 hours	>24 hours; severity based on other criteria	>1 and <7 days	Normal or abnormal
Severe	<9	>24 hours		>7 days	Normal or abnormal

--- TBI severity is determined by the highest severity across categories. ---

+Alteration of mental status must be immediately related the TBI.
**Typically normal, apart from complicated mTBI (defined above)*
GCS = Glasgow coma scale; LOC = loss of consciousness; AOC = alteration of consciousness; PTC = posttraumatic confusion; PTA = posttraumatic amnesia
Adapted from VA/DoD Guidelines, 2021

Disorders of Consciousness

Severity of TBI is a vital consideration for treatment planning, as patients with more severe injuries are expected to have greater residual cognitive and behavioral impairment. The typical course of recovery in severe TBI transitions through several phases, which can slow or stagnate at any specific stage. For this reason, graded disorders of consciousness (DOC)—defined based on physiological responsiveness, duration of LOC, and the nature of AOC—are used to classified recovery trajectory and inform rehabilitation prognosis. Definitions and diagnostic criteria for each stage are reviewed in Table 11.3 (Sherer et al. 2020).

Neuropsychologists working in inpatient settings, especially those with designated brain injury units, are often tasked with monitoring changes in mental status that would suggest a corresponding change in a patient's underlying medical status and/or a transition from one stage of consciousness to another—the latter referred to *clearance* or *emergence* from a recovery stage. The importance of accurately measuring and documenting emergence is two-fold. First, the length of time a patient remains in each stage of consciousness has direct implications on functional outcomes—with emergence an important milestone in this regard (Kowalski et al. 2021; Lucca et al. 2019; Malone et al. 2019; Whyte 2014). Second, improvements or decline in mental status may necessitate alterations to the plan of care. For example, improved mental status, particularly emergence from a PCS, warrants re-evaluation of—and possible changes to—existing behavioral management and intervention therapies. In contrast, any decline may be indicative of an underlying complication, such as hydrocephalus or an infection, that requires swift medical intervention.

Table 11.3 Stages of severe TBI

	Consciousness	Sleep/ Wake Cycles	Command Following	Cognition	Neuro-Behavioral Signs	Signs of Transition to Higher Level of Consciousness
Coma	None	Absent	Absent	None detected	None exhibited	Eye opening*
UWS	None	Present, but sleep disturbance and fluctuating arousal common	Absent	None detected	None exhibited	Visual fixation, localization to noxious stimuli, intelligible verbalizations, yes/no responses (even if inaccurate), reproducible movement to command
MCS	Partial	Present, but sleep disturbance and fluctuating arousal common	Inaccurate or inconsistent yes/no responses if present	Some observed, but difficult to assess due to fluctuations and inaccuracies.	Contingent emotional response to family or familiar stimuli	Functional communication, yes/no responses are accurate and reliable, appropriate functional object use of at least 2 common objects
PCS	Intact for internal states and external events, limited functional behavior	Present, but sleep disturbance and fluctuating arousal common	Usually consistent but may fluctuate	Can communicate but variable social appropriateness Impaired attention and memory, disorientation	Fluctuation in presentation, emotional and/or behavioral disturbance, confabulation, delusions, perceptual disturbances	Ability to attend to and process simple information, general orientation, recall of some recent events, lack of marked cognitive or behavioral fluctuations

* *While eye opening is commonly accepted as indicating transition from coma to UWS, there is no consistent evidence that this transition is associated with any improved level of consciousness.*
UWS = unresponsive wakefulness syndrome; MCS = minimally conscious state;
PCS = posttraumatic confusional state
Adapted from Sherer et al., 2020

Emergence from DOC

Patients in a coma are often not admitted for inpatient rehabilitation due to an inability to engage in rehabilitative interventions. However, poor prognosis is not necessarily inevitable for patients with prolonged DOC and many brain injury inpatient rehabilitation programs are increasingly accepting and treating patients in UWS and MCS. This is primarily due to empirical support for inpatient rehabilitation in improving long-term outcomes for these patients (Hammond et al. 2019; Lee et al. 2004). As a result, practice guidelines and competency recommendations have been developed in recent years for specialized DOC programs (Giacino et al. 2014, 2018, 2020; Seel et al. 2010).

As mental status waxes and wanes, clearance from each DOC stage should be determined via serial monitoring. While there are several validated measures for monitoring DOC, the JFK-Coma Recovery Scale-Revised (CRS-R; Giacino and Kalmar 2006; Giacino et al. 2004) has been studied more extensively than other similar measures. CRS-R was developed by neurorehabilitation professionals to objectively assist with differential diagnosis, prognosis, and treatment planning for individuals with DOC. The CRS-R is a validated tool designed to be used serially throughout the acute recovery period, with the specific frequency of administration based on clinical practice guidelines to ensure accurate diagnostic staging (Giacino et al. 2018, 2020; Sherer et al. 2020). The CRS-R contains six subscales addressing auditory, visual, motor, oromotor, communication, and arousal functions. Items are arranged hierarchically; associated with brain stem, subcortical, and cortical processes. Scoring is standardized and integrates operationally defined behaviors, with lowest scored items representing reflexive activity and highest scored items representing cognitively mediated behaviors. The CRS-R is particularly helpful in determining the presence (MCS+) or absence (MCS-) of language-mediated behavioral responses (i.e., intelligible verbalizations, command following, appropriate gestural responses, and/or functional object use) in patients in MCS (Thibaut et al. 2020). This is important as those with MCS+ show greater neural connectivity on imaging studies and have better prognosis than patients who do not evidence any receptive or expressive language functioning. An abbreviated version of the CSR-R has been recently validated, with an administration time of approximately 6 min for 5 core items (Coma Recovery Scale-Revised for Accelerated Standardized Testing [CRSR-FAST]; Bodien et al. 2023).

Management of PTC/PCS

Once cleared from an MCS, managing and monitoring PTC becomes a primary focus of rehabilitation efforts. In addition to serial monitoring, neuropsychologists in these settings contribute to interdisciplinary care by providing guidance on appropriate interventions and learning strategies at each stage of patient recovery. PTC can be a challenging phase of recovery as patients begin to demonstrate behaviors such as agitation, confabulation, delusions, and/or restlessness (Phyland et al. 2021). These behaviors can be upsetting for family members and staff. Working with the team to identify and implement strategies to help reduce or prevent these behaviors is frequently sought from inpatient neuropsychologists.

Although episodic memory is impaired in this population, this is sometimes conflated to an erroneous attribution of all learning interventions as ineffective. In fact, patients with PTC are responsive to procedural learning techniques such as errorless learning, repetition, and reinforcement contingencies that can ameliorate problematic behavior common in confused patients and maximize participation in rehabilitation therapies (Ponsford et al. 2023). Should a patient demonstrate clearance from PTC, the neuropsychologist can identify ways to adapt treatment approaches to enable integration of traditional learning techniques. Evidence-based practice guidelines are available for reference and have been updated to include recommendations for management of PTC (Fig. 11.4; Ponsford et al. 2023).

Emergence from PTC/PCS

While there is no universal criterion, emergence from PTC is traditionally determined by improvements in orientation and attention, the ability to process, learn, and recall (some) new information, and a lack of marked fluctuation in cognition or behavior (Fig. 11.5; Sherer et al. 2020). Multiple professions, including neuropsychologists, physicians, nurses, and rehabilitation therapists, regularly assess arousal, alertness, and orientation in acute care and inpatient rehabilitation environments. However, it is the neuropsychologist's role to ensure accurate and timely measurement of these factors using standardized instruments according to evidence-based practice guidelines. The Orientation Log (O-Log; Jackson et al. 1998) and the Galveston Orientation and Amnesia Test (GOAT; Levin et al. 1979) were developed specifically to

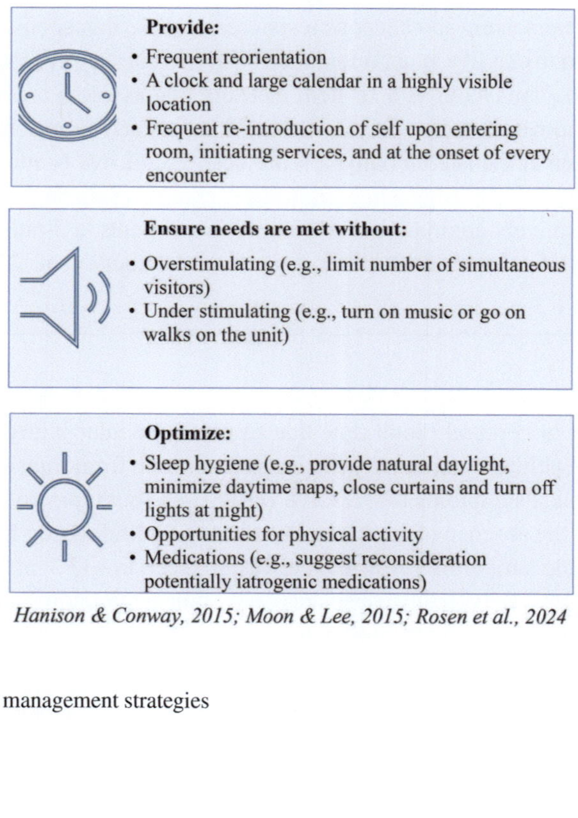

Fig. 11.4 Disorientation and confusion management strategies

Note: Emergence of PTC is determined by improvements, not necessarily complete resolution, of these key features; Sherer et al., 2020

Fig. 11.5 Emergence from posttraumatic confusion

measure orientation in rehabilitation populations with extant studies demonstrating adequate reliability and validity (Alderson and Novack 2002; Bode et al. 2000; Dowler et al. 2000; Jackson et al. 1998; Novack 2000; Novack et al. 2000). The GOAT and the O-Log are statistically similar, particularly in their assessment of clearance from PTC (Frey et al. 2007).

The O-Log is a 10-item measure scored out of 30 points that assesses orientation to time, place, and situation (frequently supplemented with questions to evaluate orientation to self). The O-Log is serially administered, and awards points (0–3) based on a patient's accuracy of answers using spontaneous responses or those in response to logical or multiple-choice cues. A cut score of 25/30 is typically used to classify intact orientation with two consecutive intact scores within 72 h representing clearance from PTC (Novack 2000). The GOAT is a 10-item measure that assesses orientation and events before and after TBI. Two consecutive scores at or above 76 on the GOAT within 72 h similarly represent clearance from PTC (Levin et al. 1979). Studies have shown that when administered daily, a score decrease of five or more points is highly likely to be followed by further declines in scores, thereby suggesting significant change (Salisbury et al. 2005). As such, while minor fluctuations (e.g., 1–2 points) are relatively common between serial assessments, a 5-point reduction warrants further investigation and consultation with the medical team for additional workup (Salisbury et al. 2005).

STROKE

Stroke is defined as a disruption of cerebral blood flow that results in cellular injury and associated clinical symptoms (Murphy and Werring 2020). The healthcare community distinguishes stroke from transient ischemic attacks (TIAs), which are characterized by stroke-like clinical symptoms that resolve within 24 h (but typically within minutes to hours). Previous diagnostic criteria for TIA included the absence of correlated neuroimaging findings on MRI or CT, but DWI sequence MRI has led to updated criteria that include temporarily restricted diffusion (seen in ~13% of TIA cases; Tsao et al. 2023).

Conversely, cellular injury from strokes typically causes residual clinical symptoms and correlated abnormal findings on structural neuroimaging. Stroke presentations are highly heterogenous and depend on the location and size of the affected vascular distribution, the presence of secondary and/or tertiary conditions, and other comorbid medical conditions. Most cerebral strokes occur in the territories of the anterior, middle, and posterior cerebral arteries, resulting in hypoperfusion of downstream distributions. Common symptom sequelae include sensory (paresis) and motor (plegia) disruptions, dysarthria, visual or hemibody neglect (usually left), visual field defects, aphasia, apraxia, and agnosia—in addition to localized cognitive deficits. Higher rates of residual psychiatric symptoms (depression, apathy, anxiety, and irritability) are also seen among stroke survivors compared to the general public (Medeiros et al. 2020).

Like TBI, a foundational grasp of functional neuroanatomy and fluency with the mechanisms of injury in stroke can help inform assessment, facilitate behavioral interventions, support adjustment to post-stroke disability, and provide education and prognostic information to the family and treatment team. Both stroke and TIA increase the risk for future cerebrovascular events and therefore prompt discussions of preventative interventions to optimize vascular health. Motivational interviewing techniques can be helpful to facilitate behavioral health changes in this regard.

Risk Factors for Stroke

Stroke mechanisms are classically divided into *hemorrhagic* and *ischemic* causes. While an extensive review of stroke risk factors goes beyond the scope of this chapter, it is notable that there are distinct and overlapping risk factors for ischemic and hemorrhagic mechanism of stroke. The greatest risk factors for all stroke types include advancing age, cardiac and vascular comorbidities (e.g., hypertension, diabetes, high cholesterol, atrial fibrillation), personal and family history of stroke or transient ischemic attack, substance use (e.g., tobacco, illicit drug use), and a sedentary lifestyle (Tsao et al. 2023). Unique risk factors become increasingly nuanced among differing cohorts, such as genetic/vascular anomalies and hypercoagulable states/conditions identified in younger cohorts of stroke patients (George 2020). Social determinants of health and cultural influences, such as access to (and trust in) medical care, create cerebrovascular risk disparities at a population-level and disproportionally disadvantaged communities (Pandian et al. 2020). For example, the risk of first stroke is nearly twice as high for non-Hispanic Black adults than White adults, and wealth status contributes to varying burden and outcomes (Tsao et al. 2023).

Stroke Types

Ischemic Stroke

According to the Center for Disease Control (CDC), most strokes in adults are ischemic in nature, accounting for nearly 87% of strokes (Virani et al. 2020). An ischemic stroke is caused by reduced cerebral blood flow resulting in hypoxygenation (or hypoxemia) of, and subsequent damage to, brain tissue within a given vascular distribution. The term *infarction* is sometimes used to describe the area of cell death associated with ischemic injury. Despite also being caused by disruptions in cerebral oxygen, primary hypoxic or anoxic brain injuries are not usually classified as ischemic strokes given differences in primary injury mechanism and associated pathological changes (Table 11.4).

Ischemic strokes can be further subdivided into embolic and thrombotic forms. *Embolic* infarcts result from an *embolism*, or a piece of matter such as a blood clot, that forms distally in another area of the body, migrates through the bloodstream, and eventually occludes one (or more) cerebral arteries. Embolic strokes are typically associated with large-vessel infarcts but may break up and cause multifocal strokes in downstream vascular distributions. Identification of the embolic etiology (e.g., atrial fibrillation, dissection) is important to prevent future strokes. In contrast, *thrombotic* infarcts result from an occlusion secondary to a local blockage (or thrombus) in the blood vessel wall, typically due to atherosclerotic plaque. These types are relatively more common.

Despite pathological mechanisms being contained with the vessel walls, patients with ischemic stroke should be monitored for hemorrhagic conversion (a secondary blood vessel rupture following ischemia), especially in the early acute hospital course. Hemorrhagic conversion occurs in approximately 32% and 20% of patients with and without intravenous thrombolytic therapy, respectively (Honig et al. 2022) and is associated with poorer prognostic outcome. As such, early identification of risk factors and symptoms of possible hemorrhagic conversion can assist medical professionals in screening for appropriateness of thrombolytic therapies based on history and presentation (e.g., history of atrial fibrillation, use of anticoagulants, larger infarct volume).

Hemorrhagic Stroke

A hemorrhagic stroke, or bleeding within the cranial cavity, is typically caused by venous abnormalities, medical comorbidities, certain medical treatments (e.g., use of blood thinners), and lifestyle factors such as tobacco and other substance use. Blood vessel abnormalities include aneurysms with or without ruptures, arteriovenous malformations, or bleeding disorders.

Table 11.4 Classification labels related to disruption in oxygen to brain

Injury	Description	Blood flow	Blood Oxygen	Pathological Mechanism
Anoxia	Complete lack of oxygen in supply to brain tissue.	Normal or Impaired	None	Permanent damage with total oxygen loss > ~4 minutes
Hypoxia	Insufficient oxygen to brain tissue despite adequate blood flow.	Normal or Impaired	Low	Resting pulse oximetry ≤ 95%
Hypoxemia	Reduced partial pressure of oxygen in arterial blood; can lead to hypoxia	Normal	Low	Ratio of arterial partial pressure of oxygen and fraction of inspired oxygen (PaO2/FiO2) = ≤300mmHg
Ischemia	Failure of perfusion of blood through cerebral vessels to brain tissue.	Decreased	Low	Permanent cerebral ischemic threshold = ~ 22ml-10ml/100g/min

Bhutta, Alghoula, & Berim, 2024; SRLF Trial Group, 2018; Yuh et al., 2017

While all strokes are considered medical emergencies, hemorrhagic strokes carry higher rates of morbidity and mortality due to the neurotoxic effects of blood product on cerebral tissue and the elevated risk of increased intracranial pressure (ICP) due to shift or mass effect of the surrounding tissue. The evolution of bleeds can persist into subacute and rehabilitative settings even after patients are presumed to be medically stable. For this reason, neuropsychologists working in these settings should be able to identify signs of ICP, such as worsening headaches, acute alterations in mental status, vision changes, papilledema (optic disc swelling), or dilated pupils. If increased ICP is suspected, neurosurgical procedures (described below) can be emergently performed.

MEDICAL MANAGEMENT OF ACQUIRED BRAIN INJURY

Acute Medical Management

Emergency medical intervention following acquired brain injury is highly beneficial for optimizing outcomes and available treatment options. While primary injuries cause local brain damage, much of the disability associated with brain injury is attributed to secondary and tertiary injuries, which are often the focus of acute care management (Vella et al. 2017). Secondary damage begins with inflammatory activity followed by disruption of the blood–brain barrier, peripheral blood cell infiltration, brain edema, and the discharge of numerous immune mediators and pro-inflammatory factors including chemotactic factors and interleukins (Stamatovic et al. 2006). Hematoma evacuation and/or thrombectomy are relatively common interventions for large hemorrhages, those that evolve over time, and in situations where there is concern for secondary edema. Early evacuations (i.e., within 8 hours) to address focal damage have been associated with reduction of poor outcomes (Kellner et al. 2021; Imoisili 2024).

Surgical interventions such as craniotomies, craniectomies, and tissue-destroying decompression (e.g., emergency removal of orbitofrontal areas given major herniation concerns) are commonly conducted to reduce injuries from edema-driven mass effect, reduce the risk of death from brainstem herniation, and to ultimately improve post-stroke prognosis. In the context of elevated intracranial pressure, conducting decompressive craniectomies relieves the pressure on the brain and by removing a large bone flap and allowing the brain to herniate. The bone flap is typically preserved, if possible, and a cranioplasty is performed once the swelling subsides. A skull can be manufactured using synthetic medical materials when this is not possible. When conducted in the early acute period post brain injury, decompressive craniectomies reduce the adverse impact of brain swelling due to edema, show promise in reducing mortality, and may have an unclear benefit on neurological outcomes (Sahuquillo and Dennis 2019; Stefano et al. 2015).

Acute medical management of stroke typically includes early bedside administration of the National Institutes of Health Stroke Scale (NIHSS; Goldstein and Samsa 1997), a measure commonly used by physicians to evaluate clinical presentation, and the severity of acute stroke to inform intervention. Along with other clinical data (e.g., CT scan), the NIHSS can quickly determine a patient's medical status and the indication for thrombolytic treatments, such as tissue plasminogen activator (tPA) or tenecteplase (TNK). Thrombolytic agents are clot-dissolving drugs administered for arterial ischemic stroke that promote more rapid reperfusion, which must be given within the first few hours since the patient was last known to be well (e.g., 3–4.5 h for tPA and possibly a longer window for TNK). The introduction of such agents in the 1970s and 1980s drastically reduced disability and supported positive functional outcomes.

Close follow-up by neuropsychologists can objectively track mental status fluctuations or declines that may inform the medical team's decision to proceed with emergent neurosurgical or medical interventions. Serial assessment tools such as the Orientation Log (O-Log), Cognitive Log (Cog-Log; Novack 2004), the Cognitive Assessment scale for Stroke Patients (CASP; Benaim et al. 2015, 2022), and the Confusion Assessment Method-Defined Delirium (3D-CAM; Marcantonio et al. 2014) have been found to have adequate sensitivity in detecting presence of bleeds, additional delirium, or other neurological factors (Jackson et al. 1998; Novack et al. 2000; Salisbury et al. 2005; Benaim et al. 2015; Alderson and Novack 2002, 2003) and offer increased standardization as compared to informal assessment.

Subacute Medical Management

Acute management is typically done in intensive care settings to ensure medical stability before the person is transferred to a subacute or rehabilitation unit (Vella et al. 2017; Salasky and Chang 2023; Stocchetti et al. 2017). By the time the person is admitted to inpatient rehabilitation, the rehabilitation team's focus shifts to developing strategies to maximize independence with functional therapies. However, cranioplasties may occur in the context of inpatient rehabilitation in some cases, particularly for those who are able to remain in rehabilitation for longer periods given that it typically takes months for swelling to improve. That said, a large majority of cranioplasties also occur on outpatient bases. Interestingly, cranioplasties conducted within the first three-to-six months (vs. longer than six months) after brain injury have been linked to positive neuropsychological outcomes, even after controlling for demographic factors (Coelho et al. 2023; Su et al. 2017; De Cola et al. 2018). As such, reinforcing the importance of close neurosurgical follow-up post discharge and assisting in managing patient and family expectations and can be of benefit if planned post discharge. The neuropsychologist can also offer pre- and post-surgical neuropsychological screening to enable detection of potential changes from pre-surgical baseline, which can facilitate care in the case of potential cognitive and medical complications.

Posttraumatic Seizure Risk

Seizure risk is higher following brain injury due to parenchymal damage, the presence of intracerebral blood, posttraumatic inflammation and associated changes, neurotransmitter imbalances, and secondary infections—the latter being particularly susceptible among patients with penetrating injuries. Seizure-associated prognostic outcomes differ among patients who experience an initial ictal event early (versus later) in their recovery period (Wilson et al. 2018). Most individuals with posttraumatic seizures experience their first seizure within the first few days post-injury. Late-onset posttraumatic seizures in the subsequent months and years following injury are less common, but possible (Laing et al. 2022; Temkin 2003; Annegers et al. 1998). Early posttraumatic seizures have been found to be associated with longer hospitalizations and poorer 24-month outcomes, including the development of posttraumatic epilepsy, and poorer functional outcomes (Laing et al. 2022).

Antiepileptic drugs (AEDs) are commonly prescribed for seizure prophylaxis in the early acute period following TBI given the increased risk of posttraumatic seizures. Prescribing patterns have changed over time, with recent guidelines offering support for AED prophylaxis in mitigating seizures that develop in early recovery, though efficacy in preventing late-onset seizures is less clear (Pingue et al. 2021; Pingue et al. 2022). This has led to variability in long-term AED use in post-TBI populations. Additionally, the mechanisms of action and side-effect profiles of AEDs may attenuate neurologic and cognitive recovery; therefore, clinical determination of the appropriateness of these medications often hinges on the risk/benefit ratio specific to each patient's circumstances (Gopalan et al. 2023; Wat et al. 2019; Fang et al. 2023; Pingue et al. 2021). Medication monitoring can also be helpful in informing possible contributions to disruptive behavior (e.g., adverse side effects from Keppra or other AEDs) or identifying patients who have clinical histories that may place them at greater risk for negative side effects (e.g., those notably agitated on admission).

Other Pharmacological Agents

Many patients benefit from pharmacological agents that enhance treatment and target problematic symptoms. For this reason, we recommend ongoing evaluation and identification of potential areas for medication intervention across phases of recovery. Medications with empirical support in ABI populations include NMDA receptor antagonists (i.e., amantadine and memantine), which offer neuroprotection by modulating post-injury excitotoxicity and secondary neuronal apoptosis (Giacino et al. 2012; Khan et al. 2021) and selective serotonin reuptake inhibitors (SSRIs), which have been found to promote mood, cognitive, and motor recovery (Bhattacharjee et al. 2018; Kakehi and Tompkins 2021). Judicious use of psychostimulants can also improve engagement in rehabilitation services by targeting slowed processing speed and attentional dysfunction following ABI (Whyte et al. 2002; Whyte et al. 2004).

However, like AEDs, other medications used to manage post-injury symptoms hold potential for deleterious cognitive and behavioral effects (Clegg and Young 2011). Familiarization with the American Geriatrics Society (AGS) Beers Criteria (2023) is recommended, which identifies potentially inappropriate medications in neurologically vulnerable adults. Potentially inappropriate medications commonly prescribed following ABI include benzodiazepines, first- and second-generation antipsychotics, agents with anticholinergic properties, and other deliriogenic agents. The neuropsychologist can assist in identifying symptoms that may be attributed to medication and informing prescribers when changes could be of benefit. For example, individuals may inadvertently remain on medications prescribed in acute care, despite an absence of continued clinical indication in subacute or rehabilitation settings. The neuropsychologist can raise concerns about a certain medication(s) hindering recovery and encourage continual evaluation of the medical necessity of medications that may contribute to attenuated progress.

ASSESSMENT CONSIDERATIONS

Evaluative Approaches

Neuropsychologists are uniquely skilled as evaluators of cognition and can offer a synthesized perspective based on integration of data gathered from the medical record, observation, and formal assessment. However, given the rapid rate of recovery in early stages, comprehensive evaluations for patients with acute ABI are not practical nor clinically efficacious. Instead, neuropsychologists in these settings utilize brief bedside assessments to characterize current mental status and track change over time. This, in turn, informs behavioral interventions that maximize treatment engagement, recovery, and patient independence. Bedside screening in the inpatient setting can also serve as a baseline to appreciate protracted cognitive recovery, giving outpatient providers an opportunity to communicate improvement—a motivator that fosters hope to patients and their families.

As patients progress in recovery, best practice dictates administration of a more robust cognitive testing battery to inform recommendations and predict functional outcomes following discharge (Kalmar et al. 2008; Hanks et al. 2008; Hannay and Sherer 1996; Pastorek et al. 2004; Roebuck-Spencer and Sherer 2008; Weeks et al. 2020). For example, as opposed to a short screening measure, administration of a more thorough memory test can offer greater nuance in detecting specific deficits at each stage of the learning and recall process. Using this information, the neuropsychologist can tailor specific recommendations for how to best accommodate the patient's deficits, while maximizing areas of cognitive strength. This could include guidance on whether the patient benefits from repeated exposure to information, the utility of incorporating written materials or visual cues to aid learning, or appropriate accommodations for impairments in memory recall (e.g., giving the patient recognition/multiple-choice cues). Similarly, performance on measures of safety awareness, insight, and problem-solving can lend psychometric support for recommendations pertaining to supervision and independently living following discharge. The neuropsychologist can also integrate data with other team members' perspectives to provide a consistent message regarding recommendations about readiness to return to functional activities (e.g., driving, work, independent living) and assistance in completing basic or instrumental activities of daily living.

Characterizing and Accommodating Deficits

Vision and Visual Cognition

Visual and visuocognitive deficits are common, particularly in brain injuries impacting posterior aspects of the brain. Presentations following acquired brain injuries may include diplopia (i.e., double vision), visual field defects (e.g., hemianopsia, quadrantanopia), cortical blindness, inattention and neglect syndromes, and visual agnosia. A visual agnosia is defined as the loss of meaning for a percept or visual object despite having intact sensory processing of the stimuli (Schoenberg and Scott 2011). While these syndromes are somewhat unusual, they are more likely to occur in the inpatient setting following acquired brain injury. Familiarity with common visual agnosia is also beneficial to the inpatient neuropsychologist (Table 11.5).

Table 11.5 Types of visual agnosia

Syndrome	Definition	Example Presentation Upon Assessment
Visual Object Agnosia	Inability to visually recognize and appreciate the meaning of objects in absence of deficit in vision, naming, or attention.	
Apperceptive Agnosia	Inability to identify objects visually despite knowing with other senses.	Unable to draw, copy, name, or match visually presented objects into groups.
Associative Agnosia	Inability to identify objects visually due to inability accessing stored information about their meaning, despite knowing with other senses.	Can draw or match visually presented objects but cannot name or demonstrate use
Simultanagnosia	Inability to simultaneously perceive multiple aspects of a single object or multiple aspects of a scene.	
Dorsal Simultanagnosia	Inability to appreciate more than one feature of an object/scene at once.	Difficulties perceiving complex visual scenes (e.g. Cookie Theft picture, overlapping stimuli); environmental difficulties (e.g., bumping into objects close together)
Ventral Simultanagnosia	Inability to appreciate more than one feature of a scene at once	Difficulties addressing more than one element on Cookie Theft, other scenic stimuli; difficulty reading full sentences but able to read a word; less severe than dorsal type.
Prosopagnosia	Inability to recognize, identify, and/or revisualize faces of familiar people despite ability to do so with other characteristics; adapted to include appreciation of inability to recognize a specific member of a class.	Difficulties identifying friends, family members, or previously known providers by face, but able to identify them by clothing, accessories, or other senses (e.g., voice, perfume, etc.); able to recognize generic dog/s but unable to identify a specific dog from the pack.

Evaluation of visuocognitive deficits following ABI include visual field examination, assessing for extinctions with simultaneous stimulation, formal cognitive testing measures (e.g., figure/clock copy), and behavioral observation (such as when patients have a specific positional preference or have difficulty spatially tracking or locating objects in their visual field). Recommendations to accommodate a visual field cut include optimization of the patient's visual field (e.g., bringing stimuli to midline, avoiding placement of objects into hemispace of a field defect), reminding patients to scan their visual field until they eventually can routinely self-initiate this activity (Pollock et al. 2019), conducting therapies in a distraction-free environment, and using corresponding visual, tactile, and auditory aids.

While there are limited evidence-based interventions for neglect syndromes (Longley et al. 2021), clinical recommendations typically include environmental changes rather than strategies initiated by the patient given the high rate of co-occurring anosognosia. Environmental modifications include positioning staff or target stimuli in the non-neglected side. Behavioral interventions such as attention-training to the neglected hemispace, when effective, can also increase patient independence.

Motor

Assessment of motor sequelae following ABI is an important part of comprehensive clinical assessment. While occupational and physical therapists traditionally assess and treat these symptoms, neuropsychologists may use observations of motor deficits to inform additional assessment and conceptualization in a different manner given their training. Understanding the extent of a patient's motor or sensory deficits can determine any necessary adaptations to the assessment and inform functional prognosis.

Paresis and Plegia: Paresis and plegia are common sequelae following acquired brain injury whereby patients experience reduced strength and functioning or paralysis to an affected limb. Paresis refers to partial loss of motor functioning for a body part, whereas plegia indicates complete paralysis. Hemiparesis may impact up to 80% of all individuals following stroke, and as many as two-thirds of all patients may evidence persistent hemiparesis at 6 months post-stroke (Lawrence et al. 2001; Lai et al. 2002). In acquired brain injury, paresis and plegia most commonly involve the contralateral hemisphere to the affected limbs, though ipsilateral impact may be observed for lesions in the lower brainstem.

Spasticity and Ataxia: Spasticity is also a common post-stroke phenomenon, seen in up to 50% of patients in the post-stroke recovery phase (Ward 2012), and often a prominent source of chronic pain, distress, and disability (Winstein et al. 2016). Ataxia is characterized by poor muscle control and impaired coordination. It is common following ABI, particularly with damage to the cerebellum and associated circuitry. Recognizing the presence of limb or truncal ataxia, or spasticity, is also an important consideration with respect to possible fall risk and ability to fully engage in treatment therapies.

Apraxia

Apraxia is a disorder of purposeful or skilled movements that is not related to basic sensory or motor deficits (Petreska et al. 2007). Apraxia is difficult to characterize given the often context-dependent presentation. For example, motor sequencing may be intact for automatic activities, with apraxia only elicited in response to command. Apraxic syndromes include ideomotor apraxia, ideational apraxia, limb-kinetic apraxia, apraxia of speech (also called verbal apraxia), and buccofacial (oral/oromotor) apraxia. Apraxia should not be explained by physical causes such as paresis, weakness debility, spasticity, ataxia, or sensory changes. To assess apraxia, patients are often asked to pantomime tool use, demonstrate oromotor or hand gestures, and complete a motor sequence. Ability to complete these tasks successfully, as well as characterization of errors, may provide insight into type of apraxia present.

Ideomotor Apraxia: Ideomotor apraxia is caused by deficits in motor planning and multistep execution, with preserved understanding of object use. Commonly pantomimed tool use to identify ideomotor apraxia may include brushing teeth, combing hair, hammering a nail, and using a pair of scissors. Both transitive (i.e., use of a tool) and intransitive (i.e., use of a body part) commands should be assessed. In ideomotor apraxia, pantomime is impaired, and individuals tend to make "body-as-object" errors such that they substitute a part of their body for the tool. For example, they may use their finger as a toothbrush, their hand as a hammer, or both fingers extended as scissors. In ideomotor apraxia, this error will persist even with demonstration with cue to imitate, and after the error is pointed out.

Ideational Apraxia: While pantomime object use is impaired in ideomotor apraxia, the mental concept of the object or item is preserved. This contrasts with ideational apraxia, where the mental concept of an object is disrupted. For this reason, ideational apraxia tends to cause more significant impairments in activities of daily living (ADLs) than ideomotor apraxia does. For example, if an individual is provided with the object and asked to demonstrate how to use it, a patient with ideomotor apraxia would often be able to perform the correct use (Cassidy 2016). With ideational apraxia, individuals may be able to name a tool (e.g., a knife) describe how to engage with tools (e.g., make a sandwich), but when asked how to pantomime making a peanut butter and jelly sandwich from start to finish, they exhibit greater difficulties. Importantly, these deficits may persist even with objects present. Tables 11.6 and 11.7 provide additional examples of how to assess and discern between ideomotor and ideational apraxia.

Limb-Kinetic Apraxia: In limb-kinetic apraxia, the largest deficits are seen in coordination of fine motor movements and often thought of as more consistent with a dexterity impairment than with other types of apraxia. Ideomotor and limb-kinetic apraxia are likely to result in fine motor 'clumsiness' and thus may interfere with activities of daily living. Typically, this clumsiness interferes with successful completion of a task given that it increases task errors, but with persistence, tasks may be able to be completed (Sunderland and Shinner 2007). Success of completing activities of daily living with ideomotor apraxia, however, may be contingent on use of an overlearned strategy; as such, if other post-injury sequelae require a new strategy to be used, task completion will likely be impaired (Goldenberg and Hagmann 1998). Table 11.8 reviews some differentiating factors between ideomotor and limb-kinetic apraxia.

Apraxia of Speech: Apraxia of speech (verbal apraxia) and buccofacial (oral/oromotor) apraxia impact speech production and are unique from aphasia. Consistent with other types of apraxia described, while these presentations involve difficulty with motor functions, they are not due to paralysis or weakness of specific muscles involved in speech or oral movement. Apraxia of speech can be confused with aphasia and dysarthria (i.e., weakness in muscles used for speech), which is further complicated by their frequent comorbidity. However, bedside assessment and collaboration with speech-language pathologists can offer the ability to differentiate. Individuals with apraxia of speech typically evidence intact ability to repeat simple speech sounds or single syllables, while those with dysarthria typically make errors. However, when asked to combine syllables together and repeat the sequence, individuals with apraxia of speech will typically make errors (Rohrer et al. 2008). Observationally, individuals with apraxia of speech tend to have inconsistency in pronunciation errors and may have distorted speech sounds given their difficulty placing their tongue, teeth, and jaw in the proper place. Contrastingly, individuals with dysarthria tend to struggle consistently with certain sounds. Automatic or overlearned speech may be intact for individuals with apraxia of speech.

Buccofacial (oral/oromotor) apraxia can be assessed by asking patients to perform oral gestures on command. This can be assessed through verbally describing the gesture or by demonstrating and asking the patient to copy the gesture; modeling of a gesture allows buccofacial apraxia to be assessed even in the presence of concern for aphasia. Common gestures assessed include opening and closing the mouth, tightening lips, inflating the cheeks, movement of the whole tongue, movement of the tip of the tongue, licking lips, and coughing. Functional movements may also be assessed, including pantomiming blowing out a candle, drinking through a straw, and blowing a kiss.

Table 11.6 Ideomotor versus ideational apraxia

	Ideomotor Apraxia	Ideational Apraxia
Summary of deficits	Difficulty with motor execution and planning despite intact ability to understand a task	Difficulty with conceptualization and sequencing tasks
Gestures / Imitation	Difficulty with motor coordination of gestures despite intact understanding of gesture	Difficulty with understanding sequence of gesture but intact ability to imitate
Object use	Deficits in execution despite intact understanding of function	Deficits in correct functional object use
Sequence / Multistep task completion	Difficulty carrying out motoric steps	Difficulty organizing steps of sequence

Table 11.7 Assessment of ideomotor and ideational apraxia

Apraxia Command	Error	Type of apraxia
(Without providing an object; gesture)		
"Show me how you would use…" A single object or tool (e.g., key / spoon / comb / pair of scissors / toothbrush, etc.)	Hand as object	Ideomotor apraxia
	Jerky, uncoordinated movements	Ideomotor apraxia (vs. Limb-kinetic apraxia)
	Use tools for correct purpose but incorrectly (e.g., attempts to write with the back of a pen)	Ideomotor apraxia
	Use object for incorrect purpose	Ideational apraxia
(Provide array of objects)		
"Show me how you would use these objects"	Hold object incorrectly but for correct purpose	Ideomotor apraxia
	Use object for incorrect purpose	Ideational apraxia
"Show me how to…" Complete a sequence (e.g., make a sandwich, make a cup of tea)	Complete steps in correct order but may use hands to mime objects (e.g.., stacking hands as pieces of bread)	Ideomotor apraxia
	Complete steps in correct order, but with incorrect demonstration of tasks	Ideomotor apraxia
	Miss steps or complete steps in incorrect order	Ideational apraxia
	Use objects for incorrect purpose (i.e., uses a piece of bread to spread peanut butter instead of knife)	Ideational apraxia

Treatment for apraxia of speech and buccofacial apraxia is typically managed by speech-language pathologists. However, many of the behavioral and environmental recommendations for aphasia may be applied to individuals with these presentations as well. Individuals with apraxia should be encouraged to slow down when trying to produce speech, and the treatment team and loved ones should provide prolonged durations to allow for a response. Individuals with verbal and oral apraxia should be encouraged to practice speaking and making speech sounds. However, the use of pictorial representations like communication boards, gestures, and nonverbal communication strategies may also be beneficial to augment communication. Additionally, since apraxia is not a disorder of language, in absence of upper extremity weakness, individuals with oral or verbal apraxia may be able to write to aid in communication.

Table 11.8 Ideomotor versus limb-kinetic apraxia

	Ideomotor Apraxia	Limb Kinetic Apraxia
Summary of deficits	Difficulty with motor execution and planning despite intact ability to understand a task	Difficulty with fine motor control or strength of limbs / hands and fingers to complete tasks, dexterity, clumsiness
Understanding / Conceptualization of tasks	Understands task, struggles with execution	Understands task, struggles with fine motor control
Core deficits	Imprecise movements, incorrect motor sequences	Fine motor dexterity and strength
Functional deficits	Gestures, tool use, sequencing tasks	Fine motor tasks

Language

Aphasia is present in approximately one-third of individuals in the acute phase following stroke (Haddad et al. 2019; Berthier 2005) and can be seen in patients with TBI—particularly those of greater severity. Accurate characterization of aphasia syndromes, tracking progress over time, and elucidating preserved cognitive and/or language functions in effort to inform communication strategies and rehabilitation progress are critical to rehabilitation post-ABI. Speech-language pathologists (SLPs) are also likely to be involved in assessment and intervention for aphasia, depending on the setting and composition of the team. As such, close collaboration between neuropsychology and SLP is encouraged to ensure consistency of feedback when communicating with the team, patient, and family.

Assessment of post-ABI language disorders includes formal measures as well as behavioral observation. At minimum, language evaluation should include tasks of fluency, repetition, and comprehension—with tests of grammar, spelling, reading, and naming often included in more comprehensive batteries. Formalized screening allows for structured evaluation of language subdomains, which assists in classification of aphasia syndromes, and systematically track progress over time. The Mississippi Aphasia Screening Test (MAST; Nakase-Thompson et al. 2002) is one such commonly used measure to assess communication and language functioning and can be found on the COMBI website. It was designed to be both brief (5–15 min administration) and repeatable, thus is well-suited for use in the inpatient setting (Nakase-Thompson 2004). The MAST consists of nine subjects: Naming, Automatic Speech, Repetition, Yes/No Accuracy, Object Recognition from a Field of Five, Following Verbal Instructions, Reading Instructions, Verbal Fluency, and Writing/Spelling to Dictation. These subjects contribute to an Expressive Language Index (50 total points), a Receptive Language Index (50 total points), and a MAST Total Score (100 total points). The MAST can be administered in full or in part, which is particularly useful for individuals who are nonverbal due to physical causes (e.g., intubation, oromotor apraxia). The Western Aphasia Battery (WAB; Shewan and Kertesz 1980) is another well-validated assessment of aphasia. The WAB was subsequently revised (WAB-R; Kertesz 2006) to include both bedside and full assessment batteries, which take approximately 15–60 min to complete, respectively. The full battery offers additional subtests in reading, writing, praxis, and construction.

Aphasia Syndromes: Aphasia syndromes are differentiated by core features, which can help guide treatment and improve outcomes. Characterization of aphasic symptoms can be determined through both formal and informal assessment. Behavioral observation of language should note any abnormalities in conversational skills and fluency, as well as any word-finding difficulties (i.e., anomia), paraphasias, neologisms, or use of overlearned or automatic phrases. Classifications are made based on patterns of relative preservation and impairment (Table 11.9) and can assist the inpatient team with localization of lesion or injury (especially in the absence of available neuroimaging), thus providing valuable clinical and prognostic information.

Table 11.9 Aphasia syndromes

	Core Characteristics	Fluency	Comprehension	Repetition	Naming
Global Aphasia	Severe impairments in both production and comprehension; limited speech production overall	Impaired	Impaired	Impaired	Impaired
Mixed Transcortical Aphasia	Severe impairments in both production and comprehension; relatively preserved repetition even when content of prompt is not understood	Impaired	Impaired	Preserved	Impaired
Broca's Aphasia	Non-fluent aphasia; slow, effortful speech with grammatical errors; characterized by short phrases or words with intact meaning	Impaired	Preserved, especially for simple sentences	Impaired	Impaired
Transcortical Motor Aphasia	Like Broca's aphasia but with notably preserved repetition, even for longer phrases and tenses	Impaired	Persevered	Preserved	Impaired
Wernicke's Aphasia	Fluent aphasia; speech is fluent but may be nonsensical, characterized by jargon, word finding difficulties, frequent paraphasias, circumlocution, and neologisms; comprehension is severely impaired	Preserved	Impaired	Impaired	Impaired
Transcortical Sensory Aphasia	Like Wernicke's aphasia but with notably preserved repetition, even if the content of the prompt is not understood	Preserved	Impaired	Preserved	Impaired
Conduction Aphasia	Characterized by difficulty with repetition and word-finding; fluent speech with phonemic errors and circumlocution; comprehension adequate for spoken and written language	Preserved	Preserved	Impaired	Impaired
Anomic Aphasia	Fluent speech with word finding difficulty and impaired naming; speech often characterized by circumlocution	Preserved	Preserved	Preserved	Impaired

Improving Communication: Once identification and classification of the aphasia is complete, providers should direct their attention to establishing and improving communication with these patients. Given that approximately half of patients with aphasia on acute stroke units initially present with global deficits (Godefroy et al. 2002), we recommend starting by establishing a reliable mechanism to communicate a dichotomous yes/no response. If a patient can vocalize, asking them to repeat "yes" and "no" may be successful. For the large subsection of patients who are nonverbal, creative problem solving (e.g., using thumbs up/down, pointing to written responses, eye blinks, communication boards, etc.) may be helpful. Once a method for responding is identified, questions should be asked in a yes/no format to establish accuracy. Additional steps include determining whether sentence complexity impacts communication accuracy and evaluating more complex language abilities (e.g., reading) as recovery progresses.

Establishing an accurate and reliable means of communication is critical for a patient's ability to interact with family, caregivers, and the medical or treatment team. Some information to consider highlighting to the team and family includes which mechanism of communication produced the most reliable/accurate response style (e.g., verbal response versus specific gesture), whether the patient may have more difficulty as sentence complexity increases, and whether there exist additional cognitive impairments (e.g., processing speed) that affect the latency to respond. Co-treatment sessions with team members also may be useful in implementing these strategies.

Learning and Memory

Understanding how a patient's learning and memory impacts treatment, including carryover from one therapy session to another, can be essential in helping the treatment team maximize the effectiveness of rehabilitation therapies. For patients with severe episodic memory impairments, strategies that capitalize on procedural memory principles can offer greater success by using repetitive motor activities or routinized behavioral patterns instead of verbal instructions or commands. As cognition improves, more formalized learning and memory assessment can be completed to evaluate both verbal and nonverbal memory. The nuance afforded by commonly used list learning and story memory tasks can be extremely helpful in providing recommendations to a team or family/collateral regarding how to best work with a patient. For example, a typical subcortical memory syndrome may be aided by reminders/recognition and providing context for future recall, whereas a primary mesial temporal deficit will not benefit from such strategies and will instead be aided by relying on compensatory strategies. Further, the relative integrity of verbal vs. nonverbal memory abilities can inform the methods with which a patient is most likely to learn and retain information. We find that employing strategies targeted to a patient's relative strengths in learning and memory will further help build rapport, foster engagement, and enhance treatment outcomes (Lequerica and Kortte 2010).

Neurobehavioral Change

Anosognosia: Anosognosia, or the limited awareness of and appreciation of deficits, is a common neurobehavioral symptom following ABI, particularly in injuries involving the frontal lobes (Prigatano and Sherer 2020). Limited insight or awareness may contribute to reduced engagement in rehabilitation given the patient's perceived lack of need for treatment or poor understanding of rationale and relevance. Furthermore, impaired awareness of deficits often has an adverse impact on daily functioning and caregiver experience (Steward and Kretzmer 2022). In the inpatient setting, medical treatment and discharge planning are dual foci of the interdisciplinary team. As such, assessment of a patient's insight, judgment, and safety awareness is fundamental in determining the extent to which the patient can participate in treatment and discharge planning. Identifying impairments in these areas can also inform future supervision needs following discharge. In inpatient settings, anosognosia is usually assessed informally through behavioral observations and interview with the patient and, when available, collateral informants.

Observations that may inform this assessment include:

- The patient is attempting unsafe behaviors in the room or neglecting rules about environmental safety (e.g., getting up unattended, not using durable medical equipment [DME] such as a helmet).
- The patient making requests for food or beverages that are not consistent with their prescribed diet (e.g., dysphasia or NPO diet).
- The patient exhibiting low frustration tolerance during discussions about restrictions (e.g., return to driving).

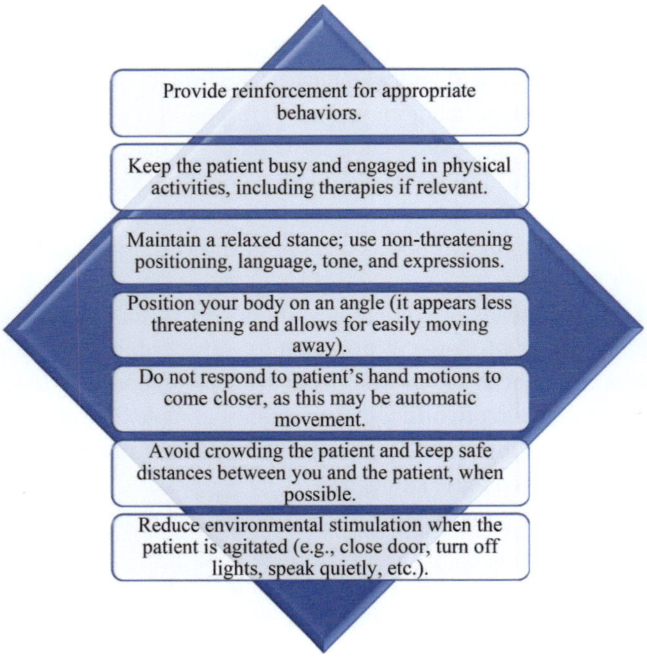

Fig. 11.6 Strategies to manage agitated behaviors

Open-ended, hypothetical questions can help elucidate the patient's anticipatory awareness of their ongoing deficits, which may reveal a deeper level of insight than simply confirming deficits when identified by a provider. Direct questioning is often beneficial (e.g., "*do you think you are currently safe to drive?*"), with follow-up as necessary (e.g., "*what concerns might you or your team have about your return to driving?*"). Hypothetical scenarios, vignettes, or role play can reveal decrements in problem-solving of safety risks that often arise following discharge (Schrijnemaekers et al. 2014).

Agitation: During early recovery following acquired brain injury, it is common for patients to experience neurobehavioral changes that adversely impact engagement and optimization of rehabilitation treatments and contribute to increased distress by patients' family and friends (Lequerica et al. 2007). Anecdotally, a patient clearing PTC may exhibit increased agitation as they regain awareness of their situation, particularly if they lack full understanding of their circumstances. In addition to serial monitoring of orientation, tracking agitation can also be done to support more objective measurement of injury recovery, behavioral regression, as well as response to behavioral and pharmacological interventions. The use of pharmacologic agents, such as atypical antipsychotics, for management of agitation and confusion is well studied and supported (Fleminger et al. 1996; Lonergan et al. 2009; McKay et al. 2021; Overshott et al. 2008). See Fig. 11.6 for strategies to manage agitated behaviors.

MANAGING COMORBIDITIES

Fatigue

Fatigue is common following ABI and can negatively impact an individual's ability to participate in treatment and/or rehabilitative care. Fatigue can manifest across physical, emotional, and/or cognitive domains, often co-occurring in acute stages of illness. Mechanisms of post-ABI fatigue are heterogenous (Lagogianni et al. 2018), including cellular and neurochemical abnormalities as well as patient-specific factors (e.g., predisposing concomitant conditions). Patients may not anticipate fatigue as an acute (and chronic) symptom following ABI—especially in less traditional realms (psychological, cognitive). Therefore, early education and detection is important to accommodate fatigue in acute recovery. Several measures, such as the Modified Fatigue Impact Scale (MFIS; Multiple Sclerosis Council for Clinical Practice Guidelines 1998; Strober et al. 2020), the Fatigue Scale for Motor and Cognitive Functions (FSMC; Penner et al. 2009), and the Fatigue Severity Scale (FSS; Krupp et al. 1989; Ozyemisci-Taskiran et al. 2019) have been developed to quantify this symptom and track change over time (Hubacher et al. 2012).

Once symptoms of fatigue have been identified, understanding contributing factors can inform effective management and interventional strategies. For example, recommending low stimulation "neurobreaks" after back-to-back therapy sessions or following a group of visitors can help decrease the likelihood of behavioral outbursts in confused individuals. Scheduled rest breaks and/or staggered therapy schedules also serve to limit fatigue levels and thereby help ensure patients are better able to engage in rehabilitations therapies that require physical and/or cognitive effort.

While fatigue often improves over time, it does not always resolve in tandem with other acute features of ABI, with up to 70% of patients with ABI reporting meaningful fatigue 5 years post-injury (Ali et al. 2022). Many patients report fatigue as a residual symptom that negatively impacts functional capacity and quality of life (QoL)—including job performance and social relationships (Cumming et al. 2016; Johansson et al. 2012). Further, fatigue following ABI is associated with greater mortality, even years after the initial injury (Glader et al. 2002; Lagogianni et al. 2018; Naess and Nyland 2013). Should a patient exhibit continued cognitive fatigue at discharge, recommendations for outpatient treatment can include interventions such as Mindfulness-Based Stress Reduction (MBSR), which has demonstrated efficacy in improving mental stamina (Johansson et al. 2012).

Sleep Disorders

Disruptions of Sleep

Insomnia and sleep disruption are common in hospital settings and serve as a point of consultation for the neuropsychologist when sleep problems begin to interfere with treatment engagement. Besides engagement in therapies, severe sleep disruptions can also negatively affect neural repair and brain injury outcomes (Silva et al. 2019; Nakase-Richardson et al. 2020b; Zunzunegui et al. 2011). Early assessment and treatment can be an important and unique role for inpatient neuropsychologists, as it is often underrecognized in hospital settings, and especially when significant cognitive and/or behavioral sequelae are present. For patients unable to accurately communicate about their sleep difficulties (e.g., pediatrics, those with aphasia or confusion), actigraphy devices can be helpful in measuring total sleep time, sleep onset metrics, and frequency of interruptions. These metrics can help inform treatment approaches, and measure treatment efficacy. First line interventions often include addressing the most prominent interfering factors within the hospital environment's control. This may mean requesting physicians place a nurse order to encourage patients to turn off electronics (e.g., phone, television, lights) after a certain time, promote keeping the patient out of bed during breaks, minimizing caffeine, and reducing daytime napping. In contrast, patients able to engage in sleep hygiene activities, will benefit from education and supportive resources (i.e., handouts, apps, online materials).

Obstructive Sleep Apnea

In addition to environmental factors that can impact sleep, obstructive sleep apnea (OSA) is prevalent following ABI (Lyons and Ryan 2015; Mathias and Alvaro 2012). Untreated OSA is associated with poorer outcomes in TBI, including greater cognitive impairment, reduced mobility, longer lengths of hospital stay, and reduced mobility (Steward et al. 2022; Silva et al. 2020). OSA often goes undiagnosed in acute and sub-acute settings, as polysomnography (PSG), the gold-standard for diagnosing sleep apnea, is not often provided in inpatient settings.

In the absence of available PSG, OSA screening measures can be helpful in prioritizing inpatient sleep medicine referrals. Several OSA screening tools were found to be effective during inpatient rehabilitation for TBI, the most well-known being the STOP-BANG questionnaire, named for the 8 domains assessed (Fig. 11.7; Chung et al. 2008). Other measures include the Multivariable Apnea Prediction Index (MAPI; Maislin et al. 1995) and the Berlin Questionnaire high-risk score (Netzer et al. 1999)—although the latter is less useful in patients with PTC (Nakase-Richardson et al. 2020a). When available, actigraphy can be used as initial screening for OSA and other sleep disorders to prioritize referrals and supplement data derived from self-report measures (Kamper et al. 2016). Due to the growing literature demonstrating the impact of untreated OSA on ABI outcomes, the American Academy of Sleep Medicine is currently developing "Clinical Practice Guidelines for the Management of Sleep-Disordered Breathing in Inpatients" to help improve access to evidence-based care for hospitalized patients.

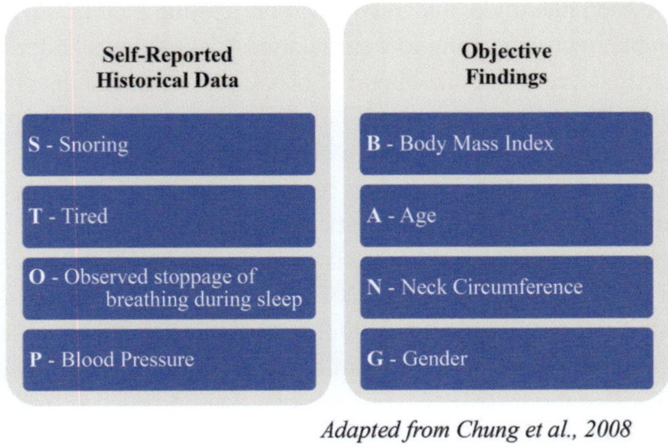

Fig. 11.7 Domains of the STOP-BANG Questionnaire

Psychiatric Symptoms

Although comprehensive review is beyond the purview of this chapter, psychiatric symptoms are common in inpatient settings. Adjustment to disability and acute illness, the nature of a hospital admission, among numerous other factors, understandably generates emotional distress in patients. For patients with ABI, these contextual risk factors are exacerbated by the structural, inflammatory, neurochemical, and metabolic changes associated with brain injury. As experts in brain-behavior relationships, inpatient neuropsychologists working with this patient population are tasked with identifying and differentiating psychiatric pathology from other post-injury sequelae, intervening to manage symptoms when appropriate, and monitoring changes in mood over time. Provision of education and prognosis to the patient and family/caregivers should be wholistic and include discussions normalizing current (and possible future) mood symptoms, as well as offer recommendations for ongoing treatment, when necessary. Below, we briefly highlight two common circumstances which place patients with ABI at risk for mood symptoms.

Post-stroke Depression

Post-stroke depression (PSD) is the most common neuropsychiatric comorbidity following stroke, with 55% of individuals experiencing depression at some point following a vascular event (Ayerbe et al. 2013). Pathophysiological processes underlying PSD are not well understood, though most experts agree it is complex and likely multifactorial. Proposed pathological mechanisms include low levels of monoamines; abnormal neurotrophic response; increased inflammation and associated dysregulation of the hypothalamic-pituitary-adrenal (HPA) axis; as well as glutamate-mediated excitotoxicity, often associated with frontal lobes, hippocampus, limbic regions, and basal ganglia projections (Medeiros et al. 2020). PSD is associated with higher mortality rates, more severe cognitive deficits, higher rates of long-term disability, lower quality of life, and increased rates of suicidal ideation as compared to individuals not experiencing depression following stroke (Medeiros et al. 2020). While the occurrence of PSD may vary over an individual's lifetime, studies have consistently demonstrated PSD rates are highest within the first year following stroke, and particularly within the first 3 months (Medeiros et al. 2020).

Inpatient neuropsychologists most frequently interact with stroke patients during the first several months following stroke and, given the impact on long-term outcomes, should be well versed in PSD assessment and treatment protocols. Empirically supported measures for PSD include the Hamilton Depression Rating-Scale (HAM-D; Hamilton 1960, 1978), Center of Epidemiological Studies—Depression Scale (CES-D; Radloff 1977) and Patient Health Questionnaire (PHQ-9; Meader et al. 2014), the latter being more efficient to administer with only nine items. In assessing PSD, providers should be sensitive to contextual factors that bear consideration of the timing and interpretation of these measures. For example, as suggested by Swartz et al. (2016), the clinical utility of screening for psychiatric symptoms in acute care may be limited due to the confusion and emotional lability common in these settings. Once acute symptoms stabilize and patient engagement improves, the transition to subacute care or inpatient rehabilitation offers an opportunity to validate gage baseline psychiatric symptoms. Care should be exercised in misattribution of adjustment reactions, poor sleep, and fatigue as post-stroke depression to ensure appropriate intervention (Medeiros et al. 2020). For example, compared to Major Depressive Disorder (MDD), patients with PSD tend to have more severe depressive symptoms, greater cognitive impairment, and

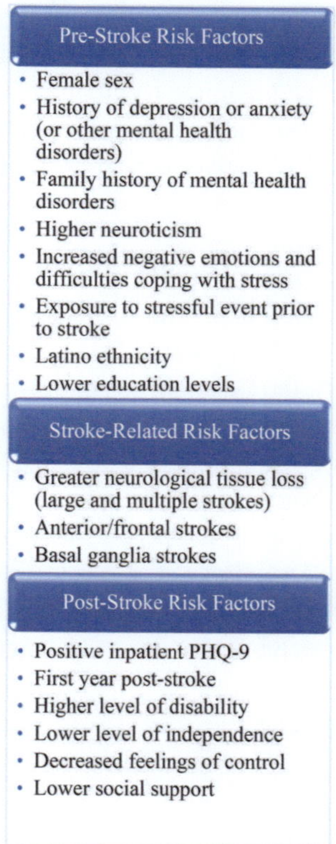

Adapted from Mederios et al., 2020

Fig. 11.8 Risk factors associated with post-stroke depression

higher rates of physical disability (e.g., sensory loss, motor/gait impairment, aphasia), necessitating a different approach to intervention. In contrast, patients with MDD tend to show greater anhedonia and sleep cycle disturbance compared to those with PSD. Other notable rule-outs when considering PSD include hypoactive delirium, dementia, bipolar disorder, apathy/abulia, and pseudobulbar affect.

Finally, while ongoing monitoring of mood is encouraged throughout a patient's admission, formal guidelines for stroke rehabilitation recommend all patients be screened for depression prior to discharge (Gittler and Davis 2018). Should screening at discharge suggest PSD, additional diagnostic evaluation including suicide risk, PSD education, and coordination of follow-up outpatient care will be needed. Further, as many patients may not experience PSD until discharged from the inpatient setting, identifying individuals at risk for developing worsening or future PSD symptomology is imperative (Fig. 11.8).

Effective PSD treatment approaches include behavioral activation, cognitive behavioral therapy (CBT), and psychoeducation, particularly when combined with pharmacological and stroke-focused interventions (e.g., physical and cognitive rehabilitation, stroke prevention, etc.; Wang et al. 2018; Medeiros et al. 2020). PSD education is the first-line intervention and should be provided to all patients and their caregivers following stroke, regardless of the presence or absence of depressive symptomology. Education should include information about stroke sequelae, associated functional and cognitive changes, PSD and associated risks, signs and symptoms of PSD onset, treatment resources, caregiver burden, and inclusion of activities that allow for socialization and quality of life. Again, for patients demonstrating PSD and/or associated risk factors, it is important to ensure follow-up mental health outpatient appointments are scheduled to provide a seamless continuity of care as patients transition home.

Aphasia and Mood

Aphasia is uniquely associated with psychological distress in post-ABI patients (Hilari et al. 2010). Deficits in language not only limit communicative expression, but also hinder access to, and utilization of, healthcare and community supports, which might otherwise mediate functional challenges and offer support for emotional distress. Working collaboratively with the medical team for medication management, as well as speech therapists to evaluate, educate and treat depression and anxiety symptoms, can be particularly beneficial. In higher functioning patients, poor frustration tolerance can be ameliorated by validation of the patient's experience, implementation of evidence-based mindfulness techniques, and compensatory strategies to improve communication. Relaxation and abdominal breathing techniques can be modeled and practiced without reliance on language abilities and neuropsychologists can educate inpatient team members on how to intervene using these techniques when patients exhibit behavioral signs of distress. Caregiver education about aphasia-related mood changes and supported communication techniques should also be provided.

Measuring Psychiatric Symptoms

Several emotional screening measures have been found to have adequate psychometric properties for use in patients with ABI (Meachen et al. 2008; Dajpratham et al. 2020; McCrory et al. 2023; Zachar-Tirado and Donders 2021). Commonly used measures include the PHQ-9 (Meader et al. 2014), the Generalized Anxiety Disorder scale (GAD-7; Spitzer et al. 2006), and the Brief Symptom Inventory (BSI-18; Derogatis 2001). When possible, measures should be selected with consideration of a patient's deficits and the measurement response structure. For example, patients with limited frustration tolerance or language impairment may benefit from shorter measures and/or those with dichotomous (vs. Likert or continuous) response options. For individuals with significant aphasia, use of self-report measures may not be suitable and reliance on pictures and/or observable behaviors may be beneficial. The Stroke Aphasic Depression Questionnaire-Hospital-10 (SADQH-10, Cobley et al. 2012) can be completed by inpatient providers based on observations over the previous week (Bennett et al. 2006). This measure is available online and has been translated into 11 languages.

PATIENT AND FAMILY EDUCATION

A final important domain in the assessment and care of patients with ABI is the role of caregivers. We recommend building rapport with family and/or other collaterals throughout a patient's hospital stay. This can include acting as a liaison for the family to the medical team or problem-solving difficulties in communication that may arise in acute care. For example, ABI not only abruptly changes a patient's life, but also disrupts a family's equilibrium. Caregivers may struggle balancing other obligations (e.g., childcare, employment) while simultaneously being present and available to the patient and treatment team. The neuropsychologist can assist both parties in addressing and providing suggestions to repair any rifts in alliance between the team and caregivers.

The content of therapeutic feedback and education will change over the course of inpatient treatment. The neuropsychologist may find themselves initially providing education on ABI, mechanisms, underlying neuropathology, expected trajectory, functional status, etc. Collaterals often anticipate a linear, upwards recovery curve, so frequent reminders of the dynamism of the process are essential, as day-to-day fluctuations and changes may be observed which should not necessarily be interpreted as indicating regression in progress. Further, caregivers may be distressed by a patient's post-ABI behavioral changes, and repeatedly explaining these behaviors in the context of the injury can reduce the perception that noxious behaviors are intentional or personal.

Providing this information allows patients and family members to better manage expectations and employ patience with the process of healing, as it can take longer and can be more complicated than initially anticipated. Prognosis is dependent on many factors, some of which may be more predictive than others. For example, Kothari and DiTommaso (2012) have proposed a protocol for relaying prognosis after severe TBI, though neuropsychologists are reminded that the expected outcome after TBI or stroke is highly variable and dependent on a myriad of factors including type and location of injury, severity, acute care management, premorbid functioning, comorbid medical conditions, demographics, and several other variables. In TBI, severity of injury, penetrating injuries, and posttraumatic seizures are related to increased risk for the development of additional disability in the years following injury (Dams-O'Connor et al. 2023) and shifting to educating both patients and family/collaterals on these possible subsequent complications as rehabilitation progresses is critical.

While practical education (e.g., medical status, residual mobility impairment) provided by medicine and rehabilitation therapists is important to third parties such as family members, this feedback is typically limited to the patient's experience. Neuropsychologists also offer a unique contribution to family education by integrating information about the patient's deficits and care needs while also addressing the emotional experience of the caregiver. It is common for family members and other collaterals to experience helplessness, grief, and anxiety as they support their loved one through the rehabilitation process. The presence of focal neurocognitive deficits like aphasia can exacerbate these feelings, as patients may experience increased frustration, which can be distressing for their support network. These emotions often crest proximal to discharge, and it is common for care providers to worry about whether they will be adequate in providing care, what to do if issues arise, and what to expect for more distal recovery. One of best ways to assist these care providers is to help equip them with resources for support, as research has shown that caregiver emotional well-being is buffered by social support (Ergh et al. 2002, 2003; Hanks et al. 2007; Ownsworth et al. 2010), which can of course influence patient outcomes.

Maintaining a constructive relationship with family is critical as patients requiring long-term care will often be discharged home (Bailey et al. 2017). Besides direct education, modeling strategies, and offering therapeutic support, the inclusion of care partners in rehabilitation therapy sessions, family conferences, and other services can help them feel like a member of the team and will help prepare the patient and caregivers for the eventual transition to discharge. Furthermore, ensuring patient and care providers' understanding of the availability of outpatient neuropsychological and psychotherapeutic services (for the patient and care partners) promotes continuity of care and may facilitate greater likelihood of follow-up with outpatient treatment and compliance with functional recommendations.

SUMMARY

In closing, the role of the inpatient neuropsychologist is dynamic and multifaceted, offering a unique perspective in addressing assessment and intervention of acquired brain injury in inpatient settings. While nominally focused on neurorehabilitation given the higher likelihood of working with ABI in that setting, the information and practice recommendations described are portable to other contexts (e.g., med/surg floors, ICU) as well. It is our hope that this chapter provided some of the central knowledge as well as routine procedures and practices needed for a neuropsychologist to work effectively in acute care and acute inpatient rehabilitation settings. Like in other contexts, the inpatient neuropsychologist must adapt to the ever-changing evolutions in medicine, treatment, and patient presentations. As such, continued consultation with other medical professionals and the empirical literature can inform the neuropsychologist and their practice.

ADDITIONAL RESOURCES

- Resources through the Brain Injury Association of America (https://biausa.org/), the Model System Knowledge Translation Center (https://msktc.org/TBI), and the American Stroke Association (https://www.stroke.org/en/) have particularly helpful information for family members and friends of those who have experience acquired brain injury.
- As communicating with individuals who have receptive or expressive language difficulties is particularly challenging, the American Congress of Rehabilitation Medicine has developed supportive resources available here https://acrm.org/resources/professional/aphasia-and-related-disorders/family-education/.
- For more information about the American Academy of Sleep Medicine development of "Clinical Practice Guidelines for the Management of Sleep-Disordered Breathing in Inpatients," please visit https://aasm.org/clinical-resources/practice-standards/guidelines-in-development/

REFERENCES

2023 American Geriatrics Society Beers Criteria® Update Expert Panel. American Geriatrics Society 2023 updated AGS Beers Criteria® for potentially inappropriate medication use in older adults. J Am Geriatr Soc. 2023;71(7):2052–81.

Adams JH, Doyle D, Ford I, Gennarelli TA, Graham DI, McLellan DR. Diffuse axonal injury in head injury: definition, diagnosis and grading. Histopathology. 1989;15(1):49–59.

Alao T, Munakomi S, Waseem M. Penetrating head trauma. In *StatPearls [Internet]*. StatPearls Publishing; 2024

Alderson AL, Novack TA. Measuring recovery of orientation during acute rehabilitation for traumatic brain injury: value and expectations of recovery. J Head Trauma Rehabil. 2002;17(3):210–9.

Alderson AL, Novack TA. Reliable serial measurement of cognitive processes in rehabilitation: the cognitive log. Arch Phys Med Rehabil. 2003;84(5):668–72.

Ali A, Morfin J, Mills J, Pasipanodya EC, Maas YJ, Huang E, et al. Fatigue after traumatic brain injury: a systematic review. J Head Trauma Rehabil. 2022;37(4):E249–57.

Annegers JF, Hauser WA, Coan SP, Rocca WA. A population-based study of seizures after traumatic brain injuries. N Engl J Med. 1998;338(1):20–4.

Ayerbe L, Ayis S, Wolfe CD, Rudd AG. Natural history, predictors and outcomes of depression after stroke: systematic review and meta-analysis. Br J Psychiatry. 2013;202(1):14–21. https://doi.org/10.1192/bjp.bp.111.107664.

Bailey EK, Nakase-Richardson R, Patel N, Dillahunt-Aspillaga C, Ropacki SA, Sander AM, et al. Supervision needs following veteran and service member moderate to severe traumatic brain injury: a VA TBI model systems study. J Head Trauma Rehabil. 2017;32(4):245–54.

Benaim C, Barnay JL, Wauquiez G, Bonnin-Koang HY, Anquetil C, Pérennou D, et al. The cognitive assessment scale for stroke patients (CASP) vs. MMSE and MoCA in non-aphasic hemispheric stroke patients. Ann Phys Rehabil Med. 2015;58(2):78–85.

Benaim C, Wauquiez G, Pérennou D, Piscicelli C, Lucas-Pineau B, Bonnin-Koang HY, et al. Cognitive assessment scale for stroke patients (CASP): a multicentric validation study. Ann Phys Rehabil Med. 2022;65(3):101594.

Bennett HE, Thomas SA, Austen R, Morris AMS, Lincoln NB. Validation of screening measures for assessing mood in stroke patients. Br J Clin Psychol. 2006;45:367–76.

Berthier ML. Poststroke aphasia: epidemiology, pathophysiology and treatment. Drugs Aging. 2005;22:163–82.

Bhattacharjee S, Axon DR, Goldstone L, Lee JK. Patterns and predictors of depression treatment among stroke survivors with depression in ambulatory settings in the United States. J Stroke Cerebrovasc Dis. 2018;27(3):563–7.

Bhutta BS, Alghoula F, Berim I. Hypoxia. In *StatPearls [Internet]*. StatPearls Publishing; 2024

Bode RK, Heinemann AW, Semik P. Measurement properties of the Galveston orientation and amnesia test (GOAT) and improvement patterns during inpatient rehabilitation. J Head Trauma Rehabil. 2000;15(1):637–55.

Bodien YG, Vora I, Barra A, Chiang K, Chatelle C, Goostrey K, et al. Feasibility and validity of the coma recovery scale-revised for accelerated standardized testing: a practical assessment tool for detecting consciousness in the intensive care unit. Ann Neurol. 2023;94(5):919–24.

Cassidy A. The clinical assessment of apraxia. In: Practical neurology, vol. 16. 16th ed; 2016. p. 317–22.

Chen W, Yao C, Li S, Huang H, Zhu Z, Chen R, et al. Cognitive impairment in diffuse axonal injury patients with favorable outcome. Front Neurosci. 2023;17:1077858.

Chung F, Yegneswaran B, Liao P, Chung SA, Vairavanathan S, Islam S, et al. STOP questionnaire. Anesthesiology. 2008;108(5):812–21.

Clegg A, Young JB. Which medications to avoid in people at risk of delirium: a systematic review. Age Ageing. 2011;40(1):23–9.

Cobley CS, Thomas SA, Lincoln NB, Walker MF. The assessment of low mood in stroke patients with aphasia: reliability and validity of the 10-item hospital version of the stroke aphasic depression questionnaire (SADQH-10). Clin Rehabil. 2012;26(4):372–81.

Coelho F, Noleto GS, Solla DJF, Martins PN, Andrade AF, Teixeira MJ, et al. Functional improvements associated with cranioplasty after stroke and traumatic brain injury: a cohort study. Brain Impairment. 2023;24(3):721–31.

Cumming TB, Packer M, Kramer SF, English C. The prevalence of fatigue after stroke: a systematic review and meta-analysis. Int J Stroke. 2016;11(9):968–77.

Dajpratham P, Pukrittayakamee P, Atsariyasing W, Wannarit K, Boonhong J, Pongpirul K. The validity and reliability of the PHQ-9 in screening for post-stroke depression. BMC Psychiatry. 2020;20:1–8.

Dams-O'Connor K, Juengst SB, Bogner J, Chiaravalloti ND, Corrigan JD, Giacino JT, et al. Traumatic brain injury as a chronic disease: insights from the United States traumatic brain injury model systems research program. Lancet Neurol. 2023;22(6):517–28.

Davis PC. Head trauma. Am J Neuroradiol. 2007;28(8):1619–21.

De Cola MC, Corallo F, Pria D, Lo Buono V, Calabrò RS. Timing for cranioplasty to improve neurological outcome: a systematic review. Brain Behavior. 2018;8(11):e01106.

Derogatis LR. Brief symptom inventory 18. Baltimore: Johns Hopkins University; 2001.

Dowler RN, Bush BA, Jackson WT. Cognitive orientation in rehabilitaton and neuropsychological outcome after traumatic brain injury. Brain Inj. 2000;14:117–23.

Ergh TC, Rapport LJ, Coleman RD, Hanks RA. Predictors of caregiver and family functioning following traumatic brain injury: social support moderates caregiver distress. J Head Trauma Rehabil. 2002;17(2):155–74.

Ergh TC, Hanks RA, Rapport LJ, Coleman RD. Social support moderates caregiver life satisfaction following traumatic brain injury. J Clin Exp Neuropsychol. 2003;25(8):1090–101.

Fang M, Liu W, Tuo J, Liu M, Li F, Zhang L, et al. Advances in understanding the pathogenesis of post-traumatic epilepsy: a literature review. Front Neurol. 2023;14:1141434.

Fleminger S, Greenwood RR, Oliver DL. Pharmacological management for agitation and aggression in people with acquired brain injury. Cochrane Database Syst Rev. 1996;2008(3)

Frey KL, Rojas DC, Anderson CA, Arciniegas DB. Comparison of the O-log and GOAT as measures of posttraumatic amnesia. Brain Inj. 2007;21(5):513–20. https://doi.org/10.1080/02699050701311026.

Fugate JE. Anoxic-ischemic brain injury. Neurol Clin. 2017;35(4):601–11.

George MG. Risk factors for ischemic stroke in younger adults: a focused update. Stroke. 2020;51(3):729–35.

Georges A. Traumatic brain injury. *StatPearls*. StatPearls Publishing Copyright © 2023, StatPearls Publishing LLC; 2023

Giacino J, Kalmar K. Coma recovery scale-revised. The Center for Outcome Measurement in brain injury; 2006. http://www.tbims.org/_combi/crs

Giacino JT, Kalmar K, Whyte J. The JFK coma recovery scale-revised: measurement characteristics and diagnostic utility. Arch Phys Med Rehabil. 2004;85(12):2020–9. https://doi.org/10.1016/j.apmr.2004.02.033.

Giacino JT, Whyte J, Bagiella E, Kalmar K, Childs N, Khademi A, et al. Placebo-controlled trial of amantadine for severe traumatic brain injury. N Engl J Med. 2012;366(9):819–26. https://doi.org/10.1056/NEJMoa1102609.

Giacino JT, Fins JJ, Laureys S, Schiff ND. Disorders of consciousness after acquired brain injury: the state of the science. Nat Rev Neurol. 2014;10(2):99–114.

Giacino JT, Katz DI, Schiff ND, Whyte J, Ashman EJ, Ashwal S, et al. Practice guideline update recommendations summary: disorders of consciousness: report of the guideline development, dissemination, and implementation Subcommittee of the American Academy of neurology; the American congress of rehabilitation medicine; and the National Institute on Disability, Independent Living, and Rehabilitation Research. Neurology. 2018;91(10):450–60. https://doi.org/10.1212/wnl.0000000000005926.

Giacino JT, Sherer M, Christoforou A, Maurer-Karattup P, Hammond FM, Long D, Bagiella E. Behavioral recovery and early decision making in patients with prolonged disturbance in consciousness after traumatic brain injury. J Neurotrauma. 2020;37(2):357–65.

Gittler M, Davis AM. Guidelines for adult stroke rehabilitation and recovery. JAMA. 2018;319(8):820–1. https://doi.org/10.1001/jama.2017.22036.

Glader EL, Stegmayr B, Asplund K. Poststroke fatigue: a 2-year follow-up study of stroke patients in Sweden. Stroke. 2002;33(5):1327–33.

Godefroy O, Dubois C, Debachy B, Leclerc M, Kreisler A. Vascular aphasias: main characteristics of patients hospitalized in acute stroke units. Stroke. 2002;3(3):702–5.
Goldenberg G, Hagmann S. Therapy of activities of daily living in patents with apraxia. Neuropsychol Rehabil. 1998;8:123–41.
Goldstein LB, Samsa GP. Reliability of the National Institutes of Health stroke scale: extension to non-neurologists in the context of a clinical trial. Stroke. 1997;28(2):307–10.
Gopalan H, et al. Use of anti-epileptic drugs for post traumatic seizure: a global survey. Ann Neurosci. 2023;30(1):26–32.
Haddad MM, Uswatte G, Mark VW, Bechtold KT. Cerebrovascular Accidents. In: Brenner LA, Reid-Arndt SA, Elliott TR, Frank RG, Caplan B, editors. Handbook of rehabilitation psychology. 3rd ed. Washington: American Psychological Association; 2019. p. 371–98.
Hamilton M. A rating scale for depression. J Neurol Neurosurg Psychiatry. 1960;23(1):56.
Hamilton M. A rating scale for depression. J Neurol Neurosurg Psychiatry. 1978;133:429–35.
Hammond FM, Giacino JT, Nakase Richardson R, Sherer M, Zafonte RD, Whyte J, et al. Disorders of consciousness due to traumatic brain injury: functional status ten years post-injury. J Neurotrauma. 2019;36(7):1136–46. https://doi.org/10.1089/neu.2018.5954.
Hanison J, Conway D. A multifaceted approach to prevention of delirium on intensive care. BMJ Qual Improv Rep. 2015;4(1)
Hanks RA, Rapport LJ, Vangel S. Caregiving appraisal after traumatic brain injury: the effects of functional status, coping style, social support and family functioning. NeuroRehabilitation. 2007;22(1):43–52.
Hanks RA, Millis SR, Ricker JH, Giacino JT, Nakese-Richardson R, Frol AB, et al. The predictive validity of a brief inpatient neuropsychologic battery for persons with traumatic brain injury. Arch Phys Med Rehabil. 2008;89(5):950–7. https://doi.org/10.1016/j.apmr.2008.01.011.
Hannay HJ, Sherer M. Assessment of outcome from head injury. In: Narayan RK, Wilberger JE, Povlishock JT, editors. Neurotrauma. New York: McGraw-Hill; 1996. p. 723–47.
Hilari K, Northcott S, Roy P, Marshall J, Wiggins RD, Chataway J, Ames D. Psychological distress after stroke and aphasia: the first six months. Clin Rehabil. 2010;24:181–90.
Hilton G. Diffuse axonal injury. J Trauma Nurs. 1995;2(1):7–12. https://doi.org/10.1097/00043860-199501000-00009.
Honig A, Percy J, Sepehry AA, Gomez AG, Field TS, Benavente OR. Hemorrhagic transformation in acute ischemic stroke: a quantitative systematic review. J Clin Med. 2022;11(5):1162.
Hubacher M, Calabrese P, Bassetti C, Carota A, Stöcklin M, Penner IK. Assessment of post-stroke fatigue: the fatigue scale for motor and cognitive functions. Eur Neurol. 2012;67(6):377–84.
Humble SS, Wilson LD, Wang L, Long DA, Smith MA, Siktberg JC, et al. Prognosis of diffuse axonal injury with traumatic brain injury. J Trauma Acute Care Surg. 2018;85(1):155–9.
Imoisili OE. Prevalence of stroke – behavioral risk factor surveillance system, United States, 2011–2022. MMWR. 2024;73:*449*.
Jackson WT, Novack TA, Dowler RN. Effective serial measurement of cognitive orientation in rehabilitation: the orientation log. Arch Phys Med Rehabil. 1998;79:718–20.
Jackson GL, Hamilton NS, Tupler LA. Detecting traumatic brain injury among veterans of operations enduring and Iraqi freedom. N C Med J. 2008;69(1):43–7.
Jain S, Iverson LM. Glasgow Coma Scale. In *StatPearls*. StatPearls Publishing; 2023
Johansson B, Bjuhr H, Rönnbäck L. Mindfulness-based stress reduction (MBSR) improves long-term mental fatigue after stroke or traumatic brain injury. Brain Inj. 2012;26(13–14):1621–8.
Kakehi S, Tompkins DM. A review of pharmacologic neurostimulant use during rehabilitation and recovery after brain injury. Ann Pharmacother. 2021;55(10):1254–66.
Kalmar K, Novack TA, Nakase-Richardson R, Sherer M, Frol AB, Gordon WA, et al. Feasibility of a brief neuropsychologic test battery during acute inpatient rehabilitation after traumatic brain injury. Arch Phys Med Rehabil. 2008;89(5):942–9. https://doi.org/10.1016/j.apmr.2008.01.008.
Kamper JE, Garofano J, Schwartz DJ, Silva MA, Zeitzer J, Modarres M, Barnett SD, Nakase-Richardson R. Concordance of Actigraphy with polysomnography in traumatic brain injury Neurorehabilitation admissions. J Head Trauma Rehabil. 2016;31(2):117–25. https://doi.org/10.1097/HTR.0000000000000215.
Kellner CP, Schupper AJ, Mocco J. Surgical evacuation of intracerebral hemorrhage: the potential importance of timing. Stroke. 2021;52(10):3391–8.
Kertesz A. Western Aphasia Battery. Revised (WAB-R). *APA PsycTests*; 2006
Khairat A, Waseem M. Epidural hematoma. In *StatPearls [Internet]*. StatPearls Publishing; 2023
Khan S, Ali AS, Kadir B, Ahmed Z, Di Pietro V. Effects of memantine in patients with traumatic brain injury: a systematic review. Trauma Care. 2021;1(1):1–14.
Kothari S, DiTommaso C. Prognosis after severe traumatic brain injury: a practical, evidence-based approach in Zasler. In: Zasler ND, Katz DI, Zafonte RD, editors. Brain injury medicine: principles and practice. Demos Medical Publishing; 2012.
Kowalski RG, Hammond FM, Weintraub AH, Nakase-Richardson R, Zafonte RD, Whyte J, Giacino JT. Recovery of consciousness and functional outcome in moderate and severe traumatic brain injury. JAMA Neurol. 2021;78(5):548–57. https://doi.org/10.1001/jamaneurol.2021.0084.
Krupp LB, LaRocca NG, Muir-Nash J, Steinberg AD. The fatigue severity scale: application to patients with multiple sclerosis and systemic lupus erythematosus. Arch Neurol. 1989;46(10):1121–3.
Lacerte M, Shapshak AH, Mesfin FB. Hypoxic brain injury. In *StatPearls [Internet]*. StatPearls Publishing; 2023
Lagogianni C, Thomas S, Lincoln N. Examining the relationship between fatigue and cognition after stroke: a systematic review. Neuropsychol Rehabil. 2018;28(1):57–116.
Lai SM, Studenski S, Duncan PW, Perera S. Persisting consequences of stroke measured by the stroke impact scale. Stroke. 2002;33(7):1840–4.
Laing J, Gabbe B, Chen Z, Perucca P, Kwan P, O'Brien TJ. Risk factors and prognosis of early posttraumatic seizures in moderate to severe traumatic brain injury. JAMA Neurol. 2022;79(4):334–41.
Lawrence ES, Coshall C, Dundas R, Stewart J, Rudd AG, Howard R, Wolfe CD. Estimates of the prevalence of acute stroke impairments and disability in a multiethnic population. Stroke. 2001;32(6):1279–84.
Lequerica AH, Kortte K. Therapeutic engagement: a proposed model of engagement in medical rehabilitation. Am J Phys Med Rehabil. 2010;89(5):415–22. https://doi.org/10.1097/PHM.0b013e3181d8ceb2.
Lequerica AH, Rapport LJ, Loeher K, Axelrod BN, Vangel SJ Jr, Hanks RA. Agitation in acquired brain injury: impact on acute rehabilitation therapies. J Head Trauma Rehabil. 2007;22(3):177–83. https://doi.org/10.1097/01.HTR.0000271118.96780.bc.

Levin HS, O'Donnell VM, Grossman RG. The Galveston orientation and amnesia test: a practical scale to assess cognition after head injury. J Nerv Ment Dis. 1979;167(11):675–84.

Lohani S, Bhandari S, Ranabhat K, Agrawal P. Does diffuse axonal injury MRI grade really correlate with functional outcome? World Neurosurg. 2020;135:e424–6.

Lonergan E, Luxenberg J, Sastre AA. Benzodiazepines for delirium. Cochrane Database Syst Rev. 2009;4:CD006379.

Longley V, Hazelton C, Heal C, Pollock A, Woodward-Nutt K, Mitchell C, et al. Non-pharmacological interventions for spatial neglect or inattention following stroke and other non-progressive brain injury. Cochrane Database Syst Rev. 2021;7

Lucca LF, Lofaro D, Pignolo L, Leto E, Ursino M, Cortese MD, et al. Outcome prediction in disorders of consciousness: the role of coma recovery scale revised. BMC Neurol. 2019;19:1–8.

Lyons OD, Ryan CM. Sleep apnea and stroke. Can J Cardiol. 2015;31(7):918–27.

Maas AI, Stocchetti N, Bullock R. Moderate and severe traumatic brain injury in adults. Lancet Neurol. 2008;7(8):728–41. https://doi.org/10.1016/s1474-4422(08)70164-9.

Macruz FBDC, Feltrin FS, Zaninotto A, Guirado VMDP, Otaduy MCG, Tsunemi MH, et al. Longitudinal assessment of magnetization transfer ratio, brain volume, and cognitive functions in diffuse axonal injury. Brain Behavior. 2022;12(3):e2490.

Maislin G, Pack AI, Kribbs NB, Smith PL, Schwartz AR, Kline LR, et al. A survey screen for prediction of apnea. Sleep. 1995;18(3):158–66.

Malone C, Erler KS, Giacino JT, Hammond FM, Juengst SB, Locascio JJ, et al. Participation following inpatient rehabilitation for traumatic disorders of consciousness: a TBI model systems study. Front Neurol. 2019;10:1314.

Marcantonio ER, Ngo LH, O'Connor M, Jones RN, Crane PK, Metzger ED, Inouye SK. 3D-CAM: derivation and validation of a 3-minute diagnostic interview for CAM-defined delirium: a cross-sectional diagnostic test study. Ann Intern Med. 2014;161(8):554–61.

Mathias JL, Alvaro PK. Prevalence of sleep disturbances, disorders, and problems following traumatic brain injury: a meta-analysis. Sleep Med. 2012;13(7):898–905.

McCrory M, Murphy DF, Morris RC, Noad RF. Evaluating the GAD-2 to screen for post-stroke anxiety on an acute stroke unit. Neuropsychol Rehabil. 2023;33(3):480–96.

McKay A, Trevena-Peters J, Ponsford J. The use of atypical antipsychotics for managing agitation after traumatic brain injury. J Head Trauma Rehabil. 2021;36(3):149–55.

Meachen SJ, Hanks RA, Millis SR, Rapport LJ. The reliability and validity of the brief symptom inventory– 18 in persons with traumatic brain injury. Arch Phys Med Rehabil. 2008;89(5):958–65. https://doi.org/10.1016/j.apmr.2007.12.028.

Meader N, Moe-Byrne T, Llewellyn A, et al. Screening for poststroke major depression: a meta-analysis of diagnostic validity studies. J Neurol Neurosurg Psychiatry. 2014;85:198–206.

Medana IM, Esiri MM. Axonal damage: a key predictor of outcome in human CNS diseases. Brain. 2003;126(3):515–30.

Medeiros GC, Roy D, Kontos N, Beach SR. Post-stroke depression: a 2020 updated review. Gen Hosp Psychiatry. 2020;66:70–80.

Mesfin FB, Gupta N, Hays Shapshak A, Taylor RS. Diffuse axonal injury. In *StatPearls*. StatPearls Publishing; 2023

Mittl RL, Grossman RI, Hiehle JF, Hurst RW, Kauder DR, Gennarelli TA, Alburger GW. Prevalence of MR evidence of diffuse axonal injury in patients with mild head injury and normal head CT findings. Am J Neuroradiol. 1994;15(8):1583–9.

Moon KJ, Lee SM. The effects of a tailored intensive care unit delirium prevention protocol: a randomized controlled trial. Int J Nurs Stud. 2015;52(9):1423–32.

Multiple Sclerosis Council for Clinical Practice Guidelines. Fatigue and multiple sclerosis: evidence-based management strategies for fatigue in multiple sclerosis: clinical practice guidelines. The Council; 1998.

Murphy SJ, Werring DJ. Stroke: causes and clinical features. Medicine. 2020;48(9):561–6.

Naess H, Nyland H. Poststroke fatigue and depression are related to mortality in young adults: a cohort study. BMJ Open. 2013;3(3):e002404.

Nakase-Richardson R, Schwartz DJ, Drasher-Phillips L, Ketchum JM, Calero K, Dahdah MN, et al. Comparative effectiveness of sleep apnea screening instruments during inpatient rehabilitation following moderate to severe TBI. Arch Phys Med Rehabil. 2020a;101(2):283–96.

Nakase-Richardson R, Schwartz DJ, Ketchum JM, Drasher-Phillips L, Dahdah MN, Monden KR, et al. Comparison of diagnostic sleep studies in hospitalized neurorehabilitation patients with moderate to severe traumatic brain injury. Chest. 2020b;158(4):1689–700.

Nakase-Thompso R. The Mississippi Aphasia Screening Test. Retrieved from The Center for Outcome Measurement in Brain Injury; 2004. http://www.tbims.org/combi/mast

Nakase-Thompson R, Manning E, Sherer M, Yablon S, Vickery C, Harris C, Dickson S. Bedside screen of language disturbance among acute care admissions: initial psychometrics of the Mississippi aphasia screening test. Arch Clin Neuropsychol. 2002;17:848.

National Academies of Sciences, Engineering, and Medicine. Characteristics of a high-quality process for determining disability resulting from traumatic brain injury. In: Evaluation of the disability determination process for traumatic brain injury in veterans. National Academies Press; 2019.

Netzer NC, Stoohs RA, Netzer CM, Clark K, Strohl KP. Using the Berlin questionnaire to identify patients at risk for the sleep apnea syndrome. Ann Intern Med. 1999;131(7):485–91.

Novack T. The orientation log. The Center for outcome measurement in brain injury; 2000. http://www.tbims.org/combi/olog

Novack T. The cognitive log. The Center for outcome measurement in brain injury; 2004. http://www.tbims.org/_combi/coglog

Novack TA, Dowler RN, Bush BA, Glen T, Scheider JJ. Validity of the orientation log, relative to the Galveston orientation and amnesia test. J Head Trauma Rehabil. 2000;15:957–61.

Overshott R, Karim S, Burns A. Cholinesterase inhibitors for delirium. Cochrane Database Syst Rev. 2008;1

Ownsworth T, Henderson L, Chambers SK. Social support buffers the impact of functional impairments on caregiver psychological well-being in the context of brain tumor and other cancers. Psycho-Oncology. 2010;19(10):1116–22.

Ozyemisci-Taskiran O, Batur EB, Yuksel S, Cengiz M, Karatas GK. Validity and reliability of fatigue severity scale in stroke. Top Stroke Rehabil. 2019;26(2):122–7.

Pandian JD, Kalkonde Y, Sebastian IA, Felix C, Urimubenshi G, Bosch J. Stroke systems of care in low-income and middle-income countries: challenges and opportunities. Lancet. 2020;396(10260):1443–51.

Pastorek NJ, Hannay HJ, Contant CS. Prediction of global outcome with acute neuropsychological testing follow- ing closed-head injury. J Int Neuropsychol Soc. 2004;10(6):807–17.

Penner IK, Raselli C, Stöcklin M, Opwis K, Kappos L, Calabrese P. The fatigue scale for motor and cognitive functions (FSMC): validation of a new instrument to assess multiple sclerosis-related fatigue. Mult Scler J. 2009;15(12):1509–17.

Peterson AB, Xu L, Daugherty J, Breiding MJ. Surveillance report of traumatic brain injury-related emergency department visits, hospitalizations, and deaths, United States. MMWR Surveill Summ. 2019;66(9):1–16.

Petreska B, Adriani M, Blanke O, Billard AG. Apraxia: a review. Prog Brain Res. 2007;164:61–83.

Phyland RK, Ponsford JL, Carrier SL, Hicks AJ, McKay A. Agitated behaviors following traumatic brain injury: a systematic review and meta-analysis of prevalence by post-traumatic amnesia status, hospital setting, and agitated behavior type. J Neurotrauma. 2021;38(22):3047–67.

Pingue V, Mele C, Nardone A. Post-traumatic seizures and antiepileptic therapy as predictors of the functional outcome in patients with traumatic brain injury. Sci Rep. 2021;11(1):4708.

Pingue V, Mele C, Biscuola S, Nardone A, Bagnato S, Franciotta D. Impact of seizures and their prophylaxis with antiepileptic drugs on rehabilitation course of patients with traumatic or hemorrhagic brain injury. Front Neurol. 2022;13:1060008.

Pollock A, Hazelton C, Rowe FJ, Jonuscheit S, Kernohan A, Angilley J, et al. Interventions for visual field defects in people with stroke. Cochrane Database Syst Rev. 2019;5

Ponsford J, Trevena-Peters J, Janzen S, Harnett A, Marshall S, Patsakos E, et al. INCOG 2.0 guidelines for cognitive rehabilitation following traumatic brain injury, part I: posttraumatic amnesia. J Head Trauma Rehabil. 2023;38(1):24–37.

Prigatano GP, Sherer M. Impaired self-awareness and denial during the postacute phases after moderate to severe traumatic brain injury. Front Psychol. 2020;11:1569.

Radloff LS. Center for epidemiological studies depression scale. Appl Psychol Meas. 1977;

Roebuck-Spencer T, Sherer M. Moderate and severe traumatic brain injury. In: Morgan JE, Ricker JH, editors. Textbook of clinical neuropsychology. Taylor & Franices; 2008.

Rohrer JD, Knight WD, Warren JE, Fox NC, Rossor MN, Warren JD. Word-finding difficulty: a clinical analysis of the progressive aphasias. Brain. 2008;131(1):8–38.

Rosen JH, Bieber E, Matta SE, Sayde GE, Fedotova NO, Devries J, et al. Hypoactive delirium: differential diagnosis, evaluation, and treatment. Primary Care Companion CNS Disorders. 2024;26(1):51618.

Sahuquillo J, Dennis JA. Decompressive craniectomy for the treatment of high intracranial pressure in closed traumatic brain injury. Cochrane Database Syst Rev. 2019;12

Salasky VR, Chang WTW. Neurotrauma update. Emerg Med Clin. 2023;41(1):19–33. https://doi.org/10.1016/j.emc.2022.09.014.

Salisbury DB, Baños JH, Novack TA, Schneider JJ. Significance of decreased orientation performance during rehabilitation. Rehabil Psychol. 2005;50(2):174.

Schoenberg MR, Scott JG. The little black book of neuropsychology: a syndrome-based approach. New York: Springer; 2011. p. 149–58.

Schrijnemaekers AC, Smeets SM, Ponds RW, van Heugten CM, Rasquin S. Treatment of unawareness of deficits in patients with acquired brain injury: a systematic review. J Head Trauma Rehabil. 2014;29(5):E9–E30.

Seel RT, Sherer M, Whyte J, Katz DI, Giacino JT, Rosenbaum AM, et al. Assessment scales for disorders of consciousness: evidence-based recommendations for clinical practice and research. Arch Phys Med Rehabil. 2010;91(12):1795–813.

Sherer M, Katz DI, Bodien YG, Arciniegas DB, Block C, Blum S, et al. Post-traumatic confusional state: a case definition and diagnostic criteria. Arch Phys Med Rehabil. 2020;101(11):2041–50.

Shewan CM, Kertesz A. Reliability and validity characteristics of the Western aphasia battery (WAB). J Speech Hear Disord. 1980;45(3):308–24.

Silva MA, Schwartz DJ, Nakase-Richardson R. Functional improvement after severe brain injury with disorder of consciousness paralleling treatment for comorbid obstructive sleep apnoea: a case report. Int J Rehabil Res. 2019;42(3):285–8.

Silva MA, Calvo D, Brennan EM, Reljic T, Drasher-Phillips L, Schwartz DJ, et al. Incidence and predictors of adherence to sleep apnea treatment in rehabilitation inpatients with acquired brain injury. Sleep Med. 2020;69:159–67.

Silverberg ND, Iverson GL, et al. The American congress of rehabilitation medicine diagnostic criteria for mild traumatic brain injury. Arch Phys Med Rehabil. 2023;104(8):1343–55.

Slazinski T, Johnson MC. Severe diffuse axonal injury in adults and children. J Neurosci Nurs. 1994;26(3):151–4. https://doi.org/10.1097/01376517-199406000-00008.

Spitzer RL, Kroenke K, Williams JB, Löwe B. A brief measure for assessing generalized anxiety disorder: the GAD-7. Arch Intern Med. 2006;166(10):1092–7.

SRLF Trial Group David, et al. Hypoxemia in the ICU: prevalence, treatment, and outcome. Ann Intensive Care. 2018;8:1–11.

Stamatovic SM, Dimitrijevic OB, Keep RF, Andjelkovic AV. Inflammation and brain edema: new insights into the role of chemokines and their receptors. In: Brain edema XIII; 2006. p. 444–50.

Stefano CD, Rinaldesi ML, Quinquinio C, Ridolfi C, Vallasciani M, Sturiale C, Piperno R. Neuropsychological changes and cranioplasty: a group analysis. Brain Inj. 2015;30(2):164–71. https://doi.org/10.3109/02699052.2015.1090013.

Steward KA, Kretzmer T. Anosognosia in moderate-to-severe traumatic brain injury: a review of prevalence, clinical correlates, and diversity considerations. Clin Neuropsychol. 2022;36(8):2021–40.

Steward KA, Silva MA, Maduri P, Tang X, Wittine L, Dams-O'Connor K, Nakase-Richardson R. Obstructive sleep apnea is associated with worse cognitive outcomes in acute moderate-to-severe traumatic brain injury: a TBI model systems study. Sleep Med. 2022;100:454–61.

Stocchetti N, Carbonara M, Citerio G, Ercole A, Skrifvars MB, Smielewski P, et al. Severe traumatic brain injury: targeted management in the intensive care unit. Lancet Neurol. 2017;16(6):452–64. https://doi.org/10.1016/s1474-4422(17)30118-7.

Strober LB, Bruce JM, Arnett PA, Alschuler KN, DeLuca J, Chiaravalloti N, et al. Tired of not knowing what that fatigue score means? Normative data of the modified fatigue impact scale (MFIS). Mult Scler Relat Disord. 2020;46:102576.

Stuss DT, Binns MA, Carruth FG, Levine B, Brandys CE, Moulton RJ, Snow WG, Schwartz ML. The acute period of recovery from traumatic brain injury: posttraumatic amnesia or posttraumatic confusional state? J Neurosurg. 1999;90(4):635–43. https://doi.org/10.3171/jns.1999.90.4.0635.

Su JH, Wu YH, Guo NW, Huang CF, Li CF, Chen CH, Huang MH. The effect of cranioplasty in cognitive and functional improvement: experience of post traumatic brain injury inpatient rehabilitation. Kaohsiung J Med Sci. 2017;33(7):344–50.

Sunderland A, Shinner C. Ideomotor apraxia and functional ability. Cortex. 2007;43(3):359–67.

Swartz RH, Bayley M, Lanctôt KL, Murray BJ, Cayley ML, Lien K, et al. Post-stroke depression, obstructive sleep apnea, and cognitive impairment: rationale for, and barriers to, routine screening. Int J Stroke. 2016;11(5):509–18.

Teasdale G, Jennett B. Assessment of coma and impaired consciousness: a practical scale. Lancet. 1974;304(7872):81–4.

Teasdale G, Jennett B. Assessment and prognosis of coma after head injury. Acta Neurochir. 1976;34(1):45–55.

Teasdale G, Maas A, Lecky F, Manley G, Stocchetti N, Murray G. The Glasgow coma scale at 40 years: standing the test of time. Lancet Neurol. 2014;13(8):844–54. https://doi.org/10.1016/s1474-4422(14)70120-6.

Temkin NR. Risk factors for posttraumatic seizures in adults. Epilepsia. 2003;44:18–20.

Thibaut A, Bodien YG, Laureys S, Giacino JT. Minimally conscious state "plus": diagnostic criteria and relation to functional recovery. J Neurol. 2020;267:1245–54.

Tsao CW, Aday AW, Almarzooq ZI, Anderson CA, Arora P, Avery CL. Heart disease and stroke statistics – 2023 update: a report from the American Heart Association. Circulation. 2023;147(8):e93–e621.

VA/DoD. Clinical practice guideline for the management and rehabilitation of post-acute mild traumatic brain injury; 2021. Retrieved from https://www.healthquality.va.gov/guidelines/Rehab/mtbi/index.asp

van Eijck MM, Schoonman GG, van der Naalt J, de Vries J, Roks G. Diffuse axonal injury after traumatic brain injury is a prognostic factor for functional outcome: a systematic review and meta-analysis. Brain Inj. 2018;32(4):395–402.

Vella MA, Crandall ML, Patel MB. Acute management of traumatic brain injury. Surg Clin. 2017;97(5):1015–30. https://doi.org/10.1016/j.suc.2017.06.003.

Virani SS, Alonso A, Benjamin EJ, Bittencourt MS, Callaway CW, Carson AP, et al. Heart disease and stroke statistics – 2020 update: a report from the American Heart Association. Circulation. 2020;141(9):e139–596.

Wang SB, Wang YY, Zhang QE, Wu SL, Ng CH, Ungvari GS, et al. Cognitive behavioral therapy for post-stroke depression: a meta-analysis. J Affect Disord. 2018;235:589–96.

Ward AB. A literature review of the pathophysiology and onset of post-stroke spasticity. Eur J Neurol. 2012;19(1):21–7.

Wat R, Mammi M, Paredes J, Haines J, Alasmari M, Liew A, et al. The effectiveness of antiepileptic medications as prophylaxis of early seizure in patients with traumatic brain injury compared with placebo or no treatment: a systematic review and meta-analysis. World Neurosurg. 2019;122:433–40.

Weeks DL, Ambrose SB, Tindall AG. The utility of the modified mini-mental state examination in inpatient rehabilitation for traumatic brain injury: preliminary findings. Brain Inj. 2020;34(7):881–8. https://doi.org/10.1080/02699052.2020.1761564.

Whyte J. Disorders of consciousness: the changing landscape of treatment. Neurology. 2014;82(13):1106–7. https://doi.org/10.1212/wnl.0000000000000276.

Whyte J, Vaccaro M, Grieb-Neff P, Hart T. Psychostimulant use in the rehabilitation of individuals with traumatic brain injury. J Head Trauma Rehabil. 2002;17(4):284–99.

Whyte J, Hart T, Vaccaro M, Grieb-Neff P, Risser A, Polansky M, Coslett HB. Effects of methylphenidate on attention deficits after traumatic brain injury: a multidimensional, randomized, controlled trial. Am J Phys Med Rehabil. 2004;83(6):401–20.

Wilson CD, Burks JD, Rodgers RB, Evans RM, Bakare AA, Safavi-Abbasi S. Early and late posttraumatic epilepsy in the setting of traumatic brain injury: a meta-analysis and review of antiepileptic management. World Neurosurg. 2018;110:e901–6.

Winstein CJ, Stein J, Arena R, Bates B, Cherney LR, Cramer SC, et al. Guidelines for adult stroke rehabilitation and recovery: a guideline for healthcare professionals from the American Heart Association/American Stroke Association. Stroke. 2016;47(6):e98–e169.

Ylioja S, Hanks R, Baird A, Millis S. Are cognitive outcome and recovery different in civilian penetrating versus non-penetrating brain injuries? Clin Neuropsychol. 2010;24(7):1097–112.

Yuh WT, Alexander MD, Ueda T, Maeda M, Taoka T, Yamada K, Beauchamp NJ. Revisiting current golden rules in managing acute ischemic stroke: evaluation of new strategies to further improve treatment selection and outcome. Am J Roentgenol. 2017;208(1):32–41.

Zachar-Tirado CN, Donders J. Clinical utility of the GAD-7 in identifying anxiety disorders after traumatic brain injury. Brain Inj. 2021;35(6):655–60.

Zunzunegui C, Gao B, Cam E, Hodor A, Bassetti CL. Sleep disturbance impairs stroke recovery in the rat. Sleep. 2011;34(9):1261–9.

Part III

Advanced Practice Considerations

Management of Cognitive and Behavioral Sequelae in Inpatient Settings

12

Jennifer Pender and Summar Reslan

ABBREVIATIONS

ABS	Agitated Behavior Scale
ACT	Acceptance and Commitment Therapy
APA	American Psychological Association
BPAD	Behavioral and Psychological Assessment of Dementia
CAM	Confusion Assessment Measure
CBT	Cognitive Behavioral Therapy
DBT	Dialectical Behavioral Therapy
MDS	Memorial Delirium Scale
MMPI-3	Minnesota Multiphasic Personality Inventory-3
OMNI-IV	OMNI-IV Personality Inventory
OBS	Overt Behaviour Scale
OQ	Outcome Questionnaire
PAI	Personality Assessment Inventory
PCOMS	Patient-Centered Outcome Measures

INTRODUCTION

Cognitive, behavioral, functional, and emotional changes experienced by patients in the inpatient setting can interfere with the provision of medical care, disrupt the care environment for others, contribute to caregiver distress, and hinder normal recovery trajectories. Neuropsychologists excel at being able to describe and address brain-behavior relationships, and, as these changes can occur following a myriad of medical, neurological, or psychiatric disorders; neuropsychologists in the inpatient setting are often asked to identify and address these changes as a critical component of the care plan (Johnson-Greene 2018; Novack et al. 2010; Soble et al. 2017). In the following chapter, we will discuss common cognitive and behavioral symptoms in the context of stabilized and altered mental status. Assessment and intervention strategies for managing these changes will also be reviewed, as cognitive and behavioral sequela differ across non-confused and confused states. Lastly, guidance will be offered on provision of holistic practice that accounts for patient characteristics, culture, and preferences (American Psychological Association [APA] 2021).

J. Pender (✉) · S. Reslan
Mental Health and Polytrauma Services, John D. Dingell Department of Veteran Affairs Medical Center, Detroit, MI, USA
e-mail: Jennifer.Pender@va.gov; Summar.Reslan@va.gov

E. K. Bailey et al. (eds.), *Inpatient Neuropsychology*, https://doi.org/10.1007/978-3-032-05880-5_12

COGNITIVE CHANGES IN NON-CONFUSED PATIENTS

Patients experiencing altered mental status have a limited ability to persistently engage in cognitive therapeutic interventions. In contrast, cognitive rehabilitation therapy administered to the non-confused patient across multiple time points shows additive, or incremental, progress over time. It is important to differentiate between the terms *cognitive training* (or cognitive remediation) and *cognitive rehabilitation* prior to reviewing symptoms, assessment, and intervention strategies. Although both are widely used interchangeably by providers and patients, the premises, goals, and outcome expectancies are different. Cognitive training (akin to cognitive remediation) consists of structured practice exercises designed to restore specific areas of cognitive functioning (Bahar-Fuchs et al. 2013). It is based on two premises. The first is that practice directly aids in brain plasticity and neural recovery. The second is that practice generalizes from training room exercises to real-world improvement. Unfortunately, neither assumption of organic recovery is supported in research examining patients with dementia (Bahar-Fuchs et al. 2013; Owen et al. 2010), traumatic brain injury (TBI; Julien et al. 2023), cancer (Fernandes et al. 2019), stroke (Gillespie et al. 2015), or movement disorders (Amatya et al. 2019). Training to test is an identified limitation with these techniques (Fernandes et al. 2019), and functional imaging studies denote changes in patterns of neural activation without evidence of morphologic regeneration (Julien et al. 2023; Weyer-Jamora et al. 2021).

Cognitive rehabilitation, in contrast, uses a variety of techniques to generate functional improvement in the settings where individuals live and work. It does not seek to restore function per se, but to minimize the cognitive and behavioral impact of an injury and maximize quality of life, independence, and safety (Amatya et al. 2019). Functional gain is facilitated through compensatory strategy instruction and environmental modification (Cicerone et al. 2005). Compensatory techniques employ residual cognitive strengths to compensate for areas of inefficiency, environmental and process modifications. This approach harnesses scientific principles of learning to identify constituent abilities and generate behavior change. Research suggests these techniques, in isolation or combination, can produce meaningful change in functional capability across a variety of conditions (Julien et al. 2023; Fernandes et al. 2019).

Observational Assessment of Cognitive Functioning

Completing neuropsychological evaluations in the inpatient setting offers unique challenges—including severe physical, sensory, cognitive, and behavioral sequelae—that often limit completion of comprehensive batteries. While traditional assessment measures are understandably contraindicated with acutely confused patients, even those who have cleared confusion may have difficulty completing comprehensive assessments due to fatigue, pain, or other various medical and psychiatric conditions. This is problematic as characterization of a patient's cognitive strengths and weaknesses once they have cleared from a confusional state is imperative to informing treatment and discharge recommendations. In these circumstances, observational data can be used to identify areas of cognitive deficit and help determine best treatment approaches. There are five cognitive domains commonly targeted in cognitive rehabilitation practice: attention, language, visual cognition, memory, and executive functioning (Holland et al. 1997; Johnstone and Stonnington 2011). Psychometric assessment allows for objective identification of both impairments and residual strengths. Clinicians should evaluate constituent abilities in target domains to individualize rehabilitation intervention (Johnstone and Stonnington 2011). Observational assessment spanning across testing and non-testing settings is of diagnostic utility and enhances ecological validity. Table 12.1 provides a select list of cognitive domains, constituent abilities, and suggestions for corresponding observational assessment.

Table 12.1 Observational assessment by cognitive domain

Cognitive Domain	Behavioral Assessment
Attention	
Arousal: Alertness & Responsiveness	Do they maintain alertness? Are they oriented? Do they respond to stimuli around them in the environment? Can they follow simple commands?
Focused Attention: Focusing on one stimuli, while ignoring others	Are they easily distractible? Are they able to maintain conversational set?
Divided Attention: Focusing on multiple stimuli at one time	Are they getting stuck on a prior conversation or task?
Sustained Attention: Focusing for extended time periods	Are they able to sit still for longer periods? Can they work on a task for several minutes without interruption?
Language	
Expressive: Speech production, repetition, naming, syntax, writing, spelling, reading	Is their speech fluent in conversation? Are they searching for words or substituting? Is their speech telegraphic or missing articles? Does it trail off? Is it organized?
Receptive: Auditory comprehension, reading, written comprehension	Are their verbal responses appropriate to the conversation? Are they able to follow simple commands? Can they make use of written materials?
Pragmatic: Use of speech in a goal directed manner, appropriate social context	Are they using language inappropriate to the setting? Can they communicate needs/wants?
Visual Cognition	
Visual Attention: Attending to all aspects of visual space	Are they missing stimuli on one side? Are they having navigation problems, or using the wall to navigate? Are they able to see multiple objects in front of them at once?

Table 12.1 (continued)

Visual Organization: Understanding and integrating part-whole relationships	Are they moving their heads to compensate consistently?
Spatial Perception: Perceiving objects in space	Are they focusing on internal details but missing the big picture? Are they able to maintain a gestalt in drawings to command?
Object Recognition: Recognizing and applying meaning to visual stimuli	Can they discriminate where an object is in front of them? Are they over or under-reaching when asked to pick something up? Are they using wholly incorrect words to describe objects?
Learning and Memory	
Explicit Memory: Verbal/Visual	Are they able to describe recent events accurately?
Encoding: Initial acquisition of information	Do they require repetition of information, questions or instructions?
Consolidation and Retrieval: Accessing and storing information	Do they improve when given hints? Are they repeating themselves in conversation?
Implicit Memory:	Can they learn new motor movements with repetition? Do they seem to recall better with the same routine?
Executive Functioning	
Initiation: Spontaneity	Is speech or movement slowed? Are there lags? Are they stuck in movement or speech?
Termination: Perseveration	Do they persist on a topic or task that has just been done?
Organization: Inefficiency	Are there careless mistakes? Are they doing things inefficiently? Do they keep redoing work? Do they have a hard time benefiting from

Table 12.1 (continued)

Abstraction: Higher-order problem-solving	corrective feedback? Are they concrete in responses?
Regulation: Inhibition Emotional control	Do they demonstrate inappropriate verbalization? Dysregulated emotion? Poor frustration tolerance?

Johnstone & Stonnington, 2011

Cognitive Compensatory Strategies

Suggestions for cognitive compensatory strategies should be domain specific and based on assessment findings (Table 12.2; Haskins 2012; Johnstone and Stonnington 2011; Sohlberg 2012). Strategies can also be broken down into external (e.g., using notebooks or electronic devices as reminders, etc.) or internal strategies (e.g., teaching patients to personally cue themselves to use a skill, etc.). The appropriateness of internal versus external compensatory strategies depends on the area of cognitive inefficiency.

In collaboration with their provider, individuals should establish mutually agreed upon goals to help promote functional progress. Early during injury recovery, one may see overall improvements in neurologic functioning and goals can reflect expected symptom recovery. Shifting away from setting expectations for symptom recovery and toward improvement in quality of life, especially for conditions with a static or more progressive course, may be beneficial. Setting appropriate expectations often results in improved treatment efficacy.

Table 12.2 Examples of cognitive compensatory strategies

Area of Cognitive Inefficiency	Examples of Possible Compensatory Strategies
Attention	‣ Minimize external distractions. + ‣ Ask individual to repeat back important information to ensure attention is focused. + ‣ Minimize situations requiring dividing attention among multiple inputs or outputs (e.g., speaking while typing). + ‣ Alternate tasks of varying interest level. + ‣ Allow extended time to complete tasks. +
Language	‣ Incorporate augmentative communication devices (e.g., communication boards) to address expressive language deficits. + ‣ Provide questions with options for yes/no responding or multiple-choice options. + ‣ Encourage use of nonverbal modalities of communication to address expressive language inefficiencies. + ‣ Utilize components of the cloze procedure to increase expression via spoken language. +
Visual Cognition	‣ If a field cut or neglect is present, direct individual to that side of their visual field by use of auditory, tactical, or visual (in the intact field) cues. + ‣ Train individual to spontaneously use visual scanning techniques. * ‣ For issues with depth perception, use brightly colored tape to improve contrast. Avoid dark environments. + ‣ Environmental modification (e.g., place all important objects [e.g., food, utensils, phone] on the side with preserved visual cognition). + ‣ Avoid environmental clutter. +

Table 12.2 (continued)

Memory	‣ Repeat information and rehearse tasks several times to promote encoding and retention. * ‣ Use alarm clocks or timers as reminders (e.g., medication administration times). + ‣ Increase the relevance of presented information (e.g., ask individual to restate information in their own words). + ‣ Provide associations between information to be remembered (e.g., place basket next to door for wallet, keys, etc.). + ‣ Encourage use of a daily planning/to do list. + ‣ Promote learning through multiple sensory modalities (e.g., see, hear, and execute task instructions). +
Executive Functioning	‣ Break down tasks into individual parts and complete said parts from easiest to most difficult. * ‣ Promote routine and structure (e.g., Saturday is for grocery shopping). * ‣ Use mnemonic coping strategies such as PDSA cycle (i.e., Plan, Do, Study, Act). *For example, ask the individual to plan an activity (PLAN), use a strategy (DO), study the outcome (STUDY), and then act accordingly in the future (ACT).* * +

Information in this table drawn from the following sources, please see original sources for comprehensive list of compensatory strategies: Johnstone & Stonnington, 2011; Haskins, 2012
** denotes an internal compensatory strategy,*
+ denotes an external compensatory strategy

BEHAVIORAL CHANGES IN NON-CONFUSED PATIENTS

Positive Disruptive Behaviors

Positive disruptive behaviors (PDBs) are multiply determined, overt behaviors that interfere with treatment, self-care, safety, and interpersonal functioning. Defined by their presence (and therefore termed *positive* behaviors), these behaviors readily capture the attention of providers, caregivers, and families. Common behaviors in this category are listed in Table 12.3. Disruptive behaviors can be mediated by neurologic, psychological, environmental, and premorbid/characterological factors. However, these predisposing and etiological contributors often interact synergistically and may mitigate or exacerbate each other when accompanied by neurologic insult.

Neurologically Mediated Positive Disruptive Behaviors

There are multiple conditions that impact the central nervous system, alter neurocognitive functioning, and contribute to disruptive behaviors. These include intracranial trauma, stroke, dementia, metabolic derangements, infectious disease, and medications/drug effects, among other medical disorders. Behavioral changes may be specific or diffuse, depending on the neurological condition and associated neurocognitive impairment. For example, neurodegenerative or

Table 12.3 Considerations of positive disruptive behaviors

Behaviors in Excess	Function of the Behavior	Neurocognitive Factors	Psychological Factors
Negative Verbalizations	• Stimulus Seeking • Attention Seeking • Stimulus Termination • Negative Emotion Reduction • Pain Communication	• Confusion • Impaired Social Awareness • Disinhibition • Perseveration	• Anxiety • Anger • Depression • Dysregulation • Delusionality/Psychosis • Personality Disorder • Discomfort/Pain
Wandering/ Elopement	• Stimulus Seeking • Escape/Safety • Negative Emotion Reduction	• Confusion • Memory Impairment • Spatial Disorientation	• Anxiety • Panic • Delusionality/Psychosis
Psychomotor Restlessness	• Stimulus Seeking • Self-Soothing • Distraction • Emotion Reduction	• Impaired Communication • Confusion • Perseveration	• Anxiety • Depression • Panic
Combativeness	• Stimulus Termination • Negative Emotion Reduction • Increased control	• Confusion • Disorientation • Language Impairment	• Anxiety • Anger • Paranoia • Discomfort/Pain
Aggression	• Stimulus Termination • Negative Emotion Reduction • Increased Control	• Language Impairment • Confusion • Disinhibition	• Anxiety • Anger • Paranoia • Dysregulation • Impulse Control Disorders • Personality Disorder • Mood Disorders

Table 12.3 (continued)

Disrobing	• Stimulus Seeking • Discomfort Reduction • Pain Communication	• Confusion • Disinhibition • Language Impairment • Impaired Social Awareness	• Personality Disorder • Discomfort/Pain
Sexually Inappropriate Behavior	• Stimulus Seeking • Attention Seeking • Biological Drive Reduction	• Impaired Social Awareness • Confusion • Disinhibition	• Grief/Loneliness • Personality Disorders • Impulse Control Disorders • Mood Disorders
Inappropriate Toileting Behaviors	• Stimulus Seeking • Discomfort Reduction	• Confusion • Disinhibition • Memory Impairment • Spatial Disorientation	• Depression • Discomfort/Pain

traumatic damage to the frontal lobes can lead to disinhibition, emotional dysregulation, stereotypy, impulsivity, sexual inappropriateness, aggression, and paranoid behaviors (Barrera Jr et al. 2013; Mitchell et al. 2009; Reilly 2001). Medications can have agitating effects, as can withdrawal from opioids, benzodiazepines, and alcohol.

Psychologically Mediated Positive Disruptive Behaviors

Anxiety, panic, depression, psychosis, grief/loss, personality disorder, and loneliness contribute to disruptive behaviors. Patients may engage in repetitive verbalizations and motor behaviors to soothe anxiety. Panic-based fight and flight can cause escape behaviors and elopement. Depression leads to negative verbalizations about self, others, and surroundings. Paranoid delusions can cause protective combative behaviors. Anger and frustration contribute to verbalizations, combativeness, and physical aggression. Loneliness, grief/loss, and social impoverishment can contribute to behaviors that facilitate social attention. Pain distress can cause agitated disrobing and toileting behaviors.

Environmentally Mediated Positive Disruptive Behaviors

Stimulation, aversive experiences, social factors, scheduling, observational learning, and approach factors can contribute to disruptive behavior. Stimulation can be either too high or too low. Sound, light, tactile, movement, and surrounding activity are all sources of stimulation. Aggressive verbalizations and disruptive behaviors can terminate perceived aversive stimuli. Conversely, when bored, individuals may engage in stimulation-seeking behavior like wandering, vocalizations, and sexual behaviors. Disruptive behaviors can generate socially reinforcing staff interaction. Observational learning can cause mimicry behaviors. Inconsistent staffing, scheduling changes, and approach without explanation can cause confusion and associated combativeness. Care without choice or perceived consent can lead to autonomy assertion and care refusal.

Negative Withdrawal Behaviors

Negative behaviors are behaviors that are characterized by their *absence* or withdrawal. They reflect deficits in behavior, and include reductions in self-care, treatment refusal, and reduced initiation/motivation. These behaviors are not overtly disruptive, so they can easily be overlooked by providers who may view the individual as non-problematic. As these behaviors persist, however, they can interfere with rehabilitation and treatment progress. They can also contribute to reduction in self-care abilities for individuals with progressive conditions where a skill cannot be easily relearned. Common negative withdrawal behaviors and associated neurocognitive, psychological, and functional constituents are found in Table 12.4.

Neurologically Mediated Negative Withdrawal Behaviors

Conditions that impact the central nervous system can lead to withdrawal behaviors. These include intracranial injury, dementia, metabolic disorders, medical disease, medication, sleep, and pain. Traumatic and acquired intracranial injury can produce amotivational or abulic presentations. Degenerative dementias are associated with care refusals, reductions in self-care, and amotivation (Boeve et al. 2022). Sedating medication and post-surgical anesthesia contribute to poor arousal and disengagement. Pain leads to treatment refusal, reduced movement, and functionality. Circadian rhythm and sleep disruption cause arousal difficulties that impede engagement. Alcohol and drug use contribute to behavioral and self-care reductions.

Psychologically Mediated Negative Withdrawal Behaviors

Depression, anxiety, pain distress, psychosis, and somatoform disorders can contribute to negative withdrawal behaviors. Depression is marked by amotivation and reduced activity. A hallmark of anxiety disorders is behavioral avoidance, where withdrawal is reinforced by reduction in anxiety. Pain-based fear causes movement and functional skill avoidance. Paranoid delusionality causes care refusal based on perceived malintent. Somatoform patients can have negative capability expectation and attention seeking behaviors.

Environmentally Mediated Negative Withdrawal Behaviors

Environmental factors that impact negative behaviors include approach, task features, poor sleep environments, reduced reinforcement, and inadequate stimulation. Care providers should be careful to identify themselves and explain tasks in advance to reduce confusion, anxiety, and care refusal. Tasks should be structured at a "just right" capability level, as tasks that are perceived as too difficult are more likely to be refused than less difficult constituent ones. Choice should be provided to facilitate perceived autonomy. The sleep environment should be reviewed, particularly in acute settings, where artificial light at night, poor sleep hygiene, limited exposure to sunlight, and nighttime disruptions occur.

Quantitative Assessment of Positive and Negative Behaviors

Some conditions have courses that are expected to demonstrate some recovery (e.g., delirium, recently acquired injury), remain static (e.g., remote TBI, stroke), or decline (e.g., neurodegenerative disorders). In the inpatient setting, treatment tracking for behavioral intervention is based on monitoring the frequency of target behaviors across several points: at baseline, before and after intervention, and at the end of treatment or inpatient admission. When possible, assessment should incorporate evaluation of neurologic/medical, neurocognitive, psychological, environmental, and functional factors. Comprehensive review, including diagnosis, behavior history, baseline functioning, imaging, laboratory studies, sensory restrictions, pain, and medications, is an important starting point. Facility staff and collateral interview can help to identify target behaviors, establish the trajectory of a behavior, articulate frequency, and clarify environmental, reinforcing, and maintaining factors.

The Overt Behavior Scale (OBS; Kelly et al. 2006) and the Agitated Behavioral Scale (ABS; Corrigan 1989) can be useful for quantifying positive disruptive behaviors. The Confusion Assessment Measure (CAM; Inouye et al. 1990) and Memorial Delirium Scale (MDS; Breitbart et al. 1997) for negative ones. The Behavioral and Psychological Assessment of Dementia

Table 12.4 Constituents of negative withdrawal behaviors

Behaviors in Absence	Function of the Behavior	Neurocognitive Factors	Psychological Factors
Care Refusal	• Stimulation Reduction • Negative Emotion Reduction • Increased Autonomy • Pain Reduction	• Confusion / Disorientation • Memory Impairment • Initiation Impairment • Language Deficits • Anosognosia	• Anxiety • Depression • Anger • Delusionality/ Psychosis
Excess Dependency	• Attention / Dependency Needs	• Confusion / Disorientation • Memory Impairment • Arousal Impairment	• Depression • Grief/Loss • Somatoform Disorders • Factitious Disorder • Personality Disorder
Social Withdrawal	• Stimulation Reduction • Negative Emotion Reduction	• Confusion / Disorientation • Impaired Processing • Impaired Communication	• Depression • Anxiety • Grief/Loss • Psychotic Disorders • Personality Traits
Decline in Self Care	• Negative Emotion Reduction • Attention / Dependency Needs • Pain Reduction	• Arousal Impairment • Anosognosia • Memory Impairment • Confusion / Disorientation • Apraxia / Motor Impairment	• Depression • Grief/Loss • Psychotic Disorders • Somatoform Disorders
Reduced Behavioral Activity	• Pain Reduction • Decreased Reinforcement	• Arousal Impairment • Confusion / Disorientation • Initiation Impairment • Impaired Planning/Sequencing	• Depression • Anxiety • Grief/Loss • Personality Traits • Psychotic Disorders

(BPAD; Schmidt and Gallo 2005) provides measurement of both positive and negative behaviors. Brief psychiatric self-report measures can be used to measure specific symptoms, contexts, and frequencies at bedside. Neuropsychological evaluation can clarify presentation validity, cognitive and psychological behavioral components. Functional analysis of behavior can identify contingencies that generate and maintain behaviors (Tables 12.3 and 12.4).

Behavioral Management Interventions

The complex and heterogeneous etiology of cognitive and behavioral changes makes it difficult to provide a uniform recommendation selecting a particular intervention. Rather, the likelihood is that multiple interventions will be utilized over time to address the changing landscape of patients' presentations, as the same behavior may serve different functions and evolve over time. Behavioral intervention for positive and negative behaviors starts with a careful assessment of the factors that are associated with, generate, and drive behaviors. Determining the underlying function of a maladaptive behavior (functional analysis) helps to identify the reason or purpose for its occurrence and to develop targeted interventions. Functional analyses should incorporate observation of contingency, environmental, internal, cultural, and approach factors associated with behavioral concerns.

When addressing behavioral changes in non-confused patients, one obviously must consider the etiology (or trigger) of the detected behavioral change (Hanley et al. 2003). Generally, behavior change will be the product of a neurologic-, environmental-, or mood-mediated process, or a combination of these. Behavioral interventions can be classified as either preventative or reactionary (Table 12.5). Prevention is the most optimal intervention option for addressing behavioral changes. Attending to, and taking note of, environmental or behavioral cues that an individual is becoming angry or frustrated is critical for preventative approaches (Labuda et al. 2018). Functional analysis of behavior as above (Tables 12.3 and 12.4) can be utilized to identify the triggers of problem behaviors (Hanley et al. 2003). Escalation can be mediated by identifying and removing antecedents of unwanted or maladaptive behavior. If removal of the triggering circumstance is not possible, the goal of the intervention shifts to identifying and maintaining factors. Although less optimal in creating sustained behavioral change, reactionary approaches are sometimes necessary to prevent harm to patients and/or providers.

An important element of effective behavioral management, consistency in executing reinforcement contingencies across care providers increases the likelihood for an overall sustained reduction in problematic behavior (Hollis et al. 2021). Extinction bursts are to be expected when implementing corrective behavioral interventions (Katz and Lattal 2020). In reaction to an intervention implemented by the care provider, a patient may temporarily increase the problem behavior in an attempt to obtain their desired outcome. By consistently responding with the same corrective behavioral intervention, the frequency of this problem behavior should eventually decrease. It is most effective when the response to the problem behavior is the same each time (i.e., don't change the corrective behavioral intervention), and the response is delivered within a few minutes of the behavior (not hours or days later).

Interventions for Neurologically Mediated Behaviors

If one suspects that an observed change in behavior (positive or negative) is the result of a neurologic change, there are some considerations before determining the appropriate intervention. The first question a provider should ask is if there is an underlying reversible cause of the behavior change that can be easily rectified (e.g., delirium contributing to overstimulation with resulting agitation). If so, the reversible cause should first be addressed in the hopes that the problematic behavioral sequelae will resolve after addressing the underlying reversible cause. If the behavioral change is the result of the injury itself (e.g., post-TBI reduced impulse control secondary to orbital frontal lobe damage), a provider should ensure the patient is medically stable before introducing any behavioral management intervention.

If a patient is deemed medically stable and still experiencing neurologically mediated behavior changes, the appropriateness of contingency-based approaches, environmental management, and/or communication-based approaches for behavior management should be explored. Contingency-based approaches, wherein behavioral changes are rewarded or reinforced to promote adaptive behavior change, are rooted in the principles of operant conditioning (Skinner 2019). Variable interval reinforcement schedules (wherein a behavior is reinforced after an unpredictable amount of time) have been shown to be the most effective for consistent and stable change toward more desirable behaviors, and that schedule is more likely to be resistant to extinction (Pritchard et al. 1976).

Communication-based approaches to behavioral management include teaching assertive communication, providing reassurance, using assertion, and implementing a primary communicator strategy. Traits of an assertive communication style include listening without interrupting, clearly stating wants and needs, and promoting patients' self-agency. Providing reas-

Table 12.5 Examples of preventative and reactionary behavioral interventions

Intervention	Definition/Example	Type of Approach
Reassurance	Explain why certain things are being done and remind patient they are safe. *Example*: Inform patients that they have recently sustained an injury and are recovering in a medical setting.	Preventative
Environmental Modification	Add or remove things from environment. *Examples*: Turn off the television, reduce number of people in room, and/or turn down the lights.	Preventative or Reactionary
Positive Reinforcement	Provide incentive (internal or external) to increase desired behavior. Catch the patient being "good" and provide praise. Avoid giving attention only when the problem behavior occurs. *Example:* Allow patient to nap after completion of all daily therapy activities.	Preventative
Assertion	Acknowledge the patient's point of view and try to arrive at a workable compromise. *Example:* State that therapy activity may contribute to heightened pain but is also contributing to functional recovery. Negotiate with patient how much therapy time they would be willing to tolerate and arrive at compromise. Add in additional tolerated therapy time daily.	Preventative
Provide Direction	Provide a clear statement about expectations for performance. *Example:* Use simple, one step instructions.	Preventative
Distraction	Redirect topic of conversation from triggering topic toward a more neutral topic. *Example:* Change topic of conversation, bring new person into conversation, or call a "time out" wherein the individual gets a break from the situation for a preplanned amount of time before returning for activity completion.	Preventative or Reactionary

Table 12.5 (continued)

Primary Communicator Strategy	One person is designated as primary communicator and comes into situation with suggestion to improve engagement. Anyone in room at the time shouldn't offer encouragement or comments unless directly asked.	Preventative
Chemical Restraints	Physician decision, but other providers can provide input based on behavioral observations. *Example:* Pharmacological agent to reduce agitation.	Reactionary
Physical Restraints	Physician decision, but other providers can provide input based on behavioral observations. *Examples:* Physical take down when combative, bilateral mitts, four bed rails, or soft wrist restraints.	Reactionary

surance or emphasizing the purpose of activities while simultaneously reminding a patient they are safe or acknowledging a patient's point of view then trying to find a workable compromise are communication-based strategies. A designated primary communicator (Table 12.5) can conduct all the talking with an aggressive, non-compliant, or emotionally dysregulated patient and propose suggestions to de-escalate a situation whilst other staff remain silent.

Use of Restraints for Behavioral Management: Chemical and physical restraints should not be the first-line strategy for managing suspected behavioral changes, but restraints are sometimes employed in situations wherein a patient's or provider's safety is in question. Restraints of any sort should attempt to be avoided as there is no therapeutic benefit or long-term positive behavioral change associated with use (Huckshorn et al. 2023). Additionally, there are pharmacological agents that the literature suggests can have a negative impact on recovery following a neurologic event (Larson and Zollman 2010; Bayley et al. 2023). That said, physical and chemical restraints are sometimes used as a time-limited, reactionary intervention when a patient is engaging in dangerous behavior that would result in harm to themselves or others if the behavior is not immediately terminated. This should not be a strategy employed in situations that are not imminently dangerous. While restraints may be effective for immediate termination of problem behavior, they are unlikely to have any contribution to long-term change (Huckshorn et al. 2023).

Interventions for Psychologically Mediated Behaviors

Mood-mediated behavior issues are also common in nonacute settings. Adjustment-related issues and/or adapting to new functional limitations (either cognitive or physical) are often contributory. A variety of variables should be taken into consideration when determining the intervention best suited to address mood-mediated changes in behavior. A thorough assessment of self-reported mood symptoms is an appropriate place to begin, as certain interventions have greater support for addressing select symptoms. For problems with emotion dysregulation and poor distress tolerance, Dialectical Behavioral Therapy (DBT; Linehan 1993) may be the intervention of choice. Decreased behavioral engagement and cognitive inflexibility can be addressed with Acceptance and Commitment Therapy (ACT; Hayes et al. 2011). Cognitive Behavioral Therapy (CBT) can reduce cognitive rumination and re-experiencing symptoms (Beck 1970). These approaches have demonstrated efficacy for addressing mood-mediated behavioral issues across a variety of neurological populations, such as ACT for TBI (Whiting et al. 2012) CBT for Brain Injury (Gallagher et al. 2019). Thus, selection of the appropriate type of treatment should be determined based on patient and provider opinion about the most distressing presenting symptom, provider expertise implementing select interventions, flexibility in adapting protocols for cognitive and behavioral deficits, and patient responsiveness to the treatment modality.

BEHAVIORAL MANAGEMENT FOR ACUTE CONFUSION

Acute confusional states are defined as abrupt fluctuations in arousal, cognition, and behavior. Acute confusion can be caused by a variety of underlying medical or metabolic conditions, medications, and procedures, as well as acquired and traumatic injuries. Common to medical settings, these disruptions in mental status can generate behavioral changes which negatively impact delivery of medical care, patient safety, recovery course, length of stay, and overall functional status. Therefore, evaluation is a crucial step in the identification, tracking, and potential remediation of delirium and confusional states.

Cognitively, acutely confused patients demonstrate dynamic alterations in attention, orientation, and memory, which can all impact behavior. Arousal changes can impact attention, causing difficulties in directing, sustaining, and shifting attention. Problems with arousal and attention, in turn, impact memory, particularly recent memory, as events cannot be encoded into memory if they are not fully apprehended. Patients demonstrate gaps in memory, and with it, confusion. Disorientation to date, time, and circumstance is common. Memory problems may result in confabulation, or agitation, when individuals in a state of confusion are unable to maintain consistent situational awareness, wherein they may misinterpret the environment around them (i.e., a nurse may be misidentified as someone trying to hurt them).

Arousal plays a unique role in the acutely confused patient and is associated with negative and positive behavioral changes. Patients with hypoarousal are more prone to negative behaviors due to reductions in arousal and responsiveness to their environment. For example, patients with hypoarousal may lack initiation in behavior, self-care, and verbalization of wants and needs. These patients often appear somnolent, with fluctuating drowsiness, and have Circadian rhythm disturbance. Providers can be apt to overlook these behavioral changes because they are not overtly disruptive, even though this can result in temporary, and in some cases, permanent functional loss in patients whose ability to relearn skills may be reduced (dementia, etc.). In contrast, hyperaroused patients may display increases in behavioral, emotional, motoric, and verbal output that interfere with care. The hyperaroused patient can appear restless, combative, agitated, emotionally labile, or delusional. These problematic behaviors serve to achieve homeostasis in arousal by reducing environmental stimulation or internal states.

Approach Strategies

The approach to provision of care is particularly important, as acutely confused patients are sensitive to both verbal and non-verbal communication. Care providers should employ non-threatening body language by moving without rushing and have their arms down at their sides, palms exposed. Differing tone of voice can play a role in patient behavior. Some agitated patients may respond better to a deeper, authoritarian voice, while alternatively, others may feel less threatened by a softer vocal presence. Should a pattern be detected, efforts should be made to adjust the provider's stimulus value (when possible).

Consistent staffing can also allow the confused patient to develop familiarity with care providers, thus improving compliance. All providers should be clearly identifiable, with nametags and standard professional attire, to reduce confusion. Prior to initiating care, providers should always tell the patient what they will be doing to ensure the confused patient understands the intended action. When there are concerns related to pain and agitation, patients can signal (verbally or non-verbally) when they are ready for the caregiver to initiate care. If a patient has difficulty comprehending auditory speech, we recommend the use of visual modeling to demonstrate the desired behavior. Additionally, rather than giving commands, which can sound like ultimatums, providers can offer the patient choices to facilitate initial behavioral engagement and autonomy. Allowing time for a slowed patient to complete tasks themselves can also increase self-agency. Excess dependency occurs in settings where patients have diminished opportunity for social interaction and engagement, as care itself becomes a source of social stimulation. In these cases, allowing for planned socialization outside of care can be beneficial.

Environmental Interventions

Environmental interventions can help remediate confusion in the delirious patient. This can be as simple as using labels (illustration or picture) to identify the patient's room and other salient locations on the unit (e.g., bathroom, nurses' station). Preventing and limiting overstimulation (i.e., routine rest periods, limiting number of individuals in room, access to quiet areas). Objects that are being used inappropriately (i.e., trash can for toilet) can be removed. Orientation can be facilitated with a visible calendar clock and sign. Patients with circadian rhythm disruptions or sundowning can benefit from scheduling changes and room modifications (like opening shades at first light) that increase exposure to natural daylight during waking hours. Conversely, superfluous sound and light sources at night should be reduced as much as possible to facilitate sleep.

Behavioral Interventions for Negative Behaviors

Causes of negative behaviors in confused patients include low arousal, somnolence, disorientation, and aversive stimulation reduction. Patients can refuse care when feeling overwhelmed, anxious, or when unable to identify staff as health providers. Intervention for negative behaviors in confused patients often include increasing stimulation, modifying activity levels, and reducing aversive stimulation (e.g., pain, anxiety, incontinence). Figure 12.1 outlines examples of negative behaviors, etiologies, and suggested interventions.

Behavioral Interventions for Positive Behaviors

Examples of positive behaviors in confused patients include combativeness, restlessness, perseveration, and agitation that often reflect responses to environmental, internal, approach, and contingency factors. Agitated behaviors include disruptive verbalizations, wandering, sexually inappropriate behaviors, disrobing, and care non-compliance. Perseverative behaviors are repetitive behaviors (noises, moaning, etc.), actions (unwrapping the same role of toilet paper, etc.), or self-stimulating behaviors (rocking, rubbing, etc.). A functional analysis and suggested interventions for example positive behaviors in confused patients are provided in Table 12.6.

Positive behaviors have multiple underlying functions. These include attention seeking, where the behavior creates an opportunity for attention or even social attention, increasing stimulation due to boredom, or decreasing stimulation to mediate hyperarousal. Wandering behaviors may have a calming-like effect and verbal/physical agitation or combativeness can serve to stop or ameliorate discomfort. Care refusals can occur when autonomy is reduced (either real or perceived) and the patient seeks to assert agency.

Behaviors that are reinforced by attention, planned rest periods, or brief removal from the environment can be used with disruptive behaviors to reduce contingency reinforcement. This is an approach where consistency is crucial for it to be effective. Planned ignoring—if the caregiver is comfortable with this, and the patient is safe—can also be an effective strategy by decreasing reinforcement for the behavior. For example, if care is stopped every time an unwanted verbalization occurs, providers can instead ignore the disruption and continue care, thus eliminating unintentional reinforcement.

For stimulation-seeking behaviors, behavioral or object substitution can replace the unwanted behavior with a more adaptive (or less disruptive) alternative. Similarly, setting substitutions should be considered in the treatment of inappropriate self-stimulating sexual behaviors (i.e., provide the patient with a private setting and explain that the setting—and not the behavior—is the problem). Targeted physical activity and allowing patients to wander safely (if not disruptive to others) may help reduce agitation and internal stimulation. When problematic behavior serves to stop, or mitigate, an unpleasant interaction for the patient, contingency change strategies can be implemented. For example, instead of discontinuing care altogether in response to an undesired behavior, care can either continue or "pause" and resume after a designated interval. Distraction, redirecting attention away from an upsetting stimulus to something more pleasant, is also an effective as a means of behavioral management with a confused patient. Using persuasion techniques, such as the "foot-in-the-door" strategy, can promote compliance. Asking a patient to complete tasks in small steps and reinforcing the successful completion of each incremental benchmark decrease the likelihood of resistance should the larger task be requested initially. For example, instead of instructing a patient to get dressed, a foot-in-the-door approach would be to start by requesting they put on socks. Providers can also offer choices, instead of commands (*"Would you like to wear the blue socks or red socks today, Mr. Jones?"*). This fosters self-agency on the part of the patient and works to diffuse power struggles.

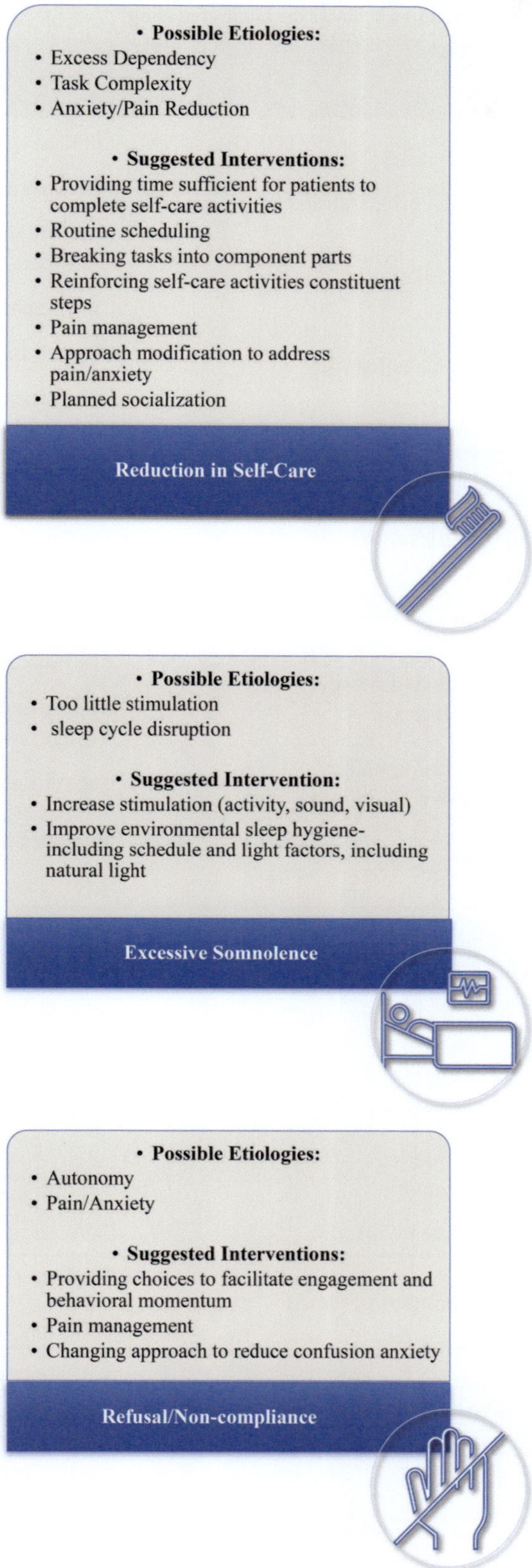

Fig. 12.1 Example behavioral interventions for negative behaviors in acute confusion

Table 12.6 Example behavioral interventions for positive behaviors in acute confusion

Function of the Behavior	Behavior Type	Intervention
Attention Seeking	Disruptive Verbalizations	▸ Planned ignoring ▸ Planned time out ▸ Reinforcement of appropriate verbal responses
	Aggressive behaviors	▸ Contingency management ▸ Planned time out ▸ Reinforcement of alternate non-aggressive behaviors
Social Interaction	Excess dependency	▸ Task modification into achievable components ▸ Verbal reinforcement ▸ Planned socialization
Stimulation Increase	Tactile, verbal, motoric, and/or sexual self-stimulation and perseverative behavior	▸ Alternate stimulation sources ▸ Appropriate environmental selection
	Environmentally destructive behaviors	▸ Behavioral substitution
	Wandering	▸ Increased activity ▸ Safety precautions
Escape/ Avoidance	Care refusal	▸ Choices
	Agitated verbalization	▸ Planned ignoring ▸ Reinforcement of appropriate verbalization
	Aggressive behaviors	▸ Planned ignoring ▸ Reinforcement of appropriate verbalization
	Toileting behaviors	▸ Medical management of constipation
	Pain behaviors	▸ Pain reduction strategies
Autonomy Increase	Refusal / Non-compliance	▸ Choices ▸ Reinforcement of engagement
Confusion	Verbal and physical agitation	▸ Routine consistency ▸ Caregiver consistency
	Wandering / Disorientation	▸ Gentle reorientation and redirection ▸ Safety precautions

TRANSDISCIPLINARY CARE

In recent years, emphasis has been placed on *transdisciplinary care*, an approach that emphasizes a more cohesive approach to patient care wherein each team member understands his/her individual role responsibilities as well as the roles of other team members to collaboratively work toward a patient's individualized functional or recovery-related goal (Reilly 2001). As team unification and consistency of intervention are vital components of behavioral management within the inpatient setting, utilizing a transdisciplinary approach will be most beneficial. Implementing effective behavioral treatment plans requires all team members to work collaboratively and consistently. It is often the neuropsychologist who is key to its development, monitoring, and ensuring team members feel competent to deliver these types of interventions. A vital role for neuropsychologists working with inpatient treatment teams is to provide diagnostic clarity to help providers understand the factors contributing to problematic behaviors. For example, being able to distinguish poor treatment engagement due to depression from other possible reasons (e.g., fear of pain, poor comprehension, or neurogenic apathy) is a unique foundational skill of neuropsychologists. Such distinctions are key to developing an individualized and effective treatment plan.

CULTURAL CONSIDERATIONS

Diversity factors, such as cultural norms and patients' past experiences, are important considerations when trying to understand behavior in both confused and non-confused patients. For example, based on these factors, some agitated patients may have difficulty with providers of a specific demeanor, race, gender, or ethnicity. For non-confused patients, who are often better able to participate in treatment planning, understanding their perception of treatment utility is critical to the overall efficacy of the intervention. This is, in part, guided by a patient's own history, which can be closely tied to diversity-related variables. Misinterpretations of the nature (or function) of a target behavior can occur when providers fail to account for a patient's culture or background (Barrera Jr et al. 2013)—possibly resulting in the selection of an inappropriate intervention. A patient who feels uncomfortable in a particular setting, for example, due to their cultural background or language of origin, may respond to feeling discomfort by exhibiting behavioral dysregulation or restlessness, symptoms that a rehabilitation professional may be called upon to address from an interventional perspective.

SUMMARY

This chapter offers a review of the common symptoms, assessments, and interventions demonstrated to be effective for the management of cognitive, emotional, and behavioral sequelae frequently encountered in inpatient settings. Emphasis was placed on cognitive compensatory strategies as the foundation of cognitive rehabilitation and behavioral interventions as evidence-based practices. Each assessment and intervention should be individualized, which makes it difficult to identify a single standardized approach for assessment or intervention. Thus, this chapter attempts to provide the dynamic empirical and personalized lens required for provider and patient success in inpatient settings.

REFERENCES

Amatya B, Khan F, Galea M. Rehabilitation for people with multiple sclerosis: an overview of cochrane reviews. Cochrane Database Syst Rev. 2019;1

American Psychological Association. Policy statement on evidence-based practice in psychology. 2021. Retrieved from https://www.apa.org/practice/guidelines/evidence-based-statement.

Bahar-Fuchs A, Clare L, Woods B. Cognitive training and cognitive rehabilitation for mild to moderate Alzheimer's disease and vascular dementia. Cochrane Database Syst Rev. 2013;6

Barrera M Jr, Castro FG, Strycker LA, Toobert DJ. Cultural adaptations of behavioral health interventions: a progress report. J Consult Clin Psychol. 2013;81(2):196.

Bayley MT, Janzen S, Harnett A, Teasell R, Patsakos E, et al. INCOG 2.0 guidelines for cognitive rehabilitation following traumatic brain injury: methods, overview, and principles. J Head Trauma Rehabil. 2023;38(1):7–23. https://doi.org/10.1097/HTR.0000000000000838.

Beck AT. Cognitive therapy: nature and relation to behavior therapy. Behav Ther. 1970;1(2):184–200.

Boeve BF, Boxer AL, Kumfor F, Pijnenburg Y, Rohrer JD. Advances and controversies in frontotemporal dementia: diagnosis, biomarkers, and therapeutic considerations. Lancet Neurol. 2022;21(3):258–72.

Breitbart W, Rosenfeld B, Roth A, Smith MJ, Cohen K, Passik S. The memorial delirium assessment scale. J Pain Symptom Manag. 1997;13(3):128–37.

Cicerone KD, Dahlberg C, Malec JF, Langenbahn DM, Felicetti T, Kneipp S, et al. Evidence-based cognitive rehabilitation: updated review of the literature from 1998 through 2002. Arch Phys Med Rehabil. 2005;86(8):1681–92.

Corrigan JD. Development of a scale for assessment of agitation following traumatic brain injury. J Clin Exp Neuropsychol. 1989;11(2):261–77.

Fernandes HA, Richard NM, Edelstein K. Cognitive rehabilitation for cancer-related cognitive dysfunction: a systematic review. Support Care Cancer. 2019;27:3253–79.

Gallagher M, McLeod HJ, McMillan TM. A systematic review of recommended modifications of CBT for people with cognitive impairments following brain injury. Neuropsychol Rehabil. 2019;29(1):1–21.

Gillespie DC, Bowen A, Chung CS, Cockburn J, Knapp P, Pollock A. Rehabilitation for post-stroke cognitive impairment: an overview of recommendations arising from systematic reviews of current evidence. Clin Rehabil. 2015;29(2):120–8.

Hanley GP, Iwata BA, McCord BE. Functional analysis of problem behavior: a review. J Appl Behav Anal. 2003;36(2):147–85.

Haskins EC. Cognitive rehabilitation manual: translating evidence-based recommendations into practice. Ed. Haskins EC. ACRM; 2012.

Hayes SC, Strosahl KD, Wilson KG. Acceptance and commitment therapy: the process and practice of mindful change. Guilford Press; 2011.

Holland D, Hogg J, Farmer J. Fostering effective team cooperation and communication: developing community standards within interdisciplinary cognitive rehabilitation settings. NeuroRehabilitation. 1997;8(1):21–9.

Hollis S, Klebine P, Nakase-Richardson R, Novack T, Reslan S. Problem behavior changes after moderate to severe traumatic brain injury. Model Systems Knowledge Translation Center; 2021. Retrieved from https://msktc.org/sites/default/files/2022-10/UndrstndBhvrChngesTBI-508.pdf

Huckshorn KA, LeBel J, Duxbury J, Hamilton B. Restraint and seclusion. In: Tasman A, Riba MB, Alarcón D, Alfonso CA, Kanba S, Ndetei DM, editors. Tasman's psychiatry. Cham: Springer; 2023. p. 1–19.

Inouye SK, van Dyck CH, Alessi CA, Balkin S, Siegal AP, Horwitz RI. Clarifying confusion: the confusion assessment method: a new method for detection of delirium. Ann Intern Med. 1990;113(12):941–8.

Johnson-Greene D. Clinical neuropsychology in integrated rehabilitation care teams. Arch Clin Neuropsychol. 2018;33(3):310–8.

Johnstone B, Stonnington HH. Rehabilitation of neuropsychological disorders: a practical guide for rehabilitation professionals. Psychology Press; 2011.

Julien A, Danet L, Loisel M, Brauge D, Pariente J, Péran P, Planton M. Update on the efficacy of cognitive rehabilitation after moderate to severe traumatic brain injury: a scoping review. Arch Phys Med Rehabil. 2023;104(2):315–30.

Katz B, Lattal KA. An experimental analysis of the extinction-induced response burst. J Exp Anal Behav. 2020;114(1):24–46.

Kelly G, Todd J, Simpson G, Kremer P, Martin C. The Overt Behaviour Scale (OBS): a tool for measuring challenging behaviours following ABI in community settings. Brain Inj. 2006;20(3):307–19.

Labuda J, Axelrod B, Windell J. Cognitive behavioral protocols for medical settings: a clinician's guide. Routledge; 2018.

Larson EB, Zollman FS. The effect of sleep medications on cognitive recovery from traumatic brain injury. J Head Trauma Rehabil. 2010;25(1):61–7.

Linehan MM. Skills training manual for treating borderline personality disorder. Guilford Press; 1993.

Mitchell PH, Veith RC, Becker KJ, Buzaitis A, Cain KC, Fruin M, et al. Brief psychosocial–behavioral intervention with antidepressant reduces poststroke depression significantly more than usual care with antidepressant: living well with stroke: randomized, controlled trial. Stroke. 2009;40(9):3073–8.

Novack TA, Sherer M, Penna S. Neuropsychological practice in rehabilitation. 2010.

Owen AM, Hampshire A, Grahn JA, Stenton R, Dajani S, Burns AS, et al. Putting brain training to the test. Nature. 2010;465(7299):775–8.

Pritchard RD, Leonard DW, Von Bergen Jr CW, Kirk RJ. The effects of varying schedules of reinforcement on human task performance. Organ Behav Hum Perform. 1976;16(2):205–30.

Reilly C. Transdisciplinary approach: an atypical strategy for improving outcomes in rehabilitative and long-term acute care settings. Rehabil Nurs J. 2001;26(6):216–20.

Schmidt KS, Gallo JL. Behavioral and psychological assessment of dementia (BPAD). Lutz: Psychological Assessment Resources; 2005.

Skinner BF. The behavior of organisms: an experimental analysis. BF Skinner Foundation; 2019.

Soble JR, Critchfield EA, O'Rourke JJ. Neuropsychological evaluation in traumatic brain injury. Phys Med Rehabil Clin N Am. 2017;28(2):339–50.

Sohlberg MM. Cognitive Rehabilitation Manual: Translating Evidence-Based Recommendations into Practice. Archives of Clinical Neuropsychology. 2012;27(8):931–2.

Weyer-Jamora C, Brie MS, Luks TL, Smith EM, Hervey-Jumper SL, Taylor JW. Postacute cognitive rehabilitation for adult brain tumor patients. Neurosurgery. 2021;89(6):945–53.

Whiting DL, Simpson GK, McLeod HJ, Deane FP, Ciarrochi J. Acceptance and commitment therapy (ACT) for psychological adjustment after traumatic brain injury: reporting the protocol for a randomised controlled trial. Brain Impairment. 2012;13(3):360–76.

Assessment of Decisional Capacity

13

Erin K. Bailey, Megan Stockamp, and Adam Fusick

ABBREVIATIONS

ABA	American Bar Association
AD	Advanced Directive
ADLs	Activities of Daily Living
AI	Artificial Intelligence
AMA	Against Medical Advice
APA	American Psychological Association
BPD	Borderline Personality Disorder
CAIPS	Capacity Assessment Instrument for People Who Misuse Substances
CAT	Capacity Assessment Tool
CAT-CAT	Compulsory Assessment and Treatment—Capacity Assessment Tool
CIS	Competency Interview Schedule
DMII	Diabetes Mellitus Type II
ED	Emergency Department
IADLs	Instrumental Activities of Daily Living
ILS	Independent Living Scale
MacCAT-T	The MacArthur Competence Assessment Tool for Treatment
MAID	Medical Aid in Dying
MMSE	Mini-Mental Status Examination
MoCA	Montreal Cognitive Assessment
NAB	Neuropsychological Assessment Battery
NPO	Nothing by Mouth (Latin: *nil* per os)
PD	Personality Disorder
PDAC	Pancreatic Ductal Adenocarcinoma
POA	Power of Attorney
SICIATRI-R	Structured Interview for Competency and Incompetency Assessment Testing and Ranking Inventory-Revised

E. K. Bailey (✉) · A. Fusick
Mental Health and Behavioral Science Service (MH/BSS), James A. Haley Veterans' Hospital, Tampa, FL, USA

Department of Psychiatry and Behavioral Sciences, Morsani College of Medicine, University of South Florida, Tampa, FL, USA
e-mail: Erin.Bailey@va.gov

M. Stockamp
Department of Psychiatry and Behavioral Sciences, Morsani College of Medicine, University of South Florida, Tampa, FL, USA

E. K. Bailey et al. (eds.), *Inpatient Neuropsychology*, https://doi.org/10.1007/978-3-032-05880-5_13

SLUMS	Saint Louis University Mental Status Examination
SNF	Skilled Nursing Facility
SUDs	Substance Use Disorders
Type 2 DM	Diabetes Mellitus Type II
QoL	Quality of Life
US	United States
VaD	Vascular Dementia
WHO	World Health Organization

INTRODUCTION

Requests for evaluations of decisional capacity are common referral questions in inpatient settings. While attending hospitalists (or other primary physicians) have some foundational knowledge in this regard, practitioners with advanced training in assessment of cognitive and mental health disorders—such as neuropsychologists and psychiatrists—are looked to as experts by medical teams when there are concerns for an individual's ability to give informed consent for treatment decisions. Although capacity evaluations utilize similar elements as other clinical exams, these specialized assessments also require fluency in integrating clinical data with medicolegal statutes and ethical standards. The critical thinking skills and empirical conceptualization emphasized in neuropsychological training, as well as a practice model that encourages nuanced assessment, make neuropsychologists particularly apt to provide consultation for decisional capacity. In some cases, this may include direct discussions with a referral source or treatment team to help foster understanding around the benefits and indications for capacity evaluation. For example, neuropsychologists may find themselves educating providers on how to identify patients most appropriate for capacity consultation to promote judicious selection of referrals and modulate workload. Similarly, distinctions between questions that can be addressed in a clinical setting versus those more appropriate for medicolegal or forensic contexts may not be readily identified by providers with less training or experience in clinical capacity. Building rapport with referring providers can not only assist in imparting knowledge and peer education about capacity but will also aid in resolving any conflict in the event of staff disagreement of capacity conclusions.

Basic Principles

Determination of medical decision-making capacity entails assessment of a person's independent ability to provide informed consent and adequately reason through a specific choice. An individual's capacity to make a specific decision is unique to the context of that decision and the time of the assessment, as capacity can vary between different decisions, or the same decision at a different time (Darby and Dickerson 2017). Guidelines to assess capacity have been established by criteria set forth by Applebaum and Grisso (Appelbaum and Grisso 1988).

These criteria consist of four main principles (Fig. 13.1):

- Communication of a clear and consistent choice
- Understanding of relevant information
- Appreciation of the situational context
- Rational manipulation and reasoning of information

Communication of Choice

Clinical capacity evaluations are inherently based on the communication of patients' thought processes, so providers should be aware of hindrances to effective communication and attempt all means necessary to compensate for these factors. For example, aphasia or memory impairment may impair verbal expression or recall of the context in which to base a decision, respectively. When provided with a memory notebook, communication picture board, or other devices, patients may demonstrate greater understanding and insight than initially thought. Other interventions to improve communication include the presence of a translator, writing material, reading glasses, or a personal amplification device for the hearing impaired. Interdisciplinary services, such as speech therapy, can assist with providing the necessary equipment or strategies to improve communication. In this way, evaluators can gain a more accurate assessment of an individual's capacity by minimizing collateral factors.

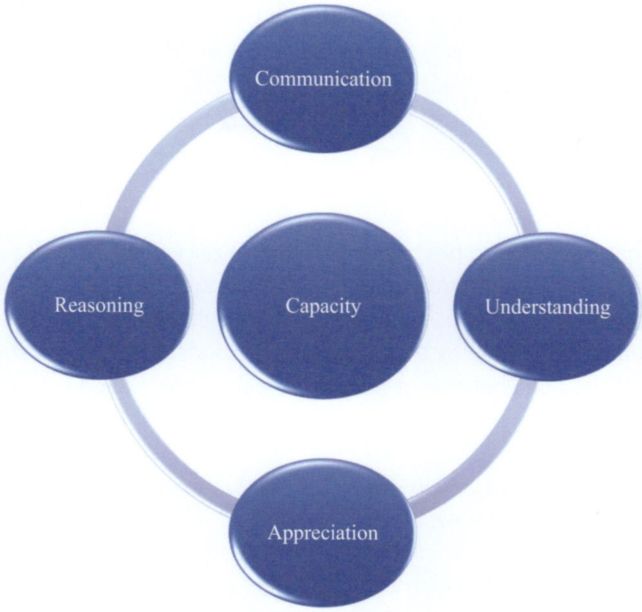

Fig. 13.1 Principles of capacity

Consistency

Consistency of choice is also an important element of communication. An individual can change their mind about a treatment decision. However, depending on the gravity of the situation, a provider should assess the reason and rationality for that decision change (Appelbaum 2007). In inpatient settings, the acuity of a medical illness may cause fluctuations in attention and orientation, such as delirium or psychosis. Patients expressing decisions in this context may lack capacity due to inconsistency of insight. However, for this same reason, hospitalized patients may find themselves making decisions in a rapidly changing context, such as worsening illnesses or availability of discharge options. Familiarization with the context of a patient's social, medical, and discharge status—as well as their goals and values—can assist in distinguishing inconsistency of choice from an individual simply changing their mind.

Understanding

To assess an individual's ability to understand relevant information, providers should first gain knowledge of the situation, including benefits/consequences of possible decisions, and have a working sense of treatment team recommendations and rationale for such. The assessing provider may request the presence of the treating provider during the evaluation to ensure relevant information is provided in a manner digestible for the patient. Cognitive difficulties in language, processing speed, or memory may require differing approaches to patient education. Neuropsychologists should advocate for patients with cognitive deficits by encouraging mindful provision of information and recommendations (using multimodalities, for example).

After ensuring patients have received proper education regarding the decision at hand, a provider can ask the individual to summarize and paraphrase the information back to the provider to gain a sense of their understanding. Providers should ask the individual to provide the information *in their own words* and correct any misunderstanding before moving on to other facets of the capacity exam.

Providers must recognize that some patients may lack intricate detail in their explanations. For example, a patient suffering from sepsis may not be able to explain the exact mechanism of infection but can simply endorse that they have an infection. This explanation may be sufficient. A patient's retention over time can also be inferred from their ability to recall information previously communicated by their treatment team. Simple questions to assess a patient's understanding of their condition can be:

- *"Why have you been admitted to the hospital?"*
- *"Please tell me in your own words what your medical team has told you about your health."*

After determining whether the patient has a basic understanding of their condition or circumstance, a provider should focus subsequent questions around the natural progression or course of possible outcomes and the proposed treatments. Examples of open-ended questions could include:

- *"What will happen to you over time with this illness?"*
- *"How does the doctor want to treat your ailment?"*

Here it is important to note that a patient simply answering "*I may get worse*" or "*I'll become sicker*" may not be sufficient. Providers will want to query for more specific outcomes. For example, if the patient has sepsis, it is true that their health may "get worse" without treatment, but this statement lacks insight into the urgency of a life-threatening condition. To ensure all relevant information is understood by the patient, a provider should ask questions surrounding the risks of the proposed treatments as well as the risks and benefits of alternative treatments, including no treatment at all. Some examples of questions include:

- *"The proposed intervention has some risks; can you tell what they are?"*
- *"What other treatment options have been presented to you?"*

Health Literacy: One important factor to consider when assessing understanding is a patient's level of baseline health literacy, or their pre-existing proficiency, exposure to, and fluidity in navigating medical environments. Studies show that providers tend to overestimate patients' health literacy, particularly for racial/ethnic minority populations. This may result in a breakdown of effective communication (Kelly and Haidet 2007). It is subsequently unsurprising that patients with low health literacy tend to have higher rates of noncompliance (Turkoglu et al. 2018; Jung and Hwang 2015; Miller 2016). For this reason, we suspect these patients may be referred for capacity examinations more often than their health-literate counterparts. Patients with lower health literacy may have less education or reduced familiarity with standard health care procedures, such as typical referral processes or normative treatment procedures (Van Der Heide et al. 2013).

Several measures have been developed to help quantify health literacy in clinical and research settings. The Newest Vital Sign, a measure of health literacy developed by Weiss et al. (2005), uses a mock nutrition label to clarify a patient's understanding of medical instructions. In this measure, patients are asked questions related to hypothetical dietary restrictions. Responses are evaluated across three core aspects of health literacy: application of words (prose literacy), numbers (numerical literacy), and forms (document literacy). *Prose literacy* refers to a person's ability to read and understand basic instructions. *Numerical literacy* requires patients to manipulate numbers in reference to certain criteria: calculating dosages of medication at certain times of day or analyzing nutritional information based on daily percentages of recommended intake, for example. *Document literacy* refers to a patient's ability to make inferences about their own circumstance when provided with health information, such as how to infer the odds or likelihood of an outcome following a medical intervention. Assessing health literacy across these three elements, whether using formal measures like The Newest Vital Sign or similar lines of questioning in a clinical interview, may help providers tailor communication to match patient's literacy levels and maximize their understanding (Kelly and Haidet 2007). Evaluation of health literacy during the informed consent process can guide interventions aimed at increasing autonomy and independence in decision-making, thereby reducing the risk of an incorrect or erroneous determination of incapacity. See Table 13.3 at the end of this chapter for suggestions to improve health literacy during informed consent.

Appreciation

Although memory is commonly identified as a chief mediating cognitive ability for decisional capacity, in our experience this domain (along with language and visuospatial skills) boasts a higher likelihood of functional responsiveness to compensatory strategies. To ensure the *least restrictive means* of oversight, steps should be taken to recommend and implement interventions which increase patients' autonomy or assist in rehabilitating aspects of decisional capacity. As mentioned previously, memory notebooks, communication boards, and other techniques can increase a patient's understanding of their current circumstance or improve communication of choice.

In contrast, executive functioning abilities, specifically insight and problem-solving, tend to be the crux of most capacity determinations. For these reasons, capacity may not necessarily align with intuitive conclusions based on behavioral observations. For example, a patient with frontotemporal dementia may perform within expectation on formal cognitive testing but may have little self-awareness of how each decision option may differentially impact outcome. To demonstrate adequate appreciation, a patient is required to discern benefits and consequences of a decision *as applied directly to their own circumstance*. Thus, it is not enough to acknowledge a general risk when that risk may be disproportionately elevated for that specific individual. In short, intact appreciation requires *insight of self*, not just of situation (Fig. 13.2).

Simple ways to assess a patient's appreciation of the decision at hand can be to ask what the patient "believes" such as:

- *"What do you believe the treatment will do?"*
- *"What do you believe will happen if you refuse treatment?"*
- *"Do you believe that you need some kind of treatment?"*

Other questions may include:

- *"What do you think the odds are of a negative outcome?"*

or

- *"What is the likelihood of this outcome?"*
- *"What can happen if you choose not to follow the doctor's recommendation?"*

Often, the more challenging capacity cases arise when there are incongruencies in this domain.

Mr. D is a 78-year-old man with severe physical disability from advanced Parkinson's Disease. He is non-ambulatory and non-verbal due to dysphagia and dysarthria and requires total bowel and bladder care. He communicates by typing out responses on an electronic board using his dominant index finger.

Mr. D has been living in his caregiver's home for the past 5 years. He was brought into the hospital by his caregiver, who feels unable to continue meeting the patient's considerable care needs. For these same reasons, he has been disqualified from available home health options.

During his admission, Mr. D refused to consider SNF placement when this option was raised by his treatment team. He expressed a value of independence "at all costs" and maintained a preference for AMA discharge into his own care, despite an absence of a safe discharge location or viable options for in-home assistance.

- Points of consideration in this example include:
  - Availability and feasibility of comprehensive in-home care
  - Prognosis for continued physical decline
  - Mr. D's insight into his limitations and ability to independently meet his own care needs
  - Mr. D's values and goals

- Mr. D was consistent in his preference for discharging into his own care, even acknowledging a risk of homelessness. While he met some standards for intact clinical capacity, his decision to obtain independence at all costs was problematic in that, despite acknowledging the increased risk of death or other serious harm should he be discharged without support, he felt this was not an *imminent* risk (due to his own health status), but rather a general hazard of independent living and (if necessary) homelessness. For example, when asked directly about the imminence and likelihood of serious harm or death should he live on the street, Mr. D stated, "stuff happens." He did not appreciate the immediacy of this risk when applied to his current situation or his own limitations in executing his preferences in a successful way. In short, he did not feel that he was at any higher risk of harm than the average individual, despite significant physical disability. Mr. D was ultimately recommended for a medical foster home to meet his need for comprehensive care in a setting that more closely aligned with his preferences for independent living.

Fig. 13.2 Capacity case example 1

Reasoning

The reasoning aspect of capacity assessments can be conceptualized as how a patient synthesizes their understanding, insight, and appreciation into their own values and goals. Providers should ask themselves the following questions:

- *Are the patient's choices congruent with their values?*
- *How does the patient's understanding guide appreciation of options based on those values?*

It is here where we strongly encourage evaluators to self-reflect on their own biases, as the salience of influences which drive a decision can be strongly tied to one's background, culture, lifestyle, or other personal factors. Differing customs and norms may conflict with the traditions and orthodoxy of Western medicine, where much emphasis is placed on maximizing intervention, even at the expense of quality of life. To avoid overly paternalistic interpretations of a patient's reasoning, it is essential to discern the motivations which underlie choice. Questions which can help deepen understanding of a patient's values include:

- *"What factors are most important to you when making this choice?"*
- *"Do you trust your medical team and why?"*

An important element of reasoning is *problem-solving*. A patient may have insight into their cognitive or functional limitations but lack adequate problem-solving for how to protect their own health and safety in this context. Some examples of questions that assess problem-solving include:

- *"How will you know when it is time to consider moving to an assisted living facility?"*
- *"How would you go about getting groceries if you were to live alone, but couldn't drive?"*
- *"What things can you do to lower your fall risk at home?"*
- *"Why do you trust your friend to be your power of attorney? How have they demonstrated trustworthiness in the past?"*

Consider the two examples below in Fig. 13.3. What additional information would be helpful in assessing each patient's reasoning and problem-solving in these contexts?

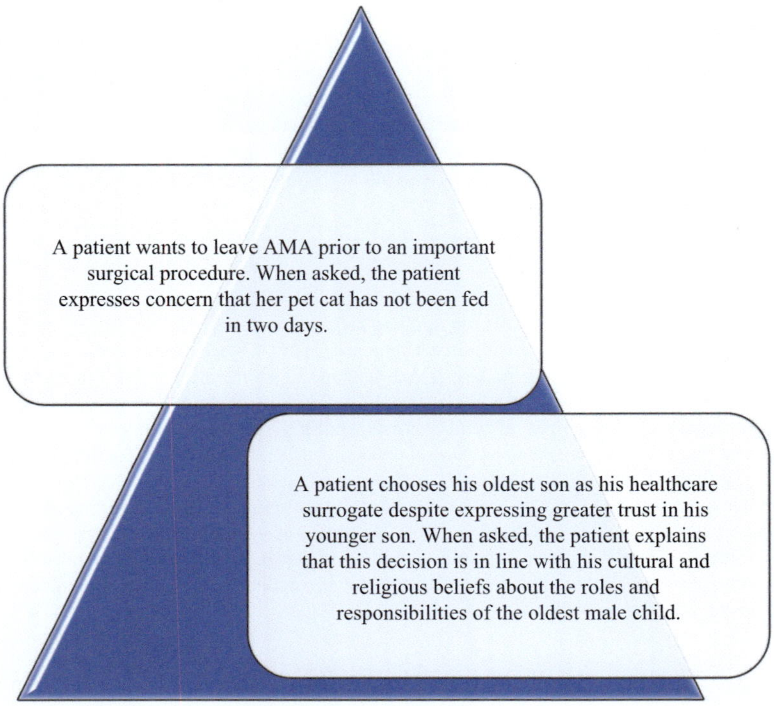

Fig. 13.3 Reasoning case examples

THE CLINICAL EVALUATION

Identifying the Referral Question

Depending on your facility's consultation process, referrals for evaluations of decisional capacity may be posed with little to no context. To avoid erroneously broad conclusions, which inadvertently limit patients' autonomy, we advise narrowing in on the specific referral question. In our experience, the specific referral question may or may not be consistent with the verbatim phrasing of the original consult. To clarify the true question at hand, conversations with team stakeholders (social work, hospitalist, family, patient, etc.) are invaluable. Common referral questions for capacity evaluations in inpatient settings may include the following:

- Informed consent for medical interventions and procedures
- Informed refusal of treatment
- Discharge decisions, including leaving against medical advice (AMA)
- Designation of a health care proxy/completion of Advanced Directive (AD)

After clarifying the referral question, the examiner will need additional information from the medical team regarding the context of that question. This will ensure the examiner accurately assesses whether the individual understands the relevant information, appreciates the situation and its consequences, and manipulates information rationally. Although many capacity referrals require in-depth exploration of a patient's decision-making, there are times when inpatient neuropsychologists may be asked to provide a capacity determination for a patient whose capacity status is obvious—such as a patient with active psychosis or delirium, or in cases where there is no evidence of, or concern for, cognitive impairment. When the referral question is unclear, we advise evaluators to reflect on "*How can I be most helpful?*" See Fig. 13.4 below for a case example.

Ms. P is an 82-year-old woman with an existing diagnosis of mild vascular dementia (VaD). She was hospitalized overnight approximately 1 month ago following a fall (no head injury) and was found to retain capacity to make independent medical decisions at that time. She was discharged home, but was recently readmitted for symptoms of nausea, fatigue, and poor appetite. Work-up was notable for an incidental finding of pancreatic ductal adenocarcinoma (PDAC), an aggressive form of cancer with a poor prognosis.

The treatment team would like guidance on whether to place a consult for re-evaluation of her medical decision-making capacity.

- Points of consideration in this example include:
  - Ms. P's existing diagnosis of mild vascular dementia
  - Recent capacity evaluation documenting intact decisional capacity for medical decisions at that time
  - New diagnosis of aggressive pancreatic cancer
  - Poor medical prognosis
  - Ms. P's values and goals

- A repeat capacity evaluation is recommended for several reasons. First, although Ms. P was found to retain medical decision-making capacity only ~1 month ago, the circumstances of her health have changed drastically since being diagnosed with terminal cancer. While significant declines in her cognitive status are not expected within that short timeframe, her ability to appreciate and reason through medical decisions may be impacted by the added complexity of a more complicated –and emotionally distressing –medical outlook. Further, re-evaluation of capacity provides an opportunity to formally document Ms. P's values and preferences with respect to end-of-life decisions anticipated in the coming months. Should Ms. P be found to retain decisional capacity for palliative/oncological decisions at present, the updated evaluation can also serve as guidance for a surrogate decision-maker should she lack capacity to make medical decisions in the future as her cancer and dementia progress.

Fig. 13.4 Capacity case example 2

Mr. S is an 85-year-old divorced man who was admitted after a witnessed fall in a grocery store. He did not sustain a head injury, but his treatment team has expressed concern about possible cognitive decline and his ability to care for himself following discharge.

Mr. S lives alone and does not have any family or caregiver support.

A consult was placed for "evaluation of discharge capacity."

- Focusing on evaluation of a patient's ability to make, or direct, discharge decisions in this scenario may result in an overly restrictive and broadly overreaching determination. The underlying question is not whether Mr. S can weight the benefits/consequences of different discharge options, but rather:
- 1) Whether or not he is cognitively impaired and,
- 2) If so, does the severity or nature of cognitive impairment raise concern for exploitation or harm should he discharge into his own care without support?
- In short *–is Mr. S a "vulnerable adult" because of his cognitive impairment?*
- Hence, the goals and focus of the evaluation would shift to first address these underlying questions, with specific emphasis on independent living. Mr. S may be asked how he currently manages basic and instrumental activities of daily living (ADLs/IADLS) and whether he uses any compensatory strategies. It would also help to gain a sense of Mr. S's values and his plan for future long-term care. For example, what are Mr. S's self-identified benchmarks for seeking additional functional assistance and are these benchmarks reasonable given his current cognitive abilities?
- Based on the conclusions of this initial exam, the evaluator would be able to provide more targeted recommendations for the team regarding feasible discharge options. Should questions remain about Mr. S's ability to direct his own discharge in this more focused context, he could then be re-referred for a formal capacity evaluation specific to discharge decision-making.

Fig. 13.5 Capacity case example 3

Any treating provider can follow capacity criteria to arrive at a formal determination of capacity to make a particular health care decision. However, neuropsychologists are often asked to render a clinical opinion given our expertise in cognitive evaluation, advanced conceptualization skills, and our ability to guide treatment planning and differential diagnosis based on clinical interview. For this reason, in addition to making determinations of capacity, neuropsychologists can be helpful in opining on a patient's cognitive prognosis, potential for future recovery of capacity, qualification/quantification of supervision or caregiving needs, and discharge disposition (see Fig. 13.5).

Discharge Capacity

Capacity to make discharge decisions is a common referral question in inpatient settings. Discharge decision-making capacity refers to the patient's ability to make informed decisions regarding the development and execution of a discharge plan following their hospitalization. To demonstrate discharge decision-making capacity, a person must display a rational understanding of their health status and how their medical conditions may impact discharge plans, in the context of their health care goals. Patients should understand the nature and availability of discharge placements in relation to their financial resources or other psychosocial limitations as well as be able to articulate their treatment team's recommendations for outpatient care. The patient must have an appreciation of how their condition applies to them, including the indication, risks, benefits, alternatives, and downsides of various discharge placement options as relating to their care needs, and make a clear, consistent choice for discharge preference that protects their basic health and safety needs (Fig. 13.6).

One common point of confusion occurs when capacity evaluations are requested for the purposes of elucidating justification for involuntary hospitalization. In these instances, the referral question is less likely to pertain to the patient's ability to make an informed decision (to refuse treatment, for example), but rather serves as an inquiry regarding their level of disability, functional status, and risk of self-neglect. As we will discuss later in this chapter, making the distinction between a patient's *ability* to make a decision and their *capability* in executing that decision can help in guiding next steps. For example, patients who retain decisional capacity but are at risk for self-neglect may benefit from establishing mechanisms for extra

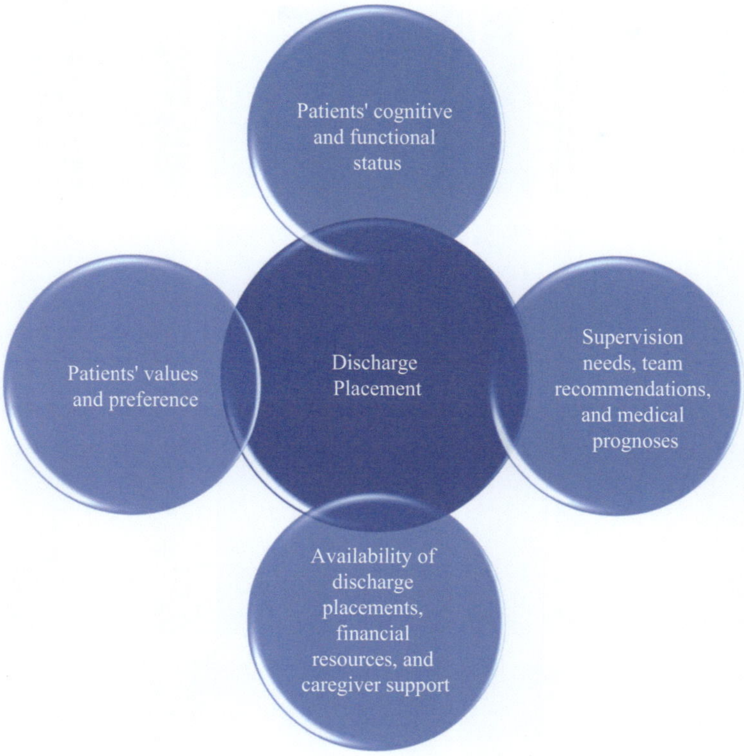

Fig. 13.6 Discharge capacity

assistance following discharge. However, treatment teams may find themselves disagreeing in cases where patients are found to lack capacity but *do not* meet the legal threshold for involuntary hospitalization (e.g., not an imminent danger to self or others)—thus generating divergence of ethical and legal principles. In cases such as these, bioethics consultation can not only guide patient care but also assist in resolving team conflict and reducing the moral distress that may arise with the ambiguity of confronting competing interests (World Health Organization [WHO] 2015; West 2020; Schneider and Bramstedt 2006).

Record Review

Capacity evaluations necessitate broader exploration of the medical record than is typically conducted for cognitive/diagnostic assessments. Examiners should familiarize themselves with the nuances of the medical treatment or procedure that forms the basis of the capacity question, as well as the risks/benefits of possible treatment alternatives. Some working knowledge of rarer diagnoses and disease processes, as well as medication effects, can be particularly useful in these circumstances. Applicable hospital policies that may not be commonly referenced in outpatient settings, such as those outlining guidelines for discharge, informed consent for emergent procedures, surrogate decision-making, and withholding of treatment should also be regularly reviewed. Documentation of chart review should be included as part of the report with a concise, longitudinal summary of the course of the inpatient admission, treatment team conclusions/recommendations, outcomes of any previously attempted interventions, as well as the concern for capacity. Any formal resources, such as institutional polices, may also be cited if used as a basis for the capacity determination.

Interview

The clinical interview is the crux of a capacity assessment. However, inpatient settings offer unique challenges, particularly with environmental and acuity factors, often not experienced in outpatient settings. For this reason, it is advisable to build rapport immediately upon entering a patient's room. Suggestions to foster rapport in inpatient settings are included in Fig. 13.7.

Fig. 13.7 Building rapport

When conducting an interview, it is important to use open-ended questions, appropriately query with follow-up questions, and to specifically address the main tenets of capacity outlined above. One key aspect of decision-making capacity is the assumption that the patient was initially provided with comprehensive information about their circumstance or options. However, it is entirely possible that an examiner finds the patient unable to convey all relevant information, because only a portion of the information, or conflicting information, was initially provided. Similarly, other situational factors may have an impact on how much information a patient later recalls, such as the time of day when the patient first met with their care team, the acuity of the presenting illness at that time, the amount of information conveyed, whether the information was repeated to the patient at any point, and the recency of the admission. Any of these factors may require the medical team to expand on or clarify their feedback and recommendations. Members of the medical treatment team can be invited to join at bedside for the initial portions of the capacity exam to provide relevant information and education to the patient.

Behavioral Observations

We believe the evaluation begins immediately upon entering a patient's room. Observations such as appearance, behaviors, affect (compared to reported mood), speech patterns, comprehension, thought processes, thought content, cognition, insight, and judgment can yield important insight about a patient's cognitive and functional status, as well as elucidate any environmental factors that influence decision-making.

Questions to consider include:

- *Do patients' responses qualitatively differ when family or other parties enter or exit the room?*
- *Is the evaluator witnessing retention of newly learned information over time?*
- *Does cognition and insight fluctuate throughout the course of the evaluation?*

Behavioral data in capacity assessment is not limited to just patient observation but can also include patient-family or patient-provider interactions. For example, it can be helpful for an evaluator to understand how patients are presented with medical information by their treatment team. Education and recommendations presented too quickly, or the use of medical jargon, may hinder a patient's understanding of their treatment options.

Formal Testing

Formal testing in capacity examinations can include cognitive measures, tests which assess specific functional skills, symptom questionnaires, and structured or semi-structured capacity interviews.

Cognitive Assessment

Administering cognitive measures as part of a broader capacity examination can be useful in several ways. First, it establishes a cognitive baseline or context for interpretation of interview responses, observed behaviors, and/or performance on functional measures. Second, cognitive testing can clarify areas of strength and weakness, which can inform compensatory strategies to increase patient autonomy. Lastly, rendering a cognitive diagnosis and proposed underlying etiology can provide useful prognostic information about cognitive status over time.

The Mini-Mental State Examination (MMSE; Folstein et al. 1983) is a widely used screening measure of broad cognition. MMSE is easily administered and scored, consisting of 11 questions that measure orientation to time and place, recall, calculation, language, and constructive ability. MMSE total scores are out of 30 possible points (Vertesi et al. 2001). Another common bedside screen is the Montreal Cognitive Assessment (MoCA; Nasreddine et al. 2005). The MoCA also has 30 total points across eight sections evaluating: visuospatial/executive function, naming, memory, attention, language, abstraction, delayed recall, and orientation. The Saint Louis University Mental Status Examination is another screening assessment that can be used at bedside to assess for cognitive impairment (SLUMS; Tariq et al. 2006). The SLUMS has similar reliability in identifying mild cognitive impairment compared to the MoCA. Since the SLUMS takes less time to administer, it may be an option for busy inpatient providers (Cummings-Vaughn et al. 2014).

A downside of using these screening instruments is the lack of sensitivity in assessing executive functioning abilities. While the MoCA does tap into executive functioning more than the MMSE or SLUMS (Nasreddine et al. 2005), we suggest supplementing bedside screens with measures specifically targeting executive functioning and/or adding functional measures (described below). Comprehensive and lengthier batteries should be used judiciously as patients are less likely to tolerate extensive testing in inpatient settings. For this reason, referrals for outpatient comprehensive neuropsychological testing following discharge are common.

Measures of Functional Abilities

Functional skills assessments are designed to mimic various activities of daily living and give (albeit imperfect) ecological validity to conclusions. These multi-subtest batteries often include mock stimuli and require a patient to execute a task while being scored on accuracy. Tasks could include writing out a mock check, paying a bill, or calculating the amount of anticipated change after a purchase. In addition to skills assessment, functional measures also commonly consist of hypothetical risk scenarios, varying in complexity, designed to elucidate a patient's problem-solving abilities. Examples include the *Health and Safety* subtest from the Independent Living Scale (ILS; Loeb 1996) or the *Judgement* subtest from Neuropsychological Assessment Battery (NAB; White and Stern 2003).

Since these measures more closely approximate real-world tasks, citing a patient's normative score or performance can lend increased support to the conclusions of the capacity exam. One important caveat to this, however, is that patients may demonstrate intact problem-solving of theoretical situations but lack insight required for application of conceptual knowledge to their own, unique situation. For this reason, it is not uncommon for a patient to score well on functional assessments although still lack capacity for decision-making within the same domain. For example, a patient may be able to identify the risk of exploitation when provided with a hypothetical example of a stranger but fails to recognize the possibility of financial exploitation from a close friend or family member. Relatedly, many questions target overlearned skills (e.g., how to call 9-1-1 in an emergency) thus possibly conflating a patient's capability for problem-solving novel situations.

Capacity-Specific Measures

There are several aids to help guide capacity interviews (Dunn et al. 2006). These measures contain standard questions and scoring guidelines that can be completed in 30 mins or less (Edelstein 2000).

Examples include:

- The MacArthur Competence Assessment Tool for Treatment (MacCAT-T; Grisso and Appelbaum 1998).
- Competency Interview Schedule (CIS; Bean et al. 1994).
- Structured Interview for Competency and Incompetency Assessment Testing and Ranking Inventory-Revised (SICIATRI-R; Kitamura 2011).
- Capacity Assessment Tool (CAT; Torroella Carney et al. 2001)

One downside to using standardized instruments is that item content often requires tailoring on the part of the examiner in order to maintain relevance for the specific situation or capacity question at hand, thus impacting true standardization. (Gurrera et al. 2007). Psychometrically, the reliability between subscales of reasoning, appreciation, and expression of a choice can unfortunately be suboptimal (Gogia and Fang 2023). The key is to consider if instrument content is relevant to the clinical question and if patients' responses to said content would yield unique insight to aid in conceptualization.

Collateral Information

Collateral informants, when available, are another important source of contextual information. As mentioned, there is great importance in establishing a patient's cognitive status when making capacity determinations. For this reason, information from family, friends, or caregivers can aid in the challenges that exist between distinguishing between delirium and other neurocognitive disorders (Gogia and Fang 2023) by providing onset and course of a patient's symptoms with potentially greater reliability. Furthermore, a consultant may only have a handful of encounters with a patient but can gain remarkable insight when talking with a family member or loved one. These individuals can help confirm if the decision being expressed by the patient is in line with their values or conversely shed light as to why there is concern for the patient expressing wishes that are out of character.

Undue Influence and Exploitation

Undue influence can be defined as manipulative influence exerted by another, which results in an individual overriding their own preferences or best interest to make a decision that simultaneously harms them but benefits the influencer. For example, despite a medical team recommending discharge to an assisted living facility (ALF), a caregiver or family member could unduly influence a patient's decision for a home discharge in order to secure benefit for themselves (continue receiving patient's social security checks, continued residence in patient's home, etc.), potentially leading to inadequate care for the patient. Exploitation, or the unfair benefit of an individual's resources at their expense, is commonly conceptualized within the realm of financial affairs. However, exploitation of medical resources can also occur, such as using or selling another's prescription medication.

Undue influence can transpire through several different mechanisms of manipulation (Quinn 2000; Blum 2003, 2005; Lichtenberg and Mast 2021; Herrmann et al. 2022; Hutton 1932; Hackard 2017; Wood and Liu 2012). When conducting capacity evaluations, it is important to assess a patient's susceptibility to undue influence, which can speak to their vulnerability to exploitation and harm. If there are elements of a patient's circumstance which raise concern for such vulnerability, such as a cognitive deficit or physical disability (Herrmann et al. 2022), we encourage documenting this in the clinic record and providing recommendations for risk mitigation. Figure 13.8. illustrates the IDEAL model for conceptualizing undue influence (Blum 2003). For additional information about undue influence and geriatric care, please refer to the text by Capezuti et al. (Capezuti et al. 2017) referenced in the 'Suggested Readings' section at the end of this chapter.

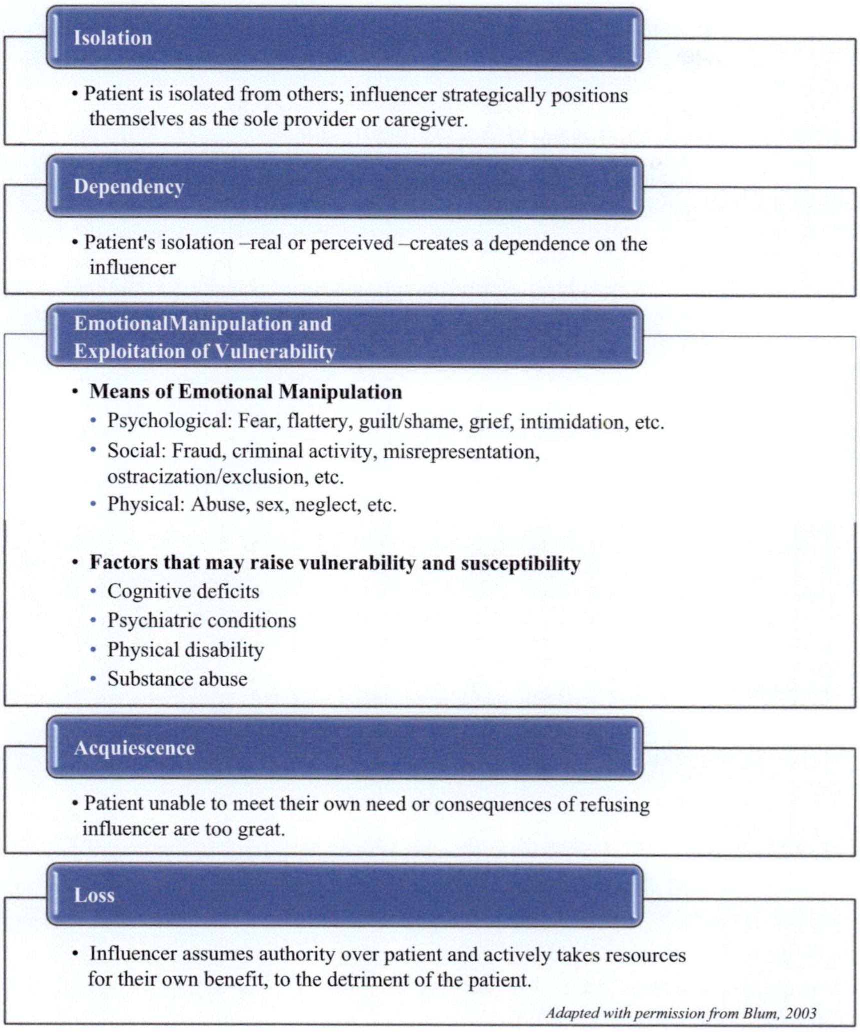

Fig. 13.8 IDEAL model of undue influence

Some points of consideration when working in inpatient settings include:

- *Who may benefit from this patient being admitted (or discharged)?*
- *Who is managing this patient's affairs during their hospitalization?*
- *Can this patient identify risk and articulate how to protect themselves from risk when given hypothetical scenarios?*

When assessing possible undue influence and/or exploitation, it is important to remember that all people retain the right to make harmful or unpopular decisions, including those against their own best interest. Similarly, incapacity must not be automatically presumed when a person has been taken advantage of or fallen victim to manipulation. Distinguishing between a lack of capacity in this regard and a mere "bad" decision requires an understanding of the patient's values, their insight into the negative consequences of their decision, and how they might re-evaluate the behavior and take steps to protect themselves in the future. Consider the examples in Fig. 13.9. What follow-up questions would help determine whether there is concern for exploitation in these instances? Think of questions that would help gauge each patient's level of vulnerability across the characteristics of exploitation outlined above.

1

Mr. G is a 68-year-old gentleman with profound vision loss. He lives alone, but his girlfriend lives nearby and assists with many functional tasks, including cooking, management of finances, and transportation to medical appointments. For this reason, she is only able to work part-time. During the evaluation, Mr. G privately discloses that his girlfriend recently purchased a new vehicle with his money. Although the couple had discussed buying a car in the past, Mr. G was not aware this was an imminent purchase. Had he known, he would have preferred to buy a used vehicle instead. Mr. G endorses feeling somewhat "uncomfortable" with his girlfriend's decision but added that he recognizes the necessity of this purchase as 1) his girlfriend would not have been able to purchase a vehicle on her income and 2) the purpose was as much for his benefit as hers.

2

Mrs. H is a 70-year-old woman with a diagnosis of mild dementia. She is widowed and lives with her daughter, who is a successful attorney, her son-in-law, and two teenaged grandchildren. Mrs. H regularly pitches in for groceries and bills. During her evaluation, Mrs. H discloses that she also frequently gives money to her grandsons for "fun" activities or to her daughter "just because." Mrs. H estimates giving several thousands of dollars to her family in the past year, over and above household expenses. When queried, she expresses feeling grateful to "support" her family in this way, stating "You can't take it with you when you die!"

Fig. 13.9 Vulnerability to exploitation

Drawing Conclusions

Integration of Capacity Elements

The four pillars of capacity outlined above are typically assessed hierarchically: with a patient's preferences and understanding used as the basis for evaluating their reasoning and appreciation. However, neuropsychologists can be particularly skilled in conducting capacity assessments in their ability to make larger inferences about cognition and functioning based on responses to individual questions across each of these four domains. Going beyond a simple summation of a patient's insight or reasoning, neuropsychologists can be helpful in providing nuanced conceptualization of *how* a patient's decision-making may impact health and safety. Thus, neuropsychologists can not only assess capacity in a traditional way but can also explain the likely cause of incapacity from a cognitive perspective as well as link the capacity determination to real-world implications and prognosis. For this reason, we encourage evaluators to be thorough in explaining why a patient's responses might be problematic with respect to autonomous integration of capacity elements decision-making on both on a theoretical and practical level.

See Fig. 13.10 for an example of how nuanced conceptualizations can add refinement and strength to a capacity determination. In this example, Summary 2 integrates the patient's responses to yield a richer conceptualization of functional status and a stronger foundation for the conclusions and recommendations.

Thresholds for Capacity

Individuals are assumed to have decisional capacity unless evidence suggests otherwise (Appelbaum and Grisso 1988). In traditional capacity evaluations, the examiner renders a binary conclusion of capacity based on the above-outlined criteria. A patient retains capacity for the decision at hand or they do not. Yet, this dualistic paradigm does not capture the often-nuanced contexts in which capacity assessments are conducted. In contrast to forensic settings, where legal determinations tend to necessitate conclusions of broad incompetency, clinical evaluations offer some flexibility in conceptualizing differing *thresholds for capacity* across various decision characteristics. Sometimes called a *sliding scale* of capacity (Mellgard and Gligorov 2022a, b), this is practically implemented by outlining the circumstances in which *complexity and urgency* of choice shape the "burden of proof" required to meet each capacity threshold. These should not be confused with *decision severity*, which can be defined as relating to the gravity of consequence (importance) of a decision. Rather, the minimum level of understanding or cognitive burden needed to adequately reason through a decision (complexity) and the imminence of adverse consequences of such decision (urgency). Examples of capacity scenarios mapped across the urgency-complexity-importance matrix are provided in Table 13.1.

Case Details

- **Understanding**: Mr. K reported basic knowledge of his medical conditions and medications, though was unable to characterize his prescribed dosages. He understands that he was brought to the hospital by police after being found confused in a shopping center. He cites his reason for hospitalization as "fatigue." Emergency Department intake lists an admitting diagnosis of diabetic ketoacidosis (DKA) with suspected poor management of DMII.
- **Choice:** Mr. K expressed a desire to discharge AMA and return to his home, where he lives alone. He expressed a value of living independently and maintaining good health.
- **Appreciation:** Mr. K had poor insight into how his noncompliance with medications contributed to his acute medical illness and hospitalization. He was unable to name any benefits of taking his medications, nor risks to his health should noncompliance continue. He was unable to give any potential consequences of leaving the hospital AMA at this time.
- **Reasoning**: Mr. K cited his reason for wanting to leave AMA as a mistrust of his doctors and healthcare establishments in general. He admits to noncompliance with medications, as he believes his doctors are poisoning him.

Summary 1

•Mr. K presents with DKA in the setting of noncompliance with DMII medications. Poor insight and appreciation of the risks/consequences of continued medication noncompliance limit his ability to make an informed decision about leaving AMA.

Summary 2

•Mr. K presents with DKA in the setting of noncompliance with DMII medications. He lives alone and, while home health assistance may be useful in helping Mr. K understand and maintain a regular medication regimen (thus meeting his goal of maintaining good health) the patient is unlikely to accept this help given his suspicion of healthcare providers and belief that his medications are poison. As such, least restrictive means of discharging with home health are not adequate for maintaining Mr. K's basic health and safety. Poor insight into the nature of his DKA and how noncompliance contributed to his current hospitalization precludes adequate problem-solving for how to prevent future hospitalizations and complications. Given medical acuity and impaired appreciation/reasoning, he is at high risk for harm should he be discharged into his own care.

Capacity Determination

Ultimately, Mr. K DOES NOT have capacity to independently make discharge decisions at this time.

Fig. 13.10 Capacity conceptualization

Table 13.1 Capacity thresholds and decision characteristics

Capacity Referral	Decision Characteristics			Capacity Threshold
	Urgency	Complexity	Importance	
Naming a family member as POA over another	Not urgent, possibly time-sensitive	Low	Moderate	**Low**
Consent to emergency surgery	Urgent	Moderate	High	**Moderate**
Choosing between two similar medications	Not Urgent	Moderate	Low	**Low-Moderate**
Consent to participate in a clinical trial	Not Urgent	Very High	Moderate	**High**
Choosing to forgo treatment of a terminal illness	Not urgent, possibly time-sensitive	Low	High	**Low**

Note that, out of these three characteristics, the cognitive complexity needed for each decision at hand most closely aligns with the capacity threshold for that decision. Therefore, understanding the complexity required is often key in establishing a criterion for conclusions of a capacity exam. Complexity can vary depending on the intricacies of the situation and the cognitive abilities of the individual. Mellgard and Gligorov (2022a, b) argue that decision severity, or the importance of the decision, is inherently subjective and thus susceptible to examiner bias of what constitutes a harm or a benefit. This reiterates how complexity is a cornerstone for determining capacity thresholds.

Directional Choice

Different decisions may require differing elements of self-reflection and contextual contingencies which vary the salience of alternative outcomes as compared to one's values and goals. Similarly, cognitive resources required for simplistic decisions do not necessarily presuppose intact capacity for higher-level tasks within the same domain. For this reason, it is possible for a patient to have differing, *task-specific* thresholds of capacity for each level of decision-making when confronted with any given decision. For example, a patient may not have the capacity to consent to a treatment but still retains capacity to select a health care surrogate who can assist in guiding treatment decisions. Similarly, a patient may retain the capacity to make small day-to-day purchases but lack capacity for complex financial investments. Task-specific capacity determinations can also be *asymmetric* in that a conclusion of capacity for a particular choice may not be consistent when presented with the opposing choice. We call this asymmetry *capacity for directional choice.*

Mellgard and Gligorov (2022a, b) summarize implications of asymmetric capacity determinations and directional choice in Fig. 13.11.

In our experience, documenting capacity for a directional choice is helpful in identifying the limitations of the capacity determination with respect to generalizability to other decisions. This can also provide a useful rubric for recommendations of when a re-evaluation of capacity may be necessary, such as if the patient were to change their mind about their decision in the opposite direction. Consider the two examples in Fig. 13.12. How and why may the capacity thresholds differ in each case?

Longitudinal Determinations

Best practice guidelines for assessment of decisional capacity traditionally emphasize the cross-sectional nature of these examinations. Assessment of capacity is typically bound to a specific decision at a singular point in time—consent to a medical procedure, for example. As such, reassessment of capacity is recommended with any change of context. This is particularly important in inpatient settings, where illness acuity can cause rapid change of a patient's cognitive status and treatment plan or where targeted medical intervention may facilitate recovery of insight and capacity. Ongoing monitoring of cognitive status is encouraged by all members of the care team to ensure appropriate attribution of incapacity across circumstance and time.

"Although such an asymmetry might seem counterintuitive, it becomes less so if refusing and accepting a treatment are construed as different decisions with each requiring different abilities. Although it might be true that the facts about diagnosis, prognosis, and treatment alternatives are shared across both decisions, there are facts related to the consequences of refusing treatment that would be distinct from the details required to accept treatment. To refuse treatment, an individual would have to be able to understand that refusal would lead to death as well as understand details about how the dying process would unfold.

Similarly, an individual accepting treatment would have to understand what is required for dialysis and the quality of life that would be maintained during treatment. Thus, refusing or accepting dialysis are distinct decisions that require the ability to understand distinct future scenarios."

Mellgard & Gligorov, 2022

Fig. 13.11 Asymmetric capacity

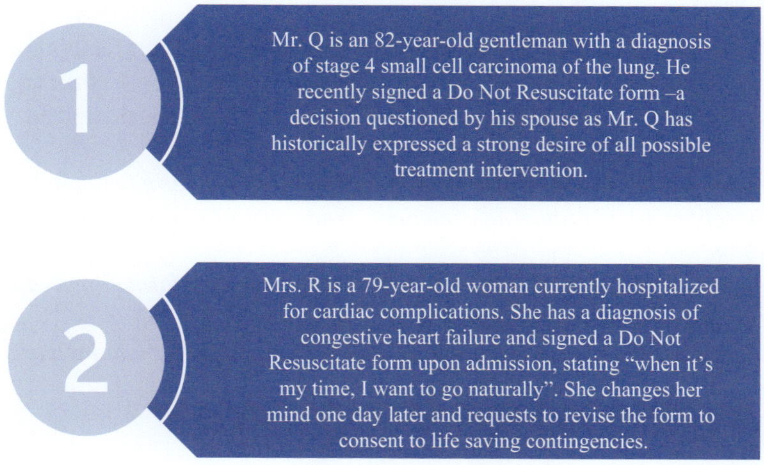

Fig. 13.12 Differences in directional choice

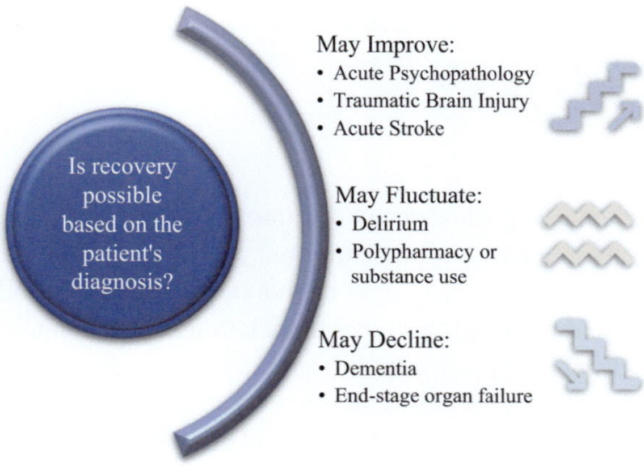

Fig. 13.13 Prognostic considerations

However, part of the role of a capacity evaluator is to provide prognostic information and impart recommendations to aid in *future* treatment planning. Consequently, a singular cross-sectional approach to capacity assessment may be disparate from the longitudinal paradigm of traditional outpatient neuropsychological evaluations. In clinical practice, it is important to weigh pragmatics with respect to applicability and relevance of prior capacity conclusions. For example, a patient with severe dementia is unlikely to regain capacity to make complex medical decisions once this capacity determination has been made. Thus, continual reassessment of capacity for every future complex medical decision may place undue burden on the patient and in this case, deferral to the patient's health care surrogate would be recommended. Ultimately, the frequency and nature of reassessment should be considered in the context of a patient's diagnosis, the clinical features of that diagnosis, and the complexity of the decision at hand (see Fig. 13.13).

An example of providing a "capacity prognosis" is in the recommendation for guardianship. Guardianship may be inappropriate for a patient recovering from delirium, where improvement is expected, but potentially applicable for a patient with severe dementia, where future deterioration of cognitive and functional status is expected given the neurodegenerative nature of the disease. Some considerations for guiding clinical indications for reassessment of capacity include:

- *Whether the capacity question related to a different domain of functioning? (financial* vs *medical)*
- *Whether the new task or decision at hand requires a greater level of cognitive burden than the initial capacity determination? (If both fall within the same domain).*
- *Whether the circumstances or context in which the decision is to be made have changed?*

Recovery of Capacity

When making determinations of incapacity, it is also important to outline reversible factors that, if addressed, may increase a patient's autonomy. For example, patients may experience *restoration* (or recovery of) their capacity with improvements in mental status and underlying health conditions. Considerations for recovery of capacity include:

- *How remediable or changeable are the factors which contributed to a determination of incapacity?*
- *Would the patient benefit from additional information? What type of assistance or support would help the patient receive that information?*
- *Does the severity of their condition prevent learning and integration of information and render any treatment recommendations ineffective in increasing understanding?*
- *Which, if any, of the patient's decision-making may be different if the patient had increased knowledge, assistance, or support?*
- *Is independence or semi-independence achievable?*

The *Assessment of older adults with diminished capacity: A handbook for psychologists* offers a comprehensive list of interventions and less restrictive alternatives to address diminished capacity and improve autonomy across numerous functional areas (American Bar Association [ABA] & American Psychological Association [APA], 2008).

ADDITIONAL CONSIDERATIONS

Mental Health Factors

Mental health conditions, whether chronic (e.g. schizophrenia) or transient (e.g. grief), can greatly influence decision-making. It is vital to not only understand a patient's psychiatric history but also their current emotional state. Symptoms of severe depression, such as apathy, anhedonia, or malaise, may limit an individual's desire or ability to forecast future consequences of immediate decisions. This is supported by imaging studies associating depression with disruptions to prefrontal and ventral striate circuity—two systems involved in executive-based, anticipatory planning and risk-reward valuation (Cáceda et al. 2014). Similarly, considerable anxiety and fear in health care settings can evoke threat-based, fight-or-flight responses that may negatively bias a patient's consideration of possible options and outcomes.

The presence of, and adherence to, psychotropic medication or other mental health treatment is essential to inform the context of a capacity assessment as well as the limits of conclusions. For example, capacity may vary widely in a patient with bipolar disorder who is intermittently compliant with antipsychotic medication, and any capacity determination made in this context should emphasize that the conclusions should not be generalized across time. We encourage nuanced deliberation and delineation of *which* cognitive impairments may be affecting the individual's capacity. For example, a patient with schizophrenia who has refractory, fixed delusions or persistent cognitive impairment despite treatment adherence may have less potential for recovery of capacity. While all attempts to improve autonomy should be explored regardless, this is important for treatment planning as this patient's discharge may require extra time and planning to procure longer-term support. Similarly, a distinction can be made between *rendering a conceptual decision* and *execution of that decision*. For a person with an intellectual disability, for example, an exam may conclude that a patient has adequate capacity to make a choice but would need assistance to execute such choice (Fig. 13.14).

Personality Disorders

Assessment of decisional capacity in persons with personality disorders (PD) presents a challenge for consultants. For example, borderline personality disorder (BPD) encompasses instability of mood, characterological disruptions of self-identity, impulse control issues, and shifting motivations which can result in actions against one's own self-interest and inconsistency of choice. While these symptoms are chronic, studies show that the manifestation of any associated cognitive dysfunction in BPD, particularly executive abilities, can be exacerbated by the presence of situational factors that heighten emotional distress in a patient population already prone to instability (Perington et al. 2020; Fertuck et al. 2006). Thus, a person who may otherwise be rational or mindful in their decision-making could exhibit greater difficulties with impulse and inhibitory control when experiencing strong emotions. Other personality disorders that present with significant cognitive rigidity and inflexibility, such as obsessive-compulsive personality disorder, may hinder assimilation and synthesis of new

Ms. Z is a 60-year-old woman with a medical history of coronary artery disease, Type 2 DM, and intellectual disability. She lives with her sister, who is her primary caregiver. Ms. Z was admitted after presenting to the ED with complaints of severe dyspnea.

Medical workup revealed pulmonary carcinoma, and chemotherapy was recommended over surgical intervention given concerns of post-surgical adherence in the context of patient's intellectual disability. A consult was placed to evaluate Ms. Z's capacity to make oncological treatment decisions.

- Points of consideration in this example include:
  - Characterization of Ms. Z's cognitive abilities and pre-admission functional status.
  - Considerations of any acute medical factors (such as possible paraneoplastic delirium) which could impact her decision-making.
  - Ms. Z's understanding of her diagnosis, prognosis, and benefits/drawbacks/alternatives of treatments.
  - Realistic complexity of pre-and post-surgical care and associated cognitive skills needed for compliance.
  - Availability of collateral support.

- Upon interview, Ms. Z identified her medical condition as "lung cancer" and the recommended treatment option as "surgery." She expressed a knowledge that, without treatment, her cancer would likely spread resulting in adverse health outcomes, including death. She had a basic, though intact, understanding of the mechanism of chemotherapy and the possible side-effects ("nausea, hair loss"). Ms. Z acknowledged the benefit of surgery being excision of her cancerous lesion as well as the limitations of surgery, stating that her cancer may still progress despite undergoing resection. While neuropsychological exam revealed cognitive deficits which may negatively impact Ms. Z's ability to independently manage all aspects of post-surgical care (as well as ongoing management of her chronic medical conditions), this risk is mitigated by the presence of a caregiver as well as Ms. Z's willingness to accept support. Ultimately, Ms. Z was found to retain capacity to choose surgical intervention over chemotherapy.

Fig. 13.14 Capacity case example 4

information into a patient's decision-making paradigm or impact compliance with newly recommended health-promoting behaviors.

Lastly, characterization of an individual's reasoning and motives can be difficult depending on various factors which impact the therapeutic relationship between the examiner and the patient (Ganzini et al. 1994). Personality disorders that include disruptions in social-emotional comportment, such as antisocial personality disorder, or that exhibit distrust or suspiciousness, such as paranoid personality disorder, can impact a patient's willingness to disclose personal values or reasoning behind a choice. While not unique to PD, determinations about capacity and any perceived threat to autonomy may initiate a power struggle and undermine the tenet of benevolence which generated the capacity referral in the first place. To help tease apart true incapacity from unconventional decision-making in patients with PD, Perington et al. (2020) suggest differentiating a patient's *willingness* to meet their own needs from their *ability* to do so. The assessing provider should consider whether social and environmental factors are contributing to the patient's decision. When possible, these factors should first be addressed to reduce a patient's distress prior to initiating a discussion of informed consent or capacity assessment. See Table 13.3 at the end of this chapter for psychiatric and cognitive considerations in informed consent.

Authenticity of Choice

In cases where mental health factors are suspected to play a role in decision-making, psychotherapeutic intervention skills, such as motivational interviewing techniques, can be particularly helpful in eliciting patients' motivations for making a particular decision. A provider can use these skills to identify underlying impetuses of a patient's valuation, goals, priorities, remediable cognitive distortions, and inaccurate assumptions. For example, a 35-year-old patient may exasperatedly claim to "*not care*" whether they live or die when asked to consent to a surgery, giving their reasoning as "*we all die anyway.*" In this case, the sincerity of their voiced preference may be clouded by severe depression and not representative of their genuine values. Conversely, after being newly diagnosed with stage 3 melanoma, a pregnant woman who strongly values parenthood may elect to postpone chemotherapy treatments until after her delivery, thus worsening her cancer prognosis. Although the consequences of her decision elevate her individual risk of harm, her decision remains in accordance with her values, preferences, and priorities. This is sometimes referred to as *authentic decision-making capacity* (Newton-Howes et al. 2019). Factors to consider when evaluating the authenticity of decision-making can include:

- Congruency between a patient's actions and their stated values.
- All-or-nothing, fatalistic, or unrealistic expectations.
- Reasoning based on inaccurate or faulty assumptions about self or others.
- Abrupt changes or complete reversal of choice without any change of context.
- Decisions out-of-character for the patient.

However, we also want to emphasize that people can make informed, autonomous decisions even in the context of severe emotional distress. Authenticity is, by definition, a term ascribed to an accepted 'truth' and, as such, is culturally bound and inherently subjective. If concerns remain regarding authenticity of a patient's decision, delaying capacity determinations to document the patient's consistency of choice and reasoning, as well as any responsiveness to mental health intervention, may help avoid inaccurate and biased conclusions (Kane et al. 2022). Building therapeutic rapport and gaining a sense of patients' personal values, beliefs, and preferences will help evaluators stay cognizant of the individual context in which capacity criteria are being applied.

Substance Use

It is not uncommon for patients with substance use disorders (SUDs) to be referred for evaluations of decisional capacity (Nassif 2019). Across hospitals, the prevalence of such cases varies, but one study at an urban hospital revealed that 41% of 300 consecutive psychiatric consultations for decisional capacity were linked to SUDs (Boettger 2015). Emphasis, however, must be made that diagnosis of a SUD does not unequivocally indicate a lack of capacity. Despite the potential for decisions rooted in addiction and a likelihood of treatment noncompliance, upholding individual autonomy remains paramount.

Determinations of incapacity in patients with SUDs require careful and nuanced consideration of the traditional capacity criteria, particularly the factors of appreciation and reasoning. To clarify ambiguous terminology, Restifo (2013) proposed two subcategories of insight: *intellectual* and *emotional*. Restifo (2013) proposed to apply these subcategories to the concepts of *judgment* and *willpower*, terms often used when deliberating addiction and other impulse-control disorders. When conducting capacity evaluations with patients with SUDs, we recommend considering both *intellectual insight*, or a person's knowledge-based awareness of the situation and possible outcomes/consequences, as well as *emotional insight*, or a person's insight that incorporates a higher degree of personal salience. *Judgement*, or the appraisal of harm/benefit, and *willpower*, defined as the ability to exert control over one's behavior and impulses, can then be examined in this context (Restifo 2013). Consider the examples in Fig. 13.15.

It can be argued that the outcomes for both Patients A and B resulted from diminished willpower, or acting incongruent with their stated intentions, despite having intact intellectual insight into their own circumstance. However, Patient B's reduced emotional insight into how his own actions preclude his preferred choice for treatment and his inaccurate (or incomplete) judgment of the motivations of his medical team when postponing previously scheduled surgeries is concerning for a lack of capacity in this regard.

When evaluating patients with SUDs, providers should consider that clinical presentations may arise from the direct effects of substances or associated complications. Since capacity assessments are time-specific, SUDs present a challenge for fluctuations of insight, problem-solving, and choice, depending on use and cognitive status. For this reason, multiple assessments at different points in time may be necessary. While these assessments can be performed with the previous questions outlined in this chapter, structured tools like the Capacity Assessment Instrument for People who Misuse Substances (CAIPS; Taylor 2014) or the Compulsory Assessment and Treatment–Capacity Assessment Tool (CAT-CAT; Kumar et al. 2022) can assist in consultations for persons with SUDs.

Patient Refusal

Like assessment of capacity in patients with SUDs, the concept of *willingness* also applies in circumstances where patients refuse to engage in capacity examinations. This can present a multitude of clinical and ethical dilemmas as providers weigh the patient's a priori presumption of autonomy with possible medical necessity of intervention (Hurst 2004). Caution should be exercised to avoid misapplication of capacity criteria in these instances, as a patient may not be expressing a clear and consistent choice based on an *inability* to do so, but rather a *volitional refusal* to participate in the evaluation. Elucidating distinctions between these two motivations can be a challenge in clinical contexts yet is vital

Patient A

- 42-year-old patient with heroin addiction and a history of three admissions for overdose.

- Upon interview, the patient:
- Acknowledges the severity of their addiction in the context of multiple overdoses.
- Has appreciation for the potential consequences of continued use, including death.
- Articulates a detailed plan for recovery and express strong motivation for treatment.

- Following discharge, the patient relapses and no-shows to their outpatient appointment at the local recovery center.

Patient B

- 66-year-old patient with a history of severe alcohol addiction admitted for surgical resection of a malignant gastrointestinal tumor. However, his surgery has been rescheduled several times as the patient is nonadherent to pre-surgical 24-hour NPO restrictions and will leave the hospital for hours at a time, only to return intoxicated.

- Upon interview, the patient:
- Expresses a good understanding of his cancer diagnosis and a consistent preference for surgery.
- Understands consequences of continued postponement of surgery (e.g., worsened prognosis, tumor growth).
- Articulates a plan to remain in the hospital until after he undergoes his surgery (re-scheduled for the next day) and is medically cleared for discharge.
- States a belief that his prior surgeries were postponed because his treatment team is prejudiced against him due to his alcoholism.
- Refuses to consider any implications of his leaving the hospital on NPO restrictions when queried.
- After finishing rounds, you observe the patient exiting the main door of the hospital.

Fig. 13.15 Substance use and capacity

to ethical practice. Appel (2023) outlines guidance for assuming consent or refusal when patients are unwilling to participate in capacity examinations and offers suggestions for encouraging patients' willingness to engage (Fig. 13.16).

Unfortunately, formal capacity determinations may be unattainable in instances where a patient's motives for disengagement are truly equivocal. However, collateral information from the medical record, behavioral observations, and/or other clinical data can be used to derive basic conclusions regarding the patient's cognitive status and level of vulnerability or risk. Consultation with colleagues, your institution's ethics committee, hospital policy, or other resources provided by practice organizations can also help shape next steps when patients refuse to participate in capacity assessment.

Culturally Competent Practice

The cultural background of a patient has a profound impact on various aspects of their life and will undoubtedly influence their decision-making in medical matters. Recall that one of the basic principles of capacity is reasoning as applied to one's values and a patient's culture is integral in this concept. Beyond just shaping the interpretation of disease, culture informs perceptions of disease origins, potential cures, and the deference given to some benefits/risks over others. These factors often lie at the core of many capacity assessments. For example, studies reveal that Asian patients see medical decision-making as a collaborative, information-sharing exchange, in contrast to American patients who may prioritize power-sharing with a focus on greater individual independence (Alden et al. 2015). However, this statement, while supported by the literature, can be too broad as membership to different subgroups affect further still one's decision making. For example, within Asian culture, research shows Japanese individuals prefer thorough, slower decision modalities whereas Koreans and Chinese persons prefer faster intuitive or rule-based modes (Yates and de Oliveira 2016). Therefore, an exhaustive understanding of every culture isn't a requirement for health care providers but an awareness and sensitivity to the fact that these beliefs can shape decision-making is essential.

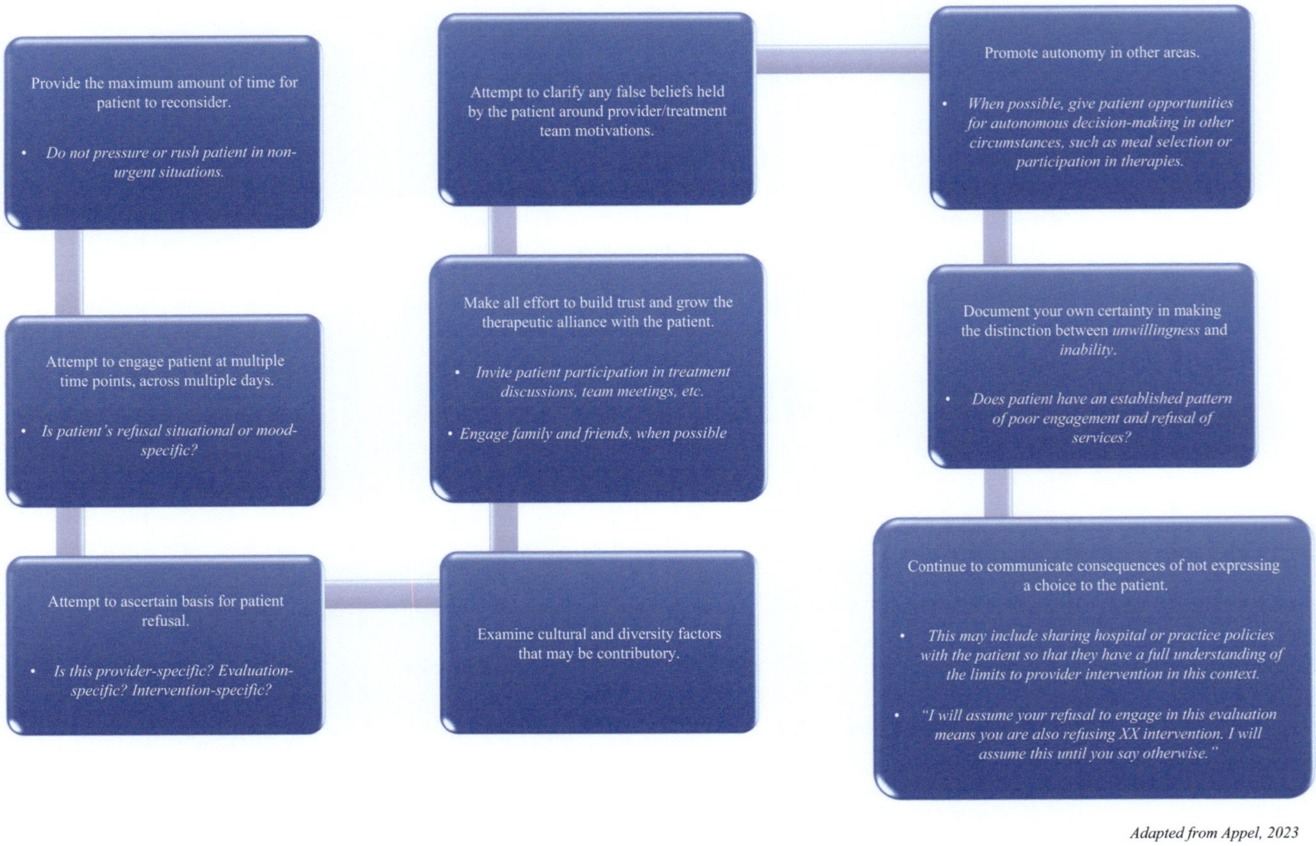

Fig. 13.16 Suggestions for volitional refusal

It is crucial not to assume that belonging to a specific ethnic or racial group implies a predetermined preference for decision-making styles and as such, providers must also acknowledge and address their own biases. Research indicates that a higher frequency of decision-making capacity referrals for Black and Hispanic patients compared to White and Asian patients (Garrett et al. 2023). One possible explanation is that the predominant White and Asian makeup of the US medical profession may not fully grasp the impact of autonomy loss on individuals from racial-ethnic communities who already harbor distrust towards the medical establishment. Such situations can strain the therapeutic relationship between patients and providers.

Further, culture is not defined solely as race or limited to the geographical region a patient was raised but also includes religion. The classic example of religion influencing medical decisions comes from Jehovah's Witnesses and their refusal of blood products (Zeybek et al. 2016). Beyond specific religions, however, it has been generally shown that patients who rate themselves higher on various dimensions of religious experience tend to favor more aggressive care specifically at the end-of-life (Geros-Willfond et al. 2016). Here again, providers are not expected to be theologians but should be able to recognize and incorporate the value system that patients follow into their capacity assessment.

Lastly, another source of potential provider bias can come from the patient's age. While it is true that an increasing age is associated with neurodegenerative illness, not every geriatric patient has neurocognitive dysfunction (Moye 2020). Despite this, the perception of an elderly patient already lacking certain faculties has been shown to impact treatment planning and can also extend to decision-making capacity (Sullivan et al. 2023). Clinicians must remember that decision-making capacity assessments are not contingent on a person's age or other characteristics (ABA/APA 2008).

Recognizing potential provider bias is paramount in not letting these misconceptions wrongfully impact patient outcomes. To safeguard the health alliance, awareness, education, and an ongoing commitment to cross-cultural communication can lead to positive patient interactions. When a provider is unsure about the significance of a patient's background and how it is impacting decision making, obtaining collateral information from a family or friend (with the patient's permission) can be a useful way to ensure you are appropriately weighing the impact of the patient's background and beliefs with the decision at hand (Petrik et al. 2015; Dickson et al. 2023). Furthermore, formal capacity training has shown to reduce the influence of bias in evaluations (Minciocchi Urban et al. 2024). Overall, the diversification of the healthcare workforce, formal cultural trainings, and other Diversity and Inclusion efforts that have been underway in the US health system aim to improve all outcomes in health care, including more appropriate and accurate determinations of decision-making capacity (Handtke et al. 2019).

Surrogacy and Supported Decision-Making

When patients are found to lack capacity to make decisions, the burden of decision-making falls to a surrogate. Federal and state laws, judicial rulings, and hospital/institutional polices will outline guidance for the nature and scope of surrogate decision-making across various settings and circumstances. The foundational assumption of surrogate decision-making is the notion of *substituted judgment*, wherein the surrogate makes decisions on behalf of the patient based on the patient's own goals, preferences, culture, and values. However, in practice, conflicts can arise between families, loved ones, and treatment teams as to who is best considering the perspectives of the patient. For this reason, patient-appointed surrogates, such as Powers of Attorney (POAs) or health care surrogates named in Advanced Directives (ADs) are preferred over court-appointed surrogates (such as a guardian) as, in theory, the patient has self-identified a trusted surrogate whose substituted judgement aligns closely with their own values and goals. Having documentation of a patient's wishes or values, as well as clinical recommendations from the treatment team, can also help alleviate the emotional burden on the surrogate of being the sole decision-maker (Sulmasy and Snyder 2010; Vig et al. 2007).

The capacity to name a health care surrogate differs from other capacities in that a patient need not have a full understanding of their own medical condition or proposed treatment, but rather, that they understand and appreciate the parameters and role of a surrogate as well as demonstrate good reasoning for their choice of surrogate (Navin et al. 2022). Consider the two case examples in Table 13.2.

A common criticism of surrogate decision-making is the dichotomous delegation of decisional responsibility: either the patient makes the decision or the surrogate. Advocacy efforts for supported decision-making argue that joint decision-making between the patient and a trusted individual should be considered as initial "least restrictive means", allowing for greater patient autonomy (Kohn et al. 2012). Neuropsychologists can assist in promoting joint decision-making by facilitating open discussions of treatment recommendations or other relevant information with the patient, family, and treatment team. Often, this can be used to support patient decision-making and avoid ruptures between the patient and team which would otherwise damage rapport and compliance. Kohn et al. (2012) also suggest that providers can work with patients to identify a "circle of support". This is defined as a group of patient-identified support persons who can offer guidance to the patient and assist the team by helping them understand and interpret the patient's words, actions, or decisions in the broader

Table 13.2 Capacity to name a health care surrogate

Referral Question: Healthcare Surrogate	Patient 1 68-year-old single man admitted for hepatic encephalopathy	Patient 2 73-year-old widowed man admitted for idiopathic altered mental status
Understanding	Patient 1 understands he is being hospitalized for "liver problems." He understands the role of a healthcare surrogate.	Patient 2 is aware he is in the hospital but does not understand why. He understands the role of a healthcare surrogate.
Choice	Patient 1 would like his 30-year-old daughter to be his surrogate decision-maker. His daughter is currently incarcerated in another state and the patient has not been in contact with her for over a year.	Patient 2 would like his 49-year-old neighbor to be his surrogate decision-maker.
Appreciation	Patient 1 had little insight into the limitations of naming a healthcare surrogate who is current incarcerated. He did not feel this would hinder communication with his healthcare team. He was unable to articulate any possible consequences to his own health, should the team be unable to reach his daughter during her incarceration.	Patient 2 demonstrated good appreciation of the benefits and drawbacks of naming various peers as his healthcare surrogate. He had insight into the demands and desired qualities of a good healthcare surrogate and why his neighbor specifically meets these qualities. For example, his neighbor is retired and readily available if needed.
Reasoning	Patient 1 states that his daughter is his only immediate family. For this reason, he feels she would respect his wishes and consider his preferences.	Patient 2 articulated a preference for his neighbor because she has been "looking out" for him since his wife died. His neighbor is familiar with his health conditions and helps set up his weekly pillbox. They often make dinner together and he "trusts" her because they are friends and go to the same church.
Conceptualization	Patient 1 has poor appreciation for how the limitations of his daughter's incarceration pose a risk to his future health and safety. Further, he did not have insight into the potentially problematic nature of his assumption that his daughter would make decisions in line with his values simple because she is family, without any prior discussion with her of his expectations for this role.	Patient 2 demonstrates good problem-solving in selecting a surrogate who will not only be able to adequately fulfill the role, but who has a demonstrated history of providing trusted assistance and support. Given they attend the same church and are close friends, it may also be reasonable to assume that their values closely align and that his neighbor would consider decisions in the patient's best interest.
Capacity Determination	Patient does NOT have capacity to name his daughter as his healthcare surrogate.	Patient 2 RETAINS capacity to name his neighbor has his healthcare surrogate.

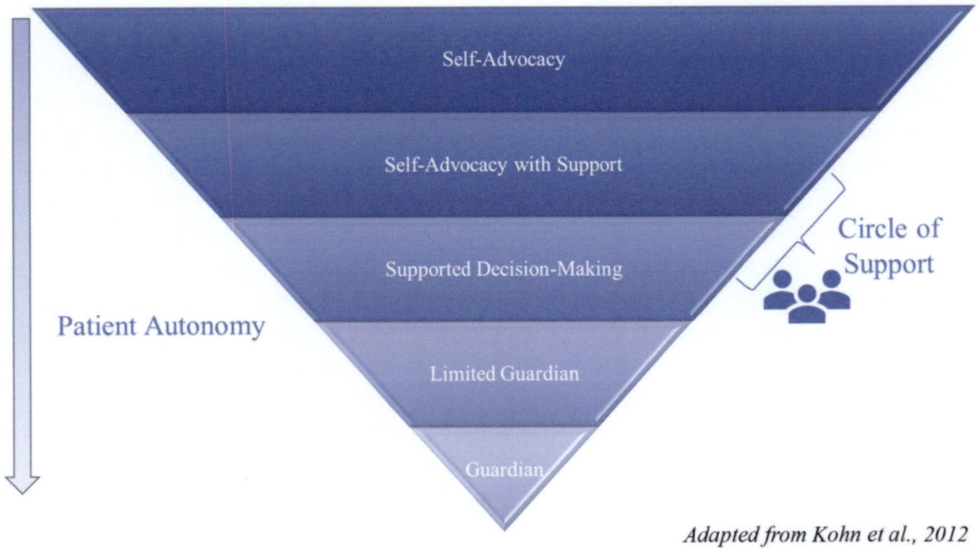

Fig. 13.17 Supported and surrogate decision-making

context of their background. A circle of support may be particularly helpful when there is a contrast between the customs of Western medicine and a patient's culturally based expectations of health care treatment. See Fig. 13.17 for an outline of surrogate and supported decision-making.

Palliative Decision-Making

Decisional capacity tends to decline toward end-of-life. In one study, 27% of patients were found to have limited capacity one month before their deaths and up to 67% in the last week of life (Kaspers et al. 2013). Those with limited capacity are more likely to die in settings outside the home and as such, evaluators on inpatient palliative care or hospice units may be asked to make end-of-life capacity determinations in circumstances with greater ethical ambiguity and moral complexity. In the United States, the notion of an inalienable right to a "good death" can be at odds with institutional and cultural prioritization of the preservation of life. For this reason, it is not uncommon to find discrepant care goals between patients, their surrogates, and treatment teams in palliative or hospice settings. Neuropsychologists can assist, not only with capacity assessments, but by weighing in on how to best integrate differing opinions in patient-centric care.

The use of Advanced Directives (ADs) can help illuminate a patient's wishes for palliative treatments and guide surrogate decision-making. Knowledge of an individual's a priori preferences for their end-of-life care can also be used as a basis for comparison when assessing their *consistency of choice* and determining if their current reasoning is out-of-character and inconsistent with their longstanding beliefs. However, examining consistency of choice in palliative settings should be interpreted with care. Some research suggests that the assumption of poor quality of life (QoL) following severe disability is not necessarily consistent with patients' lived experiences, and that the latter tends to reflect similar ranges of QoL as age-matched healthy peers (Lulé et al. 2009). Thus, it is possible a patient may opt for more aggressive treatment when confronted with a severe or terminal illness, despite previously stated preferences to the contrary. Similarly, as people approach their end-of-life, their preferences for care (or refusal of care) may shift simply because of the changing context and are not pathological nor representative of impaired capacity. If a patient's current choice does not align with their expressed past preference, evaluators should identify the timing of when ADs were completed or when discussions of a patient's preferences were conducted for guidance. For example, are previously expressed preferences or directives—made when a person was much younger or healthier—now being applied at a different phase of life or in a new circumstance with different considerations for quality of life? Such a circumstance may increase distress for surrogates or treatment teams who are now tasked with making inferences of a patient's preference in a novel context. Some studies show that patients who do change their preferences toward end-of-life tend to go from neutral or ambivalent about intervention to declining treatment (Kaspers 2013)—which again, may be at odds with provider or surrogate partiality for preserving life. Neuropsychologists working in these settings can help foster *ongoing* Advanced Care Planning discussions so that patient preferences are documented across time and with changing circumstance—ultimately leading to improved patient, family, and care team satisfaction overall (Kaspers 2013).

Medical Aid in Dying

A unique and controversial situation faced by providers in certain US states is the request for medical aid in dying (MAID). The criteria for which patients can be appropriately considered for MAID vary from state to state, and as such, providers conducting capacity evaluations in this context should be intimately familiar with state regulations, as well as scope-of-practice as outlined by state law. However, despite nuanced geographic and legal variances, there is commonality in these cases as the patient will have a terminal illness and be near the end-of-life (Price et al. 2014). When approaching these evaluations, it is important to distinguish between (1) MAID—a process in which a patient of intact capacity and terminal prognosis chooses to end *their own* life; and (2) assisted suicide—nomenclature used when death is the result of actions of a third-party, and not the patient themselves (also called euthanasia)—as the latter remains illegal in the United States even in certain states authorizing the former.

Because MAID inherently requires patient participation, capacity evaluations in this context are primarily conducted to safeguard against the impact of mental disorders (Treem 2023), although the prevalence of suicidal ideation among those with terminal illness can make it difficult to elucidate choice independent of mood (Breitbart et al. 2000). Similarly, depression may not always be appropriately recognized in patients who are requesting medical assistance with dying (Ganzini et al. 2008). This noted oversight only further underscores the necessity of completing a thorough capacity evaluation and ensuring no underlying psychiatric disorders may be impacting the patient's decision. If during your assessment, depression, or an underlying mood disorder is noted, treatment is recommended before a determination of capacity is made as there is evidence to suggest that treatment of an underlying psychiatric disorder can reduce the desire for a hastened death in this patient population (Breitbart et al. 2000).

Medicolegal Implications

In clinical settings, health care providers are obliged by law to evaluate an individual's time- and decision-specific capacity to ensure appropriate care is delivered in a timely fashion. This differs from legal determinations of competency, which are adjudicated in court. Only the court can determine whether incompetency is present, unceasing, and requires the appointment of a guardian or conservator (Garner 2019). Determinations of incompetency can be made in one or more functional domain (financial, medical, legal, etc.; Elder Justice Initiative n.d.). Specialized courts handling these matters are state-dependent and may be called probate courts, surrogate courts, or orphan courts (Elder Justice Initiative n.d.). Depending on state laws and the specifics of the case, third-party oversight of a guardian (or conservator) may be appointed by a judge for a very narrowly defined task (e.g., limited guardianship), or for very broad purposes (e.g., plenary or total guardianship). A petition for guardianship can be filed by any competent adult, including a family member, friend, or associated professional.

Forensic evaluation of competency is typically conducted by employees or contractors for the local or state courts (Elder Justice Initiative n.d.)—with each state's statutory framework dictating the specific steps of legal proceeding. Ultimately, independent findings are compiled and presented to a judge, who evaluates evidence of:

- Any medical and mental diagnoses that could affect the individual's cognition.
- Any abilities which are impaired because of these diagnoses.
- The prognosis of improvement or recovery of these abilities or diagnoses.

However, it is not uncommon for health care providers to formally testify to an individual's mental health, functional status, and cognitive abilities or for medical records to be subpoenaed as part of an evidentiary hearing or guardianship litigation. For this reason, it is important to familiarize oneself with guardianship and competency laws in your state and thoroughly document rationale for all clinical capacity determinations.

When providing documentation, we recommend including statements that address the following:

The Evaluation

- Criteria by which capacity is being determined and the specific capacity question at hand.
- Data sources included in the evaluation as well as what may *not* be represented, such as collateral information from a family member or missing medical records.
- Evidence for intact, diminished, or impaired abilities across each of the four pillars of capacity, for each functional domain assessed.
- Documentation of any prior attempts at least restrictive means and the outcome of these.

Conclusions and Recommendations

- The mechanism, and suspected etiology, that underlies impaired capacity.
- Prognosis for recovery of the above.
- Recommendations for compensatory strategies and/or contingencies in which capacity may change or through which least restrictive means may be attained.
- Documentation of relevant peer or administrative consultation or other interdisciplinary liaising.
- Disclaimer of the legal limitations for clinical capacity examinations.

See Appendix "Example Capacity Documentation" at the end of this chapter for example a capacity section to include in clinical reports.

As mentioned previously, clarifying the referral question can help differentiate clinical capacity referrals from those that are not appropriate for clinical contexts. To determine appropriateness of a consult for capacity assessment (versus a question best left to medicolegal realms), providers should reflect on how the capacity determination may impact the patient's clinical care in the present setting. If questions extend beyond this scope, or do not directly impact care planning, then the patient should be evaluated in a forensic setting. See Fig. 13.18 for an example. For additional guidance regarding the role of neuropsychology in evaluating civil capacities in legal settings, please refer to the text by Demakis (Demakis 2012) referenced in the 'Suggested Readings' section at the end of this chapter.

Artificial Intelligence in Capacity Assessment

An emerging topic in medicolegal literature is the use of Artificial Intelligence (AI). Not only is AI being integrated into clinical care (Beam and Kohane 2016; Van Hartskamp et al. 2019), but discussions have been raised regarding the utility and ethical implications of AI in obtaining informed consent and making capacity determinations (MacIntyre et al. 2023; Ng 2024). One of the biggest potential advantages is the ability to integrate unique patient data while allowing for greater standardization and reliability, thus possibly assuaging subjectivity of human evaluators (MacIntyre et al. 2023; Herman et al. 2015).

However, several pitfalls exist in the practical application of AI. As mentioned previously in this chapter, bias in capacity evaluations is not limited to inferences of whether a person retains or lacks capacity, but is also reflected in referral process itself,

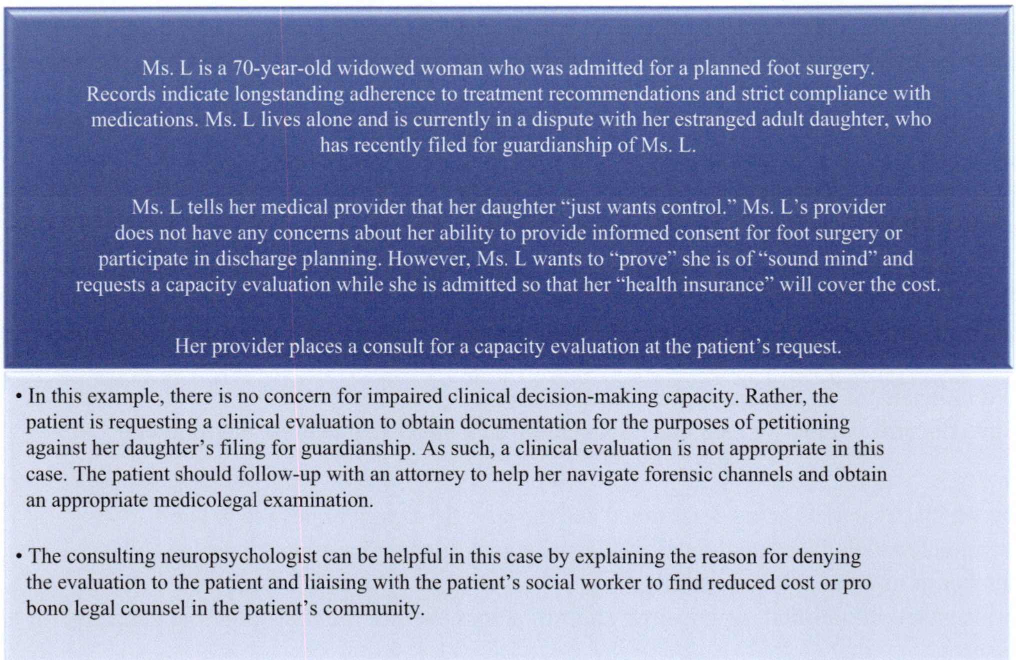

Fig. 13.18 Clinical vs medicolegal referral

where those with lower health literacy (Schulz et al. 2021; Miller 2016) or racial/ethnic minorities (Garrett et al. 2023) may be referred at disproportionately higher rates than their health literate or White counterparts. If AI algorithms are self-taught, the inherent selection bias of the so-called "normative sample" will no doubt prejudice the algorithmic logic by which an AI capacity determination is made. Another obstacle of AI in informed consent and capacity assessments is the inability to re-create and examine the proprietary deductive process of these determinations (Table 13.3; Danks and London 2017; Ng 2024). This not only has implications for evaluating the validity of conclusions, but also creates legal liability for clinical providers, who may be asked to explain their own decision-making to a judge, jury, and attorneys in forensic settings—where regulatory laws, statutes, and legal precedent are not established at the same rate as technology. While technology will continue to advance for the foreseeable future, a practitioner's approach to advanced technologies, such as generative AI, should be cautious.

Table 13.3 Clinical vs medicolegal referral

Real World Challenges to Informed Consent	Mitigation Strategies
Patient Health Literacy and Sociocultural Factors	
• Poor health literacy is more prevalent in vulnerable populations and associated with suboptimal comprehension and participation in informed consent. • Sociocultural norms affect what is considered to be locally accepted practice in informed consent.	• Use of educational tools, decision aids, improving readability of consent forms, and 'teach-back' techniques. • Cultural competency training. • Targeted health literacy programs. • Subjective ethicolegal standard of disclosure based on patient context.
Psychiatric Patients and Older Persons with Cognitive Impairment	
• At risk of losing patient autonomy in clinical decision-making due to preconceived notions on cognitive abilities and mental capacity. • More nuanced mental capacity assessments in psychiatric/cognitively impaired individuals	• Presume mentally competent until proven otherwise • Rule out transient, reversible cognitive impairment. • Overcome communication barriers. • Evaluation of psychiatric insight. • Beware of psychopathological influences • Time-and decision-specific capacity assessments
Artificial Intelligence in Clinical Care	
• Flawed decision-making due to measurement errors, selection bias, and feedback loop bias. • Ill-equipped to manage AI technologies. • 'Black box' proprietary AI models. • Accountability quandary and privacy-related concerns.	• Develop a coherent informed consent strategy • Clinical validation of AI model in real-world datasets. • Formal approval by governing authority of healthcare AI use. • AI user manual on approved clinical indication, assessment metrics, study-proven benefits, risks, data protection, and quality control measures.

Adapted with permission from Ng, 2024

Clinical Competency

There are several options for providers who wish to improve their competence and familiarity with capacity evaluations. Literature review of medicolegal, health care ethics, and mental health journals yields numerous articles discussing nuances of capacity across a spectrum of circumstances and patient populations. As a start, we provide a list of suggested readings on the topic at the end of this chapter. Formal consultative pathways are also available through ethics committees offered by practice organizations and, if applicable, setting-based resources such as your institution's office of general counsel. Using a team-approach, obtaining a second opinion, and peer-to-peer support can assist in building consensus in challenging or ambiguous cases.

SUMMARY

Informed consent is inherent to the provision of health care. When concerns exist regarding a patient's ability to understand, appreciate, and reason through options to arrive at a consistent choice, inpatient neuropsychologists may be tasked with performing evaluations of decisional capacity. Though each referral is unique, the process and criteria outlined in this chapter can assist in guiding capacity determinations across a variety of inpatient settings and circumstances. Remember, providing guidance to the treatment team on how to maximize independence is often as valuable as the capacity determination itself. To this end, individualized recommendations, such as remediation of mitigating factors (e.g., acute medical conditions, health literacy, psychiatric conditions, substance use), can improve patient autonomy (Table 13.3). Cognitive and/or functional evaluations can also elucidate areas of strength and weakness and inform tailored recommendations for supporting independence. When conducting capacity assessments, clinicians may have to manage others' expectations of their conclusions and work with the treatment team to provide the ideal balance provision of beneficent care with patients' inherent right to self-determination. Ongoing consultation with peers or practice organizations will help ensure clinical and ethical competencies. Similarly, maintaining awareness of jurisdictional laws, personal bias, and cultural sensitivities is essential for sound professional practice.

APPENDIX: EXAMPLE CAPACITY DOCUMENTATION

CAPACITY EVALUATION

The following is a CLINICAL assessment of decisional capacity. It is meant to guide clinical care and treatment decisions and does not function as a court-appointment expert assessment of legal competency. The following domains are addressed via available medical records as well as clinical evaluation, and thus, may not be exhaustive of all relevant information or directly comparable to the legal standards of the capacity in question.

ASSESSMENT OF CAPACITY TO MANAGE HEALTHCARE INDEPENDENTLY:

Medical decision-making capacity refers to the patient's ability to make informed decisions about a particular health care decision at a particular point in time. To demonstrate medical decision-making capacity, a person must display a rational understanding of his/her condition, as well as the proposed treatment, including the indication/ risks/ benefits/ alternatives/consequences of no treatment. They must have an appreciation of how their condition applies to them, and make a clear, consistent choice.

For this patient, the present exam seeks to evaluate the specific capacity to…(consent to surgery, refuse treatment, etc.)

- UNDERSTANDING: Partial

  The patient provided a partial list of their medical conditions

- CHOICE/VALUES: The patient expressed a value of independence and

- CONSISTENCY: Inconsistent

  The patient was inconsistent in making a choice to… Despite

- APPRECIATION: Impaired

  Although the patient was oriented to person and place, he provided inconsistent and often inaccurate recounts of his medical history and medications. Despite multiple recent hospitalizations for serious infections, the patient denied any

- REASONING: Impaired

  Deficits in executive functioning limited this patient's ability

  As such, this patient is vulnerable to harm because...

- POTENTIAL FOR COMPENSATION:
  (Evaluation of environmental support and the potential to compensate for knowledge/cognitive deficits, etc.)

CAPACITY DETERMINATION:

Based on the current evaluation, the patient RETAINS/DOES NOT HAVE capacity to independently make the decision to……

- DIAGNOSES/PROGNOSES:

- RECOMMENDATIONS:

REFERENCES

ABA/APA Assessment of Capacity in Older Adults Project Working Group, American Bar Association. Commission on Law, & American Psychological Association. Assessment of older adults with diminished capacity: a handbook for psychologists. American Bar Association; 2008.

Alden DL, Friend JM, Lee AY, de Vries M, Osawa R, Chen Q. Culture and medical decision making: healthcare consumer perspectives in Japan and the United States. Health Psychol. 2015;34(12):1133.

Appel JM. Anything you do not say can be used against you: volitional refusal to engage in decisional capacity assessment. J Clin Ethics. 2023;34(2):204–10.

Appelbaum PS. Assessment of patients' competence to consent to treatment. N Engl J Med. 2007;357(18):1834–40.

Appelbaum PS, Grisso T. Assessing patients' capacities to consent to treatment. N Engl J Med. 1988;319(25):1635–8.

Bathaee Y. The artificial intelligence black box and the failure of intent and causation. Harv J Law Tech. 2017;31:889.

Beam AL, Kohane IS. Translating artificial intelligence into clinical care. JAMA. 2016;316(22):2368–9.

Bean G, Nishisato S, Rector NA, Glancy G. The psychometric properties of the competency interview schedule. Can J Psychiatry. 1994;39(8):368–76.

Blum B. The "IDEAL" protocol for undue influence—an introduction. NCPJ Life Times. 2003;2(7):1–4.

Blum B. The "undue influence worksheet" and "IDEAL" protocol–an introduction. 2005.

Boettger S, Bergman M, Jenewein J, Boettger S. Assessment of decisional capacity: Prevalence of medical illness and psychiatric comorbidities. Palliat Support Care. 2015;13(5):1275–81. https://doi.org/10.1017/S1478951514001266. Epub 2014 Oct 30.

Breitbart W, Rosenfeld B, Pessin H, Kaim M, Funesti-Esch J, Galietta M, Nelson CJ, Brescia R. Depression, hopelessness, and desire for hastened death in terminally ill patients with cancer. JAMA. 2000;284(22):2907–11.

Cáceda R, Nemeroff CB, Harvey PD. Toward an understanding of decision making in severe mental illness. J Neuropsychiatry Clin Neurosci. 2014;26(3):196–213.

Capezuti E, Malone ML, Khan A, Baumann SL, editors. The encyclopedia of elder care: the comprehensive resource on geriatric health and social care. Springer; 2017.

Cullum M, Saine K, Weiner MF. Texas functional living scale. Arch Clin Neuropsychol. 2009.

Cummings-Vaughn LA, Chavakula NN, Malmstrom TK, Tumosa N, Morley JE, Cruz-Oliver DM. Veterans affairs Saint Louis University mental status examination compared with the Montreal cognitive assessment and the short test of mental status. J Am Geriat Soc. 2014;62(7):1341–6.

Danks D, London AJ. Algorithmic bias in autonomous systems. Ijcai. 2017;17(2017):4691–7.

Darby RR, Dickerson BC. Dementia, Decision Making, and Capacity. Harv Rev Psychiatry. 2017;25(6):270–278. https://doi.org/10.1097/HRP.0000000000000163.

Demakis GJ, editor. Civil capacities in clinical neuropsychology: research findings and practical applications. Oxford University Press; 2012.

Dickson KS, Gala GJ, Moseley D. The role of decision-making capacity in gathering collateral information. J Clin Ethics. 2023;34(2):123–7.

Dunn LB, Nowrangi MA, Palmer BW, Jeste DV, Saks ER. Assessing decisional capacity for clinical research or treatment: a review of instruments. Am J Psychiatry. 2006;163(8):1323–34.

Edelstein B. Challenges in the assessment of decision-making capacity. J Aging Stud. 2000;14(4):423–37.

Elder Justice Initiative. *Guardianship*. U.S. Department of Justice; n.d.. Retrieved from https://www.justice.gov/elderjustice/guardianship

Fertuck EA, Lenzenweger MF, Clarkin JF, Hoermann S, Stanley B. Executive neurocognition, memory systems, and borderline personality disorder. Clin Psychol Rev. 2006;26(3):346–75.

Folstein MF, Robins LN, Helzer JE. The mini-mental state examination. Arch Gen Psychiatry. 1983;40(7):812.

Ganzini L, Lee MA, Heintz RT. The capacity to make decisions in advance and borderline personality disorder. J Clin Ethics. 1994;5(4):360–3.

Ganzini L, Goy ER, Dobscha SK. Prevalence of depression and anxiety in patients requesting physicians' aid in dying: cross sectional survey. BMJ (Clin Res Ed). 2008;337:a1682.

Garner, B.A. (2019). Competence. In *Black's law dictionary* (11th ed). Retrieved from https://thelawdictionary.org/competence/

Garrett WS, Verma A, Thomas D, Appel JM, Mirza O. Racial disparities in psychiatric decisional capacity consultations. Psychiat Serv (Washington, DC). 2023;74(1):10–6.

Geros-Willfond KN, Ivy SS, Montz K, Bohan SE, Torke AM. Religion and spirituality in surrogate decision making for hospitalized older adults. J Relig Health. 2016;55(3):765–77.

Gogia, B. & Fang, X. (2023). Differentiating delirium versus dementia in the elderly. In: StatPearls. StatPearls Publishing, Treasure Island. Retrieved from https://www.ncbi.nlm.nih.gov/books/NBK570594/

Grisso T, Appelbaum PS. MacArthur competence assessment tool for treatment (MacCAT-T). Professional Resource Press/Professional Resource Exchange; 1998.

Gurrera RJ, Karel MJ, Azar AR, Moye J. Agreement between instruments for rating treatment decisional capacity. Am J Geriatr Psychiatry. 2007;15(2):168–73.

Hackard M. The wolf at the door: undue influence and elder financial abuse. Hackard Global Media; 2017.

Handtke O, Schilgen B, Mösko M. Culturally competent healthcare – a scoping review of strategies implemented in healthcare organizations and a model of culturally competent healthcare provision. PLoS One. 2019;14(7):e0219971.

Herrmann N, Whaley KA, Herbert DJ, Shulman KI. Susceptibility to undue influence: the role of the medical expert in estate litigation. Can J Psychiatry. 2022;67(1):5–12.

Hurst SA. When patients refuse assessment of decision-making capacity: how should clinicians respond? Arch Intern Med. 2004;164(16):1757–60.

Hutton AJ. Undue influence and fraud in wills. Dick L Rev. 1932;37:16.

Jung EY, Hwang SK. Health literacy and health behavior compliance in patients with coronary artery disease. Korean J Adult Nurs. 2015;27(3):251–61.

Kane NB, Ruck Keene A, Owen GS, Kim SY. Difficult capacity cases—the experience of liaison psychiatrists. An interview study across three jurisdictions. Front Psychiatry. 2022;13:1477.

Karel MJ, Gurrera RJ, Hicken B, Moye J. Reasoning in the capacity to make medical decisions: the consideration of values. J Clin Ethics. 2010;21(1):58–71.

Kaspers PJ. End-of-life care and preferences for (non) treatment decisions in older people during their last three months of life [PhD-thesis – research and graduation internal, Vrije Universiteit Amsterdam]. 2013. Retrieved from https://research.vu.nl/ws/portalfiles/portal/42113218/complete+dissertation.pdf

Kaspers PJ, Onwuteaka-Philipsen BD, Deeg DJ, Pasman HRW. Decision-making capacity and communication about care of older people during their last three months of life. BMC Palliat Care. 2013;12:1–10.

Kelly PA, Haidet P. Physician overestimation of patient literacy: a potential source of health care disparities. Patient Educ Couns. 2007;66(1):119–22.

Kitamura T. Structured interview for competency and incompetency assessment testing and ranking inventory-revised (SICIATRI-R). 2011.

Kohn NA, Blumenthal JA, Campbell AT. Supported decision-making: a viable alternative to guardianship. Penn St L Rev. 2012;117:1111.

Kontos N, Querques J, Freudenreich O. Capable of more: some underemphasized aspects of capacity assessment. Psychosomatics. 2015;56(3):217–26.

Kumar R, Berry J, Koning A, Rossell S, Jain H, Elkington S, Nagaraj S, Batchelor J. Psychometric Properties of a New Decision-Making Capacity Assessment Tool for People with Substance Use Disorder: The CAT-CAT. Arch Clin Neuropsychol. 2022;37(5):994–1034. https://doi.org/10.1093/arclin/acac010.

Libby C, Wojahn A, Nicolini J, Gillette G. Competency and capacity. In: StatPearls. StatPearls Publishing; 2023.

Lichtenberg PA, Mast B. Older adult financial capacity assessment webinar. Institute of Gerontology, Wayne State University; 2021. Retrieved from https://iog.wayne.edu/olderadultfinancialcapacityassessmentcourse/

Loeb PA. ILS: independent living scales manual. Tex, Psychological Corp; 1996.

Lulé D, Zickler C, Häcker S, Bruno MA, Demertzi A, Pellas F, et al. Life can be worth living in locked-in syndrome. Prog Brain Res. 2009;177:339–51.

Mellgard G, Gligorov N. Complexity, not severity: reinterpreting the sliding scale of capacity. Camb Q Healthc Ethics. 2022a;31(4):506–17.

Mellgard G, Gligorov N. Complexity, not severity: reinterpreting the sliding scale of capacity. Camb Q Healthc Ethics. 2022b;31(4):506–17.

Miller TA. Health literacy and adherence to medical treatment in chronic and acute illness: a meta-analysis. Patient Educ Couns. 2016;99(7):1079–86.

Minciocchi Urban M, Sullivan KA, Purser K. Age-bias in assessments of medical decision-making capacity: a cross-sectional experimental vignette study. Psychiat Psychol Law. 2024;

Moye J. Innovative Interprofessional Approaches to Improving Well Being for Older Adults with Interacting Mental and Physical Health Concerns. Clin Gerontol. 2020;43(4):363–364. https://doi.org/10.1080/07317115.2020.1761103.

Nasreddine ZS, Phillips NA, Bédirian V, Charbonneau S, Whitehead V, Collin I, et al. The Montreal cognitive assessment, MoCA: a brief screening tool for mild cognitive impairment. J Am Geriatr Soc. 2005;53(4):695–9.

Nassif WM. Assessing decisional capacity in patients with substance use disorders. Curr Psychiatr Ther. 2019;19:35–40.

Navin M, Wasserman JA, Stahl D, Tomlinson T. The capacity to designate a surrogate is distinct from decisional capacity: normative and empirical considerations. J Med Ethics. 2022;48(3):189–92.

Newton-Howes G, Pickering N, Young G. Authentic decision-making capacity in hard medical cases. Clin Ethics. 2019;14(4):173–7.

Ng IK. Informed consent in clinical practice: old problems, new challenges. J R Coll Physic Edinb. 2024;54(2):153–8.

Perington MR, Smith CA, Schillerstrom JE. Decisional capacity and personality disorders: substantially unable or substantially unwilling? J Forensic Sci. 2020;65(2):655–60.

Petrik ML, Billera M, Kaplan Y, Matarazzo B, Wortzel H. Balancing patient care and confidentiality: considerations in obtaining collateral information. J Psychiatr Pract. 2015;21(3):220–4.

Price A, McCormack R, Wiseman T, Hotopf M. Concepts of mental capacity for patients requesting assisted suicide: a qualitative analysis of expert evidence presented to the commission on assisted dying. BMC Med Ethics. 2014;15:1–11.

Quinn MJ. Friendly persuasion, good salesmanship, or undue influence. Marquette Elder Adv. 2000;2(4):49–56. Retrieved from http://scholarship.law.marquette.edu/elders/vol2/iss4/8

Restifo S. A review of the concepts, terminologies and dilemmas in the assessment of decisional capacity: a focus on alcoholism. Australas Psychiatry. 2013;21(6):537–40. https://doi.org/10.1177/1039856213497812. Epub 2013 Jul 24.

Schneider PL, Bramstedt KA. When psychiatry and bioethics disagree about patient decision making capacity (DMC). J Med Ethics. 2006;32(2):90–3.

Schulz PJ, Pessina A, Hartung U, Petrocchi S. Effects of objective and subjective health literacy on patients' accurate judgment of health information and decision-making ability: survey study. J Med Internet Res. 2021;23(1):e20457.

Sullivan KA, Purser K, Graham K, Parkinson L. Public awareness of legal decision-making capacity and planning instruments in dementia: implications for health care practitioners. Psychiatry Psychol Law. 2023;30(4):565–78.

Sulmasy DP, Snyder L. Substituted interests and best judgments: an integrated model of surrogate decision making. JAMA. 2010;304(17):1946–7.

Tariq SH, Tumosa N, Chibnall JT, Perry MH 3rd, Morley JE. Comparison of the Saint Louis University mental status examination and the mini-mental state examination for detecting dementia and mild neurocognitive disorder--a pilot study. Am J Geriatr Psychiatry. 2006;14(11):900–10. https://doi.org/10.1097/01.JGP.0000221510.33817.86.

Taylor LD. (2014). A tool to assess capacity to consent for treatment among homeless populations with problematic substance use (Doctoral dissertation, The University of British Columbia).

Torroella Carney M, Neugroschl J, Morrison RS, Marin D, Siu AL. The development and piloting of a capacity assessment tool. J Clin Ethics. 2001;12(1):17–23.

Treem J. Medical aid in dying: ethical and practical issues. J Adv Pract Oncol. 2023;14(3):207–11.

Turkoglu AR, Demirci H, Coban S, Guzelsoy M, Toprak E, Aydos MM, et al. Evaluation of the relationship between compliance with the follow-up and treatment protocol and health literacy in bladder tumor patients. Aging Male. 2018.

Van Der Heide I, Wang J, Droomers M, Spreeuwenberg P, Rademakers J, Uiters E. The relationship between health, education, and health literacy: results from the Dutch adult literacy and life skills survey. J Health Commun. 2013;18(sup1):172–84.

Van Hartskamp M, Consoli S, Verhaegh W, Petkovic M, Van de Stolpe A. Artificial intelligence in clinical health care applications. Interac J Med Res. 2019;8(2):e12100.

Vertesi A, Lever JA, Molloy DW, Sanderson B, Tuttle I, Pokoradi L, Principi E. Standardized mini-mental state examination. Use and interpretation. Can Family Phys. 2001;47(10):2018–23.

Vig E, Starks H, Taylor JS, Hopley EK, Fryer-Edwards K. Surviving surrogate decision-making: what helps and hampers the experience of making medical decisions for others. J Gen Intern Med. 2007;22:1274–9.
Weiss BD, Mays MZ, Martz W, Castro KM, DeWalt DA, Pignone MP, et al. Quick assessment of literacy in primary care: the newest vital sign. Annal Family Med. 2005;3(6):514–22.
West JC. What is an ethically informed approach to managing patient safety risk during discharge planning? AMA J Ethics. 2020;22(11):919–23.
White T, Stern RA. NAB, neuropsychological assessment battery: psychometric and technical manual. Psychological Assessment Resources; 2003.
Wood S, Liu PJ. Undue influence and financial capacity: a clinical perspective. Generations. 2012;36(2):53–8.
World Health Organization. (2015). Global health ethics key issues: global network of WHO collaborating centres for bioethics. Retrieved from https://www.who.int/publications/i/item/9789241549110
Yates JF, de Oliveira S. Culture and decision making. Organ Behav Hum Decis Process. 2016;136:106–18.
Zeybek B, Childress AM, Kilic GS, Phelps JY, Pacheco LD, Carter MA, Borahay MA. Management of the Jehovah's witness in obstetrics and gynecology: a comprehensive medical, ethical, and legal approach. Obstet Gynecol Surv. 2016;71(8):488–500.

Further Reading

ABA/APA Assessment of Capacity in Older Adults Project Working Group, American Bar Association. Commission on Law, & American Psychological Association. Assessment of older adults with diminished capacity: a handbook for psychologists. American Bar Association; 2008.
Appel JM. Capacity, informed consent and third-party decision-making. 2024.
Capezuti E, Malone ML, Khan A, Baumann SL, editors. The encyclopedia of elder care: the comprehensive resource on geriatric health and social care. Springer; 2017.
Demakis GJ, editor. Civil capacities in clinical neuropsychology: research findings and practical applications. Oxford University Press; 2012.
Kontos N, Querques J, Freudenreich O. Capable of more: some underemphasized aspects of capacity assessment. Psychosomatics. 2015;56(3):217–26.
Karel MJ, Gurrera RJ, Hicken B, Moye J. Reasoning in the capacity to make medical decisions: the consideration of values. J Clin Ethics. 2010;21(1):58–71.
Kane NB, Ruck Keene A, Owen GS, Kim SY. Difficult capacity cases—the experience of liaison psychiatrists. An interview study across three jurisdictions. Front Psychiat. 2022;13:1477.
Mellgard G, Gligorov N. Complexity, not severity: reinterpreting the sliding scale of capacity. Camb Q Healthc Ethics. 2022;31(4):506–17.
Ng IK. Informed consent in clinical practice: old problems, new challenges. J R Coll Phys Edinburgh. 2024;54(2):153–8.

Psychological Adjustment in Inpatient Settings

14

Daniel W. Klyce, David J. Rothman, Sarah W. Clark, and Bani Malhotra

ABBREVIATIONS

ACT	Acceptance and Commitment Therapy
APA	American Psychological Association
APAP	Auto Titrating Positive Airway Pressure
ASDI	Acute Stress Disorder Interview
ASDS	Acute Stress Disorder Scale
BDI	Beck Depression Inventory
Bi-PAP	Bilevel Positive Airway Pressure
CBT	Cognitive Behavioral Therapy
CBT-I	Cognitive Behavioral Therapy for Insomnia
CDC	Centers for Disease Control and Prevention
CPAP	Continuous Positive Airway Pressure
CPT	Cognitive Processing Therapy
C-SSRS	Columbia–Suicide Severity Rating Scale
CSF	Cerebrospinal Fluid
DSM	Diagnostic and Statistical Manual of Mental Disorders
DVPRS	Defense and Veterans Pain Rating Scale
FAIS	Factors Affecting Inpatient Sleep
GAD-7	Generalized Anxiety Disorder Scale-7 Item
HADS	Hospital Anxiety and Depression Scale
IASP	International Association for the Study of Pain
MI	Motivational interviewing
MSLT	Multiple Sleep Latency Test

D. W. Klyce (✉)
Mental Health Service, Richmond VA Medical Center,
Richmond, VA, USA

Department of Physical Medicine and Rehabilitation, Virginia Commonwealth University, Richmond, VA, USA

Psychology Service, Sheltering Arms Institute,
Richmond, VA, USA
e-mail: Daniel.Klyce@va.gov

D. J. Rothman · B. Malhotra
Department of Physical Medicine and Rehabilitation, Virginia Commonwealth University, Richmond, VA, USA
e-mail: David.Rothman@vcuhealth.org;
Bani.Malhotra@vcuhealth.org

S. W. Clark
Mental Health Service, Richmond VA Medical Center,
Richmond, VA, USA
e-mail: Sarah.Clark@va.gov

E. K. Bailey et al. (eds.), *Inpatient Neuropsychology*, https://doi.org/10.1007/978-3-032-05880-5_14

MVC	Motor Vehicle Collision
NRS	Numerical Rating Scale
NSESSS	National Stressful Events Survey Acute Stress Disorder-Short Scale
NTI	Narcolepsy Type 1
OARS	Open-Ended Questioning, Affirmation of Strengths and Progress, Reflective Listening, and Summaries to Ensure Understanding and Highlight Change Statements
OSA	Obstructive Sleep Apnea
PADSS	Paris Arousal Disorders Severity Scale
PAINAD	Pain Assessment in Advanced Dementia
PAS	Posttraumatic Adjustment Scale
PCS	Pain Catastrophizing Scale
PE	Prolonged Exposure
PHQ	Patient Health Questionnaire
PROMIS	Patient-Reported Outcomes Measurement Information System
PSEQ-2	Pain Self-Efficacy Questionnaire-2 Item
PSG	Polysomnography
PTSD	Posttraumatic Stress Disorder
QTc	Interval Corrected
RBD	REM Sleep Behavior Disorder
RBDSQ	REM Sleep Behavior Disorder Screening Questionnaire
RCSQ	Richards-Campbell Sleep Questionnaire
REM	Rapid Eye Movement
SASRQ	Stanford Acute Stress Reaction Questionnaire
SF-MPQ-2	Short-Form McGill Pain Questionnaire-2
SMHSQ	St. Mary's Hospital Sleep Questionnaire
SSI	Scale for Suicide Ideation
TENS	Transcutaneous Electrical Nerve Stimulation
TIC	Trauma-Informed Care
VSHS	Verran Snyder-Halpern Sleep Scale

INTRODUCTION

Neuropsychologists working in inpatient settings experience multiple challenges and rewards. Given the dynamic nature of these treatment environments, perennial challenges will include creating time and psychologically safe spaces to connect with patients who might feel quite vulnerable. Neuropsychologists can make a meaningful impact by humanizing and contextualizing the emotional experience of patients that might otherwise be marginalized at this level of care. Frequently, evidence-based assessment or treatment strategies will require significant adaptation or accommodation to meet the needs of the patient, team, or clinical environment. Depending on the familiarity of the consulting provider with the benefits and limitations of psychological assessment and intervention, an effective neuropsychologist may have to collaboratively shape the consult question into something more precise, practical, or achievable during the inpatient epoch of care. Neuropsychologists working in these settings can be helpful in identifying psychiatric pathology and offering guidance on intervention and differential diagnosis when these symptoms occur in the context of acute medical illness (e.g., delirium, polypharmacy, infectious processes, substance withdrawal, pain). In this chapter, we discuss adjustment-related issues, acute stress, managing risk for self-directed violence, sleep, and pain. An appendix of recommended assessment tools is included at the end of this chapter (Appendix "List of Scales and Psychometric Properties").

ADJUSTMENT-RELATED ISSUES

Consults for the evaluation and treatment of other adjustment-related issues, especially anxiety and depression, are commonly encountered by inpatient mental health providers, including neuropsychologists. It is difficult to estimate the prevalence rates of anxiety and depression among hospitalized patients, as much research regarding psychiatric disorders in hospital settings tends to be specific to certain populations, diagnoses, or medical services. A review and meta-analysis by Walker et al. (2021) included 32 studies related to the assessment of anxiety disorders among general hospital patients. Although there was considerable variability across studies, the authors estimated the prevalence of a diagnosable anxiety disorder to be ~8%, with panic disorder and generalized anxiety disorder at 3% and 5%, respectively. However, while the overall prevalence of these diagnoses was generally consistent with the general population, specific symptoms of anxiety were more commonly endorsed and had a prevalence of up to 28% in inpatient populations. The authors suggested several explanations for this discrepancy, noting that (a) anxiety symptoms associated with illness or hospitalization simply may not meet threshold for a specific diagnosis, (b) many evaluators might not deem the "excessiveness" criterion to be met in the context of acute injury or illness, and (c) it may be difficult to attribute symptoms to a primary psychiatric disorder given comorbid medical conditions. In earlier work, Walker et al. (2018) reviewed 31 studies of depression among medical and surgical patients, finding an average prevalence rate of 12%, though there was significant variability across studies ranging from 5% to 34%. Neither the 2021 review of anxiety, nor the 2018 review of depression, was able to identify reliable, population-based predictors of symptoms.

However, addressing symptoms of anxiety and depression in inpatient populations remains vitally important, as these conditions can undermine effective participation in treatment, contribute to poor outcomes, and increase healthcare utilization (Campbell 2008; Jansen et al. 2018; Pederson et al. 2016; Roy-Byrne et al. 2008). Close collaboration with the medical team is essential for any mental health provider working in the inpatient setting to distinguish psychologically based symptoms of anxiety or depression from medical or treatment factors. It can be ineffective (at best) or even dangerous (at worst) to attempt psychotherapeutic intervention when symptoms are better explained by a medical condition for which a medical or surgical intervention may be more appropriate. Otherwise, anxiety and depression, along with evidenced-based cognitive-behavioral therapies (CBT)—which can be easily adapted for brief interventions at bedside—are no doubt familiar to many neuropsychologists. For an authoritative overview of these strategies, please consider LaBuda, Axelrod, and Windell's text on CBT protocols for medical settings (LaBuda et al. 2018), as well as the validated assessment tools summarized in Appendix "List of Scales and Psychometric Properties" at the end of this chapter.

However, as identified by Walker et al. (2018, 2021), there are a range of adjustment-related issues that may not necessarily meet the diagnostic criteria for specific psychological disorders or that are not psychiatrically pathological per se but still present as problematic. In the hospital or rehabilitation setting, these symptoms can include grief, demoralization, or disengagement. The remainder of this section will focus on strategies that neuropsychologists can use to address these concerns.

Grief

Grief reactions (e.g., anger, guilt, denial, shock, yearning) are normative responses to issues such as loss of independence or agency, changes in physical functioning or health status, and interpersonal loss of, or disconnection from, sources of support. However, grief can become problematic when this undermines participation in care or interferes with the goals of rehabilitation. In their work regarding grief and loss among people with brain injuries, Niemeier and Karol (2010) illustrate several useful strategies for addressing grief that are applicable in inpatient settings, beginning with an effort to orient patients toward two key insights about the nature of rehabilitation—that (1) the goals of inpatient care may not necessarily be curative but instead focus on compensatory strategies or improved functioning, and (2) effective treatment often requires active participation from the patient. To help illustrate these points, Niemeier and Karol (2010) suggest the following strategies:

- Emphasize that grief and loss can be reduced by getting the maximal benefit out of rehabilitation therapies and treatment.
- Shift from a focus on "full recovery to normal" toward a focus on "advancement and accommodation."
- Reconceptualize role of patient from passive recipient of treatment to active participant in care.
- Prepare patients to focus on work that challenges their abilities or performance.
- Normalize seeking help or support for emotional needs.
- Practice communicating needs or preferences to providers.
- Explore and teach the rationale for treatments and therapies that may lack face validity.
- Assist families with understanding their role during a loved one's hospitalization.

Relatedly, patients can become preoccupied with roles and responsibilities outside the hospital. Although understandable, these concerns can distract from engagement in care. In these instances, it may help to "give permission" for patients to prioritize on their own health and wellness in the hospital setting and temporarily delegate outside responsibilities to others. When appropriate, patients may benefit from the message that chronic stressors were a problem *before they came into the hospital*, that these problems are not likely to resolve *while in the hospital*, and that it is in the best interest of the patient to obtain maximum benefit from their recovery and rehabilitation so as to better prepare them to reengage with these problems *after they leave the hospital.*

Demoralization

Hospitalized patients may experience demoralization for a variety of reasons. Some may be confronting a new life-changing diagnosis or disability, feeling burdensome to others, struggling with a lost sense of meaning or purpose in life, or simply feeling helpless in the hospital environment. Difficulty coping with these circumstances may be expressed as hopelessness, irritability and frustration, self-doubt, or just "giving up" (Clarke and Kissane 2002). For such patients, adapting principles of Acceptance and Commitment Therapy (ACT) into brief interventions can be helpful in addressing these symptoms (Hayes and Pierson 2005).

Described as a "third wave" of CBT, ACT is grounded in Relational Frame Theory and utilizes a situationally based, cognitive-behavioral approach that highlights how a person interacts with their environment (Hayes et al. 2001). As such, the ACT approach often resonates with individuals faced with navigating contextual stressors in the hospital environment. Indeed, there is an established evidence base for the use of ACT-informed interventions in addressing a range of issues commonly encountered by neuropsychologists in hospital settings, including anxiety, depression, substance use, and pain (Luoma et al. 2012; Ma et al. 2023; Salari et al. 2023). Additionally, there is burgeoning evidence for the benefit specifically in rehabilitation settings and in supporting adjustment to chronic illness or disability (Hoffmann et al. 2016; Kangas and McDonald 2011).

The main objective of ACT is to increase psychological flexibility, which—incidentally—is key to promoting positive adjustment to injury, illness, or disability. The goal is not to eliminate difficult feelings (acceptance), but to move purposefully toward valued behaviors (commitment). In delivering ACT, patients are encouraged to attend to unpleasant emotions and guided on ways to effectively manage their emotional experience. This, in turn, teaches patients how to experience emotional discomfort without overreacting and prepares them in confronting situations where difficult emotions are likely to occur. In general, the six principles of ACT seem well-suited as an intervention for emotional distress that accompanies any change in health and/or functional status, as well as issues that naturally arise in a hospital environment (Fig. 14.1).

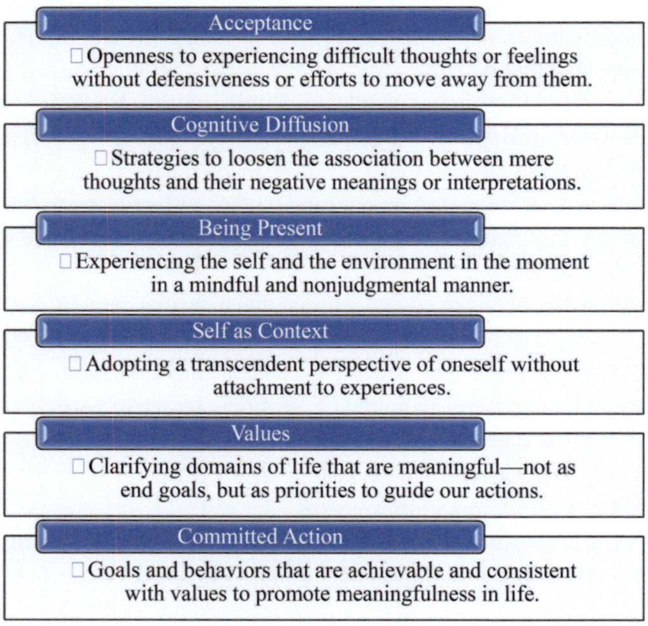

Adapted from Hayes et al., 2001; Hayes & Pierson, 2005

Fig. 14.1 Six principles of acceptance and commitment therapy

Low Engagement

Most neuropsychologists have some foundational understanding of motivational interviewing (MI), a therapeutic approach used to promote behavioral change. In the inpatient setting, principles of MI can be readily applied to challenge problematic behaviors such as disengagement or inconsistent compliance with treatment (Fig. 14.2).

The core of MI involves characterizing patients as partners in their own healthcare, ascertaining patients' own value-based reasons and resources to support change, and affirming patients' rights, responsibilities, and autonomy to enact change. In contrast to other interventions promoting health and wellness, MI offers a multi-pronged approach (Fig. 14.3). Providers of MI are tasked with expressing empathy, identifying inconsistencies between patients' stated values and discrepant behaviors, "rolling with"—or accepting—any resistance to change, and supporting patients' self-efficacy. Specific skills and techniques associated with MI include *O*pen-ended questioning, *A*ffirmation of strengths and progress, *R*eflective listening, and *S*ummaries to ensure understanding and highlight change statements (OARS). Throughout, there is an effort to elicit the patient's desire, ability, reasons, and need to change. Among hospitalized patients, these principles and techniques of MI can be applied to promote adherence to treatment recommendations, active engagement in rehabilitation and therapies, and pro-active behaviors that support continued health and wellness following discharge.

"MI is not a technique for tricking people into doing what they do not want to do. Rather, it is a skillful clinical style for eliciting from patients their own good motivations for making behavior changes in the interest of their health. It involves guiding more than directing, dancing rather than wrestling, listening at least as much as telling. The overall 'spirit' has been described as collaborative, evocative, and honoring of patient autonomy."

Rollnick & Colleagues, 2008

Fig. 14.2 Motivational interviewing

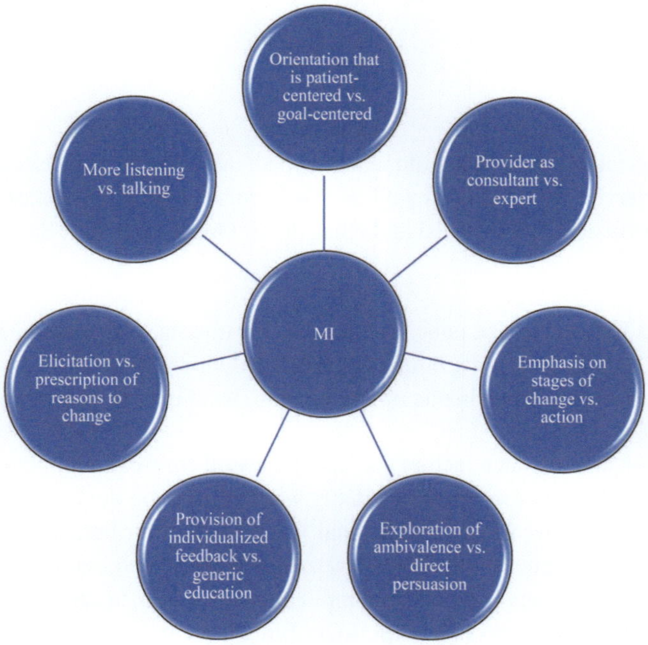

Adapted from Rollnick et al., 2008

Fig. 14.3 Approaches in motivational interviewing

ACUTE STRESS REACTIONS

The severity of medical illness or injury that necessitates a hospital admission poses a risk for developing a higher acuity of emotional distress, beyond classic symptoms of anxiety, depression, or grief. For patients who have recently endured a traumatic event or critical situation, acute stress symptoms can develop during hospitalization, either an acute stress reaction (over a few hours or days) or acute stress disorder (up to one month). Much of the research informing the conceptualization of acute stress disorder, subsequent refinement of the construct, and best practices for assessment and intervention has been conducted by Richard Bryant, whose authoritative text *Acute Stress Disorder: What It Is and How to Treat It* is strongly recommended (Bryant 2016).

Acute stress disorder was first introduced in the American Psychiatric Association's (APA) 4th Edition Diagnostic and Statistical Manual of Mental Disorders (DSM-IV; APA, 1994) as an anxiety disorder that required exposure to a traumatic event and the presence of symptoms related to dissociation, re-experiencing, avoidance, and anxious arousal. Assuming these symptoms would likely to be a precursor to posttraumatic stress disorder (PTSD), acute stress disorder was added to the DSM-IV to help facilitate access to psychological treatments for these individuals presumed to be at higher risk for PTSD (Bryant et al. 2011). However, subsequent research has not supported such a connection between acute stress reactions and PTSD (Bryant 2011). Diagnostic criteria for acute stress were reformulated for the DSM-5 (APA, 2013). The initial exposure criterion was broadened to include witnessing events, learning about events that affected a friend or family member, and repeated exposure to aversive details about an event. Instead of requiring symptoms within specific categories, the DSM-5-TR (APA, 2022) criteria allow for at least 9 out of 14 symptoms in any combination across categories of intrusion, negative mood, dissociation, avoidance, and arousal symptoms. The disturbance must result in significant distress or impairment, persist for at least 3 days but not more than 1 month, and not be attributable to substance use, another medical condition, or a brief psychotic disorder. These changes in the acute stress disorder diagnostic criteria from DSM-IV to DSM-5 appear to have resulted in improved prediction of eventual PTSD (Bryant et al. 2015). Lastly, patients may have symptoms of acute stress without meeting the full criteria for acute stress disorder. In these instances, mental health interventions can still be beneficial.

Estimations of the prevalence of acute stress disorder have been difficult to establish for multiple reasons, including the relatively brief window of time during which this diagnosis can be made, conducting research in the environments acute stress reactions are manifested, and heterogeneity of assessment across research studies. A systematic review by Ophuis et al. (2018) found rates of acute stress disorder among hospitalized patients with traumatic injuries to range from 24.0% to 24.6% within 1 week post-injury and 11.7% to 40.6% between 1 and 2 weeks post-injury.

Assessment of Acute Stress

Any assessment of psychological adjustment to trauma in the medical setting should ascertain personal strengths and coping skills that have historically served the patient, as well as risk factors for development of acute stress. Based on previous reviews and meta-analyses (Brewin et al. 2000; Fanai and Khan 2023; Ozer et al. 2003; Sareen 2014), these risk factors have been summarized in three categories:

- Pre-trauma factors such as female gender, education level, prior trauma, and mental health history
- Peri-trauma factors such as trauma severity, injury, interpersonal violence
- Post-trauma factors such as social determinants of health, pain, level of care in hospital, brain injury, and disability

Focusing on modifiable and/or maintenance factors can also be useful for treatment planning. As Bryant (2016) notes, a person who has experienced recent trauma may understandably want to focus on the trauma experience itself, with the unintended consequence of diminishing or overlooking additional factors that contribute to (or exacerbate) the person's acute stress reaction. As such, we recommend guiding the interview to other issues such as medical comorbidities (e.g., pain, side-effects of treatments) or situational contexts (e.g., legal complications, barriers to returning to work or school, community stressors/hazards, and other social determinants of health). Identifying these risk factors can help the patient understand their symptoms from a holistic point of view and emphasize the importance of monitoring symptoms over time—beyond the immediate aftermath of the traumatic event, in case long-term mental health treatment or other community-based supports and services are needed.

In addition to the clinical interview, there are multiple tools for assessing acute stress symptoms, such as:

- Acute Stress Disorder Interview (ASDI; Bryant et al. 1998)
- Acute Stress Disorder Scale (ASDS; Bryant et al. 2000)
- Posttraumatic Adjustment Scale (PAS; O'Donnell et al. 2008)
- National Stressful Events Survey Acute Stress Disorder Short Scale (NSESSS; Kilpatrick et al. 2013)
- Stanford Acute Stress Reaction Questionnaire (SASRQ; Cardeña et al. 2000)

Given the course of acute stress reactions, the timing of assessment is an important consideration. Bryant recommends waiting at least 1 week from trauma exposure to assess for manifestations of acute distress (Bryant 2016). While it may be helpful to address issues such as psychological safety in the immediate aftermath of trauma (e.g., suicidality, psychosis), further assessment during the first few dynamic days following medical trauma may not be especially productive. In the context of acute care admissions, medical instability may warrant further delay of assessment among individuals with delirium. An assessment after stabilization can add predictive power regarding risk for chronic mental health symptoms after discharge, as research indicates long-term risk for neuropsychiatric issues such as depression, posttraumatic stress, and impaired cognitive functioning following critical illness and intensive care (Jackson et al. 2014; Huang, et al. 2016; Patel et al. 2016).

Simply being admitted to a hospital environment may either exacerbate or protect individuals from an acute stress reaction. For example, factors unique to the inpatient milieu such as isolation, disrupted sleep patterns, discomfort, or invasive procedures can cause or exacerbate patients' emotional distress. Those who have been historically marginalized or mistreated in medical settings, or whose lived experiences may justify some degree of mistrust of the medical community, may experience even greater feelings of powerlessness and hardship as the nature of their medical condition now requires direct interaction with healthcare providers. In contrast, other individuals may feel safe in the relative protection of the hospital environment with ready access to care providers who can answer questions, offer reassurance, or administer treatments. Such individuals may begin to experience distress as they prepare to discharge from the hospital setting, return to a potentially stressful living situation, or encounter stimuli associated with the initial trauma (e.g., riding in a car after an accident; returning to a community where they were the victim of violence).

In addition, specific symptoms of acute stress reactions can further complicate identification and treatment. For example, individuals who experience emotional numbing may not evidence obvious outward distress, which would otherwise generate a mental health consult or be used to identify and diagnose a mental health condition. Similarly, avoidance can be a barrier to a patient's engagement in psychological assessment or willingness to participate in treatment. This is problematic as delayed intervention for acute stress during a hospital admission may place patients at increased risk for negative outcomes such as PTSD and chronic depression. For patients who would otherwise prefer to avoid mental health assessment or intervention, it is important that the interaction does not feel forced (or the patient feels coerced), as this could create resistance and further perpetuate help-rejecting or noncompliant behavior.

Psychoeducation

Psychoeducation can be a powerful intervention for individuals with acute stress. For many individuals, trauma and the experience of hospitalization are not part of everyday life. Patients may be reassured through normalization of their acute stress reaction and explanation of how these symptoms are adaptive under extreme circumstances. For example, patients with intrusive thoughts of a traumatic experience may be comforted by learning that flashbacks or intense memories are part of the brain's natural effort to process trauma and prepare a survival response for any future similar events. Similarly, conceptualizing arousal symptoms as a survival response to an extreme threat may help patients appreciate that their initial reactions in the immediate aftermath of a trauma—while distressing—are part of their brain's effort to "do its job." These reactions can be framed as adaptive in the context of acute stress, but maladaptive if symptoms persist long after the threat has subsided. Pairing psychoeducation with behavioral strategies such as relaxation or grounding techniques can empower trauma patients with knowledge and practical tools to address acute stress during the early days of hospitalization.

Psychoeducation can also include setting normative expectations for recovery and introducing patients to evidence-based treatment modalities for PTSD, such as prolonged exposure (PE) and cognitive processing therapy (CPT), should this be necessary later. Given variability in the longitudinal patterns of PTSD symptoms, preemptively reviewing these efficacious treatments for PTSD, and encouraging individuals with acute stress reactions to continue monitoring their mood after discharge, is recommended.

Trauma-Informed Care

Trauma-informed care (TIC) is a framework that healthcare providers can implement in their practice settings to guide treatment for people with a history of trauma. TIC has been identified as a potentially useful approach at various levels of care, including intensive care units and inpatient rehabilitation (Schroeder et al. 2021; Williamson and Kautz 2016). Inpatient neuropsychologists are well-positioned to assist treatment teams with implementing TIC principles and crafting recommendations that inform the patients' plans of care. TIC is based on the following principles: including (a) providing an emotionally and physically safe environment for patients, (b) promoting trustworthiness and transparency in treatment, (c) incorporating peer support, (d) enhancing collaboration and mutual ownership over the plan of care, (e) supporting patient empowerment, voice, and choice in treatment, and (f) creating awareness of cultural, historical, and gender issues that might influence the expression of a history of trauma in the medical setting.

In addition to a practical illustration of TIC principles and strategies applied to the critical care setting, Dinoso and Baudoin also provide recommendations for therapeutic communication with patients who have a history of trauma (Dinoso and Baudoin 2023). Briefly, their recommendations for therapeutic communication during patient encounters include:

- Clearly introducing yourself, your role on the treatment team, setting clear expectations for the encounter, and offering opportunities to ask questions or clarify concerns.
- Creating an environment that protects patient privacy, promotes autonomy, and allows patients to make decisions about who they do or do not want to be involved in the encounter.
- Maintaining awareness of your own physical presence, including your speech, posture/stance, affect, and movement; this includes keeping an appropriate distance and trying to stay level with the patient, instead of looming over them.
- Demonstrating that you are present in the moment, attuned to the patient, relaxed, and interested.
- Communicating reciprocally in a manner that is patient-centered, makes liberal use of active listening, and employs therapeutic silence when appropriate.
- Being mindful of touch; therapeutic touch can be a powerful tool to connect and comfort patients in the hospital setting, and patients with a history of trauma may be sensitive to being touched without first providing permission.
- Demonstrating humility around cultural factors such as gender, race, immigration status, primary language, military service, and disability status.

In addition to encouraging the application of TIC principles, neuropsychologists can liaise with the team to support other patient-centric interventions for acute stress. For example, conveyance of environmental triggers for acute stress reactions can help medical providers ensure emotionally mindful execution of medical treatment. Neuropsychologists also provide valuable observations about symptoms or behavior which can inform a physician's selection of psychotropic medication. Team education regarding non-pharmacological interventions for acute stress reactions, such as grounding or relaxation techniques, can foster encouragement and consistent use of these strategies across all therapies.

Neuropsychologists working as members of an interprofessional treatment team should capitalize on coordinated and planful opportunities for co-treatment with other providers. It is advantageous to work with team members to facilitate safe exposure to fearful stimuli—situations or sensations (e.g., pain)—associated with the traumatic experiences. For example, many patients in rehabilitation centers will be required to attempt a car transfer at some point during their treatment. Such individuals who were injured in a vehicle accident may be good candidates for co-treatment of coping strategies to manage distress while practicing vehicle transfers during physical therapy sessions. Community outings and other in vivo opportunities for exposure to challenging situations will solidify implementation of personalized coping skills following discharge.

Psychotherapeutic Interventions

CBT is the primary evidence-based psychotherapy for treating acute stress (Bryant 2016; Tol et al. 2013; Watkins et al. 2018). The main components of CBT for acute stress are similar to interventions for PTSD, such as PE and CPT (Foa et al. 2008; Resick et al. 2024). These basic components include psychoeducation, anxiety management, imaginal or in vivo exposures, cognitive restructuring, and relapse prevention. As already discussed, psychoeducation is key to empowering the patient with knowledge that normalizes their experience, helps to explain their symptoms, provides a rationale for treatment, and sets normative expectations for recovery. Anxiety management provides strategies to cope with cognitive, emotional, and—especially—physical reactions to acute stress, and could include diaphragmatic breathing or progressive muscle relaxation to engender a sense of control over symptoms of anxious arousal. Exposures, whether imaginal or in vivo, play a critical role in reducing avoidance that would otherwise maintain the acute or posttraumatic stress reaction. Exposure to feared components of the trauma experience can promote habituation and weakening of the association between these stimuli and the fear reaction. Activating the psychological or cognitive experience of the trauma through exposure also provides the patient with an opportunity to process problematic emotional reactions to otherwise innocuous triggers. Working with the provider, the patient can then start to incorporate new or corrective information that mitigates psychological distress. Relatedly, cognitive restructuring of the thoughts experienced during or around the exposures provides a framework for identifying problematic beliefs, challenging these cognitive distortions, and replacing them with more adaptive appraisals. Finally, relapse prevention strategies are important to ensure gains are maintained over time.

There are several considerations that might influence whether (and how) components of CBT for acute stress can be adapted for patients in the hospital setting. In the context of recent medical trauma, the timing of treatment is important, and the psychologist in this setting should be sensitive to whether treatment would be more beneficial once recovery has further stabilized, and a patient can engage more fully with therapy. Additionally, certain medical conditions may influence certain components of the intervention. For example, when implementing anxiety management strategies (e.g., diaphragmatic breathing; progressive muscles relaxation), it may be important to think about factors affecting the patient's respiratory status or whether deep breathing might exacerbate discomfort associated with traumatic injuries (e.g., orthopedic fractures). Finally, given that most patients have been admitted to the hospital setting with serious medical conditions and accompanied, understandable psychological stress, it is important to ensure that patients do not misinterpret therapeutic cognitive restructuring as intended to invalidate or reformulate the facts or consequences of what has happened to them. Rather, the focus should remain on critically examining the *appraisal* of these experiences and *beliefs* regarding how patients view themselves, others, and the world.

Adaptations of the treatment format are also important to consider. Evidence-based CBTs for acute stress tend to be packaged as 5 or 6 weekly sessions that include the components described above, as well as sessions dedicated to PE (Bryant 2016; Bryant et al. 2008). Acute care and inpatient rehabilitation settings present many challenges to this, including the overall length of stay during an admission, creating privacy at bedside, and finding consistent time on patient schedules that must be shared with other providers. In some settings, it may be possible to increase the frequency of encounters for a shorter overall course of psychotherapy; however, the psychologist's relationship to the treatment team may also affect what is possible to provide. For example, when the neuropsychologist is in a consultative role and only able to have one or two encounters with a patient, it may be reasonable to focus on assessment, psychoeducation, acute anxiety management, and referrals for outpatient treatment. However, when the neuropsychologist is integrated as a core member of the treatment team, providing exposure interventions, cognitive restructuring, or relapse prevention strategies over a longer length of stay may be possible.

MANAGEMENT OF SUICIDALITY

Assessing and managing suicide risk in inpatient medical settings is inherently challenging given the propensity of possible competing demands from the patient, care team, setting, and policy. Importantly, suicide risk is multifactorial and dynamic. Risk may fluctuate depending on circumstance, and ongoing assessment is needed to ensure prompt delivery of patient-centered interventions that mitigate risk factors and optimize protective factors. The following sections are not meant to comprehensively illustrate clinical decision-making with suicidal patients, but rather to describe some basic concepts for consideration when working in inpatient settings.

Patients in medical settings may experience and express suicidal ideation without meeting criteria for admission to an inpatient psychiatric setting, and similarly, such an environment may not be appropriate for their medical or rehabilitation needs. The decision to transfer a medical inpatient to secure psychiatric unit is made based on several considerations including bed availability, medical and psychiatric acuity, organizational policies, and staff competencies. For providers working at hospitals not associated with larger medical systems, it may be useful to cultivate relationships with community partners (including local law enforcement and behavioral health crisis teams) and develop knowledge of local judicial process for pursuing emergency custody or treatment orders.

However, it is possible that assessment and intervention for suicide risk and associated psychopathology is conducted in a medical or rehabilitation setting. Increasing the frequency of patient observation can enhance environmental safety in such milieus. This can be accomplished by increasing or adjusting the timing of staff rounds, relocating the patient to a room closer to the nurse's station, or assigning a one-on-one observer or "sitter" to the patient. Other options might include limiting access to potential means for self-harm and identifying environmental antecedents of suicidal ideation. For example, rigid adherence to blanket restrictions, overly cautious policy, or the traditional operating procedures of the unit may exacerbate patients' sense of powerlessness, limit access to potentially therapeutic experiences, and ultimately, be counterproductive to rehabilitation or treatment goals. For example, a patient may be prohibited from venturing off the unit due to a risk of falls—a restriction that, while well-intended, may inadvertently limit his daylight exposure and worsen insomnia and mood. In this scenario, allowing accompanied excursions outside may promote his overall well-being.

In addition, addressing patient-specific factors is critical to mitigate safety risk. For example, determining the underlying intent or purpose of apparent self-harming behaviors can clarify whether such behavior represents true suicidal intent or if behavior is being used as a means of communicating emotional distress, physical pain, or other suffering. Further, confusion and agitation that may accompany acute medical illness, such as delirium, can result in self-injurious actions that are accidental and not associated with any underlying emotional distress.

Ultimately, determining the impetus and context of self-harm behaviors can help inform appropriate intervention. There are several validated measures to assess suicide risk and structured clinical approaches to managing suicidal thoughts and behaviors. These include the Collaborative Assessment and Management of Suicide (Jobes 2016), brief cognitive behavioral therapy for suicide prevention (Bryan and Rudd 2018), and lethal means counseling (Sale et al. 2018). When intentional self-harm is expressed or observed, the inpatient neuropsychologist should conduct a safety plan, help develop coping strategies, advocate for a team-based, collaborative approach to mental healthcare that promotes the patient's values, while also addressing their medical concerns (Stanley et al. 2018).

Thus, close collaboration with other disciplines is recommended, as is consultation and consideration of facility policies. For example, a psychiatry consult could be placed to evaluate acute psychiatric needs and make recommendations for trialing or adjusting psychotropic medications, if a patient is amenable. The neuropsychologist can provide recommendations to other care members about how to best manage self-harm behaviors in inpatient settings, including ways to maintain safety and stability, as well as assist in transitioning to outpatient mental health treatment once discharged. Team education can encompass strategies for effectively communicating and treating a suicidal patient. The inpatient neuropsychologist can advocate for a team approach to reinforcing adaptive coping, offering support, and avoiding unhelpful or shaming responses. With patient permission, integration of family, friends, or spiritual support should also be considered.

SLEEP DISTURBANCE

Sleep is one of the most important and common concerns for individuals in inpatient settings. Research shows that individuals who are hospitalized tend to report worse sleep quality, reduced time spent sleeping, and more awakenings (Dobing et al. 2016). Of the factors associated with poor sleep quality, nearly 65% are related to hospital environmental factors including noise, hospital staff/timing of hospital staff entrance, and noise from hospital equipment. While a greater proportion of awakenings occur due to hospital-based factors, there also are patient factors, including pre-existing sleep disorders, emotional factors (e.g., worry about illness), and pain (Wesselius et al. 2018). There are several consequences of poor sleep while hospitalized. These include worsening motor functioning following brain injury (Fleming et al. 2020), delirium (Pisani and D'Ambrosio 2020), pain (Chouchou et al. 2014), hyperglycemia (DePietro et al. 2017), poor strength and adverse respiratory function (Elliott et al. 2021). A study on sleep quality after brain injury found that individuals with acquired brain injury experienced worse sleep quality than community-dwelling adults (Fleming et al. 2020). Disrupted sleep in those with brain injuries was further associated with poorer motor outcomes demonstrated by lower scores on a measure of functional independence during their rehabilitative stay (Fleming et al. 2020). Individuals experiencing delirium during inpatient care have also shown increased sleep-wake disruptions (Meagher et al. 2007).

In the inpatient setting, the neuropsychologist may be asked to observe behaviors, conduct a clinical assessment, and provide education to patients and families, and recommendations for behavioral management. Below, we will discuss two major sleep disorders, insomnia, and obstructive sleep apnea (OSA), as well as the assessment of sleep, and recommendations for interventions to improve sleep. Table 14.1 reviews various other sleep-related diagnoses that an inpatient neuropsychologist may encounter, such as rapid eye movement (REM) behavioral sleep disorder, non-REM parasomnia, and narcolepsy.

Table 14.1 Sleep disorders

Disorder	Diagnostic Criteria	Specifiers	Evaluations
Insomnia Disorder	• Difficulty initiating, maintaining, and returning to sleep. • Causes distress, impairment, and interference in everyday functioning. • Difficulties at least 3 times/week	• With mental disorder, medical condition, or another sleep disorder ‣ Episodic (symptoms 1-3 months) ‣ Persistent (symptoms <3 months) ‣ Recurrent (2≤ episodes within 1 year)	• Self-report sleep diary • Actigraphy or polysomnography
Narcolepsy	• Irresistible sleepiness • Loss of muscle tone, sleep paralysis • Vivid hypnagogic hallucinations • At least 3 time/week in 3 months	• Type 1: With cataplexy or hypocretin deficiency • Type 2: Without cataplexy and without/unmeasured hypocretin deficiency • Type 1 or Type 2 due to medical condition • Mild (naps 1-2 times/day), moderate (multiple naps), or severe (constant sleepiness)	• Nocturnal polysomnography for sleep latency • Cerebrospinal fluid (CSF) testing for hypocretin deficiency
Rapid Eye Movement (REM) sleep behavior disorder	• Vocalizations (talking, screaming) and/or complex motor behavior (punching, kicking, hitting) emerging from REM sleep. • Occur in the later parts of the night (<90 minutes after sleep) • Individual is alert, awake, oriented upon awakening. • Lead to distress and impairment. • Not explained by substance use, medical, or mental disorder	N/A	• Polysomnography: REM without atonia • Clinical history and established synucleinopathy diagnosis (Parkinson's disease, multiple system atrophy)

(continued)

Table 14.1 (continued)

NREM Sleep Arousal Disorder	• Sleepwalking (associated with varied complex motor behavior), sleep terrors (frequent awakening starting with a scream), with fear and signs of autonomic arousal. • Amnesia for the event, difficulty in dream imagery recall. • Cause distress and impairment. • Not explained by substance use, medical, or mental disorder	• Sleepwalking or sleep terror type • With sleep-related eating, or sleep-related sexual behavior (sexsomnia)	• Polysomnography with audiovisual monitoring • Clinical history
Obstructive Sleep Apnea	• Repeated airway obstruction during sleep (at least five per hour of sleep) and: ‣ Nocturnal breathing disturbances ‣ Daytime sleepiness, fatigues not explained by medical condition.	• Mild: Apnea hypopnea index is less than 15 • Moderate: Apnea hypopnea index is 15–30 • Severe: Apnea hypopnea index greater than 30	• Polysomnography

Insomnia

Factors altering sleep may be identified via self/other-report measures, actigraphy, or polysomnography. Specific diagnoses may be preexisting or acquired in the hospital. In the inpatient setting, the most common and impactful diagnosis is generally poor sleep or transient insomnia. Insomnia is characterized as dissatisfaction with sleep onset or maintenance, along with daytime symptoms, which include sleepiness, impaired attention, and mood disturbance that occur at least three times a week for greater than 3 months. Typically, sleep disorders are usually diagnosed based on atypical sleep duration (hypo/hypersomnolence), poor sleep quality, and abnormal movements during sleep. Of the sleep disorders, insomnia is the most common, with estimates ranging between 6 and 10% of individuals (Morin and Jarrin 2022). It is noteworthy that more than a third of individuals will experience transient insomnia at some point in their life (Morin et al. 2015). Studies reviewing the efficacy of insomnia treatment are important for multiple reasons (Mansuri et al. 2022). Recent literature has highlighted that insomnia is associated with increased risk of suicidal behavior during inpatient stay, leads to slower recovery of motor functioning among patients post stroke (Joa et al. 2017), extends the length of hospital stays among patients with acquired brain injury (Dey et al. 2021), and predicts chronic pain post burn injury (Smith et al. 2008).

Before implementing any intervention in the hospital environment, proper sleep assessment is important. Traditional assessment tools, often used in the outpatient setting, require a retrospective reflection of prior sleep quality spanning multiple weeks, and are therefore less appropriate to use in the inpatient setting. Moreover, depending on a patient's current cognitive and functional status, specific hospital-based measures that include the option of caregiver report, may be ideal (See Appendix "List of Scales and Psychometric Properties"). A recent review identified three commonly used measures: the Richards-Campbell Sleep Questionnaire (RCSQ; Richards et al. 2000), St. Mary's Hospital Sleep Questionnaire (SMHSQ;

Ellis et al. 1981), and the Verran Snyder-Halpern Sleep Scale (VSHS; Snyder-Halpern and Verran 1987). According to Hoey et al. (2014), the RCSQ is the easiest to use, has strong psychometrics, and can be used by patients and caregivers. The latter two measures provide greater levels of detail but are limited by the lack of informant report. Another recently published instrument is the Factors Affecting Inpatient Sleep scale (FAIS). This measure differs from the above in specifically identifying targets for sleep modification in hospital settings (Ye et al. 2019). These targets include emotional/physical impairment due to illness or hospitalization, sleep disturbance due to discomfort or care activities, and sleep disruption due to environmental factors. This measure can help inpatient neuropsychologists provide practical recommendations for improving sleep quality while patients are admitted to the hospital.

Inpatients diagnosed with insomnia may be treated with both pharmacologic and non-pharmacologic interventions (Morin et al. 2015). Pharmacologic agents are often prescribed in efforts to target sleep onset, improve the duration of sleep (sleep maintenance), and combat environmental factors such as the limited controllability of the hospital setting (Pavlova and Latreille 2019). However, research shows equivocal efficacy of pharmacological agents in improving sleep in this setting, with some studies suggesting adverse effects including drowsiness, dizziness, confusion, amnesia, psychosis, and sleep eating (Berry et al. 2013; Herzig et al. 2021). A systematic review by Kanji et al. (2016) found that pharmacologic agents did not promote restorative sleep and can negatively influence sleep architecture. Concerningly, sleep medications have been linked to falls in the elderly and increasing delirium (Mahoney et al. 2004). Medications such as trazodone and quetiapine are associated with prolongation of the QTc interval (i.e., abnormal prolongation of the total time *from cardiac ventricular depolarization to complete repolarization;* Indraratna et al. 2020*;* Yanowitz 2012*)*, and thus are not advised in the context of recovery of myocardial infarction and cardiac concerns (Højlund et al. 2022; Mendelson 2005).

For these reasons, non-pharmacologic interventions are encouraged to improve sleep without the adverse effects of medications. A recent systematic review and meta-analysis by Beswick et al. (2023) found generally strong evidence that non-pharmacological interventions (sleep aids, relaxation, music interventions, and manualized therapies) improved sleep quality in the inpatient setting. This study highlighted that the largest evidence base is for the use of physical sleep aids. Specifically, there was extensive evidence for using eye masks, either independent of or in combination with earplugs. Moreover, there was evidence to support the use of music interventions (e.g., on the TV or played over speakers) and relaxation techniques, especially scent-based interventions. Thus, there is support for using interventions such as sensory buffers or environmental modifications to promote feelings of relaxation and sleep.

Cognitive behavioral therapy for insomnia (CBT-I) is frequently used in outpatient settings (van der Zweerde et al. 2019), though research supporting use among hospitalized patients is limited (Beswick et al. 2023), likely due to the difficulty of implementing a comprehensive CBT-I protocol in this setting. However, certain components of CBT-I can be adapted and utilized in inpatient settings. CBT-I has multiple components, including sleep restriction, psychoeducation on sleep hygiene, stimulus control, relaxation, and cognitive therapy (Morin 2006), and much of the focus in the hospital is on sleep hygiene and relaxation-based principles.

Obstructive Sleep Apnea

Compared to the general public, hospitalized patients show higher incidence of sleep-disordered breathing (Shear et al. 2014), a spectrum of diagnoses that are distinguished by airway resistance or closure and/or diminished drive to breathe (Knauert et al. 2014). One of the more common and well-known diagnoses in this category is OSA, which is quite common, affecting approximately 17% of women and 34% of men in the general population, though the prevalence among medical inpatients is estimated at 40%. Among all patients, OSA poses risk for negative health outcomes, and this is particularly true for patients with additional medical comorbidities (Al Lawati et al. 2009). Consequences of OSA include cardiovascular issues, obesity, diabetes risk, excessive daytime sleepiness, reduced vigilance, lower quality of life, difficulty with completion of daily activities, increased rates of motor-vehicle collisions (MVCs), and increased mortality, especially when untreated (Gottlieb and Punjabi 2020).

Laboratory-administered polysomnography (PSG) is the gold standard for diagnosing OSA. PSG assesses various aspects of sleep including wake/sleep time, duration, and sleep architecture by collecting physiological data via an electroencephalogram (EEG), electrooculogram, and electromyogram. Data such as the frequency/amplitude/shape of brainwaves, limb movements, respiratory effort, heart rate, oxygen saturation, and body position are scored by sleep specialists based on standardized criteria (Berry et al. 2017). However, PSG is not often available in inpatient settings due to the associated time, cost, impact on sleep, complexity of use, reduced patient tolerance, and expertise required to synthesize these data (Delaney et al. 2021; Mantua et al. 2016).

Actigraphy, a more cost-effective alternative, measures motor activity, which can be translated into sleep and wake states (Delaney et al. 2021). Actigraphy is a method of inferring sleep/wake patterns based on an individuals' movement throughout the night (Rösler et al. 2023). Currently, actigraphy can be measured through both research-based monitors and personal devices (e.g., apple watch, fit bit). There is evidence of variability in these community available devices around specific stages of sleep though they appear to do well in assessment of sleep time (Kainec et al. 2024). This information indicates that actigraphy may be most helpful to the provider in identifying a patient's time asleep and awake, which may be useful in identifying patterns of awakenings, disturbances promoting these awakenings, and supporting non-pharmacologic treatments especially around modifying sleep behaviors. The use of actigraphy could reduce the reporting bias of patients and observers alike, with research showing there may be discrepancies between these two reports (van den Ende et al. 2022), and this may be particularly beneficial for patients with cognitive or communication disorders affecting their ability to self-report. While actigraphy has benefits, there are only modest correlations with PSG and thus this is a less accurate tool (Marino et al. 2013).

There are various treatments for obstructive sleep apnea (Pavwoski and Shelgikar 2017) such as bilevel positive airway pressure (Bi-PAP) and auto-titrating PAP (APAP), with the primary intervention being continuous PAP (CPAP). While effective, CPAP adherence ranges from 17% to 71%. Moreover, many individuals use their CPAP about 4.5 h per night, less than the recommended treatment guidelines, although this varies based on age and sex (Patel et al. 2021; Rotenberg et al. 2016; Weaver and Grunstein 2008). Individuals in the hospital also often do not obtain CPAP therapy or this may not be implemented while they are hospitalized (Sorscher and Caruso 2012). Furthermore, individuals with various diagnoses may show lower levels of adherence, including those with acquired brain injury (Silva et al. 2020). Relatedly, treatment adherence for OSA is often challenging in the hospital. This may be due to the various team, patient, and diagnosis-based factors. In the context of limited provision of PAP therapies while in the hospital, the inpatient neuropsychologist can facilitate team-based communication about a patient's previous home-based sleep routines. Furthermore, even for those on home PAP, the inpatient neuropsychologist can assess for previous levels of adherence and help to increase or promote these in the hospital.

Previous research has found that motivational interviewing and motivational enhancement therapy both may lead to short-term improvements in use, which is an important finding that could be applied in hospital settings (Rapelli et al. 2021). Based on findings highlighting lower levels of adherence among individuals with acquired brain injury, assessment of cognitive, emotional, and behavioral factors via inpatient neuropsychological assessment can be used to help tailor interventions, including external aides and reminders, a team-based approach to encourage PAP use, and inclusion of family support to increase adherence (Askland et al. 2020; Baron et al. 2022). Ultimately, inpatient neuropsychologists have an important role at multiple levels in promoting PAP adherence through motivational interviewing techniques, team-based approaches, and inclusion of supports to help increase adherence.

Overall, sleep in the inpatient setting is an important and challenging area for patients and providers alike. As noted in this chapter, brief insomnia and obstructive sleep apnea are two of the most common concerns. While self-report can be helpful in identifying at-risk patients, individuals suspected of having sleep disorders in inpatient settings often benefit from ongoing assessment, such as PSG, for a definitive diagnosis. In this area, the inpatient neuropsychologist can take a multi-pronged approach, focused on environmental modification, team-based intervention, and patient-based strategies to improve sleep and adherence with these devices, leading to overall improved outcomes. Consideration of cognitive status and behavioral/emotional factors may help tailor recommendations and interventions. In this context, an inpatient neuropsychologist is uniquely suited for this role and can be an important resource for the team (Table 14.1).

PAIN

Pain is among the most common presenting concerns in medical settings, affecting 37–84% of hospitalized patients, with 9–36% reporting severe pain (Gregory and McGowan 2016). Evidence suggests that uncontrolled pain can prolong hospitalization (Walker 2004) and neuropsychologists may be consulted to help address pain and the impact on recovery, with an emphasis on non-pharmacological options given concerns about risks (e.g., tolerance, limited efficacy) of opiate and other medications for chronic pain.

Pain is an inherently subjective experience. Although observable pain behaviors may be present, clinicians rely on patients' reports to evaluate pain quality, duration, and intensity. Underscoring the experienced quality of pain, the International Association for the Study of Pain (IASP) defines pain as "an unpleasant sensory and emotional experience associated with, or resembling that associated with, actual or potential tissue damage" (Raja et al. 2020). When addressing patients' pain, distinctions should be made between what is acute versus chronic pain (Table 14.2; Murphy et al. 2014). If

Table 14.2 Acute vs chronic pain

	Acute Pain	Chronic Pain
Duration	<3 months	>3 months
Characterization	Symptom to be cured	Condition
Etiology	Typically, has identified cause, is reflective of the body's response to injury or tissue damage	May or may not have identified cause
Course	Diminishes with healing and recovery and is response to treatment	Pain persists beyond expected healing of damage or injured tissues or despite treatment
Treatment Response	Responds well to rest	Exacerbated by rest

Adapted from Murphy et al., 2014

appropriate, providers may wish to explain the difference between "hurt versus harm": or pain signaling bodily damage (harm) and the sensation of discomfort or pain (hurt) so patients can more accurately qualify their subjective experience (Smith et al. 1998). However, providers should keep in mind that, in patients with both acute and chronic pain, the experience of one may influence or impact the experience of the other. For example, patients with chronic pain conditions may report or experience acute pain differently than patients without a chronic pain history. Similarly, suffering acute, yet transient, prior pain may impact how a patient copes with or conceptualizes their current pain condition. In this case, patients may have unrealistic expectations about the ability of chronic pain to fully remit with treatment.

Assessment of Pain

Assessing subjective pain experience can guide intervention. However, expanding upon dimensions and other contributing factors will further inform pain management. It is recommended that a comprehensive evaluation assesses all variables associated with the pain experience. These may include:

- Current and previous pain management and perceptions of their efficacy.
- Alleviating or exacerbating factors.
- Pain location, quality, and intensity.
- Interference in functioning (e.g., impact on activity, mood, sleep, relationships).
- Cognitive factors such as ignoring, ruminating, or catastrophizing.
- Personal values and beliefs about pain.
- Psychological factors including mood states, stress.
- Pain behaviors such as guarding, resting, avoidance, persistence, or requesting assistance.
- Environmental factors including responses from caregivers or loved ones.

In addition to a clinical interview, self-report measures can efficiently capture relevant aspects of a patient's pain experience and guide intervention. Within the inpatient setting, it is important to consider the impact of the patient's hospitalization and medical status when interpreting these measures. See Appendix "List of Scales and Psychometric Properties" at the end of this chapter for a summary of select self-report measures of pain.

Contextual and Cultural Confounds of Pain

Effectively providing interventions for pain in inpatient settings requires providers to validate patients' experiences of their pain and take care not to attribute all reported symptoms to somatic causes. Even if unintentional, messages that patients' pain is "all in their heads" or that their medical team has "given up" on treating their pain can undermine trust in the care team.

As such, clinicians should convey that a patient's report of pain is taken seriously. This is particularly vital when working with patients from underrepresented or marginalized groups including women or minority populations, where historically pain has been incompletely assessed and treated. For these reasons, providers should remain sensitive to cultural factors that impact how patients report and understand their pain. For example, military service members and veterans, as well as individuals from other cultural backgrounds, may view the experience of pain as a personal weakness or failure. Specifically inquiring about pain beliefs can help identify unhelpful or inaccurate appraisals that have implications for pain perception, functioning, and treatment response (Williams et al. 1994).

Interventions for Pain

Psychoeducation is a valuable mechanism for ameliorating the experience of pain. Even a single session on pain education has demonstrated clinical efficacy in improving catastrophizing of pain and associated functional limitations (Darnall et al. 2021). For example, Fletcher et al. (2016) found that pain-related fear and the resulting avoidance of movement is reduced among individuals who possess greater knowledge about pain. There are several models of pain that can aid in dispelling unhelpful beliefs about pain, increasing self-efficacy, and promoting adaptive coping. The *biopsychosocial* model views pain and other chronic conditions as a relationship among physical, psychological (affective and cognitive), and social factors (Gatchel et al. 2007). This model broadly implies that managing pain requires consideration of factors beyond medical treatment of disease or illness (Fig. 14.4). Similarly, the *fear-avoidance* model describes how fear of pain leads to certain pain behaviors, like avoidance or guarding, that increase pain experience over time (Fig. 14.5; Vlaeyen et al. 2016). In this way *kinesiophobia,* or fear of movement, may prevent participation in rehabilitation and recovery.

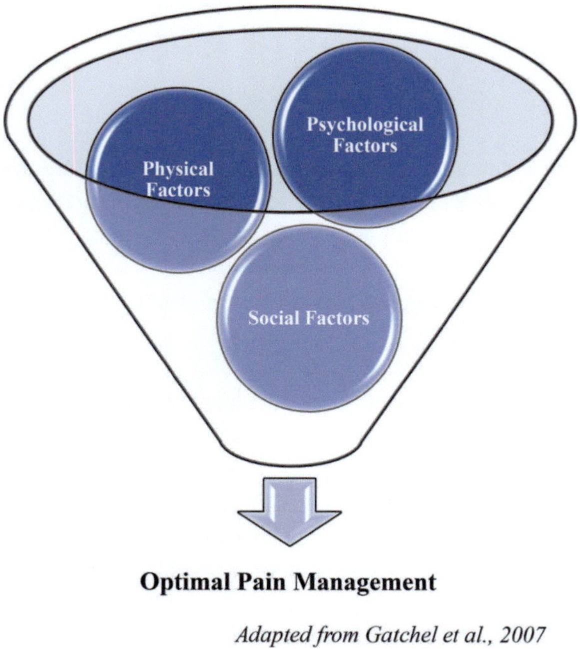

Fig. 14.4 Biopsychosocial model of pain management

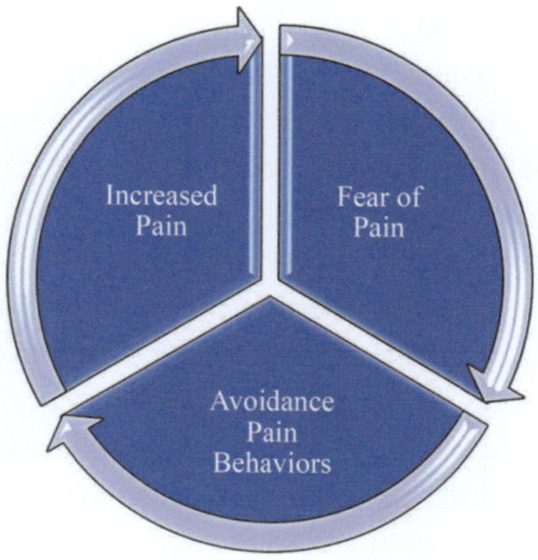

Fig. 14.5 Fear-avoidance model of pain management

Relaxation training including deep breathing, progressive muscle relaxation, and mindfulness strategies are classic pain interventions, but may require patient education to facilitate and/or improve compliance. Adaptations for medical setting or status may be needed, such as avoiding progressive muscle relaxation in patients with orthopedic injuries or ensuring guided imagery scripts are inclusive for people with pre-existing or new physical disabilities.

Cognitive Behavioral Therapy (CBT) emphasizes the interconnection of cognition, emotion, and behavior (Murphy et al. 2014). As in other CBT interventions, cognitive strategies address unhelpful thought patterns around pain whereas behavioral interventions promote more adaptive responses. Behavioral strategies may include relaxation training, pacing, and graded exposure. This can include co-treatment sessions with physical or occupational therapy, integrating CBT strategies into therapeutic activities to decrease avoidance and increase activity tolerance and participation. Co-treatment can also be an opportunity to model effective communication and responses to patient concerns. For instance, this could involve negotiating session goals (length or intensity of activity), alternating activity and rest or engagement in relaxation strategies, and identifying more accurate or helpful cognitions in vivo.

Acceptance and Commitment Therapy (ACT) integrates acceptance- and mindfulness-based interventions, emphasizing acceptance of thoughts and feelings and increased engagement in values-oriented behavior (Hayes et al. 1999b). With chronic pain, ACT emphasizes the role of decreasing attempts to control or escape painful sensations in causing distress and increasing psychological flexibility (Dahl and Lundgren 2006). Indeed, acceptance is associated with greater pain tolerance, potentially due to increased distance from internal experiences including emotions (Hayes et al. 1999a). Within inpatient settings, ACT-based interventions might emphasize mindfulness skills and clarify patients' values and goals for their medical treatment, including participation in rehabilitation therapies or effectively communicating with their care team. Helping patients identify their values and functional goals can guide rehabilitative interventions and improve compliance with treatment. In this way, proactive engagement behaviors are reinforced as aligning with the patient's own personal values, thus decreasing avoidance of perceived discomfort and fatigue which would otherwise limit patient participation.

Strategies to decrease attention and reactivity to internal pain sensations can be integrated into the patient's schedule and implemented during daily care or participation in rehabilitation therapies. For example, hypnotherapy is another non-pharmacological intervention that levies visualization and suggestive statements to redirect a patient's attention and thoughts away from pain (Jensen and Patterson 2014). While hypnotherapy can be provided during one-on-one sessions with the provider, a patient can also learn techniques for self-hypnosis, which is particularly beneficial for those who have ongoing pain or limited opportunity to schedule routine appointments. Virtual reality, another novel non-pharmacological intervention for pain management, has demonstrated efficacy in reducing pain following medical procedures and can be utilized in acute care settings (Georgescu et al. 2020).

Team-Based and Environmental Interventions

Evidence-based recommendations for pain management in hospital settings emphasize a collaborative approach (Dworkin et al. 2007; Gordon et al. 2014), including environmental, behavioral, caregiver, and team-based interventions. Importantly, non-pharmacological interventions should be introduced as a critical component of pain management rather than a substitute for medical care. Some concern has been raised that the biopsychosocial model may overweigh psychological factors when medical workup is unrevealing or otherwise the underlying pathology is not well defined or understood (Weiner 2008). Patients should not inadvertently receive messages that their pain is not credible. Similarly, providers should be aware of their own bias and not automatically assume all pain behaviors reflect secondary motives.

Clear communication from medical providers about treatment plans and goals is a simple, yet critical, recommendation. Offering explanation prior to and during care, especially for painful procedures or during early mobilization, can allay fears and set expectations. Similarly, providing rationale for therapeutic techniques and offering choices for intervention creates a collaborative working relationship that reinforces patient self-efficacy. Solicitous or punishing responses by caregivers, whether family or medical personnel, may undermine pain management efforts (McCracken 2005). Communication strategies include praise and reinforcement for adaptive coping and activity, or conversely, encouragement for appropriate pacing for patients tempted to "push through" or overdo activity. Helpful interventions to increase adherence or reinforce expectations around pain management include visual or auditory cues (e.g., timers) for pressure relief, implementation of breathing strategies, use of a transcutaneous electrical nerve stimulation device (TENS unit), or use of heat or ice.

Direct patient contact is often brief in inpatient settings and interventions should be efficient and emphasize recommendations to the consulting team. In these interactions, it may be helpful to identify skills for active pain coping that can continue beyond inpatient setting, especially in terms of promoting multimodal pain management. The optimal team-based interventions will reflect both the patient and environmental needs and resources. Where possible, monitoring of effectiveness of intervention and adjustment as needed communicates to both patients and teams that concerns are taken seriously. Consultation with social workers or other team members can assist in coordinating any necessary referrals for outpatient treatment following discharge (Hobbs 2005).

SUMMARY

In this chapter, we have discussed how neuropsychologists can deliver evidence-based assessments and intervention strategies to address a range of factors impacting psychological adjustment in inpatient settings. These are relatively common issues that can exacerbate distress and undermine the patient's engagement in care. As in any practice, neuropsychologists must distinguish between maladaptive thoughts and behaviors suitable for intervention and a wide range of potentially normative psychological reactions to illness, injury, or medical treatment. Even in the absence of problematic issues to be addressed, there is often an opportunity to reinforce positive adjustment and focus on growth. In such encounters, it is helpful to orient patients and their families toward health, wellness, and the notion that rehabilitation will require active engagement in treatment—with the potential for goals of care to range from curative to compensatory. Every encounter in this setting is an opportunity to assist someone on their path to recovery, to normalize the role of psychology in medical care, and establish the concept that all stages of treatment will involve elements of psychological adjustment.

Future directions for mental health in inpatient care might include ongoing integration of neuropsychologists within inpatient units, either as a structured consult service or integrated within an interdisciplinary team. Neuropsychologists working in inpatient care often find the desire for services greater than the available time and resources, and so future directions should consider strategies to optimize available resources. Such efforts would be bolstered by demonstration of benefits to administrators and payors, with intentional efforts to track meaningful outcomes associated with neuropsychology services for medical inpatients, at both the systems (e.g., decreasing barriers to discharge and reducing provider burnout) and patient (e.g., increasing engagement in care) levels.

APPENDIX: LIST OF SCALES AND PSYCHOMETRIC PROPERTIES

Scale	Features	Psychometrics	Clinical Interpretation
		Anxiety and Depression Scales	
Hospital Anxiety and Depression Scale (HADS; Zigmond & Snaith, 1983)	• Includes Anxiety (HADS-A) and Depression subscale (HADS-D), each with 7 items. • 4-point Likert scale (0-3). • Total scores range from 0-21 for both scales.	• Sensitivity and specificity for both subscales (0.80). • Internal Consistency: Cronbach's alpha for HADS-A varied from 0.68 to 0.93 (mean = 0.83). • Internal consistency: Cronbach's alpha for HADS-D from 0.67 to 0.90 (mean = 0.82).	• For both scales: ‣ 0-7 considered normal. ‣ 8-10 moderate symptoms. ‣ 11-21 significant symptoms. • 8+ considered a cut-off for both scales.
Patient Health Questionnaire-9 Item (PHQ-9; Kroenke et al., 2001)	• Self-report depression scale that scores 9 DSM-IV criteria. • Likert scale 0 (not at all) to 3 (nearly every day). • Total score ranges from 0-27.	• PHQ-9 score ≥ 10 has sensitivity (88%) and specificity (88%) for major depression. • Internal consistency: Cronbach's alpha (0.89). • Test-retest reliability (0.84).	• 1-4 = minimal depression. • 5-9 = mild depression. • 10-14 = moderate depression. • 15-19 = moderately severe depression. • 20-27= severe depression.
Beck Depression Inventory (BDI; Beck et al., 1961, 1996)	• BDI-2 aligned with DSM-IV symptom criteria. • 21 item self-report scale for depression. • Items grouped for somatic, affective, cognitive, and vegetative symptoms. • Total score ranges from 0-63.	• Generally, studies report high sensitivity (≥ 0.70). • Specificity varied with population and cut-off used. • Test-retest reliability range for BDI-2 (0.73 to 0.92). • Internal consistency: Cronbach's alpha (0.90).	• Clinical ranges: ‣ 0-9 = minimal depression normal. ‣ 10-18 = mild depression. ‣ 18-29 = moderate depression. ‣ 30+ = severe depression • Cut-off score varies according to type of sample.
Patient-Reported Outcomes Measurement Information System – Depression Bank (PROMIS-Depression; Cella et al., 2010)	• Assesses self-reported negative mood, views of self, social cognition, decreased positive affect and engagement. • Available for adults, pediatric, and parent proxy. • Short forms and computer adaptive test available. • Raw scores converted to T scores. Higher score indicates higher depression. • Full bank includes 28 items.	• High reliability between full item bank and short forms (0.96). • Sensitivity = 79%. • Specificity = 86%.	• General population mean T score considered 50. • Cut-off ≥53 considered optimal score to predict DSM-5 depressive disorder.

Patient-Reported Outcomes Measurement Information System – Anxiety Bank (PROMIS-Anxiety; Cella et al., 2010)	• Assesses self-reported fear, anxious misery, and somatic arousal symptoms. • Available for adults, pediatric, and parent proxy. • Short forms and computer adaptive test available. • Responses range 1 (never) to 5 (always). • Raw scores converted to T scores. Higher score indicates higher anxiety. • Full bank includes 29 items.	• Internal reliability: (above 0.93). • Full item bank and short forms are correlated (0.96). • Sensitivity = 60%. • Specificity = 74%.	• General population mean T score considered 50. • Cut-off ≥59 considered optimal score to predict DSM-5 anxiety disorder.
Generalized Anxiety Disorder Scale (GAD-7; Spitzer et al., 2006)	• 7-item, self-report anxiety scale. • Likert scale 0 (not at all) to 3 (nearly every day). • Total score ranges from 0-21.	• Sensitivity = 89%. • Specificity = 82%. • Internal reliability: Cronbach alpha (0.92). • Test-retest reliability (0.83).	• Clinical ranges: ‣ 0-4 = minimal anxiety. ‣ 5-9 = mild anxiety ‣ 10-14 = moderate anxiety. ‣ 15-21 = severe anxiety. • Scores ≥ 10 represent cut-point for identifying GAD.
	Acute Stress Reaction / Disorder		
Acute Stress Disorder Interview (ASDI; Bryant et al., 1998)	• 19-item structured clinical interview based on DSM-IV criteria. • Each item is scored dichotomously as 0 (symptom absent) or 1 (symptom present) according to DSM-IV criteria. • Total scores range from 0-19. • Higher score indicating greater symptom presence.	• Test-retest reliability: r = 0.90. • Internal consistency: r = 0.90. • Sensitivity (91%), and specificity (93%) for 65 hospitalized patients post traumatic event. • The psychometric properties of the ASDI-5 which was updated for DSM-5, have not yet been studied.	• Initial study among patients hospitalized after a traumatic event.
Acute Stress Disorder Scale (ASDS; Bryant et al., 2000)	• 19-item inventory self-report version of ASDI. • Consistent with DSM-IV. • Likert scale from 1 ("not at all") to 5 ("very much"). • Total scores range from 19-95. • Higher score indicating greater severity.	• Test-retest Reliability: r = 0.94. • Internal consistency: Cronbach's alpha 0.96. • Sensitivity (95%) and Specificity (83%) for identifying ASD against the ASD Interview on 99 civilian trauma survivors.	• The optimal cutoff score described was 56 based on hospitalized accident and assault patients.

Posttraumatic Adjustment Scale (PAS; O'Donnell et al., 2008)	• 10-item screening instrument that predicts subsequent PTSD and depression. • Likert scale ranging from 0 ("not at all") to 4 ("totally"). • PAS-PTSD score calculated by summing all 10 items. • PAS-depression score calculated by summing five of the 10 items. • Total scores range from 0-40. • Higher score indicating greater severity.	• Sensitivities of 0.82 and a specificity of 0.84 for PTSD and sensitivity of 0.72 and specificity of 0.75 for depression in hospitalized adults.	Cut off scores established with hospitalized injury patients: • PAS-PTSD cutoff score of 16. • PAS-depression cutoff score of 4.
National Stressful Events Survey Acute Stress Disorder Short Scale (NSESSS; Kilpatrick et al., 2013)	• 7-item measure assesses severity of acute stress disorder at follow up, post ASD diagnosis. • Likert scale ranging from 0 (not at all) to 4 (extremely). • Average total score is calculated by dividing the raw total score by number of items in the measure (e.g. 7). • Total scores range from 0-28. • Higher score indicating greater severity.	• Internal consistency = Cronbach's alpha was 0.91for the 9-item NSESSS-PTSD. • Convergent validity for NSESSS-PTSD established with PTSD checklist (PCL).	N/A
Stanford Acute Stress Reaction Questionnaire (SASRQ; Cardena et al. 2000)	• 30-item scale measures peritraumatic anxiety, dissociation, and acute stress disorder according to DSM-IV criteria. • Scored from (0-5) or dichotomously (0-2: 0,3-5: 1) based on symptom presence. • Scores range from 0-50. • Higher scores indicating a more severe symptom.	• Norms established among emergency rescue workers. • Internal consistency: Cronbach's alpha for 30 items ranged from 0.80 to 0.95. • Test-rest reliability was 0.69. • Convergent and discriminant validity have also been demonstrated.	• The diagnostic cutoff for ASD requires each person to assign a value of 3 or higher to at least three of the ten dissociative criteria, and a score of three or higher for at least one of symptom. clusters C, D, E, and F.
	Suicide Screening Scales		
Columbia–Suicide Severity Rating Scale (C-SSRS; Posner et al., 2011)	• Designed to assess suicidal ideation and behavior including intent and plan. • Shorter screener version available for full version. • Includes 10 categories with Yes/No responses. • Risk assessment page provides checklist of protective and risk factors.	• High sensitivity and specificity (>93%) in correctly identifying lifetime actual events and interrupted attempts. • Moderate convergent validity (r = 0.69).	• No specific clinical cutoffs because of binary responses. • Follow-up inquiries are recommended based on item endorsement for categories: ‣ 1-5 = suicide ideation. ‣ 6-10 = suicidal behavior. ‣ 1-10 = suicidal ideation and behavior.
Scale for Suicide Ideation (SSI; Beck et al., 1997)	• 19-item scale to assess suicidal intention. • Each item rated on 3-point scale (0-2). • Total scores range from 0-38.	• Internal consistency: Cronbach alpha (0.89). • Inter-judge reliability (0.83).	N/A

Sleep Disorders			
Richards-Campbell Sleep Questionnaire (RCSQ; Richards et al., 2000)	• 5 Item visual analog scale to assess sleep in critically ill patients. • Scoring: 0 (poorest quality deep) to 100 (optimum sleep). • Total score calculated by summing each score and dividing the total by five. • Lower scores indicate a poorer quality of sleep.	• Internal consistency: r = 0.90. • Significantly correlated with the sleep efficiency index in polysomnography (r = 0.58). • Sensitivity = 87%, specificity = 81% during ICU stay.	• A global score of ≥ 63.4 mm found as optimal cut-off for self-reported 'good sleep.'
St. Mary's Hospital Sleep Questionnaire (SMHSQ; Ellis et al., 1981)	• Assesses sleep quality among hospitalized patients to evaluate sleep-related issues (e.g. sleep latency, restlessness, nighttime waking, and morning alertness). • 14 Likert scale and fill-in-the -blank questions with multiple choice questions and short answers.	• Significant reliability ($p < 0.0001$) in four group (psychiatric, surgical, medical patients, and volunteers) (Kendall's tau ranging 0.70 to 0.96).	N/A
Verran Snyder-Halpern Sleep Scale (VSHS; Snyder-Halpern & Verran, 1987)	• Self-report scale to evaluate sleep disturbances and effectiveness among hospitalized patients. • Initially an eight-item scale, six additional items were added following psychometric evaluation. • Uses a 100 mm visual analog scale to evaluate sleep three nights. • Scores range from 0 to 1500; higher scores indicate better quality of sleep.	• Internal consistency of 0.82 (applied to 8 item version). • Validated among those with no history of sleep difficulties).	N/A
Epworth Sleepiness Scale (ESS; Johns, 1991)	• Eight situations self-report. • Likert scale ranging from 0 (no chance of dozing), to 3 (high chance of dozing). • Scores range from 0-24.	• Correlated significantly with multiple sleep latency test (MSLT) (r = -0.514, n = 27, $p < 0.01$). • High level of internal consistency for patients with sleep disorders: Cronbach's alpha= 0.88. • High test-retest reliability among medical students (r = 0.82). • Sensitivity of 0.94. • Specificity of 1.00.	• Cutoff of 10 for identifying sleepiness at clinical level. • Scores ranging from 16-24 considered excessively sleepy.
Narcolepsy Severity Scale (NSS; Dauvilliers et al., 2017)	• 15-item scale to evaluate narcolepsy type 1 (NT1). • 6 items evaluating frequency are rated using a six-point Likert scale. • 9 items that describe the symptom effect are rated using a four-point Likert scale. • Total scores range from 0 to 57.	• High internal consistency: Cronbach's alpha = 0.75 • Sensitivity: 61%, specificity: 81% when assessed on an independent sample of drug-free and treated patients. • Correlated with multiple sleep latency test (MSLT; r = -0.41, $p = 0.001$).	• Scores ranging from 43–57 considered very severe. • Severity groups defined with treated and untreated patients with NT1. • The cut-off value for discriminating treated and drug-free groups was 24.
REM Sleep Behavior Disorder Screening Questionnaire (RBDSQ, Stiasny-Kolster et al., 2007)	• 10 items with 13 yes/no questions. • Maximum score possible = 13.	• Good internal consistency: Cronbach's alpha = 0.885 when evaluated in RBD patients and 2 separate control groups.	• Cutoff score of 5 showed a sensitivity of 96% and a specificity of 56%.

Mayo Sleep Questionnaire (MSQ; Boeve et al., 2013)	• 16-item question scale to assess REM sleep behavior disorder (RBD) from bed partners if a behavior has been exhibited at least three times in the past. • Some questions are ignored depending on applicability.	• Sensitivity of 100% and specificity of 95% on the core question on recurrent dream enactment behavior for the diagnosis of RBD among elderly community-based sample.	• An affirmative response to question 1 is almost 100% sensitive for RBD.
Paris Arousal Disorders Severity Scale (PADSS; Arnulf et al., 2014)	• Self-report scale consisting of 3 parts: ▸ PADSS-A - Sleep behaviors - Scored as "never" = 0, "sometimes" = 1, and "often" = 2, yielding a sub-score between 0 and 34. ▸ PADSS-B - Frequency of episodes - Scored from 0 (never) to 6 (twice nightly). ▸ PADSS-C - Consequences of behaviors - scored from "never" = 0, "sometimes" = 1, and "often" = 2, yielding a sub-score between 0 and 10. • Total score ranges from 0-50.	• Good Sensitivity (83.6%), specificity (87.8%), internal consistency, and test-retest reliability (0.79) among patients with sleepwalking and sleep terror.	• The best cutoff for the total score was at 13/14 for the total scale.
STOP-BANG Questionnaire (Chung et al., 2008)	• OSA screening using: ▸ information on snoring, tiredness, observed apnea, and high blood pressure (STOP). ▸ BMI, age, neck circumference, and gender (BANG). • 8 questions based on Yes/No answers (score: 1 or 0). • Total score ranges from 0-8.	• Sensitivity for STOP-BANG based on apnea– hypopnea index cutoffs greater than 5 (83.6%), greater than 15 (92.9%) and greater than 30 (100%) among surgical patients. • Test-retest reliability: κ coefficient = 0.92.	• Score ≥3 shows high sensitivity for OSA. • Score of 5–8 indicate a high probability of moderate/severe OSA.
	Pain Disorders		
The Defense and Veterans Pain Rating Scale (DVRPS; Polomano et al., 2016)	• 5-item tool with 0-10 pain scale that incorporates impact of pain on sleep, mood, stress, and activity. • Integrates a numerical ratings scale with functional word descriptors, color coding, and drawn facial expressions matched to pain levels.	• Internal consistency: Cronbach's alpha = 0.87. • Test-retest reliability: r = 0.64 to r = 0.77.	N/A
Numerical Rating Scale (NRS; n.d.)	• Self-report scale assessing current, least, and worst intensity of pain on a 0-to-10 scales, where 0 indicates "no pain" and 10 indicates "pain as intense as you can imagine." • Response windows for least and worst can be 24 hours or seven days.	• Recommended above other pain intensity measures for chronic pain clinical trials.	• 1-2 = mild pain. • 5 = moderate pain. • 8 or above = severe pain.
Pain Assessment in Advanced Dementia (PAINAD; Warden et al., 2003)	• Utilizes observed pain behaviors to determine the level of pain and can be beneficial for patients with severe cognitive impairment or otherwise unable to report pain. • Observers rate behaviors from 0 to 2 including breathing, negative vocalization, facial expression, body language, and consolability. Each item rated from 0 to 2 based on descriptor. • Provides continuous score between 0 to 10.	• Within acute hospital setting (Dunford et al., 2022), inter-rater ICC = 0.92 at rest and 0.98 in movement. • Test-retest reliability: ICC = 0.54 at rest and 0.66 in movement. • Internal consistency: Cronbach's alpha = 0.76 at rest and 0.80 with movement. • Correlation with self-rated pain was low Kendall's tau 0.29 (p>0.001).	N/A

Pain Catastrophizing Scale (PCS; Sullivan et al., 1995)	• 13-item self-report measure. • Each item rated on 5-point scale from 0 ("not at all") to 4 ("all the time). • Scores range from 0 to 52.	Per systematic review: ‣ Internal Consistency: Cronbach's alpha = 0.92). ‣ Test-retest reliability: Spearman's ρ =0.88 for total score. • Translations available in multiple languages	• Scores of 30 or above represent clinically relevant catastrophizing.
Pain Self-Efficacy Questionnaire, 2-item (PSEQ-2; Nicholas et al., 2015)	• 2-item short form of a widely used 10-item measure of pain self-efficacy.	• Internal consistency: Cronbach's alpha = 0.76. • Test-retest reliability: r =0.76 and ICC = 0.87. • Adequate sensitivity to change and convergent validity.	• Scores <6 may be considered in need of intervention. • Scores >7 reflect adequate self-efficacy.
Short-Form McGill Pain Questionnaire (SF-MPQ-2; Trudeau et al., 2012)	• Assesses the quality of pain-related symptoms on 22 items. • Each item rated from 0 to 10 numerical rating scale. • Total score represents the average of all ratings, with subscales reflecting average of items within the subscale. • Higher subscale or total scale scores indicate greater pain intensity.	• Per a study of acute low back and radicular leg pain (Dworkin et al., 2015), Cronbach's alpha for total and subscale scores ranged from 0.77 to 0.93. • Excellent validity in acute low back and radicular leg pain. • Responsive to change. • Translated into many languages.	• Intended for both chronic and acute pain conditions.

REFERENCES

Al Lawati NM, Patel SR, Ayas NT. Epidemiology, risk factors, and consequences of obstructive sleep apnea and short sleep duration. Prog Cardiovasc Dis. 2009;51(4):285–93. https://doi.org/10.1016/j.pcad.2008.08.001.

American Psychiatric Association. Diagnostic and statistical manual of mental disorders. 4th ed. Washington, DC: American Psychiatric Publishing, Inc; 1994.

American Psychiatric Association. Diagnostic and statistical manual of mental disorders. 5th ed. Washington, DC: American Psychiatric Publishing, Inc; 2013.

American Psychiatric Association. Diagnostic and statistical manual of mental disorders. 5th ed, text revision. Washington, DC: American Psychiatric Publishing, Inc; 2022.

Arnulf I, Zhang B, Uguccioni G, Flamand M, Noël de Fontréaux A, Leu-Semenescu S, Brion A. A scale for assessing the severity of arousal disorders. Sleep. 2014;37(1):127–36. https://doi.org/10.5665/sleep.3322.

Askland K, Wright L, Wozniak DR, Emmanuel T, Caston J, Smith I. Educational, supportive and behavioural interventions to improve usage of continuous positive airway pressure machines in adults with obstructive sleep apnoea. Cochrane Database Syst Rev. 2020;4(4):CD007736. https://doi.org/10.1002/14651858.CD007736.pub3.

Baron KG, Troxel WM, Galloway S, Kharidia S, DeVettori G, Gilles A, Sundar KM, Watt MH. Couples-based interventions to promote PAP adherence among older adults: a qualitative study of patients, partners, and providers. J Clin Sleep Med. 2022;18(11):2627–34. https://doi.org/10.5664/jcsm.10180.

Beck AT, Ward C, Mendelson M, Mock J, Erbaugh J. Beck depression inventory (BDI). Arch Gen Psychiatry. 1961;4(6):561–71.

Beck AT, Steer RA, Brown GK. Beck depression inventory. 2nd ed. San Antonio: Psychological Corporation; 1996.

Beck AT, Brown GK, Steer RA. Psychometric characteristics of the scale for suicide ideation with psychiatric outpatients. Behav Res Ther. 1997;35(11):1039–46.

Berry SD, Lee Y, Cai S, Dore DD. Nonbenzodiazepine sleep medication use and hip fractures in nursing home residents. JAMA Intern Med. 2013;173(9):754–61. https://doi.org/10.1001/jamainternmed.2013.3795.

Berry RB, Brooks R, Gamaldo C, Harding SM, Lloyd RM, Quan SF, Troester MT, Vaughn BV. AASM scoring manual updates for 2017 (version 2.4). J Clin Sleep Med. 2017;13(05):665–6. https://doi.org/10.5664/jcsm.6576.

Beswick AD, Wylde V, Bertram W, Whale K. The effectiveness of non-pharmacological sleep interventions for improving inpatient sleep in hospital: a systematic review and meta-analysis. Sleep Med. 2023;107:243–67. https://doi.org/10.1016/j.sleep.2023.05.004.

Boeve BF, Molano JR, Ferman TJ, Lin S-C, Bieniek K, Tippmann-Peikert M, Boot B, St Louis EK, Knopman DS, Petersen RC, Silber MH. Validation of the Mayo Sleep Questionnaire to screen for REM sleep behavior disorder in a community-based sample. J Clin Sleep Med. 2013;9(5):475–80. https://doi.org/10.5664/jcsm.2670.

Brewin CR, Andrews B, Valentine JD. Meta-analysis of risk factors for posttraumatic stress disorder in trauma-exposed adults. J Consult Clin Psychol. 2000;68(5):Article 5. https://doi.org/10.1037/0022-006x.68.5.748.

Bryan CJ, Rudd MD. Brief cognitive-behavioral therapy for suicide prevention. Guilford Publications; 2018.

Bryant RA. Acute stress disorder as a predictor of posttraumatic stress disorder: a systematic review. J Clin Psychiatry. 2011;72(02):Article 02. https://doi.org/10.4088/JCP.09r05072blu.

Bryant RA. Acute stress disorder: what it is and how to treat it. Guilford Publications; 2016.

Bryant RA, Harvey AG, Dang ST, Sackville T. Assessing acute stress disorder: psychometric properties of a structured clinical interview. Psychol Assess. 1998;10(3):215–20. https://doi.org/10.1037/1040-3590.10.3.215.

Bryant RA, Moulds ML, Guthrie RM. Acute stress disorder scale: a self-report measure of acute stress disorder. Psychol Assess. 2000;12(1):Article 1. https://doi.org/10.1037/1040-3590.12.1.61.

Bryant RA, Mastrodomenico J, Felmingham KL, Hopwood S, Kenny L, Kandris E, Cahill C, Creamer M. Treatment of acute stress disorder: a randomized controlled trial. Arch Gen Psychiatry. 2008;65(6):Article 6. https://doi.org/10.1001/archpsyc.65.6.659.

Bryant RA, Friedman MJ, Spiegel D, Ursano R, Strain J. A review of acute stress disorder in DSM-5. Focus. 2011;9(3):Article 3. https://doi.org/10.1176/foc.9.3.foc335.

Bryant RA, Creamer M, O'Donnell M, Silove D, McFarlane AC, Forbes D. A comparison of the capacity of DSM-IV and DSM-5 acute stress disorder definitions to predict posttraumatic stress disorder and related disorders. J Clin Psychiatry. 2015;76(4):Article 4. https://doi.org/10.4088/JCP.13m08731.

Campbell FAK. Exploring internalized ableism using critical race theory. Disabil Soc. 2008;23(2):151–62. https://doi.org/10.1080/09687590701841190.

Cardeña E, Koopman C, Classen C, Waelde LC, Spiegel D. Psychometric properties of the Stanford acute stress reaction questionnaire (SASRQ): a valid and reliable measure of acute stress. J Trauma Stress. 2000;13(4):719–34. https://doi.org/10.1023/A:1007822603186.

Cella D, Riley W, Stone A, Rothrock N, Reeve B, Yount S, Amtmann D, Bode R, Buysse D, Choi S. The patient-reported outcomes measurement information system (PROMIS) developed and tested its first wave of adult self-reported health outcome item banks: 2005–2008. J Clin Epidemiol. 2010;63(11):1179–94.

Chouchou F, Khoury S, Chauny J-M, Denis R, Lavigne GJ. Postoperative sleep disruptions: a potential catalyst of acute pain? Sleep Med Rev. 2014;18(3):273–82. https://doi.org/10.1016/j.smrv.2013.07.002.

Chung F, Yegneswaran B, Liao P, Chung SA, Vairavanathan S, Islam S, Khajehdehi A, Shapiro CM. STOP questionnaire: a tool to screen patients for obstructive sleep apnea. J Am Soc Anesthesiol. 2008;108(5):812–21.

Clarke DM, Kissane DW. Demoralization: its phenomenology and importance. Aust N Z J Psychiatry. 2002;36(6):733–42. https://doi.org/10.1046/j.1440-1614.2002.01086.x.

Dahl J, Lundgren T. Acceptance and commitment therapy (ACT) in the treatment of chronic pain. Mindfulness-based treatment approaches. Clin Guide Evid Base Appl. 2006:285–306.

Darnall BD, Roy A, Chen AL, Ziadni MS, Keane RT, You DS, Slater K, Poupore-King H, Mackey I, Kao M-C. Comparison of a single-session pain management skills intervention with a single-session health education intervention and 8 sessions of cognitive behavioral therapy in adults with chronic low back pain: a randomized clinical trial. JAMA Netw Open. 2021;4(8):e2113401.

Dauvilliers Y, Beziat S, Pesenti C, Lopez R, Barateau L, Carlander B, Luca G, Tafti M, Morin CM, Billiard M, Jaussent I. Measurement of narcolepsy symptoms: the narcolepsy severity scale. Neurology. 2017;88(14):1358–65. https://doi.org/10.1212/WNL.0000000000003787.

Delaney LJ, Litton E, Melehan KL, Huang H-CC, Lopez V, Van Haren F. The feasibility and reliability of actigraphy to monitor sleep in intensive care patients: an observational study. Crit Care. 2021;25(1):42. https://doi.org/10.1186/s13054-020-03447-8.

DePietro RH, Knutson KL, Spampinato L, Anderson SL, Meltzer DO, Van Cauter E, Arora VM. Association between inpatient sleep loss and hyperglycemia of hospitalization. Diabetes Care. 2017;40(2):188–93. https://doi.org/10.2337/dc16-1683.

Dey A, Kam A, Tam A, Bayley M, Guo M. Sleep disturbance and length of stay in the setting of acquired brain injury rehabilitation. Brain Inj. 2021;35(9):1022–7. https://doi.org/10.1080/02699052.2021.1945144.

Dinoso L, Baudoin C. Using principles of therapeutic communication to enhance trauma-informed care in the critical care setting to promote positive encounters. Crit Care Nurs Clin North Am. 2023;35(2):Article 2. https://doi.org/10.1016/j.cnc.2023.02.014.

Dobing S, Frolova N, McAlister F, Ringrose J. Sleep quality and factors influencing self-reported sleep duration and quality in the general internal medicine inpatient population. PLoS One. 2016;11(6):e0156735. https://doi.org/10.1371/journal.pone.0156735.

Dunford E, West E, Sampson EL. Psychometric evaluation of the pain assessment in advanced dementia scale in an acute general hospital setting. Int J Geriatr Psychiatry. 2022;37(12):gps.5830. https://doi.org/10.1002/gps.5830.

Dworkin RH, O'connor AB, Backonja M, Farrar JT, Finnerup NB, Jensen TS, Kalso EA, Loeser JD, Miaskowski C, Nurmikko TJ. Pharmacologic management of neuropathic pain: evidence-based recommendations. Pain. 2007;132(3):237–51.

Dworkin RH, Turk DC, Trudeau JJ, Benson C, Biondi DM, Katz NP, Kim M. Validation of the short-form McGill Pain Questionnaire-2 (SF-MPQ-2) in acute low back pain. J Pain. 2015;16(4):357–66. https://doi.org/10.1016/j.jpain.2015.01.012.

Elliott R, Chawla A, Wormleaton N, Harrington Z. Short-term physical health effects of sleep disruptions attributed to the acute hospital environment: a systematic review. Sleep Health. 2021;7(4):508–18. https://doi.org/10.1016/j.sleh.2021.03.001.

Ellis BW, Johns MW, Lancaster R, Raptopoulos P, Angelopoulos N, Priest RG. The St. Mary's Hospital sleep questionnaire: a study of reliability. Sleep. 1981;4(1):93–7. https://doi.org/10.1093/sleep/4.1.93.

Fanai M, Khan MA. Acute stress disorder. In: StatPearls. StatPearls Publishing; 2023.

Fleming MK, Smejka T, Henderson Slater D, van Gils V, Garratt E, Yilmaz Kara E, Johansen-Berg H. Sleep disruption after brain injury is associated with worse motor outcomes and slower functional recovery. Neurorehabil Neural Repair. 2020;34(7):661–71. https://doi.org/10.1177/1545968320929669.

Fletcher C, Bradnam L, Barr C. The relationship between knowledge of pain neurophysiology and fear avoidance in people with chronic pain: a point in time, observational study. Physiother Theory Pract. 2016;32(4):271–6. https://doi.org/10.3109/09593985.2015.1138010.

Foa EB, Chrestman KR, Gilboa-Schechtman E. Prolonged exposure therapy for adolescents with PTSD emotional processing of traumatic experiences, Therapist Guide. Oxford University Press; 2008.

Gatchel RJ, Peng YB, Peters ML, Fuchs PN, Turk DC. The biopsychosocial approach to chronic pain: scientific advances and future directions. Psychol Bull. 2007;133(4):581.

Georgescu R, Fodor LA, Dobrean A, Cristea IA. Psychological interventions using virtual reality for pain associated with medical procedures: a systematic review and meta-analysis. Psychol Med. 2020;50(11):1795–807.

Gordon RM, Corcoran JR, Bartley-Daniele P, Sklenar D, Sutton PR, Cartwright F. A transdisciplinary team approach to pain management in inpatient health care settings. Pain Manag Nurs. 2014;15(1):426–35.

Gottlieb DJ, Punjabi NM. Diagnosis and management of obstructive sleep Apnea: a review. JAMA. 2020;323(14):1389. https://doi.org/10.1001/jama.2020.3514.

Gregory J, McGowan L. An examination of the prevalence of acute pain for hospitalised adult patients: a systematic review. J Clin Nurs. 2016;25(5–6):583–98. https://doi.org/10.1111/jocn.13094.

Hayes SC, Pierson H. Acceptance and commitment therapy. In: Freeman A, Felgoise SH, Nezu CM, Nezu AM, Reinecke MA, editors. Encyclopedia of cognitive behavior therapy. Springer; 2005. p. 1–4. https://doi.org/10.1007/0-306-48581-8_1.

Hayes SC, Bissett RT, Korn Z, Zettle RD, Rosenfarb IS, Cooper LD, Grundt AM. The impact of acceptance versus control rationales on pain tolerance. Psychol Rec. 1999a;49(1):33–47. https://doi.org/10.1007/BF03395305.

Hayes SC, Strosahl KD, Wilson KG. Acceptance and commitment therapy, vol. 6. New York: Guilford Press; 1999b.

Hayes SC, Barnes-Holmes D, Roche B. Relational frame theory: a post-Skinnerian account of human language and cognition. Springer; 2001.

Herzig SJ, Rothberg MB, Moss CR, Maddaleni G, Bertisch SM, Wong J, Zhou W, Ngo L, Anderson TS, Gurwitz JH, Marcantonio ER. Risk of in-hospital falls among medications commonly used for insomnia in hospitalized patients. Sleep. 2021;44(9):zsab064. https://doi.org/10.1093/sleep/zsab064.

Hobbs MD. The social worker on the medical transdisciplinary team. J Health Care Poor Underserved. 2005;16(2):186–91.

Hoey LM, Fulbrook P, Douglas JA. Sleep assessment of hospitalised patients: a literature review. Int J Nurs Stud. 2014;51(9):1281–8. https://doi.org/10.1016/j.ijnurstu.2014.02.001.

Hoffmann AN, Contreras BP, Clay CJ, Twohig MP. Acceptance and commitment therapy for individuals with disabilities: a behavior analytic strategy for addressing private events in challenging behavior. Behav Anal Pract. 2016;9(1):14–24. https://doi.org/10.1007/s40617-016-0105-4.

Højlund M, Andersen K, Ernst MT, Correll CU, Hallas J. Use of low-dose quetiapine increases the risk of major adverse cardiovascular events: results from a nationwide active comparator-controlled cohort study. World Psychiatry. 2022;21(3):444–51. https://doi.org/10.1002/wps.21010.

Huang M, Parker AM, Bienvenu OJ, Dinglas VD, Colantuoni E, Hopkins RO, Needham DM, National Institutes of Health, National Heart, Lung, and Blood Institute Acute Respiratory Distress Syndrome Network. Psychiatric symptoms in acute respiratory distress syndrome survivors: a 1-year National Multicenter Study. Crit Care Med. 2016;44(5):954–65. https://doi.org/10.1097/CCM.0000000000001621.

Indraratna P, Tardo D, Delves M, Szirt R, Ng B. Measurement and management of QT interval prolongation for general physicians. J Gen Intern Med. 2020;35:865–73.

Jackson JC, Pandharipande PP, Girard TD, Brummel NE, Thompson JL, Hughes CG, et al. Depression, post-traumatic stress disorder, and functional disability in survivors of critical illness in the BRAIN-ICU study: a longitudinal cohort study. Lancet Respir Med. 2014;2(5):369–79.

Jansen L, van Schijndel M, van Waarde J, van Busschbach J. Health-economic outcomes in hospital patients with medical-psychiatric comorbidity: a systematic review and meta-analysis. PLoS One. 2018;13(3):e0194029. https://doi.org/10.1371/journal.pone.0194029.

Jensen MP, Patterson DR. Hypnotic approaches for chronic pain management: clinical implications of recent research findings. Am Psychol. 2014;69(2):167.

Joa K-L, Kim W-H, Choi H-Y, Park C-H, Kim E-S, Lee S-J, Kim S-Y, Ko S-H, Jung H-Y. The effect of sleep disturbances on the functional recovery of rehabilitation inpatients following mild and moderate stroke. Am J Phys Med Rehabil. 2017;96(10):734.

Jobes DA. Managing suicidal risk: a collaborative approach. New York: Guilford Press; 2016.

Johns MW. A new method for measuring daytime sleepiness: the Epworth sleepiness scale. Sleep. 1991;14(6):540–5. https://doi.org/10.1093/sleep/14.6.540.

Kainec KA, Caccavaro J, Barnes M, Hoff C, Berlin A, Spencer RM. Evaluating accuracy in five commercial sleep-tracking devices compared to research-grade Actigraphy and polysomnography. Sensors. 2024;24(2):635.

Kangas M, McDonald S. Is it time to act? The potential of acceptance and commitment therapy for psychological problems following acquired brain injury. Neuropsychol Rehabil. 2011;21(2):250–76. https://doi.org/10.1080/09602011.2010.540920.

Kanji S, Mera A, Hutton B, Burry L, Rosenberg E, MacDonald E, Luks V. Pharmacological interventions to improve sleep in hospitalised adults: a systematic review. BMJ Open. 2016;6(7):e012108. https://doi.org/10.1136/bmjopen-2016-012108.

Kilpatrick DG, Resnick HS, Friedman MJ. Severity of acute stress symptoms—adult (national stressful events survey acute stress disorder short scale [NSESSS]). Washington, DC: American Psychiatric Association; 2013.

Knauert M, Malik V, Kamdar B. Sleep and sleep disordered breathing in hospitalized patients. Semin Respir Crit Care Med. 2014;35(05):582–92. https://doi.org/10.1055/s-0034-1390080.

Kroenke K, Spitzer RL, Williams JBW. The PHQ-9: validity of a brief depression severity measure. J Gen Intern Med. 2001;16(9):606–13. https://doi.org/10.1046/j.1525-1497.2001.016009606.x.

LaBuda J, Axelrod BN, Windell J. Cognitive behavioral protocols for medical settings: a clinician's guide. 1st ed. Routledge; 2018. https://doi.org/10.4324/9781315282411.

Luoma JB, Kohlenberg BS, Hayes SC, Fletcher L. Slow and steady wins the race: a randomized clinical trial of acceptance and commitment therapy targeting shame in substance use disorders. J Consult Clin Psychol. 2012;80(1):43.

Ma T-W, Yuen AS-K, Yang Z. The efficacy of acceptance and commitment therapy for chronic pain: a systematic review and meta-analysis. Clin J Pain. 2023;39(3):147. https://doi.org/10.1097/AJP.0000000000001096.

Mahoney JE, Webb MJ, Gray SL. Zolpidem prescribing and adverse drug reactions in hospitalized general medicine patients at a Veterans Affairs hospital. Am J Geriatr Pharmacother. 2004;2(1):66–74.

Mansuri Z, Reddy A, Vadukapuram R, Thootkur M, Trivedi C. Does winsomnia increase the risk of suicide in hospitalized patients with major depressive disorder? A nationwide inpatient analysis from 2006 to 2015. Behav Sci. 2022;12(5):117.

Mantua J, Gravel N, Spencer RM. Reliability of sleep measures from four personal health monitoring devices compared to research-based actigraphy and polysomnography. Sensors. 2016;16(5):646.

Marino M, Li Y, Rueschman MN, Winkelman JW, Ellenbogen JM, Solet JM, Dulin H, Berkman LF, Buxton OM. Measuring sleep: accuracy, sensitivity, and specificity of wrist actigraphy compared to polysomnography. Sleep. 2013;36(11):1747–55. https://doi.org/10.5665/sleep.3142.

McCracken LM. Social context and acceptance of chronic pain: the role of solicitous and punishing responses. Pain. 2005;113(1–2):155–9.

Meagher DJ, Moran M, Raju B, Gibbons D, Donnelly S, Saunders J, Trzepacz PT. Phenomenology of delirium: assessment of 100 adult cases using standardised measures. Br J Psychiatry. 2007;190(2):135–41.

Mendelson WB. A review of the evidence for the efficacy and safety of trazodone in insomnia. J Clin Psychiatry. 2005;66(4):469–76. https://doi.org/10.4088/jcp.v66n0409.

Morin CM. Cognitive-behavioral therapy of insomnia. Sleep Med Clin. 2006;1(3):375–86. https://doi.org/10.1016/j.jsmc.2006.06.008.

Morin CM, Jarrin DC. Epidemiology of insomnia: prevalence, course, risk factors, and public health burden. Sleep Med Clin. 2022;17(2):173–91. https://doi.org/10.1016/j.jsmc.2022.03.003.

Morin CM, Drake CL, Harvey AG, Krystal AD, Manber R, Riemann D, Spiegelhalder K. Insomnia disorder. Nat Rev Dis Primer. 2015;1(1):15026. https://doi.org/10.1038/nrdp.2015.26.

Murphy JL, McKellar JD, Raffa SD, Clark ME, Kerns RD, Karlin B. Cognitive behavioral therapy for chronic pain among veterans: therapist manual. Washington, DC: US Department of Veterans Affairs; 2014.

Nicholas MK, McGuire BE, Asghari A. A 2-item short form of the pain self-efficacy questionnaire: development and psychometric evaluation of PSEQ-2. J Pain. 2015;16(2):153–63. https://doi.org/10.1016/j.jpain.2014.11.002.

Niemeier J, Karol R. Overcoming grief and loss after brain injury. Oxford University Press; 2010.

O'Donnell ML, Creamer MC, Parslow R, Elliott P, Holmes ACN, Ellen S, Judson R, McFarlane AC, Silove D, Bryant RA. A predictive screening index for posttraumatic stress disorder and depression following traumatic injury. J Consult Clin Psychol. 2008;76(6):923–32. https://doi.org/10.1037/a0012918.

Ophuis RH, Olij BF, Polinder S, Haagsma JA. Prevalence of post-traumatic stress disorder, acute stress disorder and depression following violence related injury treated at the emergency department: a systematic review. BMC Psychiatry. 2018;18:311. https://doi.org/10.1186/s12888-018-1890-9.

Ozer EJ, Best SR, Lipsey TL, Weiss DS. Predictors of posttraumatic stress disorder and symptoms in adults: a meta-analysis. Psychol Bull. 2003;129(1):Article 1. https://doi.org/10.1037/0033-2909.129.1.52.

Patel MB, Jackson JC, Morandi A, Girard TD, Hughes CG, Thompson JL, Kiehl AL, Elstad MR, Wasserstein ML, Goodman RB, Beckham JC, Chandrasekhar R, Dittus RS, Ely EW, Pandharipande PP. Incidence and risk factors for intensive care unit-related post-traumatic stress disorder in veterans and civilians. Am J Respir Crit Care Med. 2016;193(12):1373–81. https://doi.org/10.1164/rccm.201506-1158OC.

Patel SR, Bakker JP, Stitt CJ, Aloia MS, Nouraie SM. Age and sex disparities in adherence to CPAP. Chest. 2021;159(1):382–9. https://doi.org/10.1016/j.chest.2020.07.017.

Pavlova KM, Latreille V. Sleep Disorders. Am J Med. 2019;132(3):292–299. https://doi.org/10.1016/j.amjmed.2018.09.021. Epub 2018 Oct 4. PMID: 30292731.

Pavwoski P, Shelgikar AV. Treatment options for obstructive sleep apnea. Neurol Clin Pract. 2017;7(1):77–85. https://doi.org/10.1212/CPJ.0000000000000320.

Pederson JL, Warkentin LM, Majumdar SR, McAlister FA. Depressive symptoms are associated with higher rates of readmission or mortality after medical hospitalization: a systematic review and meta-analysis. J Hosp Med. 2016;11(5):373–80. https://doi.org/10.1002/jhm.2547.

Pisani MA, D'Ambrosio C. Sleep and delirium in adults who are critically ill: a contemporary review. Chest. 2020;157(4):977–84. https://doi.org/10.1016/j.chest.2019.12.003.

Polomano RC, Galloway KT, Kent ML, Brandon-Edwards H, Kwon KN, Morales C, Buckenmaier III C 'Trip'. Psychometric testing of the defense and veterans pain rating scale (DVPRS): a new pain scale for military population. Pain Med. 2016;17(8):1505–19.

Posner K, Brown GK, Stanley B, Brent DA, Yershova KV, Oquendo MA, Currier GW, Melvin GA, Greenhill L, Shen S, Mann JJ. The Columbia–Suicide severity rating scale: initial validity and internal consistency findings from three multisite studies with adolescents and adults. Am J Psychiatry. 2011;168(12):1266–77. https://doi.org/10.1176/appi.ajp.2011.10111704.

Raja SN, Carr DB, Cohen M, Finnerup NB, Flor H, Gibson S, Keefe FJ, Mogil JS, Ringkamp M, Sluka KA, Song X-J, Stevens B, Sullivan MD, Tutelman PR, Ushida T, Vader K. The revised International Association for the Study of Pain definition of pain: concepts, challenges, and compromises. Pain. 2020;161(9):1976. https://doi.org/10.1097/j.pain.0000000000001939.

Rapelli G, Pietrabissa G, Manzoni GM, Bastoni I, Scarpina F, Tovaglieri I, Perger E, Garbarino S, Fanari P, Lombardi C, Castelnuovo G. Improving CPAP adherence in adults with obstructive sleep apnea syndrome: a scoping review of motivational interventions. Front Psychol. 2021;12:705364. https://doi.org/10.3389/fpsyg.2021.705364.

Resick PA, Monson CM, Chard KM. Cognitive processing therapy for PTSD. Guilford Publications; 2024.

Richards KC, O'Sullivan PS, Phillips RL. Measurement of sleep in critically ill patients. J Nurs Meas. 2000;8(2):131–44.

Rollnick S, Miller WR, Butler CC. Motivational interviewing in health care: helping patients change behavior, vol. 5. Guilford Press; 2008. p. xiv–210.

Rösler L, Van Der Lande G, Leerssen J, Cox R, Ramautar JR, Van Someren EJW. Actigraphy in studies on insomnia: worth the effort? J Sleep Res. 2023;32(1):e13750. https://doi.org/10.1111/jsr.13750.

Rotenberg BW, Murariu D, Pang KP. Trends in CPAP adherence over twenty years of data collection: a flattened curve. J Otolaryngol Head Neck Surg Le Journal D'oto-Rhino-Laryngologie Et De Chirurgie Cervico-Faciale. 2016;45(1):43. https://doi.org/10.1186/s40463-016-0156-0.

Roy-Byrne PP, Davidson KW, Kessler RC, Asmundson GJG, Goodwin RD, Kubzansky L, Lydiard RB, Massie MJ, Katon W, Laden SK, Stein MB. Anxiety disorders and comorbid medical illness. Gen Hosp Psychiatry. 2008;30(3):208–25. https://doi.org/10.1016/j.genhosppsych.2007.12.006.

Salari N, Rezaie L, Khazaie H, Bokaee S, Ghasemi H, Mohammadi M, Khaledi-Paveh B. The effect of acceptance and commitment therapy on anxiety and depression in patients with cancer: a systematic review. Curr Psychol. 2023;42(7):5694–716. https://doi.org/10.1007/s12144-021-01934-x.

Sale E, Hendricks M, Weil V, Miller C, Perkins S, McCudden S. Counseling on access to lethal means (CALM): an evaluation of a suicide prevention means restriction training program for mental health providers. Community Ment Health J. 2018;54(3):293–301. https://doi.org/10.1007/s10597-017-0190-z.

Sareen J. Posttraumatic stress disorder in adults: impact, comorbidity, risk factors, and treatment. Can J Psychiatry. 2014;59(9):Article 9.

Schroeder K, Pathak A, Sarwer DB. A call for trauma-informed intensive care. Nurs Outlook. 2021;69(5):Article 5. https://doi.org/10.1016/j.outlook.2021.06.001.

Shear TC, Balachandran JS, Mokhlesi B, Spampinato LM, Knutson KL, Meltzer DO, Arora VM. Risk of sleep apnea in hospitalized older patients. J Clin Sleep Med. 2014;10(10):1061–6.

Silva MA, Calvo D, Brennan EM, Reljic T, Drasher-Phillips L, Schwartz DJ, Kumar A, Cotner BA, Taylor DJ, Nakase-Richardson R. Incidence and predictors of adherence to sleep apnea treatment in rehabilitation inpatients with acquired brain injury. Sleep Med. 2020;69:159–67. https://doi.org/10.1016/j.sleep.2020.01.016.

Smith WB, Gracely RH, Safer MA. The meaning of pain: cancer patients' rating and recall of pain intensity and affect. Pain. 1998;78(2):123–9.

Smith MT, Klick B, Kozachik S, Edwards RE, Holavanahalli R, Wiechman S, Blakeney P, Lezotte D, Fauerbach JA. Sleep onset insomnia symptoms during hospitalization for major burn injury predict chronic pain. Pain. 2008;138(3):497–506. https://doi.org/10.1016/j.pain.2008.01.028.

Snyder-Halpern R, Verran JA. Instrumentation to describe subjective sleep characteristics in healthy subjects. Res Nurs Health. 1987;10(3):155–63. https://doi.org/10.1002/nur.4770100307.

Sorscher AJ, Caruso EM. Frequency of provision of CPAP in the inpatient setting: an observational study: "CPAP provision in the inpatient setting.". Sleep Breath Schlaf Atmung. 2012;16(4):1147–50. https://doi.org/10.1007/s11325-011-0621-y.

Spitzer RL, Kroenke K, Williams JB, Löwe B. A brief measure for assessing generalized anxiety disorder: the GAD-7. Arch Intern Med. 2006;166(10):1092–7.

Stanley B, Brown GK, Brenner LA, Galfalvy HC, Currier GW, Knox KL, Chaudhury SR, Bush AL, Green KL. Comparison of the safety planning intervention with follow-up vs usual care of suicidal patients treated in the emergency department. JAMA Psychiatry. 2018;75(9):894–900. https://doi.org/10.1001/jamapsychiatry.2018.1776.

Stiasny-Kolster K, Mayer G, Schäfer S, Möller JC, Heinzel-Gutenbrunner M, Oertel WH. The REM sleep behavior disorder screening questionnaire--a new diagnostic instrument. Mov Disord. 2007;22(16):2386–93. https://doi.org/10.1002/mds.21740.

Sullivan MJ, Bishop SR, Pivik J. The pain catastrophizing scale: development and validation. Psychol Assess. 1995;7(4):524.

Tol WA, Barbui C, van Ommeren M. Management of Acute Stress, PTSD, and bereavement: WHO recommendations. JAMA. 2013;310(5):Article 5. https://doi.org/10.1001/jama.2013.166723.

Trudeau J, Turk D, Dworkin R, Benson C, Biondi D, Kim M, Mou J, Katz N. Validation of the revised short form McGill Pain Questionnaire (SF-MPQ-2) for self-report of pain qualities in patients with acute low back pain. J Pain. 2012;13(4):S4. https://doi.org/10.1016/j.jpain.2012.01.022.

van den Ende ES, Burger P, Keesenberg M, Merten H, Gemke RJBJ, Nanayakkara PWB. Patient-nurse agreement on inpatient sleep and sleep disturbing factors. Sleep Medicine: X. 2022;4:100047. https://doi.org/10.1016/j.sleepx.2022.100047.

van der Zweerde T, Bisdounis L, Kyle SD, Lancee J, van Straten A. Cognitive behavioral therapy for insomnia: a meta-analysis of long-term effects in controlled studies. Sleep Med Rev. 2019;48:101208. https://doi.org/10.1016/j.smrv.2019.08.002.

Vlaeyen JW, Crombez G, Linton SJ. The fear-avoidance model of pain. Pain. 2016;157(8):1588–9.

Walker WC. Pain Pathoetiology after TBI: neural and nonneural mechanisms. J Head Trauma Rehabil. 2004;19(1):72.

Walker J, Burke K, Wanat M, Fisher R, Fielding J, Mulick A, Puntis S, Sharpe J, Esposti MD, Harriss E, Frost C, Sharpe M. The prevalence of depression in general hospital inpatients: a systematic review and meta-analysis of interview-based studies. Psychol Med. 2018;48(14):2285–98. https://doi.org/10.1017/S0033291718000624.

Walker J, Van Niekerk M, Hobbs H, Toynbee M, Magill N, Bold R, Hampsey E, Harriss E, Frost C, Sharpe M. The prevalence of anxiety in general hospital inpatients: a systematic review and meta-analysis. Gen Hosp Psychiatry. 2021;72:131–40. https://doi.org/10.1016/j.genhosppsych.2021.08.004.

Warden V, Hurley AC, Volicer L. Development and psychometric evaluation of the pain assessment in advanced dementia (PAINAD) scale. J Am Med Dir Assoc. 2003;4(1):9–15.

Watkins LE, Sprang KR, Rothbaum BO. Treating PTSD: a review of evidence-based psychotherapy interventions. Front Behav Neurosci. 2018;12 https://doi.org/10.3389/fnbeh.2018.00258.

Weaver TE, Grunstein RR. Adherence to continuous positive airway pressure therapy: the challenge to effective treatment. Proc Am Thorac Soc. 2008;5(2):173–8. https://doi.org/10.1513/pats.200708-119MG.

Weiner BK. Spine update: the biopsychosocial model and spine care. Spine. 2008;33(2):219–23.

Wesselius HM, Van Den Ende ES, Alsma J, Ter Maaten JC, Schuit SCE, Stassen PM, De Vries OJ, Kaasjager KHAH, Haak HR, Van Doormaal FF, Hoogerwerf JJ, Terwee CB, Van De Ven PM, Bosch FH, Van Someren EJW, Nanayakkara PWB, for the "Onderzoeks Consortium Acute Geneeskunde" Acute Medicine Research Consortium. Quality and quantity of sleep and factors associated with sleep disturbance in hospitalized patients. JAMA Intern Med. 2018;178(9):1201. https://doi.org/10.1001/jamainternmed.2018.2669.

Williams DA, Robinson ME, Geisser ME. Pain beliefs: assessment and utility. Pain. 1994;59(1):71–8.

Williamson LF, Kautz DD. Trauma-informed care is the best clinical practice in rehabilitation nursing. Rehabilitation Nursing. 2016;0:1–9. https://doi.org/10.1002/rnj.311.

Yanowitz FG. Introduction to ECG interpretation. LDS Hospital and Intermountain Medical Center; 2012.

Ye L, Owens RL, Dykes P. Individualized sleep promotion in acute care hospitals: identifying factors that affect patient sleep. Appl Nurs Res. 2019;48:63–7. https://doi.org/10.1016/j.apnr.2019.05.006.

Zigmond AS, Snaith RP. The hospital anxiety and depression scale. Acta Psychiatr Scand. 1983;67(6):361–70. https://doi.org/10.1111/j.1600-0447.1983.tb09716.x.

INDEX

E. K. Bailey et al. (eds.), *Inpatient Neuropsychology*, https://doi.org/10.1007/978-3-032-05880-5

E

P

MIX
Papier aus verantwortungsvollen Quellen
Paper from responsible sources
FSC® C105338

If you have any concerns about our products,
you can contact us on
ProductSafety@springernature.com

In case Publisher is established outside the EU,
the EU authorized representative is:
Springer Nature Customer Service Center GmbH
Europaplatz 3, 69115 Heidelberg, Germany

Printed by Libri Plureos GmbH
in Hamburg, Germany